Retrovirus Biology
and Human Disease

Retrovirus Biology and Human Disease

edited by

ROBERT C. GALLO
FLOSSIE WONG-STAAL
National Cancer Institute
National Institutes of Health
Bethesda, Maryland

CRC Press
Taylor & Francis Group
Boca Raton London New York

CRC Press is an imprint of the
Taylor & Francis Group, an **informa** business

CRC Press
Taylor & Francis Group
6000 Broken Sound Parkway NW, Suite 300
Boca Raton, FL 33487-2742

© 2009 by Taylor & Francis Group, LLC
CRC Press is an imprint of Taylor & Francis Group, an *informa* business

No claim to original U.S. Government works

ISBN-13: 978-0-824-77874-3 (hbk)
ISBN-13: 978-1-003-57370-8 (ebk)

DOI: 10.1201/9781003573708

This book contains information obtained from authentic and highly regarded sources. While all reasonable efforts have been made to publish reliable data and information, neither the author[s] nor the publisher can accept any legal responsibility or liability for any errors or omissions that may be made. The publishers wish to make clear that any views or opinions expressed in this book by individual editors, authors or contributors are personal to them and do not neces-sarily reflect the views/opinions of the publishers. The information or guidance contained in this book is intended for use by medical, scientific or health-care professionals and is provided strictly as a supplement to the medical or other professional's own judgement, their knowledge of the patient's medical history, relevant manufacturer's instructions and the appropriate best practice guidelines. Because of the rapid advances in medical science, any information or advice on dosages, procedures or diagnoses should be independently verified. The reader is strongly urged to consult the relevant national drug formulary and the drug companies' and device or material manufacturers' printed instructions, and their websites, before administering or utilizing any of the drugs, devices or materials mentioned in this book. This book does not indicate whether a particular treatment is appropriate or suitable for a particular individual. Ultimately it is the sole responsibility of the medical professional to make his or her own professional judgements, so as to advise and treat patients appropriately. The authors and publishers have also attempted to trace the copyright holders of all material reproduced in this publication and apologize to copyright holders if permission to publish in this form has not been obtained. If any copyright material has not been acknowledged please write and let us know so we may rectify in any future reprint.

Visit the Taylor & Francis Web site at
http://www.taylorandfrancis.com

Library of Congress Cataloging-in-Publication Data

Retrovirus biology and human disease / Robert C. Gallo and Flossie Wong-Staal, editors.
   p. ; cm.
  Includes bibliographical references and index.
  ISBN-13: 978-0-8247-7874-3 (alk. paper)
  ISBN-10: 0-8247-7874-X (alk. paper)
   1. Retrovirus infections--Pathogenesis. 2. Retroviruses. 3. AIDS (Disease)--Pathogenesis.
[DNLM: 1. Acquired Immunodeficiency Syndrome. 2. Retroviruses. 3. Retrovirus Infections.
QW 166 R436]
  QR201.R47R47 1989
  616.9'2--dc20

                                     89-17184

Preface

The last decade has witnessed the dawning of a new era in retrovirology, that of human retrovirology. Although the discovery of the first human retrovirus (human T-leukemia virus type 1 [HTLV-I]) in 1979 lags greatly behind the discovery of the first animal retroviruses, it was followed in rapid succession by that of other human viruses of the same lineage (HTLV-II, HTLV-V) and of an evolutionarily distinct subgroup (human immunodeficiency viruses, HIV-1 and HIV-2). The association of these viruses with a broad spectrum of diseases, including cancer, neurological disorders, and, in particular, the AIDS pandemic, makes these viruses a subject of intense interest, not only to academicians, but also to the public at large. The purpose of this book is to capture some of the essential developments, both historical and current, relating to these viruses and the diseases they cause.

Like all scientific discoveries, the discovery of human retroviruses was not made in a vacuum. It benefited from advances in technology; the availability of growth factors, culture conditions for long-term growth of hematopoietic cells in vitro, and monoclonal antibodies, and the complete arsenal of modern examples among animal retroviruses that cause leukemias, CNS disorders, and immune deficiencies. For this reason, we have included several chapters on animal retroviruses: feline leukemia virus because it was the first transmissable, pathogenic leukemia virus identified in nature, bovine leukemia virus because of its many parallels and actual kinship to the HTLVs, and the animal lentiviruses because of their relationship, albeit distant, to the HIVs. These animal retroviruses first served as models for their human counterparts, but ironically, the opposite may now be true as the progress in human retrovirology moves at a remarkable pace.

And indeed the pace has been remarkable. In the short time since the discovery of retroviruses and their link to disease, tremendous information has been uncovered relating to these viruses: their genetic structures, their tissue and cellular targets, their modes of transmission, and their cytopathology in vitro and in vivo. Many of these features are novel and complex, such that analysis of these viruses has developed into a discipline all by itself. However, the greater understanding we have of these viruses, the greater respect we have for them as truly formidable enemies. One salient feature of the human retroviruses is their ability to go undercover. Patients with adult T-cell leukemia induced by HTLV-I show no sign of virus expression in their tumor cells, even though the blueprint of the virus (proviral DNA) is invariably in safekeeping as part of the host genetic constituents. This covert presence of the virus makes therapy directed at the viral replication cycle inappropriate. It also suggests that virus infection is not only an early event (initiation), perhaps resulting in immortalization and increased proliferation of the target T-lymphocytes, but that a subsequent, irreversible event is needed for leukemic conversion. This multi-step phenomenon for ATL development may in part explain the long latency period in the infected individuals. Infection by HIV and development of AIDS are likewise interposed with long and variable periods of "latency" during which very little free virus or viral antigens can be detected. However, in contrast to the HTLV-I situation, clinical progression does correlate with or even depend on high levels of HIV expression. One might view the production of disease as a result of a usurpation of the virus's control of its own expression by exogenous factors. HIV contains a number of regulatory genes, the complex interaction of which seems to be directed at maintaining a controlled, moderate level of virus (see Chapter 10). The surprising conservation of the negative regulatory elements (acting both in *cis* and *trans*) in vivo despite strong selection against them in in vitro passage in T-cells suggests that a slow replicating virus may be preferred in an early target cell in in vivo infection. Positive and negative feedback control loops also appear to play a role in the replication cycle of HTLV (see Chapters 7 and 8). The remarkable parallels in the regulatory pathways, particularly pertaining to two genes (*tax* and *rex* for HTLV and *tat* and *rev* for HIV) may represent convergent evolution of two groups of pathogenic viruses selected for persistent infection of long-lived species (primates).

HIV has developed yet another strategy for persistent infection, namely immune evasion. The exterior envelope protein of HIV contains several hypervariable regions, and the major neutralization epitope(s) are located in one of these regions. Consequently, neutralization with hyperimmune sera is extremely type-specific. Furthermore, probably as a result of both de novo mutation and selection, neutralization-resistant variants would emerge in the course of infection. This phenomenon has been observed both in experimentally infected chimpanzees and in one accidentally infected laboratory worker. While this finding is discouraging for the prospects of vaccine development, the problem is perhaps not insurmountable. There is enough optimism that a combination of new insights and creative approaches would eventually lead to a solution (see Chapter 15).

While the availability of a vaccine for the public remains a rather distant possibility, there is the beginning of some treatment regimens for the AIDS patients (see Chapter 13). Agents that block a very key step in retrovirus replication, i.e., reverse transcription, have shown positive effects in both prolonging and improving the quality of life for these patients. Other potential agents directed at additional steps of the virus life cycle are being tested or contemplated. Having a large battery of therapeutic agents at one's disposal, it would be feasible to devise combinations with maximal potency and minimal toxicity.

We are greatly indebted to our colleagues for their contributions to this book and for their patience and tolerance throughout its assembly.

Robert C. Gallo
Flossie Wong-Staal

Contents

Contributors

William A. Blattner, M.D.   Chief, Viral Epidemiology Section, Enviromental Epidemiology Branch, National Cancer Institute, National Institutes of Health, Bethesda, Maryland

Dani P. Bolognesi, Ph.D.   Department of Surgery, Duke University Medical Center, Durham, North Carolina

Samuel Broder, M.D.   Director, Clinical Oncology Program, Division of Cancer Treatment, National Cancer Institute, National Institutes of Health, Bethesda, Maryland

A. Burny, M.D. Faculty of Agronomy, Gembloux, and Department of Molecular Biology, University of Brussels, Brussels, Belgium

Alan J. Cann, Ph.D.[*]   Post-Doctoral Fellow, Division of Hematology/Oncology, Department of Medicine, UCLA School of Medicine, Los Angeles, California

Irvin S. Y. Chen, Ph.D.   Associate Professor of Medicine, Department of Medicine/Microbiology and Immunology/Hematology-Oncology, UCLA School of Medicine, Los Angeles, California

Janice E. Clements, M.D.[†]   Associate Professor of Neurology, Molecular Biology and Genetics, The Johns Hopkins University School of Medicine, Baltimore, Maryland

Y. Cleuter University of Brussels, Brussels, Belgium

Present affiliations:
  Research Scientist, Laboratory of Molecular Biology, Medical Research Council (MRC) Centre, United Kingdom
[†]Associate Professor of Comparative Medicine, Molecular Biology and Genetics, and Neurology, Department of Comparative Medicine, The Johns Hopkins University School of Medicine, Baltimore, Maryland

Myron Essex, D.V.M., Ph.D. Chairman, Department of Cancer Biology, Harvard School of Public Health, and Chairman, Harvard AIDS Institute, Boston, Massachussetts

Anthony S. Fauci, M.D. Director, National Institute of Allergy and Infectious Diseases, National Institutes of Health, Bethesda, Maryland

Robert C. Gallo, M.D. Chief, Laboratory of Tumor Cell Biology, Division of Cancer Etiology, National Cancer Institute, National Institutes of Health, Bethesda, Maryland

Jerome E. Groopman, M.D. Chief, Division of Hematology/Oncology, and Associate Professor of Medicine, Division of Hematology/Oncology, Department of Medicine, New England Deaconess Hospital, Harvard Medical School, Boston, Massachusetts

R. Jan Gurley, M.D. Division of Hematology/Oncology, Department of Medicine, New England Deaconess Hospital, Harvard Medical School, Boston, Massachusetts

William D. Hardy, Jr., V.M.D. Associate Member and Head, Laboratory of Veterinary Oncology, Memorial Sloan-Kettering Cancer Center, New York, New York

William A. Haseltine, Ph.D. Chief, Division of Retrovirology, Dana-Farber Cancer Institute, Harvard Medical School, Boston, Massachusetts

Toshio Hattori, M.D. Associate Professor of Medicine, The Second Department of Internal Medicine, Kumamoto University Medical School, Kumamoto, Japan

Edward A. Hoover, D.V.M., Ph.D. Professor, Department of Pathology, College of Veterinary Medicine, Colorado State University, Fort Collins, Colorado

Phyllis A. Kanki, D.V.M., D.Sc. Assistant Professor of Pathobiology, Department of Cancer Biology, Harvard School of Public Health, Boston, Massachusetts

R. Kettmann Senior Researcher, Department of Molecular Biology, Faculty of Agronomy, Gembloux, and University of Brussels, Brussels, Belgium

Scott Koenig, M.D., Ph.D. Senior Staff Fellow, Laboratory of Immunoregulation, National Institute of Allergy and Infectious Diseases, National Institutes of Health, Bethesda, Maryland

M. Mammerickx, D.V.M. Faculty of Agronomy, Gembloux, and Department of Pathology for Large Animals, National Institute for Veterinary Research, Brussels, Belgium

Angela Manns, M.D., M.P.H.   Biotechnology Fellow, Viral Epidemiology Section, Enviromental Epidemiology Branch, National Cancer Institute, National Institutes of Health, Bethesda, Maryland

G. Marbaix   Doctor of Biochemistry, Laboratory of Biochemistry, University of Brussels, Brussels Belgium

Hiroaki Mitsuya, M.D.   Cancer Expert, The Clinical Oncology Program, Division of Cancer Treatment, National Cancer Institute, National Institutes of Health, Bethesda, Maryland

James I. Mullins, M.D., Ph.D.[*]   Associate Professor of Virology, Department of Cancer Biology, Harvard University School of Public Health, Boston, Massachusetts

Opendra Narayan, D.V.M., Ph.D.[†]   Professor, Department of Neurology, The Johns Hopkins University School of Medicine, Baltimore, Maryland

D. Portetelle   Faculty of Agronomy, Gembloux, and University of Brussels, Brussels, Belgium

Marvin S. Reitz, M.D.   Laboratory of Tumor Cell Biology, National Cancer Institute, National Institutes of Health, Bethesda, Maryland

Craig A. Rosen, Ph.D.   Assistant Member, Department of Molecular Oncology, Roche Institute of Molecular Biology, Nutley, New Jersey

Joseph D. Rosenblatt, M.D.   Assistant Professor of Medicine, Division of Hematology/Oncology, Department of Medicine, UCLA School of Medicine, Los Angeles, California

Motoharu Seiki, M.D.[‡]   Associate Member, Department of Viral Oncology, Cancer Institute, Kami-Ikebukuro, Toshima-ku, Tokyo, Japan

Joesph G. Sodroski, M.D.   Assistant Professor, Department of Pathology, Dana-Farber Cancer Institute, Harvard Medical School, Boston, Massachusetts

Kiyoshi Takatsuki, M.D.   Professor of Medicine, The Second Department of Internal Medicine, Kumamoto University Medical School, Kumamoto, Japan

---

Present affiliations:
[*]Professor of Virology, Department of Microbiology and Immunology, Stanford, University Medical School, Stanford, California
[†]Professor of Comparative Medicine, Department of Comparative Medicine–Retrovirus Biology Laboratories, The Johns Hopkins University School of Medicine, Baltimore, Maryland
[‡]Professor, School of Medicine, Kanazawa University, Takara-cho, Kanazawa, Japan

Ernest F. Terwilliger, Ph.D.   Instructor, Division of Human Retrovirology, Dana-Farber Cancer Institute, Harvard Medical School, Boston, Massachusetts

R. Thomas   University of Brussels, Brussels, Belgium

A. Van den Broeke   University of Brussels, Brussels, Belgium

William Wachsman, M.D., Ph.D.   Associate Professor of Medicine, Division of Hematology/Oncology, Department of Medicine, UCLA School of Medicine, Los Angeles, California

L. Willems   Chief Researcher, Faculty of Agronomy, Gembloux, and Department of Molecular Biology, University of Brussels, Brussels, Belgium

Kazunari Yamaguchi, M.D.   Associate Professor of Medicine, The Second Department of Internal Medicine, Kumamoto University Medical School, Kumamoto, Japan

Mitsuaki Yoshida, Ph.D.   Laboratory Chief and Member, Department of Viral Oncology, Cancer Institute, Kami-Ikebukuro, Toshima-ku, Tokyo, Japan

Retrovirus Biology
and Human Disease

1

Retroviruses and Human Disease
A Historical Perspective

Robert C. Gallo and **Marvin S. Reitz** *National Cancer Institute, National Institutes of Health, Bethesda, Maryland*

I. ANIMAL RETROVIRUSES AND HEMATOPOIETIC DISEASE

Hematopoietic neoplasms such as leukemia, lymphoma, and related disorders are widely distributed in the animal kingdom. They occur not only in humans and other mammals, but also in birds, reptiles, fish, and even mollusks. In most cases where the etiology of these diseases as they naturally occur is well understood, the primary event is infection with a retrovirus. Retroviruses are enveloped viruses that contain an RNA genome; a DNA copy of the RNA is synthesized by the viral DNA polymerase or reverse transcriptase, which gives this group of viruses their name. The DNA generally must be integrated into the host cell genome to establish infection, and retroviruses are generally not lytic. Infection thus tends to be permanent. Virus propagation from the infected cell can proceed through virion production, followed by infection of a new cell (horizontal transmission) or, passively, by replication of infected cells (vertical transmission).

If a retrovirus enters the germ line of a species and is thus ubiquitously transmitted vertically, it is called an endogenous virus. Endogenous viruses appear to play a very small role in naturally occurring disease, although they may recombine with retroviruses that are transmitted primarily from one host to another by horizontal transmission (exogenous viruses) to form chimeric viruses with altered biological properties. All naturally occurring pathogenic retroviruses identified thus far are exogenous.

In addition to hematopoietic neoplastic diseases, which are proliferative diseases, retroviruses also cause ablative diseases of the hematopoietic system, such as anemia and various immune disorders. In some cases, retroviruses also appear to be able to infect cells of the nervous system and cause neurological

1

disorders such as paralysis. This may reflect similarities in cell surface receptors and transcriptional and translational factors between hematopoietic cells and those of the nervous system.

Retroviruses were first isolated from chickens in the early part of this century by Ellerman and Bang from leukemias (1) and by Peyton Rous from sarcomas (2). Rous showed that the sarcoma could be transmitted by cell-free filtrates; this was the first report of an oncogenic retrovirus. Many avian retroviruses have since been described, and their associated diseases include lymphomas, myeloblastosis, erythroblastosis, and osteopetrosis (big bone disease). Many of the avian retroviruses are defective recombinants which have recombined with host cell DNA and thereby acquired cell-derived oncogenes. The recombination process alters the structure of the cellular protooncogene and, in so doing, converts it to a gene with the ability to directly transform target cells. The result is an acutely transforming virus which transforms all infected cells and which results in a polyclonal tumor. The Rous sarcoma virus is of this type.

Most of these viruses are laboratory creations, however, resulting perhaps from high multiplicities of infection during in vitro passage of the virus. Most field isolates seem to be complete, replication-competent viruses which do not directly transform cells. Their oncogenic potential, which is expressed relatively slowly and inefficiently, is thought to be mediated by insertional mutagenesis of host protooncogenes during integration of viral DNA (3,4). This results in monoclonal tumors, since it is only the rare infected cell which by chance has the relevant integration event. In addition, since some oncogenes immortalize cells and others eliminate contact inhibition, and since both properties are necessary for malignant transformation, some secondary event(s), not necessarily related to the virus, may be critical for tumorigenesis. Consequently, these slowly transforming viruses cause disease very inefficiently and generally only after a long latent phase.

The next important findings, linking retroviruses with leukemia in mammals, did not occur until the early 1950s with the pioneering work of Ludwik Gross (5), who clearly showed that murine leukemia virus, a retrovirus isolated from inbred strains of laboratory mice, could induce leukemia in neonate mice. Since that time, many different types of murine leukemia viruses, all of them related, have been identified. Some are endogenous to some inbred laboratory strains of mice, others are endogenous to all mice, and yet others are not endogenous but are horizontally transmitted. Many of the latter viruses are recombinants which have acquired genes or parts of genes from endogenous viruses, and all of the pathogenic murine leukemia viruses are horizontally transmitted (although recombination during infection with endogenous viruses may be a required step in leukomogenesis). Feral outbred mice also harbor retroviruses, including horizontally transmitted viruses which can cause lymphoma and a paralytic neurological disease. The mechanism by which the leukemogenic murine retroviruses cause neoplasia are not clear, but oncogene capture or activation by insertional mutagenesis may be involved in some cases. In addition, recombinant envelope proteins derived from two different parental viruses, as in the case of Friend and Rauscher murine leukemia viruses, can result in acutely pathogenic viruses. In this instance, it may be that the envelope inserts in or binds to the cell membrane in such a way as to cause an inappropriate signal to a surface receptor for some factor(s) regulating growth or differentiation.

The next significant set of findings concerning retroviruses and leukemia came from work in the mid-1960s by W. Jarrett and his colleagues, who found that naturally occurring hematopoietic neoplasms in outbred populations of cats were due to feline leukemia virus, a horizontally transmitted retrovirus (6). This was important because it was the first indication that transmission of these viruses occurs in a nonlaboratory setting and results in leukemia. As with other naturally occurring retroviruses, feline leukemia virus is not acutely transforming and causes leukemia only after a long latent phase and in a small percentage of infected animals. Again, the mechanism is not clear, although some evidence has been reported for insertional mutagenesis of protooncogenes, and the virus seems to have a relatively high tendency to transduce different oncogenes, creating acutely transforming viruses. This may be due to a high replication rate in infected animals. The virus commonly causes a variety of other diseases, including immunodeficiency syndrome, anemia, and spontaneous abortion. Different diseases may be caused by different strains of virus. The cat also contains related sequences that could recombine with the infecting virus to create new viruses during the course of infection.

All of the retrovirus-induced leukemias to this point involved abundant virus replication, and it was expected that this would invariably be the case. Lymphoma in cattle had long been suspected to be of viral etiology because of clustering of the disease within herds, but no virus could be identified until the late 1960s. Only when tumor cells could be successfully cultured was the bovine leukemia virus discovered (7); this was proven to be the etiological agent by seroepidemiology. It was now apparent that leukemias could have a retroviral etiology in the absence of detectable virus in the leukemic animal. In the case of bovine leukemia virus, not only is virus not expressed but neither are viral proteins or RNA, yet the viral etiology of the disease is clear.

The next significant event in the history of the study of retroviruses was the discovery by Temin and Mizutani (8) and Baltimore (9) that the viral RNA genome was reverse transcribed into DNA by the viral DNA polymerase, which was therefore called reverse transcriptase. This reversal of the normal direction of the flow of genetic information was unprecedented, and it led to the designation of viruses that utilized it as retroviruses. One extremely important consequence of its discovery was that, since it is ubiuqitous to retroviruses and since it is an enzyme, it allowed the detection of low levels of these viruses without requiring specific probes such as antibodies. Thus, in principle any retrovirus could be detected by the presence of reverse transcriptase activity.

II. THE SEARCH FOR HUMAN RETROVIRUSES

Because of the precedent of the retroviral etiology of many animal leukemias, there was considerable interest in and effort directed toward the isolation of comparable human viruses. This was particularly true following the discovery of reverse transcriptase. By the early 1970s examples of leukemias of retroviral etiology had even been extended to the gibbon apes (10-12), thus increasing expectations that human leukemia viruses would be readily identified. This turned out not to be the case, for a number of reasons. For one thing, much of this research was predicated on the studies with the previously described animal systems, where abundant virus relication and high levels of viremia were generally the case. This has notably not been the case with the human retroviruses described to date. In addition, as already mentioned in the case of bovine leukemia virus, some retroviruses can remain in a latent state even during the course of the disease unless the infected cell can be grown. The virus must be able to be grown in a proper target cell for isolation. Since the previously described animal retroviruses could all be grown in fibroblasts, it was thought (wrongly) that this would be the case with human retroviruses as well. Despite much effort, only suggestive data were obtained for the presence of human retroviruses. This included the sporadic detection of cell-associated reverse transcriptase-like activity (13,14) by various laboratories including our own and the detection of apparently exogenous (and, by implication, viral) DNA sequences in some human leukemias by Baxt and Speigleman (15). No convincing isolation of a biologically active human retrovirus was achieved, however, using the available target cell cultures.

Partly because of the notion that growth of the infected cells (i.e., hematopoietic cells) would be necessary for the successful detection of human retroviruses and that the target cell systems available for virus isolation such as fibroblasts were not appropriate, in the mid-1970s our laboratory began to look for growth factors that would support the growth of different hematopoietic cells. This effort were rewarded when Morgan et al. (16) discovered a factor in the media of phytohemagglutinin-stimulated peripheral blood cells that supported the relatively long-term growth of large quantities of T cells. We then used this factor (first called TCGF for T-cell growth factor, but now more commonly called IL-2) to establish cultures from a variety of human T-cell neoplasias (17). In the early 1980s, using these cultures in conjunction with sensitive reverse transcriptase assays, Poiesz et al. reported the isolation of HTLV-I, the first human retrovirus, from several cases of cutaneous T-cell lymphoma/leukemias (18,19), which, although first described as mycosis fungoides and Sézary syndrome, were, in retrospect, adult T-cell leukemia. HTLV-I was shown to be a horizontally transmitted virus not detectably related to other previously described retroviruses (20,21). The detailed molecular biology and epidemiology of HTLV-I are described in later chapters in this book.

Using techniques similar to those used for the isolation of HTLV-I and having antibodies to HTLV-I proteins as a probe, Kalyanaraman et al. (22) were able to isolate a virus distantly related to HTLV-I from T cells of a patient with a T-cell variant of hairy cell leukemia. This was HTLV-II, the second human retrovirus.

In the early 1980s a new and fatal epidemic disease was first noticed, which we now know as AIDS, for acquired immune deficiency syndrome. Several features pointed to an infectious etiology, including case clustering, apparent sexual transmission, and, most tellingly, apparent transmission by blood products. Several factors made retroviruses an appealing candidate. For one thing, AIDS was a T-cell disease, and all human retroviruses to date were T-cell tropic. Second animal retroviruses, notably feline leukemia virus, are commonly associated with immunosuppressive disorders (so, for that matter, is HTLV-I). Third, the disease had a rather delayed onset following presumptive transmission events, which would be consistent with a retrovirus.

Our initial hypothesis was that an HTLV-I-related virus was the AIDS agent; despite some suggestive early data, this turned out not to be the case. Using T-cell cultures and reverse transcriptase assays, however, Barre-Sinoussi et al. (23) were able to identify a virus that proved to be the actual AIDS agent. The virus, however, was highly cytotoxic for T cells and could not be grown in any quantity; in fact, it was not even immediately clear that it was a retrovirus. It was clearly not related to HTLV-I. Detailed characterizations were made possible by the identification of a replication-permissive T-cell line which was able to at least partially resist the cytotoxic effect of the virus (24,25), which has variously been called LAV or HTLV-III and is now by general agreement called HIV-1. It could then be shown that this was indeed the etiological agent for AIDS.

Using similar techniques and the available probes for HIV-1, distantly related viruses were identified in African monkeys (26) and in humans in western Africa (27,28). The latter virus is now known as HIV-2. Although it, too, appears to be associated with an AIDS-like disease, it appears far less pathogenic than HIV-1. HIV-1 and -2 are the third and fourth human retroviruses.

There are currently two groups of human retroviruses (exclusive of a recently characterized human spumaretrovirus [29]). One group consists of the distantly related HTLV-I and II and is associated with T-cell neoplastic proliferative diseases. Recently, a possible third member of this group, HTLV-V, has been reported from cutaneous T-cell lymphoma by Manzari et al. (30). The second group, which is not significantly related genetically to the first group, consists of the distantly related HIV-1 and -2, both associated with T-cell ablative diseases. Although the two groups are not related, there are some strik-ing similarities. All contain extra genes not found in most animal retroviruses. These extra genes appear to help control virus replication rates and may in part account for the low viral titers in vivo. Both groups have counterparts widely distributed in Old World monkeys, particularly in Africa. All are tropic for T cells.

That all of the human retroviruses described to date are from T cells most probably reflects our ability to grow T cells easily more than that all human retroviruses are T-cell tropic. As we gain the ability to grow large quantities of other types of hematopoietic cells, such as those in the myeloid series, it will be interesting to see which other hematopoietic diseases are etiologically linked to retroviruses. This book presents in detail much of the recent work done on the human retroviruses.

REFERENCES

1. Ellerman, V., and Bang, O. (1908). Experimentelle Keukamie bei Huhnern. *Centralbl. Bakteriol., Abt. 1 (Orig.)* 46:595–605.
2. Rous, P. (1911). Transmission of a malignant new growth by means of a cell-free filtrate. *JAMA* 56:198–204.
3. Hayward, W. S., Neel, B. G., and Astrin, S. M. (1981). Induction of lym-phocytic leukosis by avian leukosis virus: Activation of a cellular "onc" gene by promoter insertion. *Nature* 290:475–479.
4. Neel, B. G., Hayward, W. S., Robinson, H. L., Fang, J., and Astrin, S. M. (1981). Avian leukosis virus-induced tumors have common proviral integra-tion sites and synthesize discrete new RNAs: Oncogenesis by promoter insertion *Cell* 23:323–334.
5. Gross, L. (1951). "Spontaneous" leukemia developing in C3H mice follow-ing inoculation in infancy with AK-leukemic extracts, or AK-embryos. *Proc. Soc. Exp. Biol. Med.* 78:27–39.
6. Jarrett, W., Crawford, E., Martin, W., and Davie, F. (1964). Leukemia in the cat. A virus-like particle associated with leukaemia (lymphosarcoma). *Nature* 202:567–568.
7. Miller, J., Miller, L., Olson, C., and Gillette, K. (1969). Virus-like particles in phytohemagglutinin-stimulated lymphocyte cultures with reference to bovine lymphosarcoma. *J. Natl. Cancer Inst.* 43:1297–1303.
8. Temin, H., and Mizutani, S. (1970). RNA-dependent DNA polymerase in virions of Rous sarcoma virus. *Nature* 226:1211–1213.
9. Baltimore, D. (1970). Viral RNA-dependent DNA polymerase. *Nature* 226: 1209–1211.
10. Kawakami, T. G., Huff, S. D., Buckley, P. M., Dungworth, D. C., and Snyder, J. P. (1972). Isolation and characterization of a C-type virus associated with gibbon lymphosarcoma. *Nature New Biol.* 235:170–172.
11. Theilen, G. H., Gould, D., Fowler, M., and Dungworth, D. L. (1971).

C-type virus in tumor tissue of a woolly monkey (*Lagothrix* spp.) with fibrosarcoma. *J. Natl. Cancer Inst.* 47:881–886.

12. Wolfe, W. G., Deinhardt, F., Theilen, G. H., Rabin, H., Kawakami, T., and Bustad, L. K. (1971). Induction of tumors in marmoset monkeys by simian sarcoma virus type-C (*Lagothrix*): A preliminary report. *J. Natl Cancer Inst.* 47:1115–1119.

13. Goodenow, R. S., and Kaplan, H. S. (1979). Characterization of the reverse transcriptase of a type C virus produced by a human lymphoma cell line. *Proc. Natl. Acad. Sci. USA* 76:4971–4975.

14. Sarngadharan, M. G., Sarin, P. S., Reitz, M. S., and Gallo, R. C. (1972). Reverse transcriptase activity of human acute leukemic cells: Purification of the enzyme, response to AMV 70S RNA, and characterization of the DNA product. *Nature New Biol.* 240:67–69.

15. Baxt, W. G., and Spiegleman, S. (1973). Nuclear DNA sequences present in human leukemic cells and absent in normal leukocytes. *Proc. Natl. Acad. Sci. USA* 71:1309–1313.

16. Morgan, D. A., Ruscetti, F. W., and Gallo, R. C. (1976). Selective in vitro growth of T-lymphocytes from normal human bone marrow. *Science* 193:1007–1010.

17. Poiesz, B. J., Ruscetti, F. W., Mier, J. W., Woods, A. M., and Gallo, R. C. (1980). T-cell lines established from human T-lymphocytic neoplasias by direct response to T-cell growth factor. *Proc. Natl. Acad. Sci. USA* 77:6815–6819.

18. Poiesz, B. J., Ruscetti, F. W., Gazdar, A. F., Bunn, P. A., Minna, P. A., and Gallo, R. C. (1980). Detection and isolation of type C retrovirus particles from fresh and cultured lymphocytes of a patient with cutaneous T-cell lymphoma. *Proc. Natl. Acad. Sci. USA* 77:7415–7419.

19. Poiesz, B. J., Ruscetti, F. W., Reitz, M. S., Kalyanaraman, V. S., and Gallo, R. C. (1981). Isolation of a new type-C retrovirus (HTLV) in primary uncultured cells of a patient with Sézary T-cell leukaemia. *Nature* 294:268–271.

20. Gallo, R. C., Mann, D. L., Broder, S., Ruscetti, F. W., Maeda, M., Kalyanaraman, V. S., Robert-Guroff, M., and Reitz, M. S. (1982). Human T-cell leukemia-lymphoma vorus (HTLV) is in T- but not B-lymphocytes from a patient with cutaneous T-cell lymphoma. *Proc. Natl. Acad. Sci. USA* 79:4680–4684.

21. Reitz, M. S., Poiesz, B. J., Ruscetti, F. W., and Gallo, R. C. (1981). Characterization and distribution of nucleic acid sequences of a novel type C retrovirus isolated from neoplastic human T lymphocytes. *Proc. Natl. Acad. Sci. USA* 78:1887–1891.

22. Kalyanaraman, V. S., Sarngadharan, M. G., Robert-Guroff, M., Miyoshi, I., Blayney, D., Golde, D., and Gallo, R. C. (1982). A new subtype of human T-cell leukemia virus (HTLV-II) associated with a T-cell variant of hairy cell leukemia. *Science* 218:571–575.

23. Barre-Sinoussi, F., Chermann, J. C., Rey, F., Nugeyre, M. T., Chamaret, S., Gruest, J., Dauguet, C., Axler-Blin, C., Veinet-Brun, F., Rouzioux, C.,

Rosenbaum, W., and Montagnier, L. (1983). Isolation of a T-lymphotropic retrovirus from a patient at risk for acquired immune deficiency syndrome (AIDS). *Science* 220:868-870.

24. Gallo, R. C., Salahuddin, S. Z., Popovic, M., Shearer, G. M., Kaplan, M., Haynes, B. F., Palker, T. J., Redfield, R., Oleske, J., Safai, B., White, G., Foster, P., and Markham, P. D. (1984). Frequent detection and isolation of cytopathic retroviruses (HTLV-III) from patients with AIDS and at risk for AIDS. *Science* 224:500-504.

25. Popovic, M., Sarngadharan, M. G., Read, E., and Gallo, R. C. (1984). Detection, isolation, and continuous production of cytopathic retroviruses (HTLV-III) from patients with AIDS and pre-AIDS. *Science* 224:500-504.

26. Daniel, M. D., Letvin, N. L., King, N. W., Kannagi, M., Seghal, P. K., Hunt, R. D., Kanki, P. J., Essex, M., and Desrosiers, R. C. (1985). Serologic identification and characterization of a macaque T-lymphotropic retrovirus closely related to HTLV-III. *Science* 228:1201-1203.

27. Clavel, F., Guetard, D., Brun-Vezinet, F., Chamaret, S., Rey, M., Santos-Ferreira, M. O., Laurent, A. G., Dauguet, C., Katlama, C., Rouzioux, C., Klatzmann, J. L., Champalimaud, J. L., and Montagnier, L. (1986). Isolation of a new human retrovirus from West African patients with AIDS. *Science* 233:343-346.

28. Kanki, P. J., Barin, F., M'Boup, M., Allan, J. S., Romet-Lemonne, J. L., Marlink, R., McLane, M. F., Lee, T. H., Arbeille, B., Denis, F., and Essex, M. (1986). New human T-lymphotropic retrovirus related to simian T-lymphotropic virus type III (STLV-III). *Science* 232:238-240.

29. Maurer, B., Bannert, H., Darai, G., and Flugel, R. M. (1988). Analysis of the primary structure of the long terminal repeat and the *gag* and *pol* genes of the human spumaretrovirus. *J. Virol.* 62:1590-1597.

30. Manzari, V., Gismondi, A., Barillari, G., Morrone, S., Modesti, A., Albonici, L., DeMarchis, L., Fazio, V., Gradilone, A., Zani, M., Frati, L., and Santoni, A. (1987). HTLV-V: A new human retrovirus isolated in a *Tac*-negative T cell lymphoma/leukemia. *Science* 238:1581-1583.

2

Bovine Leukemia:

Facts and Hypotheses Derived from
the Study of an Infectious Cancer

A. Burny, [a,b] Y. Cleuter, [b] R. Kettmann, [a,b] M. Mammerickx, [a,c]
G. Marbaix, [b] D. Portetelle, [a,b] A. Van den Broeke, [b] L. Willems, [a,b] and
R. Thomas [b]

[a]*Faculty of Agronomy, Gembloux, Belgium*
[b]*University of Brussels, Brussels, Belgium*
[c]*National Institute for Veterinary Research, Brussels, Belgium*

I. INTRODUCTION

Bovine leukemia virus (BLV) is the etiological agent of a chronic lymphatic
leukemia/lymphoma in cows, sheep, and goats. Infection without neoplastic
transformation was also obtained in pigs, rhesus monkeys, chimpanzees, and
rabbits and observed in capybaras and water buffalo. Structurally and functionally, BLV is a relative of human T-lymphotropic viruses 1 and 2 (HTLV-I and
HTLV-II). HTLV-I induces in humans a T-cell leukemia, and its type II counterpart has been found in dermatopathic lymphadenopathy, hairy T-cell leukemia,
and prolymphocytic leukemia cases. At variance with HTLV-I, BLV has not
been associated with neurological diseases of the degenerative type.

BLV, HTLV-I, and HTLV-II show clear-cut sequence homologies. The
pathology of the BLV-induced disease—most notably, the absence of chronic
viremia, a long latency period, and lack of preferred proviral integration sites
in tumors—is similar to that of adult T-cell leukemia/lymphoma induced by
HTLV-I. The most striking feature of the three naturally transmitted leukemia
viruses is the X region located between the *env* gene and the long terminal repeat
(LTR) sequence. The X region contains several overlapping long open reading
frames. One of them, designated XBL-1, encodes a *trans*-activator function
capable of increasing the level of gene expression directed by BLV-LTR and

9

most probably involved in "genetic instability" of BLV-infected cells of the B-cell lineage. The genetic instability puts the cell into a context of fragility, ready to move along a number of stages toward full malignancy. Little is known about these events and their causes; we present some theoretical possibilities.

BLV infection has a worldwide distribution. In temperate climates the virus spreads mainly via iatrogenic transfer of infected lymphocytes. In warm climates and in areas heavily populated by hematophageous insects, there are indications of insect-borne propagation of the virus.

II. BLV GENOME AND GENE PRODUCTS

Bovine leukemia (lymphoma, lymphosarcoma) is a contagious disease induced by bovine leukemia virus (BLV), a retrovirus exogenous to the bovine species. It is a chronic disease, evolving over extended periods (1–8 years), with tumors developing in only a small number of infected animals. The same virus infects sheep, where it induces tumors with very high frequency (1).

BLV proviral genomic structure is detailed in Rice et al. (2–4), Sagata et al. (5), and Weiss et al. (6). Its salient features are: (a) the long terminal repeat (LTR) has an unusually long R region and functions as an efficient transcriptional promoter only in cells productively infected with BLV; (b) the *gag* polyprotein contains virus structural proteins p15, p24, and p12; (c) a protease, p14, is coded by an open reading frame overlapping the *gag* gene on the left and the *pol* gene to the right; *gag*, protease, and *pol* genes are in three different reading frames; (d) the *env* gene codes for a 72,000-*env* precursor (Pr 72^{env}) that is cleaved into two glycosylated envelope proteins, gp51 and gp30; (e) two overlapping open reading frames, located between *env* and the 3' LTR code for a 34-Kd and a 16-Kd protein, respectively.

BLV LTR is 531 base pairs long (7–9) and shares several peculiarities with the corresponding structures of HTLV-I and -II (10–13): (a) an energetically favored secondary structure occurs near the RNA cap site (13,13a); (b) no Proudfoot-Brownlee sequence exists near the polyadenylation site but such a sequence exists in the U_3 region; (c) HTLV-I, -II, and BLV LTRs function in a restricted range of cells. LTR sequences of HTLV-II and BLV contain cell-specific enhancer elements (14,14a); that markedly depend, in order to be active, on the presence of virus associated transacting factors. The LTR of HTLV-I promotes gene expression in uninfected cells, but the rate of transcription is greatly augmented in infected cells (15,16).

Two *gag* polyprotein precursors, 66 Kd (70 Kd) and 44 Kd, are synthesized in BLV-infected cells and in reticulocyte cell-free lysates, programmed with BLV 38S RNA and in frog oocytes microinjected with the same RNA message (17–21). The 66 Kd is the precursor to: (a) p15, myristilated fragment derived from the NH$_2$ terminus, (b) p24, the major core protein; (c) p12, the RNA interacting fragment, and (d) p14, the protease, probably synthesized via a frameshift suppression by a lysine-specific tRNA of the *gag* terminal codon (21). The 44-Kd precursor lacks the p14 protease. P10 is an additional cleavage product found in purified BLV preparations; it is an NH_2 terminal polypeptide of *gag* polyprotein. It is myristilated as well as p15, the larger and more common antigen derived from that region of the precursor.

The reverse transcriptase-endonuclease protein is made as 145-Kd *gag-pol* precursor. PR 145 indeed contains all the tryptic peptides of Pr 70^{gag}.

It is present in BLV-infected cells and in heterologous protein synthesizing systems programmed with BLV 38S RNA as an elongation product of Pr 70^{gag} (18). No information is available about the mode of synthesis and cleavage of the *gag-pol* precursor, which, in BLV, might require continuous transcription from genetic information located in three different reading frames.

Pr 72^{env} is the precursor to gp51, the external envelope glycoprotein, an and gp30 the transmembrane component. Pr 72^{env} overlaps the right-hand side of pol by 17 amino acids and is cleaved into its two components at the Arg-Val-Arg-Arg sequence at the carboxyl end of gp51. The unglycosylated homolog of Pr 72^{env} is a 47-Kd polypeptide, which is not further processed. Mouse monoclonal antibodies to gp51 made it possible to distinguish eight different epitopes, three of which (F, G, and H) are associated with the biological activities of the virus (infectivity and syncytia induction). Processing of Pr 72^{env} and biological activity are conformation-dependent, glycosylation-dependent phenomena. The neutralizing antibody-inducing sites of gp51 are subject to antigenic variations among BLV isolates of the same or different geographical origins (20,22,23). None of the presently identified variants was completely lacking F, G, and H. Moreover, no variation was found in the number and position of glycosylation sites and cysteine residues, stressing the uniformity of selection pressures exercized on BLV in all possible situations. Limited proteolytic hydrolysis of gp51 indicated that F, G, and H epitopes are located in the NH_2 moiety of the protein (Portetelle et al., in preparation).

The transmembrane glycoprotein gp30 is a highly glycosylated, 214-amino-acid polypeptide. Of its six cysteine residues, four are conserved in all type C viruses, suggesting an invariant and crucial pattern of disulfide bonding. Gp30 anchors the envelope proteins in the membrane of the infected cell and the virus particle. It is not known whether S-S bridges between gp51 and gp30 serve as linkages between the two proteins after proteolytic cleavage of the percursor or whether there are artifacts. The easy loss of gp51 from virions during purification procedures seems to favor the latter part of the hypothesis (2-5,24). Nearest neighbor relationships between components of BLV virions were investigated with chemical cross-linking and allowed the design of a structural model of the virus particle (25).

BLV genome contains two overlapping open reading frames 3' to the envelope gene. The long open reading frame (lor or XBL-I) encodes a 34-Kd protein acting as transactivator of transcription of the provirus (9,14–16,26) and presumably of some cellular genes, which is believed to be the key process in initiation of cell transformation. The short open reading frame (or XBL-II) (Sor) encodes a 16-Kd protein of presently undefined biological activity which is probably identical to p16-p18 detected by Ghysdael et al. (17,18) by in vitro translation of BLV genomic RNA or poly A$^+$ subgenomic fragments of it. P34 and p16 nonstructural BLV proteins are supposedly translated from the same double-spliced 2,1 Kb mRNA (4,27,28). Both proteins are recognized by sera of cattle in the tumor phase (29,30).

The genome structure, the nucleotide sequence of the provirus, and the size and amino acid sequence of structural and nonstructural viral proteins make BLV an obvious relative of human T-lymphotropic virus I and II and of simian T-lymphotropic virus T (Fig. 1) (2-6,9,11,13-16,26,31-34). The four viruses replicate and act via transactivation of the enhancer-promoter regions of their LTRS and of cellular genes and induce diseases with similar pathologies, notably the absence of chronic viremia, a long latency period, and a lack of preferred proviral sites of integration in tumors.

It remains to be seen whether the full coding capacity of BLV DNA has been apprehended. It may be of interest that careful scrutiny of BLV cDNA libraries derived from virus-infected cells contain cDNA inserts that correspond either to partially spliced mRNAs or to messengers with presently uncovered physiological significance.

A. Transmission of BLV

Transmission of BLV infection has been the subject of many field observations and experimental trials. Cases of natural infection are documented in cattle, sheep, capybara, and water buffalo (for a review, see Refs. 35, 36). The infection

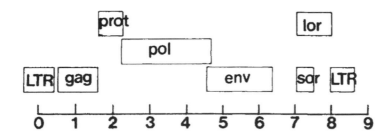

FIGURE 1   Genomic structure of BLV provirus.

can be experimentally transmitted to goats (37,38), pigs (39), rabbits (40, Altaner et al., personal communication), rhesus monkeys (41), chimpanzees (42), and buffalo (43). It has been amply documented that horizontal transmission is the rule, including the transplacental route, which amounts to 15% of infections in the offspring of BLV-positive dams (44-47) and that infected cells are the best potential vehicles of infectious BLV particles. Consequently, the concentration of BLV-infected cells in the transmitted fluid (blood in most cases) is expected to play a major role in the success or failure of BLV transmission.

The data presented in Table 1 (48) illustrate the outcome of an experiment carried out to determine the infectious dose of blood from a donor in PL. Recipients were individual sheep labeled M125-M135. Dilutions of whole blood from donor cow A (18,537 lymphocytes/μl of blood) were injected into each recipient, and seroconversion was followed with time as a marker of infection. It appears that there is a threshold amount of blood to be inoculated in order to achieve successful infection. The last infectious dose corresponded to 926 lymphocytes. Of interest too was the observation that seroconversion was delayed from day 7 to day 18 postinfection when the infectious load varied from $18,537 \times 10^3$ to 926 lymphocytes. At 49 days postinfection, antibody levels were all equal, irrespective of the number of lymphocytes administered. When the same cow was used as donor and goats as recipients, 1036 lymphocytes was the lowest dose leading to infection, an amount very similar to that required in sheep. Only twice that amount was required to achieve infection in adult cows, animals 10 times heavier than sheep or goats.

On the contrary, when the donor cow did not show persistent lymphocytosis, only the highest amount of blood (1 ml) established infection in recipient calves.

These data shed light on an important aspect of BLV transmission. Only animals in PL can be considered as highly infectious centers since minute amounts of their blood (50 nl in the documented example) transfer infection to naive sheep or goats.

Repeated observations have indicated that, in temperate climates, the rate of BLV transmission parallels the intensity of veterinary care that is dispensed to a given herd. Considering the minute amount of blood that achieved infection from PL blood (see here above and Ref. 49), it is expected that the required dose can be transmitted via needles (50,51) and dehorning gouges (52). Successful infection can also occur via PL blood-contaminated gloves used in gynecological examinations of cattle via the colorectal route (J. Evermann, personal communication), emphasizing the high permeability of the colorectal

TABLE 1 Ten Sheep Inoculated with Decreasing Numbers of Lymphocytes from Donor Cow A: ELISA (gpBLV) Titers in Sera Collected from 0 to 492 Days After Inoculation

Animal	Number of lymphocytes inoculated	ELISA (gpBLV) titers (days after inoculation)											
		0	7	11	14	18	21	25	35	49	140	311	492
M.125	18.537.000	0	20	60	540	1.620	540	1.620	43.740	43.740	43.740	43.740	43
M.126	1.853.700	0	0	20	180	4.860	14.580	14.580	14.580	43.740	43.740	131.220	131
M.127	185.376	0	0	20	60	540	1.620	4.860	14.580	43.740	43.740	43.740	131
M.128	18.537	0	0	20	20	60	60	540	43.740	43.740	43.740	53.740	131
M.129	9.268	0	0	0	60	180	540	4.860	14.580	43.740	43.740	43.740	43
M.131	1.853	0	0	0	20	0	20	60	14.580	43.740	14.580	14.580	14
M.132	926	0	0	0	0	60	180	4.860	14.580	43.740	43.740	43.740	131
M.133	185	0	0	0	0	0	0	0	0	0	0	0	0
M.134	18	0	0	0	0	0	0	0	0	0	0	0	0
M.135	1	0	0	0	0	0	0	0	0	0	0	0	0

mucosa to BLV-infected lymphocytes as was already inferred for HIV transmission in human beings.

Identification of BLV-infected wild animals in tropical regions drew attention to a possible involvement of biting insects in transmission of BLV. Observations currently made in southwestern France (Les Landes) show a positive correlation between increase of BLV-infection rate in cattle and the number of tabanids captured (A. Parodi, personal communication). Experimental transmission of BLV from a PL cow to lambs via tabanid flies has been described (53). Similar results were obtained by American investigators in Louisiana (L. D. Foil and C. J. Issel, personal communication). The latter authors could demonstrate transmission from a PL donor cow to sheep and goats with groups of 100 and 50 but not by groups of 25 and 10 horseflies (*Tabanus fuscicostatus*). On the contrary, transmission failed when the donor was a non-PL animal.

Almost ironically, the above-mentioned observations support the old practice of eliminating animals in PL in attempts to combat bovine leukemia at the herd level. The same observations may also be relevant to the possible transmission of HTLV-I via insect vectors, at least in some conditions prevalent in tropical areas (54,55). Miller et al. (55) point out: "Perhaps HTLV-I is only effectively transmitted by blood sucking insects that have had an interrupted meal on a sero-positive individual with some degree of lymphocytosis. This might then explain the apparent need for close familial contact over a prolonged period." The observations made in the BLV system support this hypothesis.

Of utmost practical importance is the repeated observation that zona pellucida-intact bovine embryos can be transferred from bovine leukemia virus-infected donors, including those bred by BLV-infected bulls (56), without risk of transmitting BLV, provided they are properly washed prior to transfer (47, 57-60). As a result, elimination of BLV infection is easily feasible even from herds with very high genetic value where culling of infected recipients would be impractical because of the high breeding potential of the animal and thus the financial loss.

B. Protection Against BLV Infection

As discussed previously, it is assumed (and demonstrated in a number of cases) that BLV transmission results from the transfer of infected cells rather than free virus. The question thus arises whether it is feasible to achieve protection via vaccination procedures against a cell-linked pathogenic agent. The status of the question has been discussed by Miller (61). From her survey of the known vaccination trials, it is obvious that very encouraging data have been obtained together with desperately poor ones. We remain, however, optimistic about the design of an efficient BLV vaccine when considering that sheep passively immunized with anti-BLV antibody can successfully resist a challenge provided they had high enough anti gp51 antibody titers (39,62). This probably means that animals able to react with efficacy against the crucial determinants of BLV gp51 do not succumb to the challenge. We have shown by inhibition of syncytia induction and neutralization of pseudotypes that the three epitopes (F, G, and H) identified on gp51 by monoclonal antibodies represent major determinants involved in the biological properties of the virus (20,22,23). Furthermore, work performed in our group has amply demonstrated that F, G, and H are conformational epitopes, glycosylation dependent, and easily denatured when handling of gp51 is carried out without the utmost care. It follows that vaccination protocols that make use of native, biologically important, determinants of gp51 constitute a prerequisite for a BLV vaccine. Experiments along these lines are under way.

III. HOST-VIRUS INTERPLAY—THE BLV TARGET CELL

A. Ab(+) Animals. The Number of Peripheral B Lymphocytes Increases. The Total Number of Circulating Lymphocytes Is Within the Normal Range

Bovine leukemia virus particles were first observed by Miller et al. (63) in short-term cultures of peripheral blood lymphocytes of BLV (+) animals in persistent lymphocytosis (see Section III.B). Numerous attempts to observe the virus in body fluids of PL animals or in tumors supposed to be of the enzootic bovine leukosis type had failed before and contributed to a large extent to make bovine leukemia a puzzling disease with obvious infectious behavior but without any observable agent. We now know that viremia can only be monitored in the first 10-12 days postinfection preceding the appearance and permanency from then on of antivirus neutralizing antibodies (64).

As the mode of action of BLV unfolds, it becomes apparent that mass production of virus particles in the host is not required. Many sites of integration of the provirus in the host DNA allow expression of p34, the trans-activating protein. In some rare cases, the latter will make the target cell conducive to full malignant transformation. If massive production of BLV is not a prerequisite, it remains true that the more permissive to BLV the B lymphocyte is, the better the chance for the animal to develop a tumor. Permissivity of sheep cells to BLV and sensitivity of the species to BLV-induced tumors make indeed a striking parallel.

The permanency of anti-BLV antibody proves the existence of a permanent antigenic stimulation, mediated via viral proteins and particles produced by lymphocytes of the B lineage and perhaps other cell types. In fact, little is known about the identity of cells involved in BLV replication in vivo. It is established that BLV persists in peripheral B lymphocytes (65) and that the proportion of B lymphocytes in the peripheral blood of BLV(+) animals increases significantly before any detectable increase in the number of circulating lymphocytes (Fossum et al., submitted; H. A. Lewin, personal communication). The PL stage is thus preceded by the polyclonal stimulation of B cells and the subsequent increase in the proportion of B cells in the blood compartment of an apparently hematologically normal animal. Moreover, refined studies with monoclonal antibodies and fluorescence-activated cell sorting have shown that: (a) some of the B cells in BLV-infected animals have altered sugar composition and seem to be arrested at an early stage of maturation; (b) a small percentage of the Ig-bearing cells of BLV(+) animals are enlarged; (c) an increased intensity of the fluorescence was observed both in the fluorescence microscope and by flow cytometry on a substantial proportion of the B lymphocytes from BLV(+) cattle compared to B lymphocytes from healthy cattle. Such observations suggest that BLV might influence the regulation of Ig expression (Fossum et al., submitted). Analogous data have been obtained by Lewin et al. (personal communication).

The BLV genome is transcriptionally repressed in both transformed and nontransformed lymphocytes in vivo (66-68). When isolated from their host and maintained in short-term cultures, PL lymphocytes and, in some cases, tumor cells (see below) express viral RNA and proteins. The mechanism by which BLV proviral transcription is repressed in vivo in B lymphocytes is not understood. It apparently involves a plasma-blocking factor (PBF) (69,70), a nonimmunoglobulin plasma protein, found primarily but not solely in BLV-infected cattle. A platelet-derived factor (PDF) has inhibitory effect on PBF (71) and exhibits high blastogenic activity on bovine lymphocytes. PDF was found in four fractions (fractions 1-4) corresponding to various molecular sizes after chromatography on Bio-gel 150. The most active fraction (fraction 4) (MW = 50,000) had physical and chemical properties similar to those of platelet-derived growth factor (PDGF). PDF and PBF acting antagonistically might be the main regulators of BLV expression in peripheral B lymphocytes, a situation at variance with the one prevailing on the HTLV-I system, where expression of the integrated provirus is controlled by antibodies to envelope glycoprotein gp51 (72).

B. Ab(+) Animals in Persistent Lymphocytosis

Persistent lymphocytosis (PL) is a polyclonal proliferation of B cells occurring in some animals in response to constant stimulation by BLV antigens. As a consequence, the B/T cell ratio is modified concurrently to an increase (or not) of the total number of circulating lymphocytes. An animal is considered to be in PL when its total lymphocyte count significantly exceeds the value recognized as normal for animals of its class of age in several successive blood tests (73,74).

As discussed here earlier, recent data (Fossum et al., submitted; H. A. Lewin, personal communication) have demonstrated that, in fact, PL prolongs and expands the phenomenon of B-cell proliferation already evident in some animals in the preceding stage of BLV infection. As expected, unusual traits of B lymphocytes already seen in some BLV-infected animals, such as modification of the glycosylation pattern of membrane components, increase in size, and increased number of surface IgM molecules, are evident when PL lymphocytes are examined by flow cytometry.

Earlier studies on heritability of susceptibility to PL led to the conclusion that PL aggregates along some families (75). Recent observations made in a herd of Shorthorn cattle (n = 117) suggested that the bovine lymphocyte antigen (BoLA) system plays a role in determining susceptibility or resistance to B-cell proliferation and lymphocytosis. Relative resistance was associated with BoLA W7 whereas susceptibility was associated with BoLA W12.3 and BoLA W8 (76). The authors concluded that subclinical progression of BLV infection is under the control of the BoLA complex and suggested that the BoLA system can be used to select for resistance to B-cell proliferation and development of lymphocytosis in BLV-infected herds. They also noted that the same BoLA alleles are involved with BLV-related traits in different herds within a breed but may differ between breeds. Considering that PL animals are the major source of infection within a herd (see Section III.A), it is highly advisable to eliminate the most sensitive haplotypes as a first step to reduce the overall rate of infection in heavily contaminated cattle populations.

Biochemical studies of numerous PL cases have shown that BLV provirus integrates at many possible sites in the cell genome. In PL, many different clones of BLV-infected cells proliferate; their propagation seems to be controlled at some times and seems to escape control at other times. PL can indeed be stabilized for a long time. It can also progress to higher values or suddenly disappear with restoration of a normal white blood cell count. No tumor cell clone seems to be present among PL lymphocytes, as techniques for in vitro cell culture that are successful in getting tumor cells to grow in vitro never led to establishment of a cell line when applied to PL lymphocytes (77; Cleuter et al., in preparation).

Identification of PL lymphocytes has been the subject of a few studies. PL cells contain BLV-infected cells, at a resting stage, incorporating very little radioactive thymidine and BLV-free, reactive cells, avidly incorporating the DNA precursor when stimulated in vitro (78,79). As mentioned earlier, the majority of PL lymphocytes are B cells, 3–10% of which are large blastoid cells, which represents an up to 30-fold increase as compared to the proportion found in control animals (Fossum et al., submitted). Flow cytometry analysis of PL lymphocytes labeled with fluoresceinated antibodies to bovine IgM showed a strong elevation of fluorescence intensity relative to that encountered with lymphocytes of normal cattle, indicating an increased number of surface immunoglobulin molecules (Fossum et al., submitted; H. A. Lewin, personal communication). It was also observed that PL lymphocytes express an altered sugar composition on their surfaces compared to noninfected animals (Fossum et al., submitted) and that induction of BLV, by short-term culture, may influence the regulation of Ig and BoLA class II antigen expression (Lewin et al., personal communication). Molecular hybridization data (80) have shown that the ratio of BLV(+) cells to BLV(-) cells (reactive cells) is roughly 1/3 to 1/4, a value in good agreement with the results of Kenyon and Piper (78,79).

In conclusion, PL or, more adequately, B-cell proliferation with or without persistent lymphocytosis is a genetically controlled manifestation of BLV infection. The expanding B-cell pool is heterogeneous with regard to the presence of BLV provirus and the stage of differentiation of the cells. Disregulation in the expression of sIg and sometimes BoLA class II antigen is obvious.

C. Ab(+) Tumor Cases

Susceptibility to tumor development varies among the three ruminant species, cattle, sheep, and goat. Sheep are the most sensitive. In one of our experimental trials, all infected animals succumbed in the tumor phase from 6 months to 6 years postinfection. Goats are the least sensitive, only two animals of 20 developed a tumor within 10 years of BLV infection. The susceptibility to tumor development in BLV-infected cattle can be qualified as intermediate. Even kept to old age, a fraction only, probably less than 10% of BLV carriers, will come down with a tumor. Irrespective of the animal species, tumors are lymphoid and involve the lymphatic system (tumorous lymph nodes) or are made of lymphoid masses of cells invading other tissues.

Liquid hybridization and Southern blot analysis using BLV-genome-specific probes showed that all BLV-induced tumors and most of the cells in individual tumors contain proviral genetic information, thus demonstrating the indispensable presence of the virus for tumor development (38,67,80). Up to four copies of proviral DNA can be found per tumor cell. Among those about 1/4 of the copies harbor deletions. We even observed tumors with a single deleted copy of the provirus per cell. In this case, however, the deletion does not cover the 5'LTR or the 3'half of the provirus. We recently cloned and sequenced such a provirus. The deletion expanded over more than 4 Kb from nucleotide 1030 in the p24 gene up to nucleotide 5332 in the middle of the gp51 gene, thus eliminating the spliced-in fragment at the end of the *pol* gene conserved in the *tat* mRNA of transactivating viruses (BLV, HTLV-I, and -II). Is the deleted provirus able to code for a functional *tat* BLV protein? We believe it is but have no proof so far. Our interpretation is that deletions occur at the provirus integration and can only be selected for by proliferation of the tumor clone if the deleted provirus expresses a functional *tat*-BLV protein. Expression of a functional *tat* protein is the mandatory prerequisite for BLV provirus to act as tumor inducer and to replicate.

Early experiments (80) showed a striking difference between PL cells and tumors. In persistent lymphocytosis, proviral DNA was found to be integrated at a large number of genomic sites in one-fourth to one-third of circulating leukocytes. In the tumor form, in contrast, proviral DNA was found to be integrated at one or very few sites in the genome of the target cell. Tumors are the results of a mono- or oligoclonal proliferation of cells, these terms referring to the site(s) of BLV integration. Integration sites, however, are not conserved from one to the other. DNAs from 25 independent hamster x bovine somatic cell hybrids were submitted to the Southern blot analysis with probes made of unique cell DNA fragments adjacent to single-copy proviruses from three different bovine tumors. It appeared that these cellular sequences, and thus the respective proviruses, belonged to three different chromosomes in the three different tumors examined (81). Previous experiments indeed had been designed to define large restriction fragments in normal cell DNA (15 and 17 Kb, respectively) corresponding to the cellular domains surrounding two proviruses in the original tumors. No rearrangement of these two cellular loci, due to the insertion of a BLV provirus, was found in 28 other BLV-induced tumors (82). We can thus safely conclude that the tumor cell can accommodate the proviral DNA sequence at many sites in the genome. This conclusion is well in line with the idea that transformation via BLV requires expression at least in the inducing phase of BLV *tat* protein. This requirement will be satisfied provided that BLV provirus resides in a region of open chromatin and presents the right signals and conformation for active transcription.

Studies of BLV expression in virus-induced tumors have led systematically to puzzling negative results. Tumor tissue taken in vivo does not express viral RNA or viral proteins. It can be argued, however, that as BLV-induced tumors are made essentially of resting cells, it was not surprising to find that the viral genes were silent. To circumvent the argument, we tried and succeeded to grow in vitro the tumor cells from four bovine and two ovine tumor cases. We could easily show (77; Cleuter et al., in preparation), via the restriction profile of the proviral sequences, that the cells growing in vitro were indeed the tumor cells. The restriction profiles were identical in the cultured cells and in the tumors they derived from. Among the so-developed six tumor cell lines, two express very little viral antigens (their viral RNA content amounts to about 0.15 molecule per cell) and four have no detectable viral RNA or viral proteins (as assessed by Western blots with high-affinity polyclonal antibody to BLV structural proteins). We are thus led to the apparent paradox that BLV is strictly mandatory as an inducer of the neoplastic process but is dispensable once the process has been switched on.

Little is known so far about the exact identity of the BLV-induced tumor cell. Flow cytometry analyses of our six tumor cell lines showed that they share the same phenotypic traits irrespective of their bovine or ovine origin. In short, they reacted with monoclonal antibodies to BoLA class I and II determinants and to heavy chains of IgM. No reactivity was encountered with monoclonals specific for T or B cells or monocytes. Fluorescein-labeled antibodies to Ig light chains did not produce any signal above background (Cleuter et al., in preparation). We tentatively concluded from these preliminary experiments that BLV-induced tumor cells, whether of bovine or ovine origin and propagated in vitro, exhibit characteristics of pre-B cells. More tumor cell lines and more antibody reagents are being developed and will allow, together with DNA probes to Ig genes, more precise identification of the differentiation stage at which BLV-induced tumor cells are blocked.

IV. FACTS AND HYPOTHESES ABOUT BLV AND CELL TRANSFORMATION

The mode of cell transformation by BLV, HTLV-I, and HTLV-II remains conjectural. Beyond any doubt, the viral *tat* gene plays a key role within a given cellular background. The virus-cell interplay leads the cell to a given stage where it can stay forever being blocked in its differentiation pathway. No doubt, the virus is necessary but by no means sufficient. Given sets of circumstances rarely encountered or rare secondary events must contribute to push the cell across barriers beyond which an irreversible state toward full malignancy has been reached. We know nothing about the nature and number of circumstances and events involved in leading a cell to transformation.

Considering that the virus is mandatory in the initial steps of the process and apparently dispensable later, we hypothesize that the transient expression of viral functions can lead to permanent expression of critical cellular genes. Two possibilities can be considered: Either the transformation process rests entirely on regulatory mechanisms without alterations of DNA (sometimes called "epigenetic processes") or alterations of the genetic material of the cell make that cell susceptible to progress toward the neoplastic state. (This may include mutations, deletions, amplifications, translocations, visible chromosomal abnormalities, etc.) Obviously, in the BLV and HTLV-I and -II systems, regulatory modifications are the results of *tat* expression, which in turn does or does not induce alterations of host cell DNA.

A. Induction of Cell Transformation via Regulatory Pathways

As far as one can tell, a viral function, presumably *tat*, switches on host functions responsible for the development of tumors, and these host functions, once initiated, remain on even though all viral functions have been switched off.

Two aspects will be briefly discussed:

1. Transactivation of host functions
2. The possible mechanisms by which a gene may be switched on by a transient signal and remain on after the signal has disappeared

That a regulatory gene can act in *trans* by emitting a diffusible regulator is one of the very bases of biological regulation, as uncovered by Jacob and Monod (83).

That these signals can be positive as well as negative has been shown by Englesberg (84) and by Thomas (85,86). More specifically, *trans*-activation has been discovered in bacterial lysogens: Silent prophage genes can be transactivated (without lifting immunology) by superinfecting the lysogen with a related heteroimmune phage (see Thomas [85,86]; Dambly et al. [87]).

How can a *transient* signal switch on a gene *permanently*? A transient signal might, of course, modify gene expression permanently by exerting an appropriate transmissable modification of DNA structure (see below). However, the same result can also be reached without any change at the level of DNA structure, provided the gene is (or is controlled by) a regulatory gene exerting (directly or indirectly, via other genes forming with it a positive feedback loop) a positive control on its own expression. Let us first consider such a positive loop in the absence of the above-mentioned signal. Either the regulatory product is below its threshold of efficiency ("absent") and since it is necessary to its own synthesis it will not be further synthesized and remain indefinitely absent (positive loop "off"), or the regulatory product is already above its threshold of efficiency ("present") and it will be further synthesized and remain indefinitely present (positive loop "on"). Now, if the loop was initially off but it is switched on by a signal, it will remain on permanently even though the signal itself is transient. A concrete example is that of gene cI of bacteriophage. This gene is switched on by the positive regulator cII (signal); since cI exerts a positive control on its own synthesis, once lit (under cII control) it remains so (under its own control) even though cII expression is now repressed as well as other viral genes (by the cI product!) (88).

Many features of our system would be consistent with the idea that the transient expression of *tat* permanently switches on a positive loop made of host genes, of which one or more switch off the viral genes (including *tat*) and one or more are involved in the development of the tumor.

B. Induction of Cell Transformation via Alterations of DNA

Abnormalities in the structure of chromosomes (translocations) or in their number (aneuploidy, hyperdiploidy, trisomy) are frequently observed in human leukemias and lymphomas (89). Aneuploidy and structural modifications have also been reported in BLV-induced bovine leukemia (for a review, see Burny et al. [35]) and are observed in cultured tumor cells from cow and sheep in the tumor phase (Popescu et al., in preparation). The major question that arises is whether chromosomal aberrations are a primary event in tumorigenesis, leading to activation of oncogenes, of growth factor or growth factor receptor genes, of differentiation genes, or of metal ion regulating genes (90) or rather are secondary to the transformation process and reflect an adaptation of the cancer cells to tissue culture conditions or to abnormalities of metabolic pathways. Solving this problem is by no means simple.

Observations made in a number of myeloid leukemia cases in humans indicate that these diseases have a multistep pathogenesis, with clonal proliferation of marrow stem cells *preceding* acquisition of distinctive chromosomal, morphological, and clinical abnormalities. The first step, as hypothesized by Fialkow and Singer (91) is characterized by genetic instability and confers a proliferative advantage to a cell clone. Other steps induce chromosomal abnormalities in the offspring of this clone and adaptation to abnormal metabolic conditions.

Chromosomal imbalances are also frequently observed in solid tumors (92) such as endometrial adenocarcinomas in humans. They show several recurrent trisomies or tetrasomies: 1 (long arm), 10, 2, 7, 12, 3, and possibly X (long arm). These chromosomes or chromosome segments carry the majority of the genes coding for enzymes of glucidic metabolism, citrate cycle, and initial steps of nucleotide synthesis (93). These anomalies are interpreted as reflecting a disturbance of metabolic pathways whose enzymes are coded by housekeeping genes and thus an adpative modification rather than a causative event. In favor of this interpretation is the repeated observation that no chromosomal anomalies are detected in the less advanced stages of endometrial carcinomas (94).

Aneuploidy and pseudodiploidy are also currently observed in B-cell lymphomas of baboons thought to be induced by the concomitant infection of the animal by an HTLV-I-related, T-cell tropic virus and a herpes-type agent, which is B-cell tropic. Karyotypic anomalies are frequently observed in fresh lymphoma cells for a given set of chromosomes, namely chromosomes 20, 1, 4, 10, and X, which, interestingly, are the evolutional homologs of human chromosomes 14, 1, 6, 8, and X, the most frequently involved in numerical and structural aberrations in human malignant lymphomas (95).

It is tempting to consider, at this writing, that BLV-*tat* gene product (p34) is the key molecule that introduces genetic instability in the cell to be transformed. P34 probably up- or downregulates expression of a number of genes. (For example, the high level of expression of BoLA class II molecules in our BLV-induced cultured tumor cells is striking.) As explained earlier, expression of *tat* may induce continuous expression of positively autoregulated cellular genes. Otherwise, *tat* expression may lead directly or indirectly to deregulation of expression of some critical genes such as histone genes, a condition known to initiate in yeast an imbalance of histone class proteins, which, in turn, induces chromosomal abnormalities (96). Such karyotypic rearrangements may well markedly affect expression of some oncogenes, differentiation genes, metal ion–regulating genes, growth factor and growth factor receptor genes. Known examples are the *myc* and *abl* oncogenes in Burkitt lymphoma and Philadelphia chromosome-positive chronic myelogenous leukemia, respectively. Putatively new oncogenes *bcl-1* and *bcl-2* are located on chromosomes 11q13 and 18q21, respectively, and are thought to be activated in B-cell lymphomas. (for review see Klein and Klein [97]; Goldman and Harnden [89]). Consequences of *tat* expression may also be the abnormal ploidies as examplified here earlier.

We are thus led to conclude that *tat* expression is a primary event whose neoplastic effect may be mediated via genetic instability. The malignant state, the final stage in the multistep process for B-cell lymphomagenesis (98), is maintained owing to positive regulatory loops without karyotypic alterations (see Section IV.A) or via permanent modifications of the genome materialized by chromosomal abnormalities (see B here above). If this is so, it is understandable that no expression of the BLV provirus is required to maintain the transformed state. The proviral information plays a major role which is mandatory but not sufficient. As soon, however, as a critical step in the tumorigenic process has been reached, the provirus is of no use; it is dispensable. We do not know the identity of this critical step, nor do we appreciate the range of possibilities included in this step. We have, indeed, several cell clones that do not grow well in vitro. It is possible that a number of cells escape the leukemic block, differentiate, and die while others remain neoplastic and divide. Should this hypothesis be verified experimentally, it would show that there are degrees in the intensity of transformation. What are these degrees? What molecules are involved and what are the regulatory circuits in which they play a role?

Karyotypic analysis of our six BLV-induced tumor cell lines shows that profound rearrangement of one X chromosome is a common event (Popescu et al., in preparation). Comparisons between fresh, stimulated tumor cells and their cultured counterparts will demonstrate whether or not the chromosomal abnormalities preexist in the original tissue. If the chromosomal aberrations are the result of in vitro culture, it might suggest that BLV-induced tumors are conditioned neoplasms (99-101), progressing to autonomous behavior via chromosome rearrangements.

The fruitful work performed in the last few years on *trans*-activating retroviruses is shedding light on initial steps of cell transformaion. The BLV, HTLV-I, and HTLV-II systems should provide basic clues about the chain of events that lead T- and B-committed lymphocytes from normal to neoplastic behavior.

ACKNOWLEDGMENTS

The work performed in the authors' laboratory was helped financially by the Fonds Cancérologique de la Caisse Générale d'Epargne et de Retraite and by the Ministry of Agriculture. R. Kettmann and G. Marbaix are Maître de Recherche and L. Willems is Aspirant of the Fonds National de la Recherche Scientifique; A. Van den Broeke is a Fellow of the Lady Tata Memorial Trust.

REFERENCES

1. Mammerickx, M., Dekegel, D., Burny, A., and Portetelle, D. (1976). Study on the oral transmission of bovine leukosis to the sheep. *Vet. Microbiol.* 1: 347–350.
2. Rice, N. R., Stephens, R. M., Couez, D., Deschamps, J., Kettmann, R., Burny, A., and Gilden, R. V. (1984). The nucleotide sequence of the *env* gene and post-*env* region of bovine leukemia virus. *Virology* 138:82–93.
3. Rice, N. R., Stephens, R. M., Burny, A., and Gilden, R. V. (1985). The *gag* and *pol* genes of bovine leukemia virus: Nucleotide sequence and analysis. *Virology* 142:357–377.
4. Rice, N. R., Stephens, R. M., and Gilden, R. V. (1987). Sequence analysis of the bovine leukemia virus genome. In: *Enzootic Bovine Leukosis and Bovine Leukemia Virus* (A. Burny and M. Mammerickx, eds.), pp. 115–144. Boston, Martinus Nyhoff.
5. Sagata, N., Yasunaga, T., Tsuzuku-Kawamura, J., Ohishi, K., Ogawa, Y., and Ikawa, Y. (1985). Complete nucleotide sequence of the genome of bovine leukemia virus: Its evolutionary relationship to other retroviruses. *Proc. Natl. Acad. Sci. USA* 82:677–681.
6. Weiss, R., Varmus, H., and Coffin, J. (eds.) (1984). *RNA Tumor Viruses.* Cold Spring Harbor, NY, Cold Spring Harbor Laboratory.
7. Tsimanis, R., Bichko, V., Dreilina, D., Meldrais, J., Lozha, V., Kukaine, R., and Gren, E. (1983). The structure of cloned 3'-terminal RNA region of bovine leukemia virus. *Nucl. Acids Res.* 11:6079–6087.
8. Couez, D., Deschamps, J., Kettmann, R., Stephens, R. M., Gilden, R. V., and Burny, A. (1984). Nucleotide sequence analysis of the long terminal repeat of integrated bovine leukemia provirus DNA and of adjacent viral and host sequences. *J. Virol.* 49:615–620.
9. Derse, D., Diniak, A. J., Casey, J. W., and Deininger, P. L. (1985). Nucleotide sequence and structure of integrated bovine leukemia virus long terminal repeats. *Virology* 141:162–166.
10. Seiki, M., Hattori, S., and Yoshida, M. (1982). Human adult T-cell leukemia virus: Molecular cloning of the provirus DNA and the unique terminal structure. *Proc. Natl. Acad. Sci. USA* 79:6899–6902.
11. Seiki, M., Hattori, S., Hirayama, Y., and Yoshida, M. (1983). Human adult

T-cell leukemia virus: Complete nucleotide sequence of the provirus genome integrated in leukemia cell DNA. *Proc. Natl. Acad. Sci. USA* 80: 3618–3622.

12. Shimotohno, K., Golde, D. W., Miwa, M., Sugimura, T., and Chen, I. S. Y. (1984). Nucleotide sequence analysis of the long terminal repeat of human T-cell leukemia virus type II. *Proc. Natl. Acad. Sci. USA* 81:1079–1083.

13. Sodroski, J. G., Rosen, C. A., and Haseltine, W. A. (1984). *Trans*-acting transcriptional activation of the human T lymphotropic virus long terminal repeat in infected cells. *Science* 225:381–384.

13a. Sodroski, J., Trus, M., Perkins, D., Patarca, R., Wong-Staal, F., Gelmann, E., Gallo, R., and Haseltine, W. A. (1984). Repetitive structure in the long-terminal-repeat element of a type II human T-cell leukemia virus. *Proc. Natl. Acad. Sci USA* 81:4617–4621.

14. Derse, D., and Casey, J. W. (1986). Two elements in the bovine leukemia virus long terminal repeat that regulate gene expression. *Science* 231: 1437–1440.

15. Rosen, C.A., Sodroski, J. G., Kettmann, R., and Haseltine, W. A. (1986). Activation of enhancer sequences in type II human T-cell leukemia virus and bovine leukemia virus long terminal repeats by virus-associated trans-acting regulatory factors. *J. Virol.* 57:738–744.

16. Rosen, C. A., Sodroski, J. G., Willems, L., Kettmann, R., Campbell, K., Zaya, R., Burny, A., and Haseltine, W. A. (1986). The 3' region of bovine leukemia virus genome encodes a *trans*-activator protein. *EMBO J.* 5:2585–2589.

17. Ghysdael, J., Kettmann, R., and Burny, A. (1978). Translation of bovine leukemia virus genome information in heterologous protein synthesizing systems programmed with virion RNA and in cell-lines persistently infected by BLV. *Ann Rech. Vet.* 9:627–634.

18. Ghysdael, J., Kettmann, R., and Burny, A. (1979). Translation of BLV virion RNAs in heterologous protein-synthesizing systems. *J. Virol* 29: 1087–1098.

19. Ghysdael, J., Bruck, C., Kettmann, R., and Burny, A. (1984). Bovine leukemia virus. In: *Current Topics in Microbiology and Immunology* (P. K. Vogt and H. Koprowski, eds.), pp. 1–19. Berlin, Springer Verlag.

20. Bruck, C., Rensonnet, N., Portetelle, D., Cleuter, Y., Mammerickx, M., Burny, A., Mamoun, R., Guillemain, B., Van der Maaten, M., and Ghysdael, J. (1984). Biologically active epitopes of bovine leukemia virus glycoprotein gp51: Their dependence on protein glycosylation and genetic variability. *Virology* 136:20–31.

21. Yoshinaka, Y., Katoh, I., Copeland, T., Smythers, G. W., and Oroszlan, S. (1986). Bovine leukemia virus protease: Purification, chemical analysis and in vitro processing of gag precursor polyproteins. *J. Virol.* 57:826–832.

22. Bruck, C., Mathot, S., Portetelle, D., Berte, C., Franssen, J. D., Herion, P., and Burny, A. (1982). Monoclonal antibodies define eight independent antigenic regions on the bovine leukemia virus (BLV) envelope glycoprotein gp51. *Virology* 122:342–352.

23. Bruck, C., Portetelle, D., Burny, A., and Zavada, J. (1982) Topographical analysis by monoclonal antibodies of BLV gp51 epitopes involved in viral functions. *Virology* 122:353-362.
24. Schultz, A. M., Copeland, T. D., and Oroszlan, S. (1984). The envelope proteins of bovine leukemia virus: Purification and sequence analysis. *Virology* 135:417-427.
25. Uckert, W., Wunderlich, V., Ghysdael, J., Portetelle, D., and Burny, A. (1984). Bovine leukemia virus (BLV)–A structure model based on chemical cross-linking studies. *Virology* 133:386-392.
26. Rosen, C. A., Sodroski, J. G., Kettmann, R., Burny, A., and Haseltine, W. A. (1985). Trans-activation of the bovine leukemia virus long terminal repeat in BLV-infected cells. *Science* 227:321-323.
27. Mamoun, R., Astier-Gin, T., Kettmann, R., Deschamps, J., Rebeyrotte, N., and Guillemain, B. (1985). The px region of the bovine leukemia virus is transcribed as a 2,1 kilobase mRNA. *J. Virol.* 54:625-629.
28. Sagata, N., Tsuzuku-Kawamura, J., Nagayoshi-Aida, M., Shimizu, F., Imagawa, K. I., and Ikawa, Y. (1985). Identification and some biochemical properties of the major X BL gene product of bovine leukemia virus. *Proc. Natl. Acad. Sci. USA* 82:7879-7883.
29. Yoshinaka, Y., and Oroszlan, S. (1985). Bovine leukemia virus post-envelope gene coded protein: Evidence for expression in natural infection. *Biochem. Biophys. Res. Commun.* 131:347-354.
30. Willems, L., Bruck, C., Portetelle, D., Burny, A., and Kettmann, R. (1986). Expression of a cDNA clone corresponding to the long open reading frame XBL-I of the bovine leukemia virus. In: *Modern Trends in Human Leukemia VII* (R. Neth, M. F. Greaves, and R. C. Gallo, eds.). Berlin, Springer Verlag, in press.
31. Poiesz, B. J., Ruscetti, F. W., Gazdar, A. F., Bunn, P. A., Minna, J. D., and Gallo, R. C. (1980). Isolation of type C retrovirus particles from cultured and fresh lymphocytes of a patient with cutaneous T-cell lymphoma. *Proc. Natl. Acad. Sci. USA* 77:7415-7419.
32. Kalyanaraman, V. S., Sarngadharan, M. G., Robert-Guroff, M., Miyoshi, I., Blayney, D., Golde, D., and Gallo, R. C. (1982). A new subtype of human T-cell leukemia virus (HTLV–II) associated with a T-cell variant of hairy cell leukemia. *Science* 218:571-573.
33. Seiki, M., Hikikoshi, A., Taniguchi, T., and Yoshida, M. (1985). Expression of the pX gene of HTLV-I. General splicing mechanism in the HTLV family. *Science* 288:1532-1534.
34. Oroszlan, S., Copeland, T. D., Henderson, L. E., Stephenson, J. R., and Gilden, R. V. (1979). Amino terminal sequence of bovine leukemia virus major internal protein: Homology with mammalian type C virus p30s. *Proc. Natl. Acad. Sci. USA* 76:2996-3000.
35. Burny, A., Bruck, C., Chantrenne, H., Cleuter, Y., Dekegel, D., Ghysdael, J., Kettmann, R., Leclercq, M., Leunen, J., Mammerickx, M., and Portetelle, D. (1980). Bovine leukemia virus: Molecular biology and epidemiology In: *Viral Oncology* (G. Klein, ed.), pp. 231-289. New York, Raven Press.

36. Marin, C., de Lopez, N., de Alvarez, L., Castanos, H., Espana, W., Leon, A., and Bello, A. (1982). Humoral spontaneous response to bovine leukemia virus infection in zebu, sheep, buffalo and capybara. *Current Topics Vet. Med. Animal Sci.* 15:310-320.

37. Olson, C., Kettmann, R., Burny, A., and Kaja, R. (1981). Goat lymphosarcoma from bovine leukemia virus. *J. Natl. Cancer Inst.* 67:671-675.

38. Kettmann, R., Mammerickx, M., Portetelle, D., Grégoire, D., and Burny, A. (1984). Experimental infection of sheep and goat with bovine leukemia virus: Localization of proviral information in the target cells. *Leukemia Res.* 8:937-944.

39. Mammerickx, M., Portetelle, D., and Burny, A. (1981). Experimental cross-transmission of bovine leukemia virus (BLV) between several animal species. *Zentralb. Veterinaermed. B.* 28:69-81.

40. Burny, A., Cleuter, Y., Couez, D., Dandoy, C., Gras-Masse, H., Gregoire, D., Kettmann, R., Mammerickx, M., Marbaix, G., Portetelle, D., Tartar, A., Van den Broeke, A., and Willems, L. (1985). Bovine leukemia virus (BLV) as a model system for human lymphotropic virus (HTLV) and HTLV as a model for BLV. In: *Proceedings of the XIIth Symposium for Comparative Research on Leukemia and Related Diseases* (F. Deinhardt, ed.), pp. 336-348.

41. Schödel, F., Hahn, B., Hübner, R., and Hochstein-Mintzel, V. (1986). Transmission of bovine leukemia virus (BLV) to immunocompromised monkeys: Evidence for persistent infection. *Microbiologica* 9:163-172.

42. Van der Maaten, M. J., and Miller, J. M. (1976). Serological evidence of transmission of bovine leukemia to chimpanzees. *Vet. Microbiol.* 1:351-357.

43. Persechino, A., Montemagno, F., and D'Amore, L. (1984). Sulla recettivia del buffalo al virus della leucosi bovina enzootica. II. Prova di infezione sperimentale. *Atti Soc. Ital. Buiatria* 15:497-498.

44. Piper, C. E., Abt, D. A., Ferrer, J. F., and Marshak, R. R. (1975). Seroepidemiological evidence for horizontal transmission of bovine, C-type virus. *Cancer Res.* 35:2714-2716.

45. Piper, C. E., Ferrer, J. F., Abt, D. A., and Marshak, R. R. (1979). Postnatal and prenatal transmission of the bovine leukemia virus under natural conditions. *J. Natl. Cancer Inst.* 62:165-168.

46. Van der Maaten, M. J., Miller, J. M., and Schmerr, M. J. (1981). In utero transmission of bovine leukemia virus. *Am. J. Vet. Res.* 42:1052-1054.

47. Digiacomo, R. F., Studer, E., Evermann, J.F., and Evered, J. (1986). Embryo transfer and transmission of bovine leukosis virus in a dairy herd. *J. Am. Vet. Med. Assoc.* 188:827-828.

48. Mammerickx, M., Portetelle, D., De Clercq, C., and Burny, A. (1986). Experimental transmission of bovine leukemia virus to cattle, sheep and goats: Infectious doses of blood and incubation period of the infection. *Leukemia Res.*, in press.

49. Van der Maaten, M. J., and Miller, J. M. (1977). Susceptibility of cattle to bovine leukemia infection by various routes of exposure. In: *Advances in*

Comparative Leukemia Research (P. Bentvelzen, J. Hilgers, and D. S. Yohn, eds.), pp. 29–32. New York, Elsevier/North Holland Biomedical Press.

50. Wilesmith, J. W. (1979). Needle transmission of bovine leukosis virus. *Vet. Rec.* 104:107.

51. Evermann, J. F., Digiacomo, R. F., Ferrer, J. F., and Parish, S. M. (1986). Transmission of bovine leukosis virus by blood inoculation. *Am. J. Vet. Res.* 47:1885–1887.

52. Digiacomo, R. F., Darlington, R. L., and Evermann, J. F. (1984). Natural transmission of bovine leukaemia virus in dairy calves by dehorning. *Can. J. Comp. Med.* 49:340–342.

53. Ohshima, K., Okada, K., Numakunai, S., Yoneyama, Y., Sato, S., and Takahashi, K. (1981). Evidence on horizontal transmission of bovine leukemia virus due to blood-sucking tabanid flies. *Jpn. J. Vet. Sci.* 43:79–81.

54. Greaves, M. F., and Miller, G. J. (1986). Are haemotophagous insects vectors for HTLV-I? In: *Modern Trends in Human Leukaemia VII* (R. Neth, M. F. Greaves, and R. C. Gallo, eds.). Berlin, Springer Verlag, in press.

55. Miller, G. J., Pegram, S. M., Kirkwood, B. R., Beckles, G. L. A., Byam, N. T. A., Clayden, S. A., Kinlen, L. J., Chan, L. C., Carson, D. C., and Greaves, M. F. (1986). Ethnic composition, age, sex and the location and standard of housing as determinants of HTLV-I infection in an urban trinidadian community. *Int. J. Cancer,* in press.

56. Monke, D. R. (1986). Noninfectivity of semen from bulls infected with bovine leukosis virus. *J. Am. Vet. Med. Assoc.* 188:823–826.

57. Olson, C., Rowe, R. F., and Kaja, R. (1982). Embryo transplantation and bovine leukosis virus: Preliminary report. In: *Fourth International Symposium on Bovine Leukosis,* Bologna, 1980 (O. C. Straub, ed.), pp. 361–369. The Hague, Martinus Nyhoff.

58. Parodi, A. L., Manet, G., Vuillaume, A., Crespeau, F., Toma, B., and Levy, D. (1983). Transplantation embryonnaire et transmission de l'agent de la leucose bovine enzootique. *Bull. Acad. Vet. France* 56:183–189.

59. Kaja, R. W., Olson, C., Rowe, R. F., Stauffacher, R. H., Strozinski, L. L., Hardie, A. R., and Bause, I. (1984). Establishment of a bovine leukosis virus-free dairy herd. *J. Am. Vet. Med. Assoc.* 184:184–185.

60. Hare, W. C. D., Mitchell, D., Singh, E. L., Bouillant, A. M. P., Eaglesome, M. D., Ruckerbauer, G. M., Bielanski, A., and Randall, G. C. B. (1985). Embryo transfer in relation to bovine leukemia virus control and eradication. *Can. Vet. J.* 26:231–234.

61. Miller, J. (1986). Bovine leukemia virus vaccine. In: *Animal Models of Retrovirus Infection and Their Relationship to AIDS* (L. A. Salzman, ed.), pp. 421–430. New York, Academic Press.

62. Kono, Y., Arai, K., Sentsui, H., Matsukawa, S., and Itohara, S. (1986). Protection against bovine leukemia virus infection in sheep by active and passive immunization. *Jpn. J. Vet. Sci.* 48:117–125.

63. Miller, J. M., Miller, L. D., Olson, C., and Gillette, K. G. (1969). Virus-like

particles in phytohemagglutinin-stimulated lymphocyte cultures with reference to bovine lymphosarcoma. *J. Natl. Cancer Inst.* 43:1297–1305.

64. Portetelle, D., Bruck, C., Burny, A., Dekegel, D., Mammerickx, M., and Urbain, J. (1978). Detection of complement-dependent lytic antibodies in sera from bovine leukemia virus-infected animals. *Ann. Rech. Vet.* 9:667–674.

65. Paul, P. S., Pomeroy, K. A., Johnson, D. W., Muscoplat, C. C., Handwerger, B. S., Soper, F. F., and Sorensen, D. K. (1977). Evidence for the replication of bovine leukemia virus in the B lymphocytes. *Am. J. Vet. Res.* 38:873–876.

66. Kettmann, R., Marbaix, G., Cleuter, Y., Portetelle, D., Mammerickx, M., and Burny, A. (1980). Genomic integration of bovine leukemia provirus and lack of viral RNA expression in the target cells of cattle with different responses to BLV infection. *Leukemia Res.* 4:509–519.

67. Kettmann, R., Deschamps, J., Cleuter, Y., Couez, D., Burny, A., and Marbaix, G. (1982). Leukemogenesis by bovine leukemia virus: Proviral DNA integration and lack of RNA expression of viral long terminal repeat and 3' proximate cellular sequences. *Proc. Natl. Acad. Sci. USA* 79:2465–2469.

68. Marbaix, G., Kettmann, R., Cleuter, Y., and Burny, A. (1981). Viral RNA content of bovine leukemia virus-infected cells. *Mol. Biol. Rep.* 7:135–138.

69. Gupta, P., and Ferrer, J. F. (1982). Expression of bovine leukemia virus genome is blocked by a nonimmunoglobulin protein in plasma from infected cattle. *Science* 215:405–407.

70. Gupta, P., Kashmiri, S. V. S., and Ferrer, J. F. (1984). Transcriptional control of the bovine leukemia virus genome: Role and characterization of a nonimmunoglobulin plasma protein from bovine leukemia virus-infected cattle. *J. Virol.* 50:267–270.

71. Tsukiyama, K. (1985). Control of expression of bovine leukemia virus genome: Effect of plasma blocking factor and platelet-derived growth factor. *Jpn. J. Vet. Sci.* 33:101.

72. Tochikura, T., Iwahashi, M., Matsumoto, T., Koyanagi, Y., Hinuma, Y., and Yamamoto, N. (1985). Effect of human serum anti-HTLV antibodies on viral antigen induction in in vitro cultured peripheral lymphocytes from adult T-cell leukemia patients and healthy virus carriers. *Int. J. Cancer* 36:1–7.

73. Bendixen, H. J. (1963). Leukosis enzootic bovis. In: *Diagnostik, Epidemiologi, Bekaempelse.* Copenhagen, Carl F. Mortensen.

74. Burny, A., and Mammerickx, M. (eds.) (1987). *Enzootic Bovine Leukosis and Bovine Leukemia Virus.* Boston, Martinus Nyhoff.

75. Abt, D. A., Marshak, R. R., Kulp, H. W., and Pollock, R. J., Jr. (1970). Studies on the relationship between lymphocytosis and bovine leukosis. *Bibl. Haematol.* 36:527–536.

76. Lewin, H. A., and Bernoco, D. (1986). Evidence for BoLA-linked resistance and susceptibility to subclinical progression of bovine leukaemia virus infection. *Animal Genet.* 17:197–207.

77. Kettmann, R., Cleuter, Y., Gregoire, D., and Burny, A. (1985). Role of the 3' long open reading frame region of bovine leukemia virus in the maintenance of cell transformation. *J. Virol.* 54:899-901.
78. Kenyon, S. J., and Piper, C. E. (1977). Cellular basis of persistent lymphocytosis in cattle infected with bovine leukemia virus. *Infect. Immun.* 16:891-897.
79. Kenyon, S. J., and Piper, C. E. (1977). Properties of density gradient-fractionated peripheral blood leukocytes from cattle infected with bovine leukemia virus. *Infect. Immun.* 16:898-903.
80. Kettmann, R., Cleuter, Y., Mammerickx, M., Meunier-Rotival, M., Bernardi, G., Burny, A., and Chantrenne, H. (1980). Genomic integration of bovine leukemia provirus: Comparison of persistent lymphocytosis with lymph node tumor form of enzootic bovine leukosis. *Proc. Natl. Acad. Sci. USA* 77:2577-2581.
81. Gregoire, D., Couez, D., Deschamps, J., Heuertz, S., Hors-Cayla, M. C., Szpirer, J., Szpirer, C., Burny, A., Huez, G., and Kettmann, R. (1984). Different bovine leukemia virus-induced tumors harbor the provirus in different chromosomes. *J. Virol.* 50:275-279.
82. Kettmann, R., Deschamps, J., Couez, D., Claustriaux, J. J., Palm, R., and Burny, A. (1983). Chromosome integration domain for bovine leukemia provirus in tumors. *J. Virol.* 47:146-150.
83. Jacob, F., and Monod, J. (1961). Genetic regulatory mechanisms in the synthesis of proteins. *J. Mol. Biol.* 3:318-356.
84. Englesberg, E., Irr, J., Power, J., and Lee, N. (1965). Positive control of enzyme synthesis by gene C in the L-arabinose system. *J. Bacteriol.* 90:946-957.
85. Thomas, R. (1966). Control of development in temperate bacteriophages. I. Induction of prophage genes following heteroimmune super infection. *J. Mol. Biol.* 22:79-95.
86. Thomas, R., (1970). Control of development in temperate bacteriophages. III. Which prophage genes are and which are not *trans*-activable in the presence of immunity? *J. Mol. Biol.* 49:393-404.
87. Dambly, C., Couturier, M., and Thomas, R. (1968). Control of development in temperate bacteriophages. II. Control of lysozyme-synthesis. *J. Mol. Biol.* 32:67-81.
88. Eisen, H., Brachet, P., Pereira da Silva, L., and Jacob, F. (1970). Regulation of repressor expression in λ. *Proc. Natl. Acad. Sci. USA* 66:855-862.
89. Goldman, J. M., and Harnden, D. G. (eds.) (1986). *Genetic Rearrangements in Leukaemia and Lymphoma.* London, Churchill Livingstone.
90. Van den Berghe, H., and Mecucci, C. (1986). Some karyotypic aspects of human leukemia. In: *Modern Trends in Human Leukemia VII* (R. Neth, M. F. Greaves, and R. C. Gallo, eds.). Berlin, Springer Verlag, in press.
91. Fialkow, P. J., and Singer, J. W. (1985). Tracing development and cell lineages in human hemopoietic neoplasia. In: *Leukemia* (I. L. Weissman, ed.), pp. 203-222. Dahlem Koferenzen 1985. Berlin, Springer Verlag.
92. Dutrillaux, B., Muleris, M., and Gerbault-Seureau, M. (1986). Imbalance

of sex chromosomes, with gain of early-replicating X, inhuman solid tumors. *Int. J. Cancer* 37:475–479.

93. Dutrillaux, B., and Couturier, J. (1986). Chromosome imbalances in endometrial adenocarcinomas: A possible adaptation to abnormal metabolic pathways. *Ann. Genet.* 29:76–81.

94. Fujita, H., Wake, N., Kutsuzawa, T., Ichinoe, K., Hreshchysshyn, M. M., and Sandberg, A. A. (1985). Marker chromosomes of the long arm of chromosome 1 in endometrial carcinoma. *Cancer Genet. Cytogenet.* 18: 283–293.

95. Markaryan, D. S., and Popova, E. A. (1986). Cytogenetics of monkey malignant lymphomas. A comparison with cytogenetics of human malignant lymphomas. In: International Conference on Primary-Localized Haematopoietic Tissue Tumors of the Non-human Primates and the Ways of Their Generalization. Sukhumi, USSR. Abstract, p. 9.

96. Meeks-Wagner, D., and Hartwell, L. H. (1986). Normal stoichiometry of histone dimer sets is necessary for high fidelity of mitotic chromosome transmission. *Cell* 44:43–52.

97. Klein, G., and Klein, E. (1985). *myc*/Ig juxtaposition by chromosomal translocations: Some new insights, puzzles and paradoxes. *Immunol. Today* 6:208–215.

98. Gordon, J., Aman, P., Rosen, A., Ernberg, I., Ehlin-Henricksson, B., and Klein, G. (1985). Capacity of B-lymphocytic lines of diverse tumor origin to produce and respond to B-cell growth factors: A progression model for B-cell lymphomagenesis. *Int. J. Cancer* 35:251–256.

99. Furth, J. (1953). Conditioned and autonomous neoplasms: A review. *Cancer Res.* 13:477–492.

100. Foulds, L. (1958). The natural history of cancer. *J. Chronic Dis.* 8:2–37.

101. Klein, G., and Klein, E. (1986). Conditioned tumorigenicity of activated oncogenes. *Cancer Res.* 46:3211–3224.

3

Biology of Feline Retroviruses

William D. Hardy, Jr. *Memorial Sloan-Kettering Cancer Center, New York, New York*

I. INTRODUCTION

There are three subfamilies of retroviruses (Table 1): (a) Oncoviriniae, (b) Lentivirinae, and (c) Spumavirinae (1). Oncoviruses are the cancer-inducing leukemia, sarcoma, and carcinoma viruses such as the feline leukemia viruses (FeLVs), feline sarcoma viruses (FeSVs), avian leukosis viruses (ALVs), murine leukemia viruses (MuLVs), and murine mammary tumor viruses (MuMTVs). Lentiviruses, long mostly ignored because they caused rare diseases of goats and sheep in far-off lands, are now very important because the human AIDS viruses (human immunodeficiency viruses, HIV-1 and -2) have been shown to be members of this subfamily. In addition, a new feline lentivirus, which induces AIDS in cats, was discovered and was named the feline T-lymphotoropic lentivirus (FTLV) (2). Howver, the FTLV was recently renamed the feline immunodeficiency virus (FIV) Spumaviruses or foamy viruses occur in many animals, cause inapparent infections, and are not known to cause any disease. Pet cats are infected with all three subfamilies of retroviruses and there are now two feline retroviruses, from different subfamilies, that causes AIDS in cats, FeLV and FIV (Tables 1 and 2).

FeLV was discovered in Scotland in 1964 from a cat that lived with several cats (cluster household) that had developed lymphosarcoma (LSA) (3). At that time it was thought that all retroviruses were endogenous viruses that were transmitted genetically (vertically) as a Mendelian trait (4). However, by observing pet cats living in their natural household environments (5–8), and later confirmed in laboratory experiments (9), it was found that FeLV is an exogenous retrovirus that is transmitted contagiously among cats. It is now known that all disease-inducing retroviruses of animals and humans, except those of inbred laboratory mice, are exogenous and are contagiously transmitted among members of their species (1,10,11). At present 2%, or almost 1 million cats, of the

33

TABLE 1 Classification of Retroviruses

Subfamilies
  A. Oncovirinae
    1. Leukemia and sarcoma viruses
      a. Avian—ALV, ASV
      b. Murine—MuLV, MuSV
      c. Feline—FeLV, FeSV, RD-114
      d. Bovine—BLV
      e. Baboon—BaEV
      f. Wolly monkey—SSV
      g. Gibbon ape—GaLV
      h. Monkey—STLV–I
      i. Human—HTLV–I, HTLV–II
    2. Carcinoma viruses
      a. Murine—MMTV
      b. Monkey—MPMV
    3. Immunosuppressive viruses
      a. Feline—FeLV–FAIDS
      b. Monkey—SRV
  B. Lentivirinae
    1. Anemiagenic
      a. Equine—EIAV
    2. Neurotropic
      a. Ungulates: Visna and Maedi viruses
    3. Immunosuppressive
      a. Feline—FIV (FTLV)
      b. Bovine—BIV
      c. Monkey—SIV (STLV–III)
      d. Human—HIV-1, -2 (HTLV–III/LAV, LAV-2)
  C. Spumavirinae: Foamy viruses
      a. Feline—FeSFV
      b. Many species—
      c. Human—HuSFV

---

estimated 50 million pet cats in the United States are infected with FeLV (12).

The elucidation of the unique biology of feline retroviruses has often pointed toward new directions and concepts for the study of retroviruses of all animals including humans. For example, FeLV was the first naturally occurring retrovirus to be shown to be commonly transmitted in a contagious manner (5–9); FeLV was the first "oncogenic" retrovirus to be shown not to only cause cancer, but, more often, to cause degenerative diseases such as aplastic anemias

TABLE 2 Feline Retroviruses

1. Subfamily Oncovirinae	
A. Endogenous viruses	Genetically transmitted
1) RD–114	Xenotropic–does not replicate in cats, no known disease association
2) FeLV-related sequences	Full-length and shorter sequences Cannot be induced to replicate Recombine with exogenous FeLV-A to form FeLV-B and FeLV-C
B. Exogenous viruses	Spread contagiously
1) FeLVs	
Subgroup A	Ecotropic–found in all infected cats
Subgroup B	Amphotropic–found in 50% of infected cats
Subgroup C	Amphotropic–found rarely, less than 1%
2) Defective FeLV-*myc*	Found in 14% of FeLV-infected thymic LSAs Recombinant proviruses
3) Defective FeLV–FAIDS	Experimentally induces AIDS
4) FeSVs	
11 well-characterized isolates	Recombinants between FeLV and cellular oncogenes
2. Subfamily Lentivirinae	
A. FIV (feline immunodeficiency virus)	Induces AIDS syndrome
3. Subfamily Spumavirinae	
A. FeSFV (feline syncytium-forming virus)	Causes no known disease

Xenotropic = grows in heterologous (noncat) cells only; ecotropic = grows in homologous (cat) cells only; amphotropic = grows in homologous and heterologous cells, broad host range.

and immunosuppressive diseases (5,8,13–20); FeLV was the first retrovirus for which a simple and practical diagnostic test for viral infection, an immuno-fluorescent antibody test for FeLV antigens, was developed (21–25); FeLV was the first retrovirus for which an effective program of control and prevention of spread was developed and used worldwide (22–24); and FeLV was the first retro-virus for which a vaccine was developed and is now in use (26,27).

Studies of the biology and basic virology of FeLV and the bovine leukemia virus (BLV), conducted in the United States, Scotland, and Europe during the 1960s and 1970s, laid much of the groundwork that would prove so relevant for the discovery of human retroviruses (28–30). During those years, Gallo and his associates collaborated with several of the groups studying feline and bovine retroviruses. These collaborations, especially those that involved retrovirus-induced "virus-negative" leukemias, helped to develop the concepts and tech-niques that were eventually so important in the isolation of the first human retroviruses.

Present studies of feline retroviruses include: molecular characterization of defective *myc*-containing FeLVs; search for and characterization of new onco-genes in new isolates of feline sarcoma viruses; and studies of the recombination of exogenous FeLV-A with endogenous FeLV-related sequences that generate highly pathogenic viruses. These studies may lead to a better understanding of the general mechanisms by which retroviruses induce transformation and im-munosuppression and thus may contribute significantly to the treatment and prevention of human cancers and AIDS.

In this chapter I will review the current knowledge of the biology of all three subfamilies of feline retroviruses and relate this information to the biology of human retroviruses.

II. ONCOVIRUSES

There are two major groups of feline Oncovirinae (oncogenic retroviruses or oncoretroviruses): The endogenous and exogenous viruses (Table 2 and 3).

A. Virology

I will review the virology and some of the molecular biology pertaining to the unique biology of feline retroviruses, but a more detailed description of the molecular biology of the feline leukemia viruses can be found in chapter 4, "Molecular Biology of Feline Leukemia Viruses" (31).

1. Endogenous Feline Oncoviruses

a. **RD-114 Virus** The RD-114 virus is an endogenous xenotropic virus of domestic cats which is only distantly related to FeLV (32–34). Multiple com-plete copies of the RD-114 viral genomes are found in all cat cells, but the virus

TABLE 3 Origin of Feline Retroviruses

Virus	Source of donor of virus	How virus is spread among cats
Oncoviruses		
FeLV	Rat ancestor	Contagiously
enFeLV	Mouse ancestor	Genetically
RD-114	Old World monkey	Genetically
Lentiviruses		
FIV (FTLV)	Not determined	Contagiously
Spumaviruses		
FeSFV	Not determined	Contagiously

does not replicate in cats, although viral mRNA and p30 are expressed in stimulated leukemic and nonleukemic lymphoid tissues regardless of their FeLV status (35). Sera from healthy pet cats and cats with various diseases do not contain antibodies to RD-114 viral proteins, which indicates that cats are immunologically tolerant to RD-114 proteins and/or that these antigens are rarely expressed (36). RD-114 is closely related to the baboon endogenous virus (BaEV), and it is likely that RD-114 and BaEV originated together (37). The progenitor virus probably arose as a result of horizontal transmission from one species to the other or via horizontal transmission from a third species into ancestral cats and baboons at approximately the same time (Table 3) (38). RD-114 is not associated with any known feline disease and there is no evidence that these endogenous sequences recombine with exogenous FeLV to produce recombinant retroviruses.

  b. **Endogenous FeLV-Related Sequences (enFeLV)** The cellular DNA of uninfected domestic cats and their close *Felis* relatives, but not of their more distantly related *Felis* species, also contains sequences that are partially homologous to the RNA of exogenous horizontally transmitted FeLVs (39–44). A portion of the exogenous FeLV genome is not endogenous to uninfected cats cells (44). The endogenous FeLV-related sequences (enFeLVs) are found not only in specific pathogen-free domestic cats (*Felis catus*), but also in closely related Felidae including the jungle cat (*Felis chaus*), the European wildcat (*Felis sylvestris*), and the sand cat (*Felis margarita*), all originating from the Mediterranean basin (38). Other species of *Felis* from sub-Saharan Africa, Southwest Asia, and North and South America lack enFeLV. Only the cellular DNA of rodents, in particular rats, contains related virogene sequences (38). The lack of enFeLV in most Felidae and the presence of related sequences in rodents suggest that enFeLVs were acquired by cats, via transspecies infection with a virus of rodent origin, subsequent to the initial Felidae divergence but prior to that of the four positive *Felis* species.

It is interesting that those *Felis* species which contain sequences related to RD-114 virus also contain enFeLV sequences, especially since each group of viral genes were derived from a distinctly different group of animals, i.e., primates and rodents. The cellular control of enFeLV and RD-114 virogenes is quite different (38). The FeLV-related virogenes have not been induced from cat cells in culture, but the closely related exogenous FeLVs are able to replicate and cause disease in cats. In contrast, the RD-114 virogenes can be readily induced from normal "virus-negative" cat cells but are generally restricted from replicating in cat cells (33).

The enFeLV sequences are arranged as multiple (8–12 copies per haploid genome), discreet copies in a nontandem fashion and are conserved among the tissues of the same cat but vary among different cats (42). There are full-length and truncated enFeLV sequences in cat cells, although these sequences are not inducible as infectious virus (38,43–45). A region (U3) of the LTRs of the enFeLV and exogenous FeLV is different (46,47). Although there are some full-length endogenous sequences, many of the sequences are shorter than the exogenous cloned infectious FeLVs, which are 8.5–8.7 kbp in length (45,48). The shorter enFeLV sequences have a large deletion (3.3–3.6 kbp) in the *gag-pol* region and a 0.7–1.0 kbp deletion in the *env* region. Sequence analysis of the genomes of the three subgroups of FeLV, FeLV-A, -B, and -C, has shown that FeLV-B and -C arise through recombination of FeLV-A with enFeLV *env* sequences to form *env* recombinant MCF-like FeLVs (31,39,47,49-51).

2. Exogenous Infectious Oncoviruses (FeLVs)

As has been previously discussed, the exogenous infectious FeLVs probably originated through cross-species infection of an endogenous rodent, rat or ancestor of the rat, retrovirus into ancestors of the pet cat (Table 3) (37,38,52). The contagiously transmitted FeLV is a replication-competent, chronic leukemia virus which does not possess an oncogene and induces monoclonal lymphosarcomas after a long latent period. The FeLV genome consists of the 5'-*gag-pol-env*-3' genes flanked by two long terminal repeats (LTRs).

a. FeLV Subgroups There are three subgroups of FeLV (FeLV-A, -B, and -C) that are identified by their envelope gp70 molecules (Table 2) (53,54). FeLV-A is ecotropic, has the most restricted host range, and thus grows almost exclusively in cat cells. FeLV-A is always present in infected pet cats either alone (50%), in combination with FeLV-B (49%), or together with FeLV-B and FeLV-C (1%) (20,55). FeLV-B has the widest host range (is amphotropic) and can replicate in cat, mink, hamster, dog, pig, bovine, monkey, and human cells. FeLV-C is also amphotropic but has an intermediate host range and can replicate in cat, dog, mink, guinea pig, and human cells (53,54). FeLV-C is only rarely found in pet cats, less than 1%, even though many cats have antibodies to the gp70 of FeLV-C, which suggests that FeLV-C is formed by recombination with FeLV-A and is immunogenic in cats (51,56). The reason why FeLV-C cannot be isolated from these cats is not known but may be due to suppression of viral replication by the neutralizing antibody. There is an 85% sequence homology between the genomes of the three FeLV subgroups (57), although more divergence has been noted by T1 oligonucleotide analysis (58).

There is no clear association of any subgroup of FeLV with any specific naturally occurring disease. However, under experimental conditions, the Rickard strain of FeLV-A induces mainly thymic LSAs, a variant of FeLV-A causes a feline acquired immune deficiency syndrome (FAIDS), and several isolates of FeLV-C induce erythroid hypoplasia (aplastic anemia) (16,17,19,50, 51,59-61). It is now known that FeLV-A gives rise to FeLV-B and FeLV-C viruses, de novo, via recombination of its genome with the enFeLV-B or enFeLV-C related *env* sequences that are present in the DNA of all cats (see Chapter 4, "Molecular Biology of Feline Leukemia Viruses") (31,45,47,48,51, 62).

b. **Defective FeLV-*myc* Recombinant Proviruses** Defective FeLV proviruses containing the *myc* oncogene have been found in the DNA of some (14%) pet cats with naturally occurring FeLV-positive thymic LSAs (63-66). A c-*myc* rearrangement in one FeLV-positive thymic LSA and a c-*myc* amplification in an FeLV-negative B-cell alimentary LSA was also found (63). The novel *myc* sequences were encapsidated in functional viruses. These observations suggest that *myc*-containing FeLVs may be transmitted contagiously among pet cats and may induce some forms of thymic LSAs. The FeLV-*myc* defective viruses were rescued by replicating FeLV and were found to rapidly induce lymphosarcoma when inoculated into kittens (67,68) and to partially transform feline fibroblasts in culture (69). The etiological role of these acute transforming FeLVs in nature is not known since more than half of all feline lymphosarcomas do not contain such viruses.

c. **FeLV-FAIDS Viruses** In nature more pet cats die of FeLV-induced immunosuppression than die of lymphosarcoma or leukemia. Recently, a specific FeLV variant, termed FeLV-FAIDS, has been isolated and shown to induce feline AIDS (FAIDS) in 100% of SPF cats (50,59). The FeLV-FAIDS variant is replication defective and is associated with replication competent FeLVs (see Chapter 4, "Molecular Biology of Feline Leukemia Viruses").

d. **Feline Sarcoma Viruses (FeSVs)** Feline sarcoma viruses (FeSVs) are replication-defective acute transforming viruses that possess transforming genes, viral oncogenes (v-*oncs*), acquired through recombination of the FeLV genome with single-copy cellular c-*onc* (proto-oncogenes) genes (70-72). FeSVs are generated de novo in individual infected cats; they do not appear to be transmitted contagiously; and they induce neoplasms with a short latent period in animals and transform tissue culture cells of various species including human cells (70,71,73).

Pet cats are an excellent source of viral oncogene containing acute transforming viruses because approximately 1 million pet cats are chronically viremic with the replication-competent FeLV. To date 11 naturally occurring FeSVs with seven different oncogenes have been isolated from pet cats (Table 4). The study of oncogenes appears to offer great promise to our understanding of the cause of cancer. The increasing number of viral-transduced oncogenes, presently at about 30, which are shared by viruses isolated from the same species as well as those isolated in viruses from different species implies that the number of oncogenes is limited.

The 11 well characterized replication-defective FeSV isolates (Table 4) contain the v-*fes*, v-*sis*, v-*abl*, v-*fgr*, v-k-ras, and v-*kit* oncogenes. All FeSV-infected cats are also infected with replication-competent helper FeLV (70,73,74).

Experimentally FeSV, like FeLV, appears to be able to induce FeSV-negative tumors in cats and in heterologous species such as monkeys and dogs (73–80). Some FeSV-transformed cells can retain their FeSV provirus but revert to a normal morphology in vitro (81). FeSV can persist in the blood for up to 3 months in experimentally inoculated cats that have regressed their sarcomas. In

TABLE 4 Feline Sarcoma Viruses

FeSV isolate		Oncogene	Protein product	Oncogene type
Snyder-Theilen	ST-FeSV	*fes*	P85gag-fes	Tyrosine kinase
Gardner-Arnstein	GA-FeSV	*fes*	P95gag-fes	Tyrosine kinase
Hardy-Zuckerman 1	HZ1-FeSV	*fes*	P96gag-fes	Tyrosine kinase
Susan-McDonough	SM-FeSV	*fms*	gp170gag-fms	CSF-1 receptor
Hardy-Zuckerman 5	HZ5-FeSV	*fms*	ND	CSF-1 receptor[a]
Parodi-Irgens	PI-FeSV	*sis*	P76gag-sis	Platelet-derived growth factor
Gardner-Rasheed	GR-FeSV	*fgr*	P70gag-fgr	Tyrosine kinase/ actin
Theilen-Pedersen 1	TP1-FeSV	*fgr*	P83gag-fgr	Tyrosine kinase/ actin[a]
Hardy-Zuckerman 2	HZ2-FeSV	*abl*	P98gag-abl	Tyrosine kinase
Noronha-Youngren	NY-FeSV	Ki-*ras*	ND	Adenyl cyclase G protein[a]
Hardy-Zuckerman 4	HZ4-FeSV	*kit*	P80gag-kit	Tyrosine kinase

ND = not determined.
[a]Assumed although not yet determined.

a small number of these FeSV- and FeLV-negative cats, about 6%, reucrrent sarcomas develop which contain integrated FeSV proviruses (82–84). The FeSV-negative fibrosarcomas that develop in cats that were once infected with FeSV are similar to the FeLV-negative lymphosarcomas that develop in cats free of FeLV but previously infected with the virus. The recurrent fibrosarcoma cell cultures derived from FeSV-free cats do not express FeSV or FeLV antigens during initial passages, but after several passages, FeSV and FeLV are expressed (82). This suggests that virus production is suppressed in vivo by an antiviral immunity and that the antisarcoma immunity is not as competent, which allows recurrent virus-negative sarcomas to develop in cats, monkeys, and dogs. However, there is no evidence that FeSV induces virus-negative fibrosarcomas in FeLV–FeSV uninfected pet cats (73).

v-fes FeSVs: ST-, GA-, and HZ1-FeSVs. The Snyder-Theilen (ST)-FeSV, Gardner-Arnstein (GA)-FeSV, and Hardy-Zuckerman-1 (HZ1)-FeSV possess the fes oncogene (71,75,85–87). The fps oncogene of the Fujinami avian sarcoma virus (ASV) was found to be homologous with the fes oncogene of the FeSVs (88,89). This was the first demonstration that acute transforming retroviruses from different species are able to transduce homologous c-oncs. The fes oncogene is the most prevalent acute transforming retroviral oncogene and is found in three FeSVs and five ASVs (70). The gag-fes product exhibits a tyrosine-specific protein kinase activity (90).

In addition to inducing fibrosarcomas in experimentally inoculated kittens, the GA–FeSV can induce melanomas when inoculated into the eye or intradermally (91-93) and the ST–FeSV can transform murine pre-B cells (94). Thus the fes oncogene, and probably other oncogenes, can transform cells of different germ line origin.

v-fms FeSVs: SM- and HZ5- FeSVs. The Susan McDonough (SM) FeSV and the Hardy-Zuckerman-5 (HZ5)-FeSV possess the v-fms oncogene (95,96). The genome organization of the 8.6-kb HZ5-FeSV provirus is 5'-gag-fms-pol-env-3' (95). HZ5- and SM-FeSVs display indistinguishable in vitro transformed cell focus morphology. The protein products of the SM–FeSV genome are somewhat unusual and consist of a primary-translation, 155,000-dalton gag-fms protein which is glycosylated to yield a gp170 gag-fms protein. The gp170 protein is then cleaved to yield gp120 and gp140 fms proteins and a p60 gag protein (97-99). Unlike other transforming proteins, the v-fms protein products are extensively glycosylated (97).

The protein product of the protooncogene c-fms is the receptor for monocyte-macrophage proliferation which is designated colony-stimulating factor-1 (M–CSF-1) (100-102). C-fms is expressed during macrophage differentiation in normal cats, mice, and humans, and normal fibroblasts produce CFS-1 (101, 102). Since v-fms contains the entire outer domain of the M–CSF-1 receptor, SM- and HZ-5-FeSV-infected fibroblasts may acquire the ability to overrespond to M–CSF-1 by producing v-*fms* protein which then binds M–CSF-1, which may continually stimulate these cells to proliferate by an autocrine loop.

v-abl *FeSV*: *HZ2-FeSV*. The Hardy-Zuckerman-2 (HZ2)-FeSV contains a v-*onc* that is homologous to the v-*abl* of the Abelson MuLV (A—MuLV) (85, 103). The A–MuLV was isolated from a steroid-treated, Moloney-MuLV–infected mouse that developed lymphosarcoma, and the virus can transform B, T, and myeloid cells (104,105). No fibrosarcomas have been induced by the A–MuLV, whereas the HZ2-FeSV can induced multicentric fibrosarcomas in kittens (103).

The HZ2-FeSV *gag-abl* 98,000-dalton polyprotein has a protein kinase activity and is found with the cytosol and membrane fractions of transformed cells (103,106,107). A significant amount of the 98K polyprotein is found on the cytoplasmic aspect of the plasma cell membranes and has properties of an integral membrane protein (103,106).

v-sis *FeSV*: *PI-FeSV*. The v-*onc* sequences of the Parodi-Irgens (PI)-FeSV are homologous to the v-*sis* sequences of the Woolly monkey simian sarcoma virus (SSV) (108,109). Both viruses display indistinguishable transformed cell focus morphology even though the v-*sis* sequences in these viruses are expressed with different strategies. The SSV genome structure is 5'-*gag-pol-env-sis-env*-3', whereas the PI-FeSV genome structure is 5'-*gag-sis-env*-3'.

The 76K protein product of the PI–FeSV v-*sis* does not exhibit protein kinase activity and has extensive homology with platelet-derived growth factor (PDGF), a growth factor produced in platelets which stimulates fibroblast proliferation in wound healing (110–112). Thus, an overproduction of PDGF by *sis*-transformed fibroblasts might lead to excessive proliferation of cells with PDGF receptors, such as fibroblasts, and result in an autocrine stimulation and neoplastic growth.

v-fgr *FeSV*: *GR-FeSV and TP-1-FeSV*. The Gardner-Rasheed (GR)-FeSV and the Theilen-Pedersen-1 (TP1)-FeSV possess the unique oncogene designated v-*fgr* (113–115). The GR–FeSV can transform a large variety of mammalian cell lines. Protein kinase activity has been detected in the GR–FeSV p70 polyprotein, and in addition to the kinase activity, v-*fgr* has additional sequences that are homologous with the actin gene (113,116).

v-kit *FeSV*: *HZ4-FeSV*. The Hardy-Zuckerman-4 (HZ4)-FeSV has a unique oncogene that has been designated v-*kit* (117). The HZ4-FeSV genome is 5'-*gag-kit-pol-env*-3' which encodes a p80 *gag-kit* polyprotein with protein kinase activity. There is a 58% homolgy of the v-*kit* and v-*fms* sequences. The v-*fms* sequence is 2.9 kb long, while v-*kit* is 1.1 kb, and the v-*kit* segment corresponds to the inner cellular domain of the v-*fms* sequences. Thus, like c-*fms*, c-*kit* could encode a transmembrane receptor that is involved in signal transduction (117). To date the v-*kit* oncogene has been found in only one feline acute transforming retrovirus.

v-K-ras *FeSV*: *NY-FeSV*. The Noronha-Youngren (NY)-FeSV v-*onc* is homologous to the v-*ras* of the Kirsten murine sarcoma virus (118,119). No studies of the *gag-onc* protein or the biology of this virus have yet been reported.

It is interesting that the oncogenes *src*, *ros*, and *yes* have been isolated only from avian retroviruses and have not been found in any of several dozen mammalian retroviruses. In contrast, the *fms*, *kit*, and *fgr* oncogenes have been found only in feline retroviruses (117). This apparent specificity may be due to viral recombinogenic sequences, functional constraints for the expression of transforming proteins, or different tissue tropisms of feline and avian retroviruses.

The domestic pet cat is an excellent animal from which to isolate oncogene-containing acute transforming viruses because approximately 1 million pet cats are chronically viremic with the replication-competent chronic leukemia virus, FeLV. The continual replication of FeLV in cat cells for long periods of time (3–5 years) increases the chances that the rare transduction of a c-*onc* will occur. The increasing number of viral-transduced oncogenes, presently at about 30, which are shared by viruses isolated from the same species as well as isolated in viruses from different species predicts that the number of oncogenes is limited.

e. **FeLV Genome** FeLV is a replication-competent chronic leukemia virus which does not possess an oncogene. The FeLV genome consists of the $5'$-*gag-pol-env*-$3'$ genes flanked by two long terminal repeats (LTRs) (see Chapter 4).

f. **FeLV Proteins** The FeLV proteins are produced in great excess in infected cats and free viral proteins are found in the plasma, tissue fluids, and membranes and the cytoplasm of infected cells. Most of these proteins are never packaged into viral particles (21,120). Nine proteins are encoded by the FeLV genome (Table 5) and include, based on a newly proposed nomenclature, the *gag* gene internal viral structural proteins: p15 (matrix protein, MA), p12 (unknown) function, p27 (capsid protein, CP), and p10 (nucleocapsid protein, NC); the *pol* gene enzymes: p14 (protease, PR), p80 (reverse transcriptase, RT), p46 (integration protein, IN); and the *env* gene envelope proteins: gp70 (surface protein, SU) and p15E (transmembrane protein, TM) (1,10,11,121). The new nomenclature proposed for proteins common to all retroviruses is based on their biological function, enzymatic activity, and/or virion location (121).

Detection of FeLV antigens in the cytoplasm of peripheral blood leukocytes by indirect immunofluorescent antibody (IFA) tests or as soluble antigens in the plasma by enzyme-linked immunosorbent assays (ELISA) has been used in the study of the occurrence and control of FeLV in pet cats (5,8,20–24,122). A positive IFA test indicates persistent, usually lifelong (in 97% of IFA-positive cats), viremia and shedding of the virus in the saliva (5,12,21,123,124).

g. **Feline Oncornavirus-Associated Cell Membrane Antigen (FOCMA)** FOCMA was initially defined, by an IFA test using antibody present in pet cat sera, as an antigen detected on cell membranes of cultured feline FL74 LSA cells which produce all three subgroups of FeLV (125,126). Cats with high titers of

TABLE 5   Feline Leukemia Virus Proteins

Viral gene	Viral protein	Protein[125] two-letter designation	Function or enzyme activity or location in virus
gag 5'	p15	MA	Matrix protein
	p12	?[a]	Unknown
	p27	CA	Capsid protein
3'	p10	NC	Nucleocapsid protein
pol 5'	p14[b]	PR	Protease
	p80	RT	Reverse transcriptase
3'	p46[b]	IN	Integration protein
env 5'	gp70	SU	Surface protein
3'	p15E	TM	Transmembrane protein

[a]Unknown function—no name given yet.
[b]Assumed from analogous MuLV protein, not yet identified for FeLV.

FOCMA antibody are resistant to FeSV-induced fibrosarcomas and FeLV-induced LSAs (20,125,127,128). FOCMA antibody and FeLV-induced virus neutralizing (VN) antibody are discordant in many cats (20,129-131). This suggested that FOCMA is not an FeLV structural antigen, at least not an FeLV-A or -B structural antigen. FOCMA is found on the cell membranes of all feline T-cell and B-cell LSA cells irrespective of their FeLV status, on FeSV-induced fibrosarcoma cells, and on FeLV-infected erythroid and myeloid leukemic cells (129-135). In contrast, FOCMA is not found on normal feline lymphocytes, even those productively infected with FeLV-A and -B. Thus, FOCMA is an FeLV- and FeSV-induced tumor-specific antigen. The finding of FOCMA on FeLV-negative LSA cells indicated that FOCMA is a marker of FeLV leukomogenesis (129-135). FOCMA, isolated from FeLV-positive and FeLV-negative LSA cells, was found to be a 70K protein which, by peptide mapping, was different from the gp70 molecules of FeLV-A and -B (129-131).

FOCMA was reported to be identical to the gp70 of FeLV-C, the FeLV subgroup that is rarely ever found in infected pet cats (136). However, further studies found that FOCMA is related to, but is distinguishable from, FeLV-C gp70 (131). FOCMA proteins were purified from lysates of FeLV-negative and -positive primary feline LSAs by immunoaffinity column chromatography, and the *Staphylococcus aureus* V8 protease partial digest maps of FOCMA p70s and FeLV-C gp70, antigens in lysates of FeLV-positive LSAs and negative LSAs competed, but only to a maximum of approximately 50%, which indicated that FeLV-C gp70 and FOCMA were related but they were serologically distinguishable (131).

To further clarify the relationship between FOCMA and FeLV-C gp70, the degree of concordance in the occurrence of FOCMA antibody and of FeLV-C neutralizing antibody in the sera of 409 cats was studied. Only 66 of 96 (68.8%) cats with FOCMA antibody also had neutralizing antibody to FeLV-C. Similarly, 66 of 86 (76.7%) cats with neutralizing antibody to FeLV-C had antibody to FOCMA (118). These serological data further indicate that FOCMA is related to, but is serologically distinguishable from, FeLV-C gp70.

Many cats viremic with FeLV-A have neutralizing antibodies to FeLV-C, even though FeLV-C cannot be isolated from their plasma (20,137). This suggests that FeLV-A transforms some lymphocytes and may recombine with endogenous FeLV sequences to produce the FeLV-C gp 70-FOCMA recombinant molecule. FOCMA p70 proteins appear to be a family of proteins that are structurally related to FeLV-C gp70 but are serologically distinct from the viral protein. Thus FOCMA may be a recombinant FeLV-C gp70 molecule with additional endogenous cellular determinants. However, these findings do not alter the previous conclusion that FOCMA is an FeLV- or FeSV-induced tumor-specific antigen.

B. Mechanisms of FeLV-Induced Leukemogenesis

The molecular (genetic) mechanisms by which oncogenic retroviruses induce leukemia are not known. Two distinct types of leukemogenic retroviruses exist: (a) chronic leukemia viruses and (b) acute leukemia viruses. Acute leukemia viruses possess oncogenes transduced from cellular protooncogene information and induce leukemia rapidly. Chronic leukemia viruses, in contrast, do not possess oncogenes and induce leukemia only after a long latent period.

The best example of a chronic leukemia virus causing tumors occurs in mice, where MCF-MuLVs are generated as recombinants between ecotropic MuLV and endogenous *env* sequences related to xenotropic viruses (138-140). Generation of MCF viruses appears to be the proximal event in the induction of AKR lymphomas. Subgroup FeLV-B and a Moloney virus-derived MuLV MCF virus show a striking homology in the nucleotide sequences of their envelope genes (62). The homologies are in the substituted (presumably xenotropic) portion of the MuLV envelope genes. It is now known that FeLV-B and -C arise from recombination of FeLV-A with enFeLV *env* sequences and these recombinant viruses may play a proximal role in the formation of lymphoid tumors in cats (49,62).

Acute transforming oncogene containing FeLVs have been discovered in some cats with T-cell LSAs. These defective *FeLV-myc*-containing viruses, are recombinant viruses derived from FeLV and c-*myc* (63-65). Since less than 50% of the LSAs in cats have defective FeLV-*myc* viruses, there must be several ways by which FeLV can induce LSAs.

C. Pathogenesis of FeLV Infection

FeLV is transmitted contagiously by intimate prolonged contact through the saliva to the ocular, oral, and nasal membranes of uninfected cats (8,123,141, 142). As many as 2×10^6 infectious FeLV per milliliter occur in the saliva of infected cats. The virus replicates initially in lymphocytes of the local lymph nodes of the head and neck, but most infected cats reject the virus at this early stage and become immune (12,20,142). However, in cats that do not reject the virus, it spreads to the bone marrow where it replicates to high titers in all nucleated cells of the myeloid and erythroid series. Infected leukocytes and platelets released from the infected bone marrow, or virus in the plasma, spread the infection throughout the cat's body. Within a few weeks the virus spreads to the salivary glands, oral mucosa, and respiratory epthelial cells, from where it is shed. FeLV is also transmitted in utero to unborn fetuses and through the milk of infected queens (8,12). Although the period of time from FeLV infection to disease development is highly variable, 83% of infected healthy cats die within 3.5 years from onset of FeLV-induced diseases (143).

Ninety-seven percent of cats that have widespread replication of FeLV in their bone marrow remain persistently, lifelong, infected. Conversely, 3% of infected cats do reject their virus infection and rid themselves of all virus-replicating cells (12,20). This observation has important implications for humans infected with retroviruses as it may mean that some people may be able to reject HIV or HTLV infection and thus may not remain persistently infected or develop a retroviral disease. The immune system is the key to FeLV rejection, and cats that have rejected FeLV develop high titers of neutralizing antibody. However, unfortunately in humans, neutralizing antibody and retroviruses coexist and no person has been shown to have a protective immunity to any human retrovirus. In contrast to humans infected with HIVs or HTLVs, cats persistently infected with FeLVs do not produce detectable free neutralizing antibody to the virus with which they are infected (20).

D. Latent FeLV

Cats that have rejected the virus and become immune can harbor latent, unexpressed, nonreplicating FeLV in a small number of mononuclear cells of the bone marrow (144,145). Latent FeLV can be reactivated from the bone marrow cells by treatment with corticosteroids or by stimulating bone marrow cultures with corticosteroids or *S. aureus* Cowan I (144). Several months after rejection of replicating FeLV, latent virus usually can no longer be reactivated from most cats (146). This may also imply that people who reject their retroviral infections may not have the potential for reactivation of latent retroviruses for their entire lives.

E. Consequences of FeLV Exposure

A vigorous immunological response to FeLV exposure is essential in order to resist infection. Macrophages are the first line of defense against FeLV infection, and both humoral and cellular immune responses are critical for virus rejection. Exposure to FeLV does not invariably result in persistent infection and development of an FeLV disease. Only about 10% of persistently infected pet cats develop LSA, whereas more than 50% die of the immunosuppressive effects of the virus (5,12,14,20,127,143,147,148). Twenty-eight percent of cats exposed to FeLV are persistently infected, 42% are immune to the virus, whereas the remaining 30% are neither immune nor infected (Table 6) (12,20).

Macrophages are essential in the initial defense against FeLV infection, and experimentally, agents that depress macrophage responses make cats more susceptible to infection with the virus (149). An adequate antibody response to the viral envelope gp70 will result in neutralization of the virus and clearance from the body. An inappropriate antibody response in some cats against the FeLV *gag* antigens p15, p12, p27, and p10 is not beneficial and, in persistently infected cats, may lead to the formation of immune complexes, immunosuppression, and the development of immune complex glomerulonephritis (Table 7) (12,56,148,150-152).

F. Epidemiology of FeLV

1. Occurrence of FeLV

During the 1960s, many veterinarians observed the clustering of cases of LSA in unrelated cats living in the same households, which suggested the contagious transmission of FeLV (5,6,153). In 1970 a simple and accurate immunofluorescent antibody test (IFA) for FeLV antigens was developed which enabled epidemiological surveys of FeLV infection to be performed in the United States and in Europe (14,20-25, 123,124,147,154). These studies showed that infected cats

TABLE 6 Consequences of Exposure to FeLV

FeLV exposure	Permanent FeLV status	FeLV immune response	Percentage of cats
Not exposed	Not infected	Not immune	30
Exposed	Not infected	Immune	42[a]
Exposed	Infected	Not immune	28

[a]Cats were transiently infected.

TABLE 7 Circulating Immune Complexes in Healthy Cats

FeLV status	Number tested	Amount of CIC in serum (μg/ml)[a]
Uninfected	32	10–240
Infected	9	480–12,000

[a]Modified Raji cell radioimmunoassay.

are found mainly in exposure households such as multiple-cat households. Twenty-eight percent of healthy FeLV-exposed cats were found to be persistently infected, whereas only 1–2% of household pet cats with no known exposure were found to be infected (Table 8) (5,12,20). Less than 1% of stray cats and 2% of shelter cats were found to be infected.

2. Outcome of Persistent FeLV Infection

Persistent FeLV infection shortens a cat's expected life-span. Eighty-three percent of healthy FeLV-infected cats die within 3.5 years, compared to only 16% of FeLV-uninfected cats living in the same households (143). Most of these deaths are caused by FeLV-induced immunosuppression and development of FAIDS. Approximately five times more FeLV-infected cats die of nonneoplastic diseases than do cats living in FeLV-free households.

3. FeLV Testing Methods

A specific, sensitive, rapid IFA test for detection of FeLV antigens in leukocytes in the peripheral blood was introduced into veterinary medicine in 1973 (5,21, 22). A positive IFA test indicates FeLV infection but is not diagnostic of any disease. FeLV can be isolated from 97.5% of IFA-positive cats (Table 9), and an IFA positive cat is shedding the virus in its saliva (5,21,134). Ninety-seven percent of IFA-positive cats remain infected for life, whereas 3% are able to reject the virus and become immune. FeLV cannot be isolated from 98% of IFA-negative cats. There is an excellent correlation between the IFA test result for FeLV and the ability to isolate the virus.

In-hosptial enzyme-linked immunosorbent assays (ELISAs) for FeLV have been introduced into veterinary medicine, and comparative studies of the IFA and ELISA tests have been performed (Table 8) (12,25,122,155). In general, there is fairly good agreement, 86.7%, between an ELISA negative test and the IFA test. However, there is only a 40.8% agreement between ELISA positive tests and the IFA test. Thus, more than half of the positive ELISA tests are incorrect. Positive FeLV ELISA tests should always be confirmed with an IFA test, and similarly, all HIV-positive ELISA antibody tests should be confirmed with a Western blot test.

G. Immune Response to FeLV Infection

1. Immune Response to FeLV

The outcome of an FeLV-infected cat depends mainly on its humoral immune response to the viral gp70, virus neutralizing (VN) antibody, and to the FeLV-induced tumor specific antigen, FOCMA (20,127,137,156). A sufficient antibody titer to the FeLV gp70 is all that is required to protect a cat against both FeLV infection and FeLV-induced diseases, except FeLV-negative LSAs. Cats with protective titers (1:10 or greater) of VN antibody have rejected the virus and are resistant to subsequent viral infection (20).

Seroepidemiological studies have shown that unexposed pet cats do not have VN antibodies to FeLV subgroups A and B, whereas about 42% of exposed cats have FeLV-A and -B neutralizing antibodies (Table 6) (12,20,137). Between 1% and 4% of uninfected stray cats have VN antibodies to FeLV-A and -B, whereas no persistently FeLV-infected cats have protective VN antibody to FeLV-A and -B. However, there is an unexplained high prevalence of VN antibodies to FeLV-C in all populations of cats, especially in FeLV-A and -B viremic cats (45%), most of which (98%) have no VN antibodies to FeLV-A and -B (12,20,137). The reason for an immune response to FeLV-C in these cats in not known.

2. Immune Response to the Feline Oncornavirus-Associated Cell Membrane Antigen (FOCMA)

Infected cats with protective titers (1:32 or greater) of FOCMA antibody, in the absence of protective titers of VN antibody, are resistant to the development of LSA or other FeLV-induced tumors but are not resistant to the development of FeLV infection and nonneoplastic FeLV diseases such as FeLV-FAIDS (20, 125,127,133). FOCMA antibody is found mainly in cats exposed to FeLV

TABLE 8 Occurrence of FeLV in Healthy Pet and Stray Cats

Environment	Percent FeLV-infected
Multiple-cat households	
Not exposed to FeLV	0
Exposed to FeLV	28
Single-cat households	
No known exposure to FeLV	1
Stray cats	
Unknown FeLV exposure	1
Shelter-adopted cats	
Unknown FeLV exposure	2

TABLE 9   Comparison of IFA FeLV Test and Other Test Methods

Immunofluorescent antibody test result	Number of cats tested	Number of cats FeLV isolated in tissue culture	Number of cats ELISA positive	Percent agreement
Tissue culture isolation comparison			·	
Negative	153	3	ND	98
Positive	121	118	ND	97.5
ELISA comparison				
Negative	661	ND	88	86.7
Positive	7819	ND	3193	40.8

(38.4%), only rarely in unexposed cats, and never in SPF cats (12,20,125,127). Twenty-five percent of FeLV-infected cats have protective titers of FOCMA antibody. FOCMA antibody has even been used therapeutically to induce remission of LSAs in pet cats (157).

H.   Prevention of the Spread of FeLV

An IFA test and removal program has been used successfully in veterinary medicine for the past 16 years (12,22-24). In this program, all cats are tested for FeLV and any infected cats are removed from the household. After the infected cats have been removed, the uninfected cats are quarantined in the household and are retested 3 months later. The 3-month retest is needed because the incubation period of FeLV infection can be as long as 3 months. If any of the initially uninfected cats are found to be FeLV positive in the second FeLV test, they are removed and a third test is done 3 months later. When all cats test negative in two consecutive tests, done 3 months apart, the cats in the household are free of FeLV infection. This program has been able to reduce the spread of FeLV 40-fold compared to households that did not remove infected cats from contact with uninfected cats (22).

It is perplexing to find that many of the same reservations that were proposed, in the early 1970s, by some cat owners and veterinarians in an attempt to prevent the testing of cats for FeLV are now being proposed to prevent the widespread testing of humans for HIV (158-162). However, 15 years later it is now a well-accepted part of good preventive veterinary medicine to test cats for FeLV in order to prevent the spread of the virus.

I.   FeLV Vaccine

Several different types of FeLV vaccines have been studied, including a low-dose live-virus vaccine (12,118,163), killed virus vaccines (26,164), and a recombinant

FeLV-vaccinia virus vaccine (165). The live-virus vaccine produced extremely high VN antibody titers but one of the nine vaccinated cats developed a virus-negative LSA (12). One of the killed FeLV vaccines enhanced FeLV infection rather than protecting cats from infection and the recombinant FeLV-vaccinia vaccine was not immunogenic in cats (26,165).

The first successful retrovirus vaccine was developed for protection of cats against FeLV infection and is a "subunit" vaccine (26,27). By manipulating the growth conditions of FL74 LSA cells, which produce FeLV-A, -B, and -C, the viral antigens and FOCMA are solubilized and released into the supernatant fluids of the culture medium and can be harvested for the vaccine. The antigens consist of FeLV p27, p12, p10, gp70, and p15E (166). The vaccine preparation is inactivated to ensure that there is no infectious FeLV in the product. Clinical trials of the efficacy of the vaccine, performed by the manufacturer, demon-strated an 80% protection against virus challenge (166). However, another study of the immunogenicity and efficacy of the vaccine reported poorer protection (167). Only 50% of vaccinated cats developed good antibody to FeLV gp70-related antigens and the response to virus challenge was disappointing.

The FeLV vaccine will not reverse FeLV infection in a viremic cat. All cats that are to be vaccinated should be tested for FeLV at the time of the first dose of vaccine, and those found to be infected should be removed from the house-hold or strictly isolated away from uninfected cats. Employing the FeLV test and removal program along with vaccination may markedly reduce the spread of FeLV among pet cats.

J. FeLV Diseases

FeLV replicates best in rapidly dividing lymphoid, myeloid, mucosal, and epi-thelial cells and can induce proliferative (neoplastic) and degenerative (blasto-penic) diseases in these cells (Table 10) (12,21). Before the introduction of the FeLV vaccine, FeLV-induced diseases were collectively the leading cause of death among pet cats from infectious causes (12,14,148,156,168,169,170,171). It is still too early to know whether the vaccine has significantly lowered the occurrence of FeLV diseases.

Lymphosarcoma (LSA) and the lymphadenopathy syndrome are the proliferative diseases of lymphocytes caused by FeLV, whereas thymic atrophy and general lymphoid depletion are the degenerative diseases of lymphocytes (13,17,148,170). Proliferative diseases of erythrocyte precursors caused by FeLV are erythremic myelosis and erythroleukemia, whereas the degenerative erythrocyte disease aplastic anemia (erythroblastopenia) occurs far more often in pet cats than the proliferative diseases (16,19,50,60,169,170). The most frequent clinical manifestations of FeLV infection is severe immunosuppression which results in the development of secondary opportunistic infections and death (18,148,168,172). This syndrome is called the FeLV-induced feline

TABLE 10   Feline Leukemia Virus Diseases

Cell type	Proliferative diseases (neoplastic)	Degenerative diseases (blastopenic)
Lymphocytes	Lymphosarcoma Lymphadenopathy syndrome (distinctive peripheral lymph node hyperplasia)	Thymic atrophy Lymphopenias Feline acquired immune deficiency syndrome (FAIDS)
Bone marrow cells Primitive mesenchymal cell	Reticuloendotheliosis	—
Erythroblast	Erythremic myelosis Erythroleukemia	Erythroblastosis (regenerative anemia) Erythroblastopenia (aplastic anemia) Pancytopenia
Myeloblast	Granulocytic leukemia	Myeloblastopenia-enteritis syndrome (panleukopenia-like disease)
Megakaryocyte	Megakaryocytic leukemia	Thrombocytopenia
Fibroblast	Myelofibrosis	—
Osteoblast	Medullary osteosclerosis Osteoochondromatosis	— —
Intestine	—	Enteritis
Kidney	—	FeLV immune complex glomerulonephritis
Uterus	—	Abortions and resorptions
Fibroblasts, skin	FeSV-induced multicentric fibrosarcomas	—

acquired immune deficiency syndrome (FeLV–FAIDS) and is very similar to its human counterpart (172).

1. Lymphocyte Diseases

   a. **Proliferative Lymphocyte Diseases**

   Lymphosarcoma (LSA). Domestic cats have the highest incidence of spontaneous LSA of any animal, with 200 cases occurring yearly per 100,00 cats at risk (12,168,173). One-third of all cat tumors are hematopoietic tumors and 90% of these are lymphoid tumors (173). Multicentric LSA, in which the tumor localizes in internal organs and lymph nodes, is the most common anatomical form of

LSA in cats (Table 11). Approximately 30% of the cats with LSA have leukemic blood profiles, and these cats are classified as having multicentric LSA (168,169, 174). Thymic LSA is the second most common form, followed by alimentary LSA, which localizes in the gastrointestinal tract and usually is FeLV-negative (132,134,169,174).

Most feline LSAs are T-cell tumors although B-cell LSAs occur frequently in the gastrointestinal tract (132,134,135). Leukemic blood profiles are uncommon in cats with lymphoid tumors, and about two-thirds of cats with LSA have nonregenerative anemias; almost all of these cats are FeLV-positive (169,174).

FeLV-Negative Lymphosarcomas. Thirty percent of cats with LSA are FeLV negative (Tables 11 and 12), and no FeLV antigens can be detected nor can infectious FeLV be isolated from the tumor tissue (132–134). However, the FeLV-induced FOCMA is present on the membranes of both FeLV-positive and -negative LSA cells and strongly indicates a role for FeLV as the cause of these tumors (134,135).

Cats with FeLV-positive LSAs are young, less than 7 years old, and usually have T-cell multicentric or thymic LSAs, whereas cats with FeLV-negative LSAs are older, over 7 years of age, and usually have alimentary B-cell LSAs (Tables 11 and 12) (132,134,135,175). FeLV-negative LSAs develop most frequently in cats living in households where FeLV-infected cats live (6,12,20,21,175). A large epidemiological study found that cats that developed FeLV-negative LSAs were exposed to FeLV as often as cats that developed FeLV-positive LSAs (132,134).

No FeLV proviral sequences, above the level of endogenous FeLV sequences present in all uninfected cat cells, are present in the FeLV-negative LSA tumor tissues, which suggests an indirect mechanism of leukemogenesis in these cases (42). However, additional FeLV sequences can be found in non-LSA tissues, most often bone marrow cells, in 60% of these cats, which indicates that cats with FeLV-negative LSAs were previously infected with FeLV and that integrated exogenous FeLV sequences exist in some non-LSA tissues. This finding and the observation that, in healthy cats with FOCMA antibody, the titers remain high for many years suggest that there is continual FeLV stimulation. Latent FeLV can be reactivated from the bone marrow of healthy FeLV immune cats and from the bone marrow, but not from the FeLV-negative LSA cells, of 80% of cats with FeLV-negative LSAs (144,145). Thus it appears that FeLV induces both FeLV-positive and FeLV-negative LSAs in cats.

FeLV can also induce FeLV-negative LSAs in dogs (176). Puppies inoculated with FeLV during the first day of life developed FeLV-negative LSAs, and those inoculated in utero developed FeLV-positive LSAs. It appears that FeLV is capable of altering the genome of lymphocytes to induce transformation but is not required to persist in the cellular genome to maintain the transformed state. Such a mechanism of tumorigenesis is common to FeLV, radiation, and chemical carcinogens. These observations in cats with commonly occurring

TABLE 11 Classification of Feline Lymphosarcomas

Anatomical site	Percent of total	Percent FeLV positive	Average age of cats FeLV+/FeLV-	Cell of origin (%)			
				T	B	Null	Mixed
Multicentric	44	80	3.3/5.8	54	13	4	29
Thymic	38	77	2.1/3.8	93	0	0	7
Alimentary	15	23	5.0/8.9	20	60	0	20
Unclassified	3	38	5	0	0	0	100[a]

[a]Only one tumor studied for cell of origin.

TABLE 12 Comparison of FeLV-Positive and -Negative Feline
Lymphosarcomas

Category	FeLV-positive LSAs	FeLV-negative LSAs
Anatomical form	Multicentric form Thymic form	Most often alimentary or skin form
T- or B-cell origin	Usually T-cell	Usually B-cell
Age	Young—under 7 years	Old—over 7 years
FeLV antigens	Present	Absent
FeLV genome in LSAs	Present	Absent
Infectious FeLV	Present	Absent from LSA cells Reactivatable from: bone marrow cells, not from LSA cells
FOCMA	Present	Present
Exposure to FeLV	Exposed	Exposed

leukemogenic, retrovirus-induced, virus-negative lymphoid tumors suggest that retrovirus-negative, retrovirus-induced lymphoid tumors may also occur in humans and the search for genetic evidence of these viruses should include non-tumor tissues such as bone marrow.

Lymphadenopathy Syndrome (LAS). A syndrome of peripheral lymph node enlargement, lymphadenopathy syndrome (LAS), also called distinctive peripheral lymph node hyperplasia, occurs in FeLV-infected young cats (6 months to 2 years of age) (18,148,177). This syndrome is very similar to the lymphadenopathy syndrome that precedes the development of AIDS in humans. Clinically, half of the cats appear normal; the others develop fever, lethargy, anorexia, and hepatosplenomegaly. The mandibular lymph nodes are commonly involved, but popliteal, visceral, and other nodes can also be affected. Architecture of the lymph nodes is distorted, and the sinuses and follicles are difficult to discern. The paracortex becomes filled and expanded with macrophages, lymphocytes, immunoblasts, and plasma cells and causes distortion of the lymphoid follicles. This syndrome has a variable outcome: In some cats the lymph nodes return to normal size, some cats progress to develop LSA years later, and most cats progress to develop FeLV–FAIDS. The pathogenesis of LAS is not known, but inappropriate stimulation by persistent FeLV replication in folliclar germinal centers is the probable cause.

b. **Degenerative Lymphoid Diseases** As mentioned previously, the most frequent and most devastating consequence of FeLV infection is a profound immunodeficiency syndrome. FeLV replicates to high titers in lymphoid cells and often induces severe depletion or dysfunctions of these cells (Tables 10 and 13).

Thymic Atrophy. Thymic atrophy is a degenerative lymphoid disease of T lymphocytes of the thymus gland of FeLV-infected young cats (13,18,148). Depletion of thymic lymphocytes and lymphocytes in other lymphoid organs occurs, probably owing to viral lymphocytosis, resulting in a deficient cell-mediated immune response and rendering infected kittens susceptible to opportunistic infectious microorganisms. Many kittens with thymic atrophy become cachectic, develop bronchopneumonia or enteritis, and usually die of these diseases in the first 3 months of life.

Lymphoid Atrophy. Generalized peripheral lymphoid hyperplasia occurs early in most FeLV-infected cats but usually progresses to generalized lymphoid atrophy, lymphopenia, and death from opportunistic infections (13,148,170).

FeLV-Induced Feline Acquired Immune Deficiency Syndrome (FeLV–FAIDS) Two feline retroviruses, FeLV and the feline immunodeficiency virus (FIV), can induce almost identical feline AIDS (FAIDS) (2). In FeLV-induced FAIDS, FeLV replicates in lymphoid and myeloid cells of the immune system and can cause degenerative blastopenic diseases (lymphopenias and neutropenias) involving these cells (12,18,21,148,168,178). FeLV-infected cats often develop immune cell deficiencies, characterized by drastic reductions in lymphocyte and neutrophil numbers (12,148,178), and immune cell dysfunctions consisting of cutaneous anergy (15), reduced T-cell blastogenic responsiveness (179,180) depressed helper T-cell function as evidenced by impaired antibody

TABLE 13 FeLV-Induced Feline Acquired Immune
Deficiency Syndrome, FAIDS

I. Immune cell deficiencies
   A. Lymphoid depletions
      1. Thymic atrophy—kittens
      2. General lymphoid depletion—adults
   B. Myeloid depletion
      1. Neutropenias—myeloblastopenia syndrome
II. Immune cell dysfunctions
   A. Deficient cell-mediated immune response
      1. Cutaneous anergy—decreased allograph rejection
   B. Deficient antibody-mediated immune response
      1. To threshold antigen stimulation
   C. Deficient neutrophil functions
III. Pathogenic antibody immune-mediated disease
   A. Immune complex glomerulonephritis
IV. Complement deficiency

TABLE 14 Occurrence of Opportunistic Infectious Diseases in FeLV-Infected FAIDS Cats

Disease	Percent FeLV-infected
Chronic stomatitis and gingivitis	48
Skin sores and recurrent abscesses	34
Viral infections: feline infectious peritonitis	46
Upper respiratory disease and pneumonias	55
Chronic generalized infections	57
	45

production (148,181), and neutrophil dysfunctions since they are less able (four-fold) to produce antibody to threshold doses of antigens than uninfected cats (148,181).

Many more pet cats die from FeLV-induced FAIDS than from FeLV-induced neoplastic diseases (18,148). About half of the cats with feline infectious peritonitis (feline coronavirus); half of the cats with chronic oral inflammation, gingivitis, and necrotizing stomatitis; one-third of the cats with chronic cutaneous or deep dermal abscesses and nonhealing lesions of the skin; half of the cats with chronic upper respiratory disease and pneumonia; half of the cats with chronic generalized infections (septicemias and pyothorax); and 87% of kittens with thymic atrophy are infected with FeLV (Table 14) (13,18,148). Many of these cats have unremitting spirochete, coliform, staphylcoccal, and steptococcal infections. Common secondary infections include: (a) viruses: feline herpesvirus (rhinotracheitis), feline coronavirus (feline infectious peritonitis virus); (b) pathogenic fungal infections: *Candida, Cryptococcus,* and *Aspergillus* species; and (c) protozoa: *Toxoplasma* and *Hemobartonella* species (155).

FeLV may induce immunosuppression by one of the following mechanisms: (a) By the process of viral replication and budding from lymphocyte and neutrophil cell membranes, FeLV may cause lysis or sensitize cells to cell-mediated immune destruction. (b) FeLV soluble circulating antigens may cause immunosuppression by themselves or as constituents of immune complexes. In this regard, purified p15E has been reported to decrease in vitro blast transformation by 45-92% (179,182). (c) Circulating immune complexes (CICs) are immunosuppressive, and CICs composed of whole infectious FeLV, FeLV gp70, p27, p15, and p15E occur in FeLV-infected pet cats (Table 7) (56,148,150). In addition, when CICs were therapeutically removed by ex vivo immunosorption on *S. aureus* Cowans I columns, several FAIDS cats showed clinical improvement (183,184). (d) The complement system is also affected in cats infected with

FeLV. In one study all FeLV-infected cats with LSA and 50% of FeLV-infected healthy cats were hypocomplementemic (185). The hypocomplementemia observed in FeLV-infected cats probably contributes to the generalized immuno-suppression.

As discussed previously, neutrophils and neutrophil progenitors are always infected in cats with persistent FeLV infection. The consequences of neutrophil infection are effector inhibition and reduction of their numbers. Persistent, transient, and cyclical neutropenias occur in FeLV-infected cats and many progress to develop FeLV–FAIDS (5,14,18,148,186).

There are numerous similarities between human AIDS and FeLV-induced FAIDS in pet cats (172,187). In cats and humans the syndromes are character-ized by lymphopenias, reduced lymphocyte blastogenesis, cutaneous anergy, reduced numbers of T cells, impaired antibody response, and the occurrence of secondary infectious diseases (Tables 15 and 16) (172,188–190). However, there are significant differences between FeLV and HIVs, and it is now apparent that the new feline lentiretrovirus, FIV, is a more appropriate viral model for human AIDS (2). FeLV is in the retrovirus subfamily Oncovirinae and the two presently known human AIDS viruses HIV-1 and HIV-2 and FIV are members of the subfamily Lentivirinae (1). Infection of cats with FeLV is determined by the detection of viral antigens in the blood, whereas determination of HIV infec-tions in humans relies on the detection of antibodies to HIV proteins (21,29, 191).

TABLE 15   Immune Parameters of Cats and Humans with AIDS

FeLV-FAIDS	AIDS

1. Cellular immunity
  a. Reduction in numbers and
    functions of T lymphocytes
	1. Cellular immunity
	  a. Reduction of helper (OKT-4)
	    T lymphocytes but have
	    normal numbers of suppressor
	    (OKT-8) T lymphocytes

b. Lymphopenias common	b. Lymphopenias common

c. Reduced lymphocyte blasto-
    genesis
	c. Reduced lymphocyte blasto-
	    genesis

d. Cutaneous anergy	d. Cutaneous anergy

2. Humoral immunity
  a. Impaired antibody response
    to:
    1. Sheep RBCs—low dose
    2. Synthetic polypeptide
      antigen
	2. Humoral immunity
	  a. Impaired antibody response
	    to:
	    1. New antigens
	    2. Have antibody to infecting
	      microorganisms

TABLE 16   Secondary Intercurrent Infections of Cats and Humans with AIDS

FeLV-FAIDS	AIDS
1. Diseases	1. Diseases
a. Lymphadenopathy	a. Lymphadenopathy
b. Pneumonia and upper respiratory diseases	b. Pneumonia—*Pneumocystis carinii*
c. Toxoplasmosis *Candida* Cryptococcosis	c. Toxoplasmosis Mucosal *Candida* Cryptococcosis
d. Skin sores and infections	d. Kaposi's sarcoma
e. Viral diseases Herpesvirus Coronavirus (FIP)	e. Viral diseases Herpesvirus Hepatitis B virus

2. Erythroid Diseases

FeLV replicates in all nucleated erythroid cells in the bone marrow and can induce neoplastic or blastopenic diseases of these cell types (Table 10) (12,18, 168).

    a. Erythroid Neoplastic Diseases   FeLV replicates well in erythroid progenitor cells, which still have nuclei, but as these erythroid cells mature the nucleus is extruded and thus the FeLV provirus and the ability of FeLV to replicate is lost. FeLV-induced erythroid diseases occur in nucleated erythroid progenitor cells in which the FeLV provirus is present (18,169). FeLV rarely induces erythremic myelosis and erythroleukemia in infected cats. Feline erythroid neoplasms are similar to those that occur in chickens and in mice infected with oncogene-containing acute transforming retroviruses (192,193).

    Reticuloendotheliosis.   FeLV-induced reticuloendotheliosis occurs very rarely in pet cats (169). Primitive mesenchymal pluripotent stem cell proliferation can lead to an accumulation of these cells, which show no recognizable progression to a more differentiated cell type. Cats with reticuloendotheliosis are usually anemic, and the neoplastic cells appear to be closely related to both the erythroid and the granulocytic myeloid precursor cells.

    Erythremic myelosis.   The occurrence of abnormally high numbers of proliferating nucleated erythroid cells without significant concurrent proliferation of granulocytes is termed erythremic myelosis (169). Thus, erythremic myelosis is a proliferative disease of only the erythroid precursor cells. The disease does not occur commonly in pet cats and is characterized by a severe anemia and marked variations in the numbers and morphology of the nucleated erythrocytes. The anemia is nonregenerative, and even though there are nucleated erythrocytes, there is a normal or reduced number of reticulocytes, indicating a block in maturation from the early nucleated erythrocyte to the reticulocyte.

Erythroleukemia. No clear distinction between erythremic myelosis and erythroleukemia exists. However, for the purpose of classifying myeloproliferative diseases, the distinction between the two diseases is based on the presence of myeloblasts (granulocytic leukocyte precursors) along with abnormal nucleated erythrocytes in the peripheral blood of cats with erythroleukemia, whereas only neoplastic erythroid cells are present in cats with erythremic myelosis (169). Most cats with erythroleukemia have a profound nonregenerative anemia, and even though nucleated erythrocytes are present, there is a normal or low reticulocyte count, indicating a block in the process of erythroid cell maturation. The half-life of the erythrocytes is reduced by 50% in cats with this disease. Proliferating erythroid and myeloid neoplastic cells are found in the blood, bone marrow, and various organs, such as the spleen, liver, and lymph nodes.

b. Erythroid Blastopenic Diseases Erythroid blastopenic diseases occur far more often in FeLV-infected pet cats than do erythroid neoplastic diseases (18,168,169). There are three types of FeLV-induced anemias: (a) erythroblastosis (aplastic or regenerative anemia), (b) erythroblastopenia (nonregenerative anemia), and (c) pancytopenia (18,19,168). FeLV-A has been experimentally shown to induced nonfatal transient erythroblastosis, whereas several FeLV-C isolates have induced fatal aplastic anemias (erythroblastopenias) (19,60). FeLV-induced anemias can result from hemolysis, bone marrow dysplasia, or erythroid aplasia. The number of platelets can be severely reduced in some infected cats.

The mechanism by which FeLV causes degenerative bone marrow diseases is unknown, but three possibilities exist. The first possibility is that FeLV causes lysis by damaging the infected cell's membrane, possibly by the process of budding. The second possibility is that FeLV may antigenically alter the cell membrane, by budding or by inducing FOCMA, and the altered membrane may then be lysed by antibody to FeLV or to FOCMA or lysed by killer lymphocytes. Finally, FeLV may affect effector cells such as erythropoietin-producing cells, which then may not produce sufficient erythropoietin, resulting in erythroid depletion. This mechanism seems unlikely, however, since erythropoietin levels were found to be increased in most cats with erythroid aplasia (194). FeLV-induced degenerative bone marrow diseases occur far more commonly than FeLV-induced myeloproliferative diseases.

Cats are more susceptible to anemias than most other species because their erythrocytes have a shorter life-span, 70-80 days, compared to the erythrocyte life-span of 120 days for most other species. The three distinct types of primary FeLV-induced anemias—(a) FeLV erythroblastosis (regenerative anemia), (b) FeLV erythroblastopenia (aplastic or nonregenerative anemia), and (c) FeLV pancytopenia—often develop in sequence, beginning with erythroblastosis, which leads to erythroblastopenia and finally to pancytopenia and death. These phases indicate that FeLV has an initial stimulatory effect on the erythroid cells followed by a degenerative effect.

FeLV Erythroblastosis (Regenerative Anemia). FeLV regenerative anemia, erythroblastosis, is not as common a disease entity as FeLV erythroblastopenia or FeLV pancytopenia. Only about 15% of FeLV-infected anemic pet cats have regenerative anemias (14,18). Immature erythrocytes are rushed out of the marrow in response to the FeLV anemia, and there is an increase in the number of reticulocytes and nucleated red blood cells in the blood and bone marrow. Extramedullary hematopoiesis occurs in the spleen and liver. Cats with non-FeLV-induced regenerative anemias usually have a good prognosis, whereas FeLV-infected cats with regenerative anemias do not have good prognoses since many of them eventually develop the fatal erythroblastopenia or pancytopenia syndromes and some even develop myeloproliferative disease or lymphosarcoma. Experimentally, FeLV-A causes regenerative anemias and not the more common and lethal nonregenerative anemias (19).

FeLV Erythroblastopenia (Aplastic or Nonregenerative Anemia). FeLV erythroblastopenia, also known as pure red cell aplasia, erythroid aplasia, or aplastic anemia, is the most common form of FeLV-induced anemia and occurs more often than any FeLV-induced neoplasm (169). Sixty-eight percent of cats with this form of anemia are FeLV-positive (12,14,169). Aplastic anemia is a progressive and fatal degenerative disease of only the erythroid cells. There is a hypoplasia of the bone marrow erythroid elements and a normocytic and normochromic anemia.

Several groups have shown that FeLV-C can induce aplastic anemias in SPF kittens (16,17,60,195,196). However, since all pet cats that develop non-regenerative anemias are infected with FeLV-A and since FeLV-A probably recombines with enFeLV sequences and other non-enFeLV endogenous cellular sequences to form FeLV-C, it is likely that FeLV-A is responsible for all forms of FeLV-induced anemias that develop in pet cats (see Chapter 4, "Molecular Biology of Feline Leukemia Viruses") (31).

Although serum erythropoietin concentrations are high, 10-20 times normal values, in cats with aplastic anemias, ferrokinetic data show decreased erythropoiesis (194,197). Marrow culture studies also revealed low numbers of BFU-E and CFU-E, but normal numbers of granulocyte-macrophage progenitors remained (60,197). FeLV infection apparently impairs the ability of feline marrow to respond physiologically to anemia.

FeLV Pancytopenia. FeLV pancytopenia is the second most common form of primary FeLV myelodegenerative anemia. In this disease, all hemato-poietic cells, erythroid, myeloid, and megakaryocytes, are lacking and there is a normocytic, normochromic nonregenerative anemia, a leukopenia, and a decreased number of platelets. Many of these cats have recurrent secondary diseases due to their low leukocyte counts (18). FeLV pancytopenia can lead to myelofibrosis before death in some cats.

Pathogenesis of FeLV Anemias. Experimentally the different subgroups of FeLV cause different types of anemias. FeLV-A and FeLV-AB induce transient nonfatal macrocytic normochromic anemias (erythroblastosis) with extensive splenic extramedullary hematopoiesis (16,19), whereas FeLV-C induces a fatal aplastic or nonregenerative normocytic normochromic anemia in SPF cats (16,17,60,195,196). Recently, FeLV-C has been routinely isolated from pet cats with aplastic anemia (155). In kittens inoculated with FeLV-B alone no anemias developed. Thus two of the three FeLV subgroups have been found capable of inducing anemias.

The pathogenesis of aplastic anemia is probably determined by the two FeLV envelope proteins, gp70 and p15E (198,199). As mentioned previously, the gp70s external knob glycoproteins are responsible for subgroup FeLV-A, -B, and -C specificities, host ranges, and disease determinants. FeLV-C has been shown to induce rapidly progressive aplastic anemias in SPF cats. The onset of anemia coincides exactly with the appearance of FeLV-C in the plasma. FeLV-C infection destroys two classes of erythroid progenitor cells, colony-forming units—erythroid (CFU-E)—and burst-forming units—erythroid (BFU-E) cells. Ten days after exposure to FeLV-C the BFU-E are infected and begin to disappear from the bone marrow, never to return (60,61). CFU-E inhibition occurs 3–6 weeks after exposure and marks the initiation of the fatal anemia (61,200, 201). The FeLV envelope spike transmembrane (TM) p15E protein can suppress lymphocyte functions and also affects BFU-E and CFU-E adversely. The development of BFU-E and CFU-E is inhibited when uninfected cat bone marrow cells are exposed, in vitro, to infectious FeLV, to inactivated FeLV, or to the purified TM (p15E) (198,199).

In summary, FeLV-A, the most commonly occurring and contagious FeLV subgroup found in all pet cats, may give rise, by recombination with endogenous FeLV-C env sequences and additional non-enFeLV sequences, to the anemiogenic FeLV-C subgroup, which then induces fatal anemias in some cats.

3. Myeloid Diseases

FeLV replicates in all precursor and differentiated myeloid cells in the bone marrow and can induce proliferative or degenerative diseases of these cells (18,21).

a. Myeloid Neoplastic (Myeloproliferative) Diseases FeLV induces myeloproliferative diseases, reticuloendotheliosis, myelogenous (usually neutrophilic) leukemia, megakaryotic leukemia, myelofibrosis, and osteosclerosis in pet cats (169,171). These diseases occur relatively rarely in cats and are similar to those induced by acute transforming oncogene-containing retroviruses in chickens.

The term "myeloproliferative" was first used in 1951 to indicate abnormal proliferation of a variety of bone marrow cells that leads to severe anemia and often terminates in granulocytic leukemia (202). Thus, myeloproliferative diseases (MPD) are a group of primary bone marrow neoplastic disorders which may involved any one or a combination of two or more cell types that originate in the bone marrow (202). The primitive mesenchymal cell of the bone marrow gives rise to erythroblasts, myeloblasts, megakaryoblasts, osteoblasts, and fibroblasts, and FeLV replicates in all of these nucleated cells and can apparently transform all of these cell types except eosinophils (21,169).

There are four stages of MPDs in cats (Table 10). The first stage, erythremic myelosis, is characterized by marked hyperplasia of erythroid cells of the bone marrow. In the second stage there is a mixed erythroid and granulocytic precursor proliferation called erythroleukemia. In the third stage the major proliferative cell is the myeloblast. The fourth stage, in some cats with MPD, is characterized by the presence of erythroid and granulocytic leukocyte precursor cells in the blood and spleen together with a proliferation of cancellous bone and/or fibrous tissues in the bone marrow resulting in medullary osteosclerosis or myelofibrosis. This stage may represent the terminal stage of feline MPD, in which the bone marrow is replaced with cancellous bone and fibrous tissue. The various MPDs may be considered diseases in their own right or different stages of the overall MPD disease entity.

Cats with MPD usually have profound normocytic normochromic, nonregenerative anemias with either neoplastic erythroid or granulocytic myeloid precursor cells, or a mixture of both types of cells, in the blood and bone marrow (202). There is usually extramedullary hematopoiesis in the spleen, liver, and lymph nodes, which, if severe enough, causes splenomegaly, hepatomegaly, or lymphadenopathy. Platelets are often reduced in numbers, resulting in bleeding disorders, and occasionally giant, abnormal platelets are present.

Myeloid Neoplastic Diseases—Myelogenous (Granulocytic) Leukemias. Granulocytic leukocytes (neutrophils and basophils) can be transformed by FeLV in the cat, but myelogenous leukemias are far less common than lymphoid malignancies. FeLV can replicate in normal and neoplastic neutrophils and in normal eosinophils (134).

Neutrophilic leukemia is rare, although it is the most common type of myelogenous leukemia in cats. In cats with neutrophilic leukemia the total leukocyte number is usually increased, and neutrophilic peroxidase-positive myelocytes, progranulcytes, and myeloblasts are present in the blood (203,204). There is usually splenomegaly, hepatomegaly, and variable lymphadenopathy. All cats with the disease are severely anemic even though large numbers of nucleated erythrocytes are present in some cats. Eosinophilic leukemia is the only MPD of cats not known to be induced by FeLV. The disease occurs very rarely in pet cats and is characterized by an overproduction of eosinophils with immature forms present in the blood and various tissues.

Megakaryocytic leukemia is extremely rare in cats (133). Cats with this disease have large numbers of bizarre platelets in the peripheral blood, have increased numbers of megakaryocytes in the bone marrow and spleen, are severely anemic, and have hepatosplenomegaly.

Myelofibrosis. Myelofibrosis is an end stage of MPD, but most FeLV-infected cats with MPD die before they develop myelofibrosis. Cats with myelofibrosis are severely anemic, have extensive replacement of the bone marrow with fibrous tissue (fibroblasts), have very few remaining erythroid or myeloid cells in the marrow, and have fibrosis in the liver (169).

Osteochondromatosis and Medullary Osteosclerosis. Osteochondromatosis or multiple cartilaginous exostoses is a benign proliferative disease of bone which occurs in humans, dogs, horses, and cats (205). In humans, dogs, and horses, but not in cats, the disease appears to have a hereditary basis and multiple osteochondromas are usually found both in the metaphyseal regions of long bones and in flat bones such as the scapulae, ribs, pelvis, and vertebrae. The growths usually cease when the growth plates close in young adult people, dogs, and horses. However, in cats, osteochondromatosis appears to be different since the disease occurs in mature cats about 2 years old whose growth plates have already closed and affects mainly the flat bones, the scapulae, pelvis, ribs, and skull, rather than the long bones. Medullary osteosclerosis developed in 12 of 13 kittens that became anemic after experimental infection with FeLV (201). C-type viral particles were seen in osteocytes, osteoblasts, and megakaryocytes, and FeLV antigen was present in the peripheral blood leukocytes of these cats.

Osteochondromatosis and medullary osteosclerosis may be induced by FeLV stimulation or transformation of periosteal fibroblasts or medullary osteocytes and osteoblasts resulting in excess cartilage, bone, or fibrous tissue proliferation. Myelofibrosis and medullary osteosclerosis represent the final stages of reactive bone marrow cells in FeLV-induced MPD.

b. Myeloid Blastopenic Diseases Blastopenic myeloid diseases occur more commonly in infected pet cats than do myeloproliferative diseases.

FeLV-Myeloblastopenia-Enteritis Syndrome (MES) (Panleukopenia-like Syndrome). FeLV-myeloblastopenia-enteritis syndrome (MES) (panleukopenia-like syndrome) was previously termed FeLV myeloblastopenia syndrome without the notation of enteritis (18,168,206). However, a syndrome of FeLV-induced enteritis without the concomitant panleukopenia has been described which appears different from this syndrome. MES is characterized by severe panleukopenia and enteritis with dysentery which can lead to severe anemia and prostation due to the rapid blood loss. Erosion of the epithelium of the tips of the small intestinal villi permits opportunistic infections to enter, resulting in septicemia and death (18,168,172). FeLV antigens and replicating virus are present in the intestinal epithelial cells and lymphocytes in the lamina propria. Although this syndrome resembles panleukopenia (feline distemper-feline parvovirus), it occurs in FeLV-infected cats that are immune to the panleukopenia virus. FeLV-infected healthy cats that are stressed by such things as cat fights or hospitalization often develop this syndrome 2-3 weeks after the stress. There is a very low white blood cell count and the characteristic bone marrow finding is hypoplasia of the granulocytic leukocytes. Lymphoid depletion and hemorrhagic necrosis occur in the mesenteric, cecal, colonic, and sublumbar lymph nodes. For comparison, see the FeLV-enteritis syndrome in Section II.J.5.d.

4. Megakaryocyte Disease

Megakaryocytes are the first cells in the bone marrow to become infected with FeLV after the virus reaches the marrow (21). The platelets that bud from infected megakaryocytes contain FeLV antigens, and budding virus has been observed in their membranes although it is not known whether the virus is replicated in circulating platelets that have no nuclei or is "frozen" in the platelet membranes after the platelets are budded off of the megakaryocytes. Platelet infection is not without consequences, and platelets in FeLV-infected cats may be too large (macrothombocytosis), have an abnormal morphology, or be reduced in numbers (thrombocytopenia) (207,208). Platelets from FeLV-infected cats have decreased function and decreased life-spans (207,208). Mean platelet half-lives are reduced in infected cats by about 50% (11.9 hr compared to 21.5 hr for platelets from uninfected cats). Bleeding disorders occur only rarely even though many FeLV-infected cats have abnormal platelets.

5. Other FeLV Diseases

   a. Abortion and Resorption Syndromes FeLV can cause fetal abortions or resorptions late in gestation. The virus has been detected in two-thirds of cats with a history of abortions or fetal resorptions (18,168). FeLV has been shown experimentally to induce fetal death, resorption, placental involution, and abortion in the middle trimester (209,210). Viremic queens can give birth to infected kittens even after having aborted previous litters (18). Transmission of the virus to fetal or newborn kittens is transplacental and transcolostral (8,209, 210). Also, queens with latent FeLV infections will occasionally bear kittens with congenital or perinatal FeLV infections (144,211). Focal mammary reactivation of latent FeLV and secretion of infectious virus occurs in these latently infected queens (211). Most retroviruses of animals and humans are transmitted in utero or via the milk, and, in fact, health authorities in Japan are now recommending that Japanese women should not nurse their newborn infants in areas of the country where HTLV-I is endemic.

   Although the pathogenesis of FeLV-induced abortions and resorptions is unknown, infection of pregnant mice with MuLV leads to fetal death (212). The MuLV provirus inserts itself into the alphal collagen gene in early mouse embryos, which prevents the synthesis of collagen. The blood vessels and connective tissues are weakened and the fetus hemorrhages fatally. FeLV has also been shown to infect and reduce the viability of hamster fetuses (213). Thus, FeLV may cause fetal mortality in FeLV-infected queens by a similar mechanism.

b. FeLV Neurological Syndrome A neurological syndrome, similar to the neurological syndrome observed in MuLV-infected wild mice, occurs in FeLV-infected pet cats and is characterized by posterior paresis leading to paralysis or tetraplegia (18,168,214). These cats usually show progressive fore- and hindleg weakness leading to eventual paralysis. However, FeLV has not yet been proven to cause this syndrome under experimental conditions. A similar neurological syndrome characterized by paralysis also occurs in humans infected with HTLV-I and has been named HTLV-associated myelopathy (HAM) and tropical spastic paraparesis (TSP) (215,216). A neurological disorder is also associated with HIV-1 infections in humans and is characterized by encephalopathy and dementia (217).

c. FeLV Immune Complex Glomerulonephritis Continual formation and circulation of immune complexes is necessary for disease production, and the proportion of antibody and antigen in immune complexes influences their deposition in glomerular capillaries (152). Small immune complexes, which form in antigen excess, are more nephrotoxic than large immune complexes. Persistently FeLV-infected pet cats have a lifelong viremia which is ideal for the formation and deposition of immune complexes in glomeruli and the induction of glomerulonephritis (56,148,150,151,168). FeLV replication produces a continuous supply of soluble viral antigens over a long period of time, and these antigens occur in excess of antibody, thus encouraging the formation of the more nephrotoxic small immune complexes (8,20).

Fifty percent of persistently viremic cats have circulating immune complexes consisting of whole infectious FeLV complexed with cat IgG (148,172). FeLV antigens, antibody (IgG), and complement deposited in the glomeruli have been found in 25% of healthy FeLV-infected cats but none of these catas had glomerulonephritis (148,172,218). Although there are many chronically viremic pet cats, clinical glomerulonephritis is not a frequent outcome of FeLV infection.

d. Enteritis FeLV-induced enteritis is a recently described syndrome which appears to be distinct from the previously described FeLV myeloblastopenia (panleukopenia-like) syndrome (18,148,172,206). In the FeLV enteritis syndrome, hemorrhagic vomitus, diarrhea, anemia, and lymphoid hyperplasia are prominent. In contrast to the FeLV-myeloblastopenia-enteritis syndrome, cats with FeLV enteritis had intestinal lesions in the crypts, not the tips of the villi; they were most often not panleukopenic; and no hemorrhagic lymphadenopathy was noted. The distinctions between these two syndromes are slight and may not be meaningful but rather may represent different stages of the same disease.

e. **Cachexia** Many FeLV-infected cats fail to thrive and become cachetic (155). The cachexia can be very severe with as much as 50–60% loss in body weight. Lesions include severe muscle atrophy, especially in the temporal muscles, and serious atrophy (myxomatous degeneration) of pleural and peritoneal fat stores. Although the pathogenesis of cachexia is not known, anorexia and depression certainly contribute to the malnourished state. Similarly, human AIDS patients become extremely cachetic; the disease is described as "slims disease" in many parts of Africa. Humans with AIDS often have deficiencies of selenium and other trace elements and develop increases in tumor necrosis factor or cachetin (219). The nutritional status of FeLV-infected cats with cachexia has not been studied.

6. FeSV-Induced Tumors of Pet Cats

Fibrosarcomas account for 6–12% of all cat tumors (73,74). FeSVs induce multicentric fibrosarcomas of young (average age 3 years) pet cats, whereas the more common single fibrosarcomas that occur in older cats (average age 10 years) are not associated with FeSVs (73,74). FeSV-induced fibrosarcomas are usually poorly differentiated and are more invasive than the non-FeSV-induced fibrosarcomas.

There is a histological difference between the FeSV-positive multicentric fibrosarcomas of young cats and the FeSV-negative solitary fibrosarcomas of older cats (73,75,87,96). The solitary fibrosarcomas are usually compact, well-differentiated, slowly invasive tumors which contain considerable amounts of collagen and reticulum. Mitotic activity is usually modest and thus the tumors are slow growing and can reach considerable size. In contrast, the multicentric FeLV-positive fibrosarcomas are less compact (multiple lesions), less well-differentiated, more invasive tumors which contain less collagen and reticulum. The fibrosarcoma cells are often invasive into surrounding tissues and grow rapidly. The tumors are usually pleomorphic, containing fusiform, polygonal, and giant fibroblasts with numerous mitotic figures. Frozen sections of these tumors contain FeLV–FeSV antigens in the tumor cell cytoplasm.

Although it is clear that FeLV is spread contagiously among cats, there is no evidence to date that FeSV is also transmitted contagiously. From detailed epidemiological studies of cats living in households where other cats developed FeSV-induced fibrosarcomas, it can be concluded that FeSVs are rarely, if ever, spread contagiously, but rather are generated de novo in FeLV-infected cats.

III. LENTIVIRUSES

A. Feline Immunodeficiency Virus (FIV)

Retroviruses are an ever-expanding family of RNA-containing viruses of animals and humans. A new feline lentivirus, which induces AIDS in cats, has recently been discovered and was initially named the feline T-lymphotropic lentivirus (FTLV) (2). Thus, cats are infected with representatives of all three subfamilies of retroviruses and there are now two feline retroviruses, from different subfamilies, that causes AIDS in cats, FeLV and FIV. FIV was recently discovered in a cattery in northern California which was free of FeLV but which had an outbreak of FAIDS in several cats living together in the same pen. Sero-epidemiological studies, in progress, show that the virus has a worldwide distribution (220). Morphologically, FIV differs from FeLV in that it has a rod-shaped versus an oval nucleoid. Unlike FeLV, which replicates in all nucleated cells of the lymphoid and myeloid series, FIV is highly T-lymphotropic. However, like all retroviruses, FIV results in a persistent, lifelong infection in those cats that do not mount an appropriate immune response and clear the virus after exposure. FIV, like FeLV, appears to be transmitted mainly via the saliva through biting.

Similarly to people infected with the AIDS viruses and cows infected with the bovine leukemia virus, cats infected with FIV produce antibodies to the viral proteins that coexist with the virus. The antibody immune response against FIV is not sufficient to clear the virus from the cat's body and a persistent infection develops. FIV antibody can be detected by a fluorescent antibody test using FIV-infected cat T lymphocytes as a target for the test or by ELISA and Western blot assays (2). This method of detection of FIV infection differs from those used to detect FeLV infection, where a fluorescent antibody test or an ELISA test detects the viral antigens in the absence of antibody to FeLV. In order to diagnose FIV-FAIDS, a diagnosis of a secondary disease syndrome must be made in an FeLV-test-negative cat that is positive for FIV antibody. It is still not known whether many cats with FAIDS are coinfected with FeLV and FIV, and if so, which virus is mainly responsible for the immunosuppression.

FIV may have been a recently introduced virus into cats and, as was found for the human AIDS virus, probably originated from another species. Since only about 50% of pet cats with chronic immunosuppressive syndromes (FAIDS) are infected with FeLV, FIV may be responsible for some or most of the non-FeLV-caused immunosuppression observed in pet cats. FIV may represent a major new naturally occurring feline pathogen, and veterinarians should be aware of this possibility and be prepared to institute control measures to prevent the spread of this virus. Until an FIV vaccine can be developed, such control measures will include FIV test and removal programs.

B. FIV-FAIDS

The clinical signs of FIV-FAIDS include generalized lymphadenopathy, fever, leukopenia, conjunctivitis, gingivitis, periodontitis, rhinitis, emaciation, diarrhea, and pustular dermatitis. The clinical signs (Table 17) of FeLV- and

TABLE 17 Comparison of FeLV and FIV

FeLV	FIV
1. Subfamily: Oncovirinae	1. Subfamily: Lentiviriniae
2. Pancytotropic—bone marrow and lymphoid cells	2. Highly T-lymphocyte tropic
3. Causes FAIDS, anemia, cancer, and neurological disorders	3. Causes FAIDS, anemia, and neurological disorders
4. Morphology: oval nucleoid	4. Morphology: rod-shaped nucleoid
5. Mg^{2+}-dependent reverse transcriptase	5. Mn^{2+}-dependent reverse transcriptase
6. Does not coexist with antibody	6. Coexists with antibody
7. Spread via saliva	7. Spread via saliva (biting)
8. Moderately contagious	8. Poorly contagious
9. Detected by detection of viral antigens	9. Detected by detection of viral antibodies

FIV-FAIDS are almost identical (2,18,172,221). One feature that clinically distinguishes the two diseases somewhat is that FIV-FAIDS cats develop a more severe gingivitis with extreme redness and swelling of the gingiva.

IV. SPUMAVIRUSES

A. Feline Syncytium-Forming Virus (FeSFV)

The feline synytium-forming virus (FeSFV), also known as the feline syncytial virus, feline foamy virus, and feline syncytia-forming virus, is a member of the subfamily Spumavirinae (222). Spumaviruses are commonly found in many species and have not been strongly linked with specific diseases. FeSFV has a typical retrovirus morphology, and in tissue culture the virus causes fusion of adjacent cells into large, multinucleated cells or synctia. Although the virus is considered nononcogenic, FeSFV has been reported to cause transformation of kidney cells in vitro (223).

FeSFV has been recovered from the tissues of many diseased and healthy cats. The incidence of FeSFV infection varies greatly with the geographical area and the living environments of the cats. The infection rates increase dramatically after 1 year of age and reach 50% by 4 years of age (222). Like all animals infected with lentiviruses, cats with antibody against FeSFV virus are infected.

Cats can become infected with FeSFV by several routes. In utero transmission occurs in 25–50% of kittens born to FeSFV-infected queens (222).

The virus is also shed in high levels from the oropharynx, but cat-to-cat contact transmission in housing experiments was infrequent (222,224). The virus is found more commonly in older, free-roaming, outdoor cats, which suggests that bite transmission may be the major route for spread of the virus. After infection the virus becomes latent in virtually all tissues but can be found in complete infectious form only in secretions of the oropharynx. Infectious replicating FeSFV is not present in whole blood, plasma, or red blood cell pack. However, the buffy coat from peripheral blood is a rich source of virus, even though replication of the virus is suppressed in vivo. Latent virus can be readily recovered after only several passages of cells in tissue culture.

As mentioned previously, FeSFV has been isolated from healthy cats and cats with various nonneoplastic and neoplastic diseases. One study found a "statistical link" of the virus with myeloproliferative diseases and chronic progressive polyarthritis (222). However, a direct etiological link to these diseases has not been established since none of the cats inoculated with FeSFV developed these or any disease. If FeSFV does cause any diseases, they must be unusual and occur at a low frequency a long time after infection.

V. SUMMARY

Pet cats are infected with all three subfamilies of retroviruses and develop a wide variety of retroviral diseases ranging from lymphoid cancers to feline AIDS. Feline retroviruses and their diseases offer an excellent comparative model for the study of the control and treatment of human retroviruses and human retroviral diseases. In addition, the continual search for new oncogene-containing feline sarcoma and leukemia viruses may yield new and important oncogenes whose normal cellular homologs may elucidate new growth-regulatory molecules.

ACKNOWLEDGMENTS

I thank E. E. Zuckerman for her excellent and devoted technical help during the past 14 years which enabled the collection of much of the data cited in this chapter. I also thank my colleagues and friends L. J. Old, M. Essex, R. Gallo, F. Wong-Staal, H. Snyder, Jr., G. Geering, P. Besmer, G. MacEwen, P. Hess, H. Nauts, J. Wolfe, E. Hoover, J. Rojko, O. Jarrett, F. deNoronha, C. F. Jones, C. Grant, D. Armstrong, J. Gold, B. Polsky, J. Whitehead, G. Robinson, and the many veterinarians for their help and encouragement. Special thanks go to J. Rojko for help in the preparation of this chapter. I also thank the National Institutes of Health, Leukemia Society of America, Cancer Research Institute, New York State Department of Health, and Society of Memorial Sloan-Kettering Cancer Center for funding this work.

REFERENCES

1. Teich, N. (1982). Taxonomy of retroviruses. In: *RNA Tumor Viruses*. R Weiss, N Teich, H Varmus, and J Coffin, eds.), pp. 25–207. New York, Cold Spring Harbor Laboratory.
2. Pedersen, N. C., Ho, E. W., Brown, M. L., and Yamamoto, J. K. (1987). Isolation of a T-lymphotropic virus from domestic cats with an immuno-deficiency-like syndrome. *Science* 235:790–793.
3. Jarrett, W. F. H., Crawford, E. M., Martin, W. B., and Davie, F. (1964). A virus-like particle associated with leukemia (lymphosarcoma). *Nature* 202: 567–569.
4. Huebner, R. J., and Todaro, G. J. (1969). Oncogenesis of RNA tumor viruses as determinants of cancer. *Proc. Natl. Acad. Sci. USA* 64:1087–1094.
5. Hardy, W. D., Jr., Old, L. J., Hess, P. W., Essex, M., and Cotter, S. (1973). Horizontal transmission of feline leukaemia virus. *Nature* 244:266–269.
6. Brodey, R. S., McDonough, S., Frye, F. L., and Hardy, W. D., Jr. Epidemiology of feline leukemia. In: *Comparative Leukemia Research* (R. M. Dutcher, ed.), pp. 333–342. Karger, Basel.
7. Essex, M., Cotter, S. M., Sliski, A. H., Hardy, W. D., Jr., Stephenson, J. R., Aaronson, S. A., and Jarrett, O. (1977). Horizontal transmission of feline leukemia virus under natural conditions in a feline leukemia cluster house-hold. *Int. J. Cancer* 19:90–96.
8. Hardy, W. D., Jr., Geering, G., Old, L. J., deHarven, E., Brodey, R. S., and McDonough, S. (1969). Feline leukemia virus; occurrence of viral antigen in the tissues of cats with lymphosarcoma and other diseases. *Science* 166: 1019–1021.
9. Jarrett, W. F. H., Jarrett, O., Mackey, L., Laird, H., Hardy, W. D., Jr., and Essex, M. (1973). Horizontal transmission of leukemia virus and leukemia in the cat. *J. Natl. Cancer Inst.* 51:833–841.
10. Hardy, W. D., Jr. (1983). Naturally occurring retroviruses (RNA tumor viruses). I. *Cancer Invest.* 1:67–83.
11. Hardy, W. D., Jr. (1983). Naturally occurring retroviruses (RNA tumor viruses). II. *Cancer Invest.* 1:163–174.
12. Hardy, W. D., Jr. (1981). The feline leukemia virus. *J. Am. Animal Hosp. Assoc.* 17:951–976.
13. Anderson, L. J., Jarrett, W. F. H., Jarrett, O., and Laird, H. M. (1971). Feline leukemia-virus infection of kittens: Mortality associated with atrophy of the thymus and lymphoid depletion. *J. Natl. Cancer Inst.* 47: 807–817.
14. Cotter, S. M., Hardy, W. D., Jr., and Essex, M. (1975). The association of the feline leukemia virus with lymphosarcoma and other disorders. *J. Am. Vet. Med. Assoc.* 166:449–454.
15. Perryman, L. E., Hoover, E. A., and Yohn, D. S. (1972). Immunologic reactivity of the cat: Immunosuppression in experimental feline leukemia. *J. Natl. Cancer Inst.* 49:1357–1365.

16. Hoover, E. A., Kociba, G. J., Hardy, W. D., Jr., and Yohn, D. S. (1974). Erythroid hypoplasis in cats inoculated with feline leukemia virus. *J. Natl. Cancer Inst.* 53:1271–1276.

17. Hoover, E. A., Olsen, R. G., Hardy, W. D., Jr., Schaller, J. P., and Mathes, L. E. (1976). Feline leukemia virus infection: Age related variation in response of cats to experimental infection. *J. Natl. Cancer Inst.* 57:365–369.

18. Hardy, W. D., Jr. (1981). Feline leukemia virus non-neoplastic diseases. *J. Am. Animal Hosp. Assoc.* 17:941–949.

19. Mackey, L. J., Jarrett, W., Jarrett, O., and Laird, H. (1975). Anemia associated with feline leukemia virus infection in cats. *J. Natl. Cancer Inst.* 54:209–217.

20. Hardy, W. D., Jr., Hess, P. W., MacEwen, E. G., McClelland, A.J., Zuckerman, E. E., Essex, M., Cotter, S. M., and Jarrett, O. (1976). Biology of feline leukemia virus in the natural environment. *Cancer Res.* 36:582–588.

21. Hardy, W. D., Jr., Hirshaut, Y., and Hess, P. (1973). Detection of the feline leukemia virus and other mammalian oncornaviruses by immunofluorescence. In: *Unifying Concepts of Leukemia*, (R. M. Dutcher and L. Chieco-Bianchi, eds.), pp. 778–799. Basel, Karger,

22. Hardy, W. D., Jr., McClelland, A. J., Zuckerman, E. E., Hess, P. W., Essex, M., Cotter, S. M., MacEwen, E. G., and Hayes, A. A. (1976). Prevention of the contagious spread of feline leukaemia virus and the development of leukaemia in pet cats. *Nature* 263:326–328.

23. Hardy, W. D., Jr., McClelland, A. J., Hess, P. W., and MacEwen, G. (1974). Veterinarians and the control of feline leukemia virus. *J. Am. Animal Hosp. Assoc.* 10:367–372.

24. Weijer, K., and Daams, J. H. (1978). The control of lymphosarcoma/leukaemia and feline leukaemia virus. *J. Small Animal Pract.* 19:631–637.

25. Jarrett, O., Golder, M. C., and Weijer, K. (1982). A comparison of three methods of feline leukaemia virus diagnosis. *Vet. Rec.* 110:325–328.

26. Olsen, R. G., Hoover, E. A., Mathes, L. E., Heding, L. D., and Schaller, J. P. (1976). Immunization against feline oncornavirus disease using a killed tumor cell vaccine. *Cancer Res.* 36:3642–3646.

27. Olsen, R. G., Lewis, M., Mathes, L. E., and Hause, W. (1980). Feline leukemia vaccine, efficacy testing in a large multicat household. *Feline Pract.* 10:13–16.

28. Poiesz, B. G., Ruscetti, F. W., Gazdar, A. F., Bunn, P. A., Minna, J. D., and Gallo, R. C. (1980). Detection and isolation of type C retrovirus particles from fresh and cultured lymphocytes of a patient with cutaneous T-cell lymphoma. *Proc. Natl. Acad. Sci USA* 77:7415–7419.

29. Gallo, R. C., Salahuddin, S. Z., Popovic, M., Shearer, G. M., Kaplan, M., Haynes, B. F., Palker, T. J., Redfield, R., Oleske, J., Safai, B., White, G., Foster, P., and Markham, P. D. (1984). Frequent detection and isolation of cytopathic retroviruses (HTLV-III) from patients with AIDS and at risk for AIDS. *Science* 224:500–503.

30. Wong-Staal, F., and Gallo, R. C. (1985). Human T-lymphotropic retroviruses. *Nature* 317:395–403.
31. Mullins, J. I., and Hoover, E. A. (1989). Molecular biology of feline leukemia viruses. In: *Retrovirus Biology: An Emerging Role in Human Diseases* (R. C. Gallo and F. Wong-Staal, eds.). New York, Marcel Dekker.
32. Fischinger, P. G., Pebbles, P. T., Nomura, S., and Haapala, D. K. (1973). Isolation of RD-114-like oncornavirus from cat cell lines. *J. Virol* 11:978–985.
33. Livingston, D. M., and Todaro, G. J. (1973). Endogenous type-C virus from a cat cell clone with properties distinct from previously described feline viruses. *Virology* 53:142–151.
34. McAllister, R. M., Nicolson, M., Gardner, M. B., Rongey, R. W., Rasheed, S., Sarma, P. S., Huebner, R. J., Hatanaka, M., Oroszlan, S., Gilden, R. V., Kabigting, A., and Vernon, L. (1973). RD-114 comparison with feline and murine type C viruses released from RD cells. *Nature* 242:75–78.
35. Niman, H. L., Stephenson, J. R., Gardner, M. B., and Roy-Burman, P. (1977). RD-114 and feline leukemia virus genome expression in natural lymphomas of domestic cats. *Nature* 226:357–360.
36. Mandel, M. P., Stephenson, J. R., Hardy, W. D., Jr., and Essex, M. (1979). Endogenous RD-114 virus of cats: Absence of antibodies to RD-114 envelope antigens in cats naturally exposed to the feline leukemia virus. *Infect. Immunol.* 24:282–285.
37. Todaro, G. J. (1980). Interspecies transmission of mammalian retroviruses. In: *Viral Oncology* (G. Klein, ed.), pp. 291–309. New York, Raven Press.
38. Benveniste, R. E., Sherr, C. J., and Todaro, G. J. (1975). Evolution of type C viral genes: Origin of feline leukemia virus. *Science* 190:886–888.
39. Baluda, M. A., and Roy-Burman, P. (1973). Partial characterization of RD-114 virus by DNA–RNA hybridization studies. *Nature* 244:59–62.
40. Benveniste, R. E., and Todaro, G. J. (1973). Homology between type-C viruses of various species as determined by molecular hybridization. *Proc. Natl. Acad. Sci. USA* 70:3316–3320.
41. Okabe, H., Twiddy, E., Gilden, R. V., Hatanaka, M., Hoover, E. A., and Olsen, R. G. (1976). FeLV-related sequences in DNA from a FeLV-free cat colony. *Virology* 69:798–801.
42. Koshy, R., Wong-Staal, F., Gallo, R. C., Hardy, W. D., Jr., and Essex, M. (1979). Distribution of feline leukemia virus DNA sequences in tissues of normal and leukemic domestic cats. *Virology* 99:135–144.
43. Bolognesi, D. P., Montelaro, R. C., Frank, H., and Schafer, W. (1978). Assembly of type C oncornaviruses: A model. *Science* 199:183–186.
44. Okabe, H., DuBuy, J., Gilden, R. V., and Gardner, M. B. (1978). A portion of the feline leukemia virus genome is not endogenous in cat cells. *Int. J. Cancer* 22:70–78.
45. Soe, L. H., Shimizu, R. W., Landolph, J. R., and Roy-Burman, P. (1985). Molecular analysis of several classes of endogenous feline leukemia virus elements. *J. Virol.* 56:701–710.

46. Casey, J. W., Roach, A., Mullins, J. T., Burck, K. B., Nicolson, M. O., Gardner, M. B., and Davidson, N. (1981). The U3 portion of the feline leukemia virus identifies horizontally acquired proviruses in leukemic cats. *Proc. Natl. Acad. Sci. USA* 78:7778-7782.

47. Mullins, J. I., Casey, J. W., Nicolson, M. D., Burck, K. B., and Davidson, N. (1981). Sequence arrangement and biological activity of cloned feline leukemia proviruses from a virus productive human cell line. *J. Virol.* 38: 688-703.

48. Soe, L. H., Devi, B. C., Mullins, J. I., and Roy-Burman, P. (1983). Molecular cloning and characterization of endogenous feline leukemia virus sequences from a cat genomic library. *J. Virol.* 46:829-840.

49. Stewart, M. A., Warnock, M., Wheeler, A., Wilkie, N., Mullins, J. I., Onions, D. E., and Neil, J. C. (1986). Nucleotide sequences of a feline leukemia virus subgroup A envelope gene and long terminal repeat and evidence for the recombinational origin of subgroup B viruses. *J. Virol.* 58:825-834.

50. Overbaugh, J., Donahue, P. R., Quackenbush, S. L., Hoover, E. A., and Mullins, J. I. (1988). Molecular cloning of a feline leukemia virus that induces fatal immunodeficiency disease in cats. *Science* 239:906-910.

51. Overbaugh, J., Riedel, N., Hoover, E. A., and Mullins, J. I. (1988). Transduction of endogenous envelope genes by feline leukemia virus in vitro. *Nature* 332:731-734.

52. Todaro, G. J., Tevethia, S., and Melnick, J. (1974). Isolation of an RD-114 related type-C virus from feline sarcoma virus-transformed baboon cells. *Intervirology* 1:399-404.

53. Jarrett, O., Laird, H. M., and Hay, D. (1973). Determinants of the host range of feline leukaemia viruses. *J. Gen. Virol.* 20:169-175.

54. Sarma, P. S., and Log, T. (1973). Subgroup classification of feline leukemia and sarcoma viruses by viral interference and neutralization tests. *Virology* 54:160-169.

55. Jarrett, O., Hardy, W. D., Jr., Golder, M. C., and Hay, D. (1978). The frequency of occurrence of feline leukemia virus subgroups in cats. *Int. J. Cancer* 21:334-337.

56. Snyder, H. W., Jr., Jones, F. R., Day, N. K., and Hardy, W. D., Jr. (1982). Isolation and characterization of circulating feline leukemia virus-immune complexes from plasma of persistently infected pet cats removed by ex vivo immunosorption. *J. Immunol.* 128:2726-2730.

57. Levin, R., Ruscetti, S. F., Parks, W. P., and Scolnick, E. M. (1976). Expression of feline type-C virus in normal and tumor tissues of the domestic cat. *Int. J. Cancer* 18:661-671.

58. Rosenberg, Z., Pedersen, F. S., and Haseltine, W. A. (1980). Comparative analysis of the genomes of feline leukemia viruses. *J. Virol.* 35:542-546.

59. Mullins, J. T., Chen, C. S., and Hoover, E. A. (1986). Disease-specific and tissue-specific production of unintegrated feline leukemia virus variant DNA in feline AIDS. *Nature* 319:332-336.

60. Onions, D., Jarrett, O., Testa, N., Frassoni, F., and Toth, S. (1982). Selective effect of feline leukaemia virus on early erythroid precursors. *Nature* 296:156-158.

61. Testa, N. G., Onions, D., Jarrett, O., Frassoni, F., and Eliason, J. F. (1983). Haematopoietic colony formation (BFU–E, GM–CFC) during the development of pure red cell hypoplasia induced in the cat by feline leukemia virus. *Leukemia Res.* 7:103–116.

62. Elder, J. H., and Mullints, J. T. (1983). Nucleotide sequence of the envelope gene of Gardner-Arnstein feline leukemia virus B reveals unique sequence homologies with a murine mink cell focus-forming virus. *J. Virol.* 46:871–880.

63. Neil, N. C., Hughes, D., McFarlane, R., Wilkie, N. M., Onions, D. E., Lees, G., and Jarrett, O. (1984). Transduction and rearrangement of the *myc* gene by feline leukaemia virus in naturally occurring T-cell leukaemias. *Nature* 308:814–820.

64. Mullins, J. T., Brody, D. S., Binari, R. C., Jr., and Cotter, S. M. (1984). Viral transduction of *c-myc* gene in naturally occurring feline leukaemias. *Nature* 308:856–858.

65. Levy, L. S., Gardner, M. B., and Casey J. W. (1984). Isolation of a feline leukaemia provirus containing the oncogene *myc* from a feline lymphosarcoma. *Nature* 308:853–856.

66. Stewart, M. A., Forrest, D., McFarlane, R., Onions, D., Wilkie, N., and Neil, J. C. (1986). Conservation of the *c-myc* coding sequence in transduced feline v-*myc* genes. *Virology* 154:121–134.

67. Onions, D. E., Lees, G., Forrest, D., and Neil, J. C. (1987). Recombinant feline viruses containing the *myc* gene rapidly produce clonal tumors expressing T-cell antigen receptor gene transcripts. *Int. J. Cancer* 40: 40–45.

68. Forrest, D., Onions, D., Lees, G., and Neil, J. C. (1987). Altered structure and expression of *c-myc* in feline T-cell tumors. *Virology* 158:194–205.

69. Bonham, L., Lobelle-Rich, P. A., Henderson, L. A., and Levy, L. S. (1987). Transforming potential of a *myc*-containing variant of feline leukemia virus in vitro in early-passage feline cells. *J. Virol.* 61:3072–3081.

70. Besmer, P. (1983). Acute transforming feline retroviruses. *Curr. Topics Microbiol. Immunol.* 107:1–27.

71. Frankel, A. E., Gilbert, J. H., Porzig, F. J., Scolnick, E. M., and Aaronson, S. A. (1979). Nature and distribution of feline sarcoma virus nucleotide sequences. *J. Virol.* 30:821–827.

72. Bishop, J. M. (1981). Enemies within, the genesis of retrovirus oncogenes. *Cell* 23:5–6.

73. Hardy, W. D., Jr. (1981). The feline sarcoma viruses. *J. Am. Animal Hosp. Assoc.* 17:981–996.

74. Hardy, W. D., Jr. (1980). The biology and virology of the feline sarcoma viruses. In: *Feline Leukemia Virus* (W. D. Hardy, Jr., M. Essex, and A. J. McClelland, eds.) pp. 79–118. New York, Elsevier.

75. Gardner, M. B., Arnstein, P., Rongey, R. W., Estes, J. D.,Sarma, P. S., Rickard, C. F., and Huebner, R. J. (1970). Experimental transmission of feline fibrosarcoma to cats and dogs. *Nature* 226:807–809.

76. Theilen, G. H., Snyder, S. P., Wolfe, L. G., and Landon, J. L. (1970). Biological studies with viral induced fibrosarcomas in cats, dogs, rabbits,

and nonhuman primates. In: *Comparative Leukemia Research 1969* (R. M. Dutcher, ed.), pp. 393–400. Basel, Karger.

77. Gardner, M. B., Arnstein, P., Johnson, E., Rongey, R. W., and Charman, H. P. (1971). Feline sarcoma virus tumor induction in cats and dogs. *J. Am. Vet. Med. Assoc.* 158:1046–1054.

78. Theilen, G. H. (1971). Continuing studies with transmissable feline fibrosarcoma virus in fetal and newborn sheep. *J. Am. Vet. Med. Assoc.* 158:1040–1045.

79. Pearson, L. D., Snyder, S. P., and Aldrich, C. D. (1973). Oncogenic activity of feline fibrosarcoma virus in newborn pigs. *Am. J. Vet. Res.* 34:405–409.

80. Deinhardt, F., Wolfe, L. G., Theilen, G. H., and Snyder, S. P. (1970). ST-feline fibrosarcoma virus: Induction of tumors in marmoset monkeys. *Science* 167:881.

81. Donner, L., Turek, L. P., Ruscetti, S. K., Fedele, L. A., and Sherr, C. J. (1980). Transformation defective mutants of feline sarcoma virus which express a product of the viral *src* gene. *J. Virol.* 35:129–140.

82. Pedersen, N. C., Johnson, L., and Theilen, G. H. (1984). Biological behavior of tumors and associated retroviremia in cats inoculated with Snyder-Theilen fibrosarcoma virus and the phenomenon of tumor recurrence after primary regression. *Infect. Immunol.* 43:631–636.

83. Aldrich, C. D., and Pedersen, N. C. (1974). Persistent viremia after regression of primary virus-induced feline fibrosarcomas. *Am. J. Vet. Res.* 35:1383–1387.

84. Noronha, F., Grant, C. K., Lutz, H., and Keyes, A. (1983). Circulating levels of feline leukemia and sarcoma viruses and fibrosarcoma regression in persistently viremic cats. *Cancer Res.* 43:1663–1668.

85. Hardy, W. D., Jr., Zuckerman, E., Markovich, R., Besmer, P., and Snyder, H. W. (1982). Isolation of feline sarcoma viruses from pet cats with multicentric fibrosarcomas. In: *Advances in Comparative Leukemia Research 1981* (D. S. Yohn and J. R. Blakeslee, eds.) pp. 205–206. North Holland, Elsevier.

86. Snyder, H. W., Jr., Singhal, M. C., Zuckerman, E. E., and Hardy, W. D., Jr. (1984). Isolation of a new feline sarcoma virus (HZ1-FeSV). Biochemical and immunological characterization of its translation product. *Virology* 132:205–210.

87. Snyder, S. P., and Theilen, G. H. (1969). Transmissable feline fibrosarcoma. *Nature* 221:1074–1075.

88. Fujinami, A., and Inamoto, K. (1914). Ueber geschwulste bei japanischen haushuhnern insbesodere uber einen transplatablen tumor. *Zeitschr. Krebsforsch.* 14:94–119.

89. Shibuya, M., Hanafusa, T., Hanafusa, H., and Stephenson, J. R. (1980). Homology exists among the transforming sequences of avian and feline sarcoma viruses. *Proc. Natl. Acad. Sci. USA* 77:6536–6540.

90. Snyder, H. W., Jr. (1982). Biochemical characterization of protein kinase activities associated with transforming gene products of the Snyder-Theilen and Gardner-Arnstein strains of feline sarcoma virus. *Virology* 117:165–172.

91. McCullough, B., Schaller, J., Shadduck, J. A., and Yohn, D. S. (1973). Induction of malignant melanomas associated with fibrosarcoma in gnotobiotic cats inoculated with Gardner feline fibrosarcoma virus. *J. Natl. Cancer Inst.* 48:1893–1896.

92. Niederkorn, J. Y., Shadduck, J. A., Albert, D., Essex, M. (1980). Humoral antibody response to FOCMA in feline sarcoma virus-induced ocular melanoma. In: *Feline Leukemia Virus* (W. D. Hardy, Jr., M. Essex, and A. J. McClelland, eds.), pp. 181–185. New York, Elsevier.

93. Shadduck, J. A., Albert, D. M., and Niederkorn, J. Y. (1981). Feline uveal melanomas induced with feline sarcoma virus: potential model of the human counterpart. *J. Natl. Cancer Inst.* 67:619–626.

94. Pierce, J. H., and Aaronson, S. A. (1983). In vitro transformation of murine pre-B lymphoid cells by Snyder-Theilen feline sarcoma virus. *J. Virol.* 46:993–1002.

95. Besmer, P., Lader, E., George, P., Bergold, P. J., Qui, F-H., Zuckerman, E. E., and Hardy, W. D., Jr. (1986). A new acute transforming feline retrovirus with *fms* homology specifies a c-terminally truncated version of the c-*fms* protein that is different from SM-feline sarcoma virus v-*fms* protein. *J. Virol.* 60:194–203.

96. McDonough, S. K., Larsen, S., Brodey, R. S., Stock, N. D., and Hardy, W. D., Jr. (1971). A transmissable feline fibrosarcoma of viral origin. *Cancer Res.* 31:953–956.

97. Anderson, S. J., Furth, M., Wolff, L., Ruscetti, S. K., and Sherr, C. J. (1982). Monoclonal antibodies to the transformation specific glycoprotein encoded by the feline retroviral oncogene v-*fms*. *J. Virol.* 44:696–702.

98. Barbacid, M., Lauver, A. V., and Devare, S. G. (1980). Biochemical and immunological characterization of polyproteins coded for by the McDonough, Gardner-Arnstein, and Snyder-Theilen strains of feline sarcoma virus. *J. Virol.* 33:196–207.

99. Reynolds, F. H., Van de Ven, W. J. M., Blomberg, J., and Stephenson, J. R. (1981). Differences in mechanism of transformation by independent feline sarcoma virus isolates. *J. Virol.* 38:1084–1089.

100. Sherr, C. J., Rettenmier, C. W., Sacca, R., Roussel, M. F., Look, A. T., and Stanley, D. R. The c-*fms* proto-oncogene product is related to the receptor for the mononuclear phagocyte growth factor, CSF-1. *Cell* 41:665–676.

101. Sacca, R., Stanley, E. R., Sherr, C. J., and Rettenmier, C. W. (1986). Specific binding of the mononuclear phagocyte colony-stimulating factor CSF-1 to the product of the v-*fms* oncogene. *Proc. Natl. Acad. Sci. USA* 83:3331–3335.

102. Sariban, E., Mitchell, T., and Kufe, D. (1985). Expression of the c-*fms* protooncogene during human monocyte differentiation. *Nature* 316:164–166.

103. Besmer, P., Hardy, W. D., Jr., Zuckerman, E. E., Bergold, P., Lederman, L., and Snyder, H. W., Jr. (1983). The Hardy-Zuckerman 2-FeSV, a new feline retrovirus with oncogene homology to Abelson-MuLV. *Nature* 303:825–828.

104. Abelson, H. T., and Rabstein, L. S. (1970). Lymphosarcoma, virus-induced thymic independent disease in mice. *Cancer Res.* 30:2213–2222.

105. Goff, S. P., and Baltimore, D. (1982). The cellular oncogene of the Abelson murine leukemia virus genome. In: *Advances in Viral Oncology* (G. Klein, ed.), pp. 127–139. New York, Raven Press.

106. Lederman, L., Singhal, M. C., Besmer, P., Zuckerman, E. E., Hardy, W. D., Jr., and Snyder, H. W., Jr. (1985). Immunological and biochemical characterization of HZ2 feline sarcoma virus and Abelson leukemia virus translation products. *J. Gen. Virol.* 66:2057–2063.

107. Bergold, P. J., Wang, J. Y. J., Hardy, W. D., Jr., Littau, V., Johnson, E., and Besmer, P. (1987). Structure and origins of the HZ2-feline sarcoma virus. *Virology* 158:320–329.

108. Besmer, P., Snyder, H. W., Jr., Murphy, J. E., Hardy, W. D., Jr., and Parodi, A. (1983). The Parodi-Irgens feline sarcoma virus and simian sarcoma virus have homologous oncogenes, but in different contexts of the viral genomes. *J. Virol.* 46:606–613.

109. Irgens, K., Wyers, M., Moraillon, A., Parodi, A., and Fortuny, V. (1973). Isolement d'un virus sarcomatogene feline a partir d'un fibrosarcome spontane due chat, etude du pouvoir sarcomatogene in vivo. *C. R. Acad. Sci. Paris* 276:1783–1786.

110. Doolittle, R. F., Hunkapiller, M. W., Hood, L. E., Devare, S. G., Robbins, K. C., Aaronson, S. A., and Antoniades, H. N. (1983). Simian sarcoma virus *onc* gene, *v-sis*, is derived from the gene (or genes) encoding a platelet-derived growth factor. *Science* 221:275–277.

111. Robbins, K. C., Antoniades, H. N., Devare, S. G., Hunkapiller, M. W., and Aaronson, S. A. (1983). Structural and immunological similarities between simian sarcoma virus gene product(s) and human platelet-derived growth factor. *Nature* 305:605–608.

112. Waterfield, M. D., Scrace, G. T., Whittle, N., Stroobant, P., Johnsson, A., Wasteson, A., Westermark, B., Heldin, C-H., Huang, J. S., and Deuel, T. F. (1983). Platelet-derived growth factor is structurally related to the putative transforming protein p28^{sis} of simian sarcoma virus. Nature 304:35–39.

113. Naharro, G., Dunn, C. Y., and Robbins, K. C. (1983). Analysis of the primary translational product and integrated DNA of a new feline sarcoma virus, GR-FeSV. *Virology* 125:502–507.

114. Rasheed, S., Barbacid, M., Aaronson, S., and Gardner, M. B. (1982). Origin and biological properties of a new feline sarcoma virus. *Virology* 117:238–244.

115. Ziemiecki, A., Hennig, D., Gardner, L., Ferdinand, F. J., Friis, R. R., Bauer, H., Pedersen, N. C., Johnson, L., and Theilen, G. H. (1984). Biological and biochemical characterization of a new isolate of feline sarcoma virus: Theilen-Pedersen (TP1-FeSV). *Virology* 138:324–331.

116. Naharro, G., Robbins, K. C., and Reddy, E. P. (1984). Gene product of *v-fgr onc*: Hybrid protein containing a portion of actin and a tyrosine-specific protein kinase. *Science* 223:63–66.

117. Besmer, P., Murphy, J. E., George, P. C., Qui, F. H., Bergold, P. J., Snyder, H. W., Jr., Brodeur, D., Zuckerman, E. E., and Hardy, W. D., Jr. (1986). A new acute transforming feline retrovirus and relationship of its oncogene *kit* with the protein kinase gene family. *Nature* 320:415–421.

118. Hardy, W. D., Jr. (1985). Feline retroviruses. In: *Advances in Viral Oncology*. Vol. 5 (G. Klein, ed.), pp. 1–34. New York, Raven Press.

119. Youngren, S. D., and deNoronha, F. (1983). A transmissable feline sarcoma virus containing sequences related to the K-*ras* onc gene. In: *RNA Tumor Viruses* (E. Scolnick, and N. Hopkins, eds.), p. 33. New York, Cold Spring Harbor Laboratory.

120. Yoshiki, T., Mellors, R. C., Hardy, W. D., Jr., and Fleissner, E. (1974). Common cell surface antigen associated with mammalian C-type RNA viruses. *J. Exp. Med.* 139:925–942.

121. Leis, J., Baltimore, D., Bishop, J. M., Coffin, J., Fleissner, E., Goff, S. P., Oroszlan, S., Robinson, H., Skalka, A. M., Temin, H. M., and Vogt, V. (1988). Standardized and simplified nomenclature for proteins common to all retroviruses. *J. Virol.* 62:1808–1809.

122. Kahn, D. E., Mia, A. S., and Tierney, M. M. (1980). Field evaluation of Leukassay F, an FeLV detection test kit. *Feline Pract.* 10:41–45.

123. Francis, D. P., Essex, M., and Hardy, W. D., Jr. (1977). Excretion of feline leukemia virus by naturally infected pet cats. *Nature* 269:252–254.

124. Gwalter, R. H. (1981). FeLV-testmethoden, If order ELISA. *Kleintier Praxis* 26:23–28.

125. Essex, M., Klein, G., Snyder, S. P., and Harrold, J. B. (1971). Correlation between humoral antibody and regression of tumours induced by feline sarcoma virus. *Nature* 233:195–196.

126. Thelien, G. H., Kawakami, T. G., Rush, J. D., and Munn, R. J. (1969). Replication of cat leukaemia virus in cell suspension cultures. Nature 222:589–590.

127. Essex, M., Sliski, A., Cotter, S. M., Jakowski, R. M., and Hardy, W. D., Jr. (1975). Immunosurveillance of naturally occurring feline leukemia. *Science* 190:790–792.

128. Essex, M., Cotter, S. M., Stephenson, J. R., Aaronson, S. A., and Hardy, W. D., Jr. (1977). Leukemia, lymphoma and fibrosarcoma of cats as models for similar diseases of man. In: *Origins of Human Cancer* (H. H. Hiatt, J. D. Watson, and J. A. Winston, eds.), pp. 1197–1214. New York, Cold Spring Harbor Laboratory.

129. Snyder, H. W., Jr., Hardy, W. D., Jr., Zuckerman, E. E., and Fleissner, E. (1978). Characterization of a tumour-specific antigen on the surface of feline lymphosarcoma cells. *Nature* 275:656–658.

130. Snyder, H. W., Jr., Phillips, K. J., Hardy, W. D., Jr., Zuckerman, E. E., Essex, M., Sliksi, A. H., and Rhim, J. (1980). Isolation and characterization of proteins carrying the feline oncornavirus-associated cell-membrane antigen. *Cold Spring Harbor Symp. Quant. Biol.* 44:787–799.

131. Snyder, H. W., Jr., Singhal, M. C., Zuckerman, E. E., Jones, F. R., and Hardy, W. D., Jr. (1983). The feline oncornavirus-associated cell membrane

antigen (FOCMA) is related to, but distinguishable from, FeLV–C gp70. *Virology* 131:315–327.

132. Hardy, W. D., Jr., Zuckerman, E. E., MacEwen, E. G., Hayes, A. A., and Essex, M. (1977). A feline leukaemia virus- and sarcoma virus-induced tumour-specific antigen. *Nature* 270:249–251.

133. Hardy, W. D., Jr., Zuckerman, E. E., Essex, M., MacEwen, E. G., and Hayes, A. A. (1978). Feline oncornavirus-associated cell-membrane antigen, an FeLV-and FeSV-induced tumor-specific antigen. In: *Differentiation of Normal and Neoplastic Hematopoietic Cells.* (B. Clarkson, P. A. Marks, and J. E. Till, eds.), pp. 601–623. New York: Cold Spring Laboratory.

134. Hardy, W. D., Jr., McClelland, A. J., Zuckerman, E. E., Snyder, H. W., Jr., MacEwen, E. G., Francis, D., and Essex, M. (1980). Development of virus non-producer lymphosarcomas in pet cats exposed to FeLV. *Nature* 288: 90–92.

135. Hardy, W. D., Jr., McClelland, A. J., Zuckerman, E. E., Snyder, H. W., MacEwen, E. G., Francis, D. P., and Essex, M. (1980). Immunology and epidemiology of feline leukemia virus nonproducer lymphosarcomas. In: *Viruses in Naturally Occurring Cancers.* (M. Essex, G. Todaro, and H. ZurHausen, eds.), pp. 677–697. New York, Cold Spring Harbor Laboratory.

136. Vedbrat, S. S., Rasbeed, S., Lutz, H., Gonda, M. A., Ruscetti, S., Gardner, M. B., and Prensky, W. (1983). Feline oncornavirus-associated cell membrane antigen: A viral and not a cellularly coded transformation-specific antigen of cat lymphomas. *Virology* 124:445–461.

137. Russell, P. H., and Jarrett, O. (1978). The occurrence of feline leukemia virus neutralizing antibodies in cats. *Int. J. Cancer* 22:351–357.

138. Chattophadhyay, S. K., Cloyd, M. W., Linemeyer, D. L., Lander, M. R., Rands, E., and Lowy, D. R. (1982). Cellular origin and role of mink cell focus-forming viruses in murine thymic lymphomas. *Nature* 295: 25–31.

139. Hartley, J. W., Wolford, N. K., Old, L. J., and Rowe, W. P. (1977). A new class of murine leukemia virus associated with development of spontaneous lymphomas. *Proc. Natl. Acad. Sci. USA* 74:789–792.

140. Elder, J. H., Gautsch, J. W., Jensen, F. C., Lerner, R. A., Hartley, J. W., and Rowe, W. P. (1977). Biochemical evidence that MCF murine leukemia viruses are envelope (*env*) gene recombinants. *Proc. Natl. Acad. Sci. USA* 74:4676–4680.

141. Gardner, M. B., Rongey, R. W., Johnson, F. V., De Journett, R., and Huebner, R. J. (1971). C-type virus particles in salivary tissue of domestic cats. *J. Natl. Cancer Inst.* 47:561–565.

142. Rojko, J. L., Hoover, E. A., Mathes, L. E., Olsen, R. G., and Schaller, J. P. (1979). Pathogenesis of experimental feline leukemia virus infection. *J. Natl. Cancer Inst.* 63:759–768.

143. McClelland, A. J., Hardy, W. D., Jr., and Zuckerman, E. E. (1980). Prognosis of healthy feline leukemia virus infected cats. In: *Feline Leukemia*

Virus (W. D. Hardy, Jr., M. Essex, and A. J. McClelland, eds.), pp. 121–126. New York, Elsevier.

144. Rojko, J. L., Hoover, E. A., Quackenbush, S. L., and Olsen, R. G. (1982). Reactivation of latent feline leukaemia virus infection. *Nature* 298:385–388.

145. Post, J. E., and Warren, L. (1980). Reactivation of latent feline leukemia virus. In: *Feline Leukemia Virus* (W. D. Hardy, Jr., M. Essex, and A. J. McClelland), pp. 151–155. New York, Elsevier.

146. Pedersen, N. C., Meric, S. M., Ho, E., Johnson, L., Plucker, S., and Theilen, G. H. (1984). The clinical significance of latent feline leukemia virus infections in cats. *Feline Pract.* 14:32–48.

147. Gardner, M. B., Brown, J. C., Charman, H. P., Stephenson, J. R., Rongey, R. W., Hauser, D. E., Diegmann, F., Howard, E., Dworsky, R., Gilden, R. V. (1977). FeLV epidemiology in Los Angeles cats: Appraisal of detection methods. *Int. J. Cancer* 19:581–589.

148. Hardy, W. D., Jr. (1982). Immunopathology induced by the feline leukemia virus. In: *Springer Semin. Immunopathol.* (G. Klein, ed.), pp. 75–105. New York, Springer-Verlag.

149. Hoover, E. A., Rojko, J. L., Wilson, P. L., and Olsen, R. G. (1981). Determinants of susceptibility and resistance to feline leukemia virus infection. I. Role of macrophages. *J. Natl. Cancer Inst.* 67:889–898.

150. Day, N. K., O'Reilly-Felice, C., Hardy, W. D., Jr., Good, R. A., and Witken, S. S. (1980). Circulating immune complexes associated with naturally occurring lymphosarcoma in pet cats. *J. Immunol.* 126:2363–2366.

151. Jakowski, R. M., Essex, M., Hardy, W. D., Jr., Stephenson, J. R., and Cotter, S. M. (1980). Membranous glomerulonephritis in a household of cats persistently viremic with feline leukemia virus. In: *Feline Leukemia Virus.* (W. D. Hardy, Jr., M. Essex, and A. J. McClelland, eds.), pp. 141–149. New York, Elsevier.

152. Weksler, M. E., Ryning, F. W., and Hardy, W. D., Jr. (1975). Immune complex disease in cancer. *Clin. Bull.* 5:109–113.

153. Schneider, R., Frye, F. L., Taylor, D. O. N., and Dorn, C. R. (1967). A household cluster of feline malignant lymphoma. *Cancer Res.* 27:1316–1322.

154. Lutz, H., Pedersen, N. C., Harris, C. W., Higgins, B. S., and Theilen, G. H. (1980). Detection of feline leukemia virus infection. *Feline Pract.* 10:13–23.

155. Rojko, J. L., and Hardy, W. D., Jr. (1989). Feline leukemia virus and other retroviruses. In: *The Cat: Diseases and Clinical Management.* (R. G. Sherding, ed.). pp. 229–332. New York, Churchill Livingstone.

156. Essex, M. (1980). Feline leukemia and sarcoma viruses. In: *Viral Oncology* (G. Klein, ed.), pp. 205–229. New York, Raven Press.

157. Hardy, W. D., Jr., MacEwen, E. G., Hayes, A. A., and Zuckerman, E. E. (1980). FOCMA antibody as specific immunotherapy for lymphosarcoma of pet cats. In: *Feline Leukemia Virus* (W. D. Hardy, Jr., M. Essex, and A. J. McClelland, eds.), pp. 227–233. New York, Elsevier.

158. Kleinman, S. (1985). Screening for HTLV–III antibodies: The relation between prevalence and positive predictive value and its social consequences. *JAMA* 253:3395–3397.

159. Handsfield, H. H. (1985). Screening for HTLV–III antibody. *N. Engl. J. Med.* 313:888.

160. Kolata, G. (1986). New blood test raises thorny issues. *Science* 233: 149–150.

161. Nuttall, P., Pratt, R., Nuttall, L., and Daly, C. (1986). False-positive results with HIV ELISA kits. *Lancet* 2:512.

162. Swinbanks, D. (1987). AIDS tests upset haemophiliacs. *Nature* 327:8.

163. Pedersen, N. C., Theilen, G. H., and Werner, L. L. (1979). Safety and efficacy studies of live- and killed-feline leukemia virus vaccines. *Am. J. Vet. Res.* 40:1120–1126.

164. Jarrett, W., Mackey, L., Jarrett, O., Laird, H., and Hood, C. (1974). Antibody response and virus survival in cats vaccinated against feline leukemia. *Nature* 248:230–232.

165. Gilbert, J. H., Pedersen, N. C., and Nunberg, J. H. (1987). Feline leukemia virus envelope protein expression encoded by a recombinant vaccinia virus: Apparent lack of immunogenicity in vaccinated animals. *Virus Res.* 7:49–67.

166. Mastro, J. M., Lewis, M., Mathes, L. E., Sharpee, R., Tarr, M. J., and Olsen, R. G. (1986). Feline leukemia vaccine: Efficacy, contents and probable mechanism. *Vet. Immunol. Immunopathol.* 11:205–213.

167. Pedersen, N. C., and Ott, R. L. (1985). Evaluation of a commercial feline leukemia virus vaccine for immunogenicity and efficacy. *Feline Pract.* 15: 7–20.

168. Hardy, W. D., Jr. (1980). Feline leukemia virus diseases. In: *Feline Leukemia Virus* (W. D. Hardy, Jr., M. Essex, and A. J. McClelland, eds.), pp. 3–31. New York, Elsevier.

169. Hardy, W. D., Jr. (1981). Hematopietic tumors of cats. *J. Am. Animal Hosp. Assoc.* 17:921–940.

170. Hoover, E. A., Perryman, L. E., and Kociba, G. J. (1973). Early lesions in cats inoculated with feline leukemia virus. *Cancer Res.* 33:145–152.

171. Herz, A., Theilen, G. H., and Schalm, O. W. (1970). C-type virus in bone marrow cells of cats with myeloproliferative disorders. *J. Natl. Cancer Inst.* 44:339–348.

172. Hardy, W. D., Jr., and Essex, M. (1986). FeLV-induced feline acquired immune deficiency syndrome (FAIDS): A model for human AIDS. In: *Acquired Immune Deficiency Syndrome, Progress in Allergy 37* (E. Klein, ed), pp. 353–376. Basel, Karger.

173. Dorn, C. R., Taylor, D. O. N., Schneider, R., Hibbard, H. H., and Klauber, M. R. (1968). Survey of animal neoplasms in Alameda and Contra Costa Counties, California. II. Cancer morbidity in dogs and cats from Alameda County. *J. Natl. Cancer Inst.* 40:307–318.

174. Crighton, G. W. (1969). Lymphosarcoma in the cat. *Vet. Rec.* 84:329–331.

175. Francis, D. P., Cotter, S. M., Hardy, W. D., Jr., and Essex, M. (1979). Comparison of virus-positive and virus-negative cases of feline leukemia and lymphoma. *Cancer Res.* 39:3866–3870.

176. Rickard, C. G., Post, J. E., Noronha, F., and Barr, L. M. (1973). Interspecies infection by feline leukemia virus: Serial cell-free transmission in dogs of malignant lymphomas induced by feline leukemia virus. In: *Unifying Concepts of Leukemia* (R. M. Dutcher and L. Chieco-Bianchi, eds.), pp. 102–112. Basel, Karger.

177. Moore, F. M., Emerson, W. E., Cotter, S. M., and DeLellis, R. A. (1986). Distinctive peripheral lymph node hyperplasia of young cats. *Vet. Pathol.* 23:386–391.

178. Hoover, E. A., Rojko, J. L., and Olsen, R. G. (1980). Host-virus interactions in progressive versus regressive feline leukemia virus infection in cats. In: *Viruses in Naturally Occurring Cancers* (M. Essex, G. Todaro, and H. zurHausen, eds.), pp. 635–651. New York, Cold Spring Harbor Laboratory.

179. Cockerell, G. L., and Hoover, E. A. (1977). Inhibition of normal lymphocyte mitogenic reactivity by serum from feline leukemia virus-infected cats. *Cancer Res.* 37:3985–3989.

180. Cockerell, G. L., Hoover, E. A., Krakowka, S., Olsen, R. G., and Yohn, D. S. (1976). Lymphocyte mitogen reactivity and enumeration of circulating B- and T-cells during feline leukemia virus infection in the cat. *J. Natl. Cancer Inst.* 57:1095–1099.

181. Trainin, Z., Wernicke, D., Ungar-Waron, H., and Essex, M. (1983). Suppression of the humoral antibody response in natural retrovirus infections. *Science* 220:858–859.

182. Mathes, L. E., Olsen, R. G., Hebebrand, L. C., Hoover, E. A., and Schaller, J. P. (1978). Abrogation of lymphocyte blastogenesis by a feline leukemia virus protein. *Nature* 274:687–689.

183. Jones, F. R., Yoshida, L. H., Ladiges, W. C., and Kenny, M. A. (1980). Treatment of feline leukemia and reversal of FeLV by ex vivo removal of IgG. *Cancer* 46:675–684.

184. Snyder, H. W., Jr., Singhal, M. C., Hardy, W. D., Jr., and Jones, F. R. (1984). Clearance of feline leukemia virus from persistently infected pet cats treated by extracorporeal immunoadsorption is correlated with an enhanced antibody response to FeLV gp70. *J. Immunol.* 132:1538–1543.

185. Kobilinsky, L., Hardy, W. D., Jr., and Day, N. K. (1979). Hypocomplementemia associated with naturally occurring lymphosarcoma in pet cats. *J. Immunol.* 122:2139–2142.

186. Pedersen, N. C., Theilen, G., Keane, M. A., Fairbanks, L., Mason, T., Orser, B., Chen, D-H., and Allison, C. (1977). Studies of naturally transmitted feline leukemia virus infection. *Am. J. Vet. Res.* 38:1523–1531.

187. Hardy, W. D., Jr. (1984). Feline leukemia virus as an animal retrovirus model for the human T-cell leukemia virus. In:*Human T-Cell Leukemia/ Lymphoma Viruses* (R. C. Gallo, M. Essex, and G. Gross, eds.), pp. 35–43. New York, Cold Spring Harbor Laboratory.

188. Center for Disease Control, Task Force on Kaposi's Sarcoma and Opportunistic Infections (1982). Epidemiologic aspects of the current outbreak of Kaposi's sarcoma and opportunistic infections. *N. Engl. J. Med.* 306: 248–252.

189. Gottlieb, M., Schroff, R., Schanker, H. M., Weisman, J. D., Fan, P. T., Wolf, R. A., and Saxon, A. (1981). *Pneumocystis carinii* pneumonia and mucosal candidiasis in previously healthy homosexual men. *N. Engl. J. Med.* 305:1425–1431.

190. Masur, H., Michelis, M. A., Greene, J. B., Onorato, I., Vande Stouwe, R. A., Holzman, R. S., Wormser, G., Brettman, L., Lange, M., Murray, H. W., and Cunningham-Rundles, S. (1981). An outbreak of community-acquired *Pneumocystis carinii* pneumonia. Initial manifestation of cellular immune dysfunction. *N. Engl. J. Med.* 305:1431–1438.

191. Sarngadharan, M. C., Popovic, M., Bruch, I., Schupbach, J., and Gallo, R. C. (1984). Antibodies reactive with human T-lymphotropic retroviruses (HTLV–III) in the serum of patients with AIDS. *Science* 224:506–508.

192. Roussel, M., Saule, S., Lagrou, C., Rommens, C., Beug, H., Graf, T., and Stehelin, D. (1979). Three new types of viral oncogenes of cellular origin specific for haemotopoietic cell transformation. *Nature* 281:452–455.

193. Scolnick, E. M. (1982). Hyperplastic and neoplastic erythroproliferative diseases induced by oncogenic murine retroviruses. *Biochem. Biophys. Acta* 651:273–283.

194. Kociba, G. J., Lange, R. D., Dunn, C. D. R., and Hoover, E. A. (1983). Serum erythropoietin changes with feline leukemia virus-induced erythroid aplasia. *Vet. Pathol.* 20:548–552.

195. Riedel, N., Hoover, E. A., Gasper, P. W., Nicolson, M. O., and Mullins, J. I. (1986). Molecular analysis and pathogenesis of the feline aplastic anemia retrovirus, feline leukemia virus C–SARMA. *J. Virol.* 60:242–250.

196. Riedel, N., Hoover, E. A., Dornsife, R. E., and Mullins, J. I. (1988). Pathogenic and host range determinants of the feline aplastic anemia retrovirus. *Proc. Natl. Acad. Sci USA* 85:2758–2762.

197. Wardrop, K. J., Kramer, J. W., Abkowitz, J. L., Clemons, G., and Adamson, J. W. (1986). Quantitative studies of erythropoiesis in the clinically normal, phlebotomized, and feline leukemia virus-infected cat. *Am. J. Vet. Res.* 47:2274–2277.

198. Rojko, J. L., Cheney, C. M., Gasper, P. W., Hamilton, K. L., Hoover, E. A., Mathes, L. F., and Kociba, G. J. (1986). Infectious feline leukemia virus is erythrosuppressive in vitro. *Leukemia Res.* 10:1193–1199.

199. Wellman, M. L., Kociba, G. J., Lewis, M. G., Mathes, L. E., and Olsen, R. G. (1984). Inhibition of erythroid colony-forming cells by a 15,000 dalton protein of feline leukemia virus. *Cancer Res.* 44:1527–1529.

200. Abkowitz, J. L., Holly, R. D., and Adamson, J. W. (1987). Retrovirus-induced feline pure red cell aplasia: The kinetics of marrow failure. *J. Cell Physiol.* 132:571–577.

201. Hoover, E. A., and Kociba, G. J. (1974). Bone lesions in cats with anemia induced by feline leukemia virus. *J. Natl. Cancer Inst.* 53:1277–1284.

202. Dameshek, W. (1951). Some speculations on the myeloproliferative syndromes. *Blood* 6:372.

203. Case, M. T. (1970). A case of myelogenous leukemia in the cat. *Zentralbl. Vet. Med.* 17:273–277.

204. Fraser, C. J., Joiner, G. N., Jardine, J. H., and Gleiser, C. A. (1974). Acute granulocytic leukemia in cats. *J. Am. Vet. Med. Assoc.* 165:355–359.

205. Pool, R. R., and Harris, J. M. (1975). Feline osteochondromatosis. *Feline Pract.* 5:24–30.

206. Reinacher, M. (1987). Feline leukemia virus-associated enteritis—A condition with features of feline panleukopenia. *Vet. Pathol.* 24:1–4.

207. Boyce, J. T., Kociba, G. J., Jacobs, R. M., and Weiser, M. G. (1986). Feline leukemia virus-induced thrombocytopenia and macrothrombocytosis in cats. *Vet. Pathol.* 23:16–20.

208. Jacobs, R. M., Boyce, J. T., and Kociba, G. J. (1986). Flow cytometric and radioisotopic determinations of platelet survival time in normal cats and feline leukemia virus-infected cats. *Cytometry* 7:64–69.

209. Hoover, E. A., Rojko, J. L., and Quackenbush, S. L. (1983). Congenital feline leukemia virus infection. *Leukemia Rev. Int.* 1:7–8.

210. Hoover, E. A. (1984). Experimental studies of the pathogenesis of feline leukemia virus infection. In: *RNA Tumor Viruses, Oncogenes, Human Cancer and AIDS: On the Frontiers of Understanding.* (P. Furmanski, J. C. Hager, and M. A. Rich, eds.), pp. 267–288. Boston, Martinus Nijhoff.

211. Pacitti, A. M., Jarrett, O., and Hay, D. (1986). Transmission of feline leukemia virus in the milk of a non-viraemic cat. *Vet. Rec.* 118:381–384.

212. Jahner, P., and Haenisch, R. (1985). Chromosomal position and specific demethylation in enhancer sequences of germ-line transmitted retroviral genomes during mouse development. *Mol. Cell Biol.* 5:2212–2220.

213. Chapman, A. L., Weitlauf, H. M., and Bopp, W. (1974). Effect of feline leukemia virus on transferred hamster fetuses. *J. Natl. Cancer Inst.* 52:583–585.

214. Gardner, M. B., Henderson, B. E., Officer, J. E., Rongey, R. W., Parker, J. C., Oliver, C., Estes, J. D., and Huebner, R. J. (1973). A spontaneous lower motor neuron disease apparently caused by indigenous type C RNA virus in wild mice. *J. Natl. Cancer Inst.* 51:1243–1254.

215. Osame, M., Usuku, K., Izumo, S., Ijichi, N., Amitani, H., Igata, A., Matsumoto, M., and Tara, M. (1986). HTLV–I associated myelopathy, a new clinical entity. *Lancet* 1:1031–1032.

216. Vernant, J. C., Maurs, L., Gessain, A., Barin, F., Gout, O., Delaporte, J. M., Sanhadji, K., Buisson, G., and de-The, G. (1987). Endemic tropical spastic paraparesis associated with human T-lymphotropic virus type I: A clinical and seroepidemiological study of 25 cases. *Ann. Neurol.* 21:123–130.

217. Snider, W. B., Simpson, D. M., Kielsen, S., Gold, J. W. M., Metroka, C. E., and Posner, J. B. (1983). Neurological complications of acquired immune deficiency syndrome: Analysis of 50 patients. *Ann. Neurol.* 14:403–418.

218. Glick, A. D., Horn, R. G., and Holscher, M. (1978). Characterization of feline glomerulonephritis associated with viral-induced hematopoietic neoplasms. *Am. J. Pathol.* 92:321–332.

219. Dworkin, B. M., Rosenthal, W. S., Wormser, G. P., and Weiss, L. (1986). Selenium deficiency in the acquired immunodeficiency syndrome. *J. Parenteral Nutr.* 10:405–407.

220. Ishida, T., Washizu, T., Toriyabe, K., Motoyoshi, S., Tomoda, I., and Pedersen, N. C. (1989). Feline immunodeficiency virus infection in cats of Japan. *J. Am. Vet. Med. Assoc.* 194:221–225.

221. Hardy, W. D., Jr. (1988). Feline T-lymphotropic lentivirus: Retrovirus-induced immunosuppression in cats. *J. Am. Animal Hosp. Assoc.* 24: 241-243.

222. Pedersen, N. C. (1987). Feline syncytium-forming virus infection. In: *Diseases of the Cat.* (J. Holzworth, ed.), pp. 268–272. Philadelphia, Saunders.

223. McKissick, G. E., and Lamont, P. H. (1970). Characteristics of a virus isolated from a feline fibrosarcoma. *J. Virol.* 5:247–257.

224. Ellis, T. M., MacKenzie, J. S., Wilcox, G. E., and Cook, R. D. (1979). Isolation of feline syncytia-forming virus from oropharyngeal swabs of cats. *Aust. Vet. J.* 55:202–203.

4

Molecular Aspects of Feline Leukemia Virus Pathogenesis

James I. Mullins* *Harvard University School of Public Health, Boston, Massachusetts*

Edward A. Hoover *Colorado State University College of Veterinary Medicine, Fort Collins, Colorado*

I. INTRODUCTION

Feline leukemia viruses (FeLV) are naturally occurring, contagiously transmitted type C retroviruses (oncovirus subfamily) of domestic cats. Study of FeLV provided the first evidence that horizontally transmitted retroviruses were the predominant cause of leukemia in nature (1-5). Naturally occurring FeLV-associated fibrosarcomas and lymphomas have also been a rich and important source for the identification of viral oncogenes (6,7). Furthermore, study of FeLV provided early evidence that antiproliferative diseases appeared to be caused by retrovirus infections (8-11) and provided an important paradigm, perceived by Gallo, Essex, and colleagues, in seeking a retroviral etiology for AIDS (12,13). Thus, FeLV has provided unique contributions to our understanding of the role of retroviruses in naturally occurring disease.

FeLVs are endemic in free-roaming, urban domestic cats, and it has been estimated that up to 50% become infected at some point in their lifetime (14). Up to 30% of these cats become persistently viremic (progressors), particularly those which are repeatedly exposed or exposed at an early age, and remain generally virus neutralizing-antibody negative (15-19). Once established, this progressor (persistent viremic) state rarely is reversed and mortality is nearly 100% within 3 years after diagnosis (20-22). The majority of these cats succumb

---

*_Present affiliation_: Stanford University Medical School, Stanford, California

to degenerative diseases such as anemia or immunodeficiency. However, a substantial minority develop leukemia or lymphoma, myeloproliferative disease, fibrosarcoma, or degenerative neurological syndromes (Figure 1) (17,20). The majority of adult cats exposed to FeLV experience regressive infection, curtail virus replication, and develop serum antibody to the virus. Some of this latter population harbor latent FeLV infection, which persists for months to years in bone marrow cells (23). Latent FeLV infections may on occasion be reactivated, transmitted congenitally, or be involved in the genesis of "virus-negative" leukemia or aplastic anemia (22-24).

Molecular analyses of FeLV genomes, their gene products, and the genetic mechanisms underlying the development of FeLV-associated diseases have been conducted in several laboratories over the past decade, utilizing both naturally occurring and experimentally induced infections. Some insight has been provided into the divergent mechanisms of disease induction for most of the diseases described in Figure 1, yet the proposed mechanisms remain incompletely understood.

An important question in the genesis of FeLV-induced disease is whether the particular outcome of infection is determined by a host-specific response, by other intercurrent infections or synertistic factors, by specific viral variants, or by some combination of these factors. Clearly, oncogene-bearing FeLV and feline sarcoma viruses (FeSV) play an important role in the development of tumors, but it is not clear what features of the parental virus are required for the formation of these recombinants or whether certain FeLV are more or less recombinogenic. Furthermore, in the case of induction of aplastic anemia and fatal immunodeficiency syndrome, it has been shown that specific, non-oncogene-bearing FeLV variants can induce these diseases when inoculated into genetically outbred cats. These findings indicate that the genetic determinants for specific pathogenicity can reside within these viral genomes. This chapter focuses on the genetic analysis of several naturally occurring FeLV isolates, their pathogenic potential in vitro and in vivo, and elucidation of the viral determinants of disease induction. Oncogene transduction by FeLV is discussed in Chapter 3 and will be only briefly summarized here.

II. GENOME ORGANIZATION AND GENE PRODUCTS

FeLV has a typical type C oncovirus genomic structure containing the *gag, pol,* and *env* open reading frames found in all retroviruses (Figure 2). None of the accessory genes associated with the human and simian lentiviruses and the human, simian, and bovine oncornaviruses have so far been detected. Although the possibility that some of the short open reading frames encode additional proteins has not been addressed experimentally, analysis of the relevant sequences fails to provide compelling evidence for other open reading frames following methionine codons on either strand (Figure 2). Furthermore, the only viral mRNA species found in FeLV-infected cells are the full-length, 8.2-Kb

Fate of Viremic Cats   Virus Association

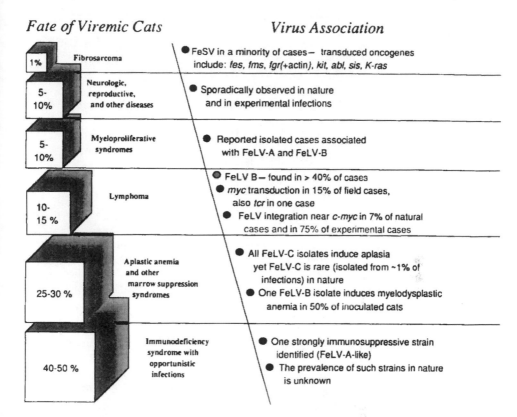

FIGURE 1 Consequences of FeLV infection. Percentages are taken from Ref. 20, W. D. Hardy, Jr. (personal communication), and E. A. Hoover (unpublished observations). Diseases are discussed in the text and in Chapter 3. Feline sarcoma virus (FeSV) isolates, molecular clones, and sequences are described in the following references: *fes*, three isolates (122–124,130,131,133, 134); *fms*, two isolates (126,127,134); *abl* (135); *sis* (133,136); *kit* (134); *fgr* + actin (137,138); *K-ras* (F. deNoronha, personal communication). FeLV-*myc* recombinant viruses (nine isolates) are described in Refs. 50,110–112,139,140. The FeLV-*tcr* isolate is described in Ref. 141. The remaining viruses are discussed in the text.

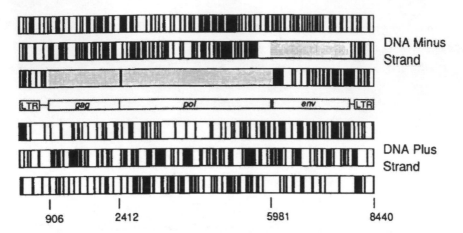

906   2412   5981   8440

FIGURE 2 FeLV genome structure. The position of stop codons is shown for
each of the six possible reading frames. The shaded regions indicate the coding
regions of the known FeLV proteins. The nucleotide positions shown at the
bottom of the figure correspond to the beginning of the *gag, pol,* and *env* polyproteins, respectively. Data taken from Ref. 25.

genome message and the spliced 3.0-Kb envelope message (Mullins, unpublished
observations).

FeLV proteins are expressed initially as part of *gag, gag-pol,* or *env* polyproteins and cleaved into the predicted mature products (depicted in Figure 3).
gag and *pol* are encoded by the same translational reading frame, *pol* products
being expressed as a result of incorporation of a glutamine residue at the TAG
stop codon within the *pol*-protease gene near its N-terminus (25,26) (Figure 3).
Assignment of coding regions for the known FeLV gene products is made by
comparison of the predicted sequences from DNA sequence analysis (25,27)
with the complete peptide sequence of the basic nucleic acid binding protein
(p10) (28), the N-terminal sequence of the major core protein (29,30), and by
alignment with the known murine virus proteins (31). Although the function of
the various FeLV *gag* and *pol* proteins has not been extensively investigated,
these proteins share substantial homology with peptides of the murine leukemia
viruses (25,27) for which some functions have been elucidated (32) (Figure 3).

III. STRUCTURE AND ORIGIN OF ENDOGENOUS FeLV–RELATED
SEQUENCES

Although feline leukemia viruses are horizontally transmitted exogenous retroviruses, uninfected domestic cat cells possess approximately 10 copies of FeLV-
cross-reactive sequences per haploid genome (33–35), referred to as endogenous

FeLV (enFeLV) (36,37). Endogenous FeLV sequences were presumably introduced into cats from a rodent progenitor in relatively recent evolutionary time, and at approximately the same time as RD114, an unrelated endogenous xenotropic retrovirus (38–41). Multiple copies of enFeLV proviruslike elements exist in cat cells (34,35,39,42,43) containing most, but not all, of the sequences found in the infectious viral genome (44). enFeLV sequences have never been shown to produce infectious virus (37,45), although some copies possess a colinear, 8.5-Kb genome structure while other copies possess substantial internal deletions (37,46,47). DNA hybridization studies and sequence analysis have revealed that enFeLV elements contain LTRs with U3 regions that are highly divergent from infectious FeLV (37,48,49) and therefore are not detected by DNA hybridization in uninfected cats using a U3 sequence probe derived from exogenous FeLV (48,50,51).

The enFeLV LTR is present at 10-fold higher copy number in uninfected cat cell DNA than are complete enFeLV proviruses (48). Some of these elements are at different sites in the genome of different animals, suggesting that they correspond to relatively mobile genetic elements (37).

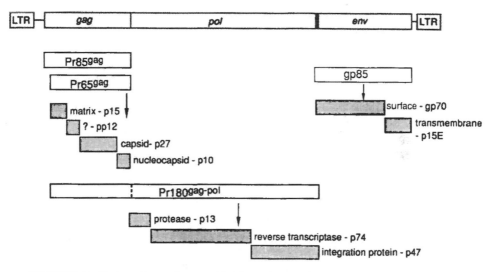

FIGURE 3 Deduced and known FeLV proteins. The coding segments for each protein are shown at the top of the figure. Precursor polyproteins are lightly shaded, mature proteins are heavily shaded. The observed or, in the case of protease and integration proteins, deduced molecular weights of the mature proteins are listed adjacent to their expected functions (32,58,142,143). The functions of some of the *gag* and *pol* viral proteins have not been independently verified; rather they are identified by virtue of their gene order and homology to their murine virus counterparts (25,27,58).

A homology search of the GENBANK nucleic acid sequence database reveals that the enU3 sequence shows a high degree of sequence similarity with the LTRs of the murine retrovirus (MuLV) MCF group. The LTRs of the ecotropic murine and feline retroviruses form a distinct second group (37). When the enFeLV LTR sequence is aligned with those of 12 FeLVs and representatives of each known class of MuLV, aside from about 140 nucleotides conserved among all groups of viruses, only 32 nucleotides of the 410 base pair enU3 could be uniquely aligned with at least two FeLV sequences (37). By contrast, 52 nucleotides could be uniquely aligned with at least two MuLV sequences (37).

However, six short regions of strong sequence conservation (~90–100% identity) are shared by enU3 and the other LTRs (Figure 4), at least three of which are recognized as important in virus integration and transcription (52). These include the 11-bp terminal inverted repeat (IR) sequence involved in virus integration, a 20-bp region terminating with the CCAAT box, and an 8-bp sequence including the TATA box, both involved in transcriptional regulation (Figure 4). The other three conserved regions include a 26-bp sequence preceding the TATA box, a 17-bp sequence near the 5′ end of U3, which immediately

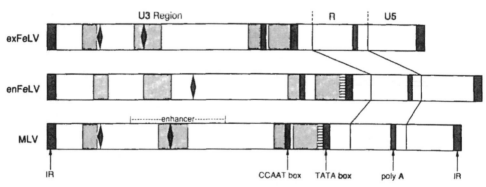

FIGURE 4 Schematic representation of feline and murine LTR sequences. The exogenous FeLV (exFeLV) sequence is representative of all known isolates (see Table 1) and taken from Ref. 25, endogenous FeLV-related sequence (enFeLV) is from Ref. 37, and murine leukemia virus (MLV) from Ref. 144. Black boxes are described in the text and correspond to regions of perfect homology between the three sequences, whereas darkly shaded regions correspond to regions of strong sequence conservation. Black diamonds correspond to positions of potential core enhancer sequences (53). Lightly shaded regions correspond to regions of strong sequence conservation between the two feline sequences. U3 sequences are derived from the 3′ end of the viral RNA genome, U5 sequences are derived from the 5′ end, and R sequences correspond to the terminal repetition in viral RNA.

precedes, but does not include, the consensus potential core enhancer sequence TGG T/A T/A A/T (53) (each occurrence of this sequence is indicated by a diamond in Figure 4), and a 40-bp sequence central within U3 which also includes the aforementioned core sequence. The functional requirements for these sequences have recently been investigated in transient expression of a linked reporter gene, and the region surrounding the conserved ecotropic murine virus core enhancer (54) has been shown to function as an enhancer (49). The position of the potential core enhancer sequence is not conserved between enFeLV and infectious feline and murine viruses (Figure 4). However, three independent enFeLV 5'-LTRs have been tested and shown to promote and enhance transcription (47,49).

In contrast to the U3 region, R and U5 LTR sequences are virtually identical in enFeLV and exFeLV-LTR sequences (37,49). A similar database search revealed that in contrast to enU3, the enR-U5 segments are more closely related to exogenous FeLVs than to any of the murine sequences.

The DNA sequence and deduced enFeLV protein sequence of the 5' portion of one clone extending from the LTR through p15*gag* and the *env*-LTR region of another clone reveal it to be closely related to exFeLV. Both *gag* and gp70 sequences have frame-shift and nonsense mutations that are consistent with the lack of infectivity of each provirus examined (37,49). Interestingly, enFeLV p15E, like that of the endogenous xenotropic murine viruses, has an extended C-terminal sequence (37). These short extensions share 6/12 base sequence identity. Furthermore, the N-terminal halves of mature FeLV p15E genes are more closely related to xenotropic MuLV, whereas the C-terminal half is more closely related to ecotropic MuLV (37). These data, combined with analysis of gp70 protein similarities (see below), strongly argue that enFeLV corresponds to an evolutionary remnant of transmission of a progenitor MCF-like virus that gave rise to present-day MuLV and exFeLV.

IV. FeLV SUBGROUPS AND THEIR PREVALENCE IN NATURAL INFECTIONS

FeLV isolates have to date been assigned to three subgroups, A, B, and C, based on their ability to interfere with infection by other FeLV isolates and on their susceptibility to neutralizing antibodies (55,56), properties associated with the virus-encoded extracellular glycoprotein or surface protein (gp70) (57,58). FeLV-A is found in all natural isolates and is efficient in establishing viremia in cats of all ages (25,59–62). FeLV-B, in the absence of FeLV-A, is generally unable to establish viremia in other than newborn animals (59–62), but is reportedly found (along with FeLV-A) in approximately 40% of all infections. FeLV-B has been reported in a higher proportion of cats with tumors. However, as will be discussed below, its true prevalence is unclear, as is the contribution of subgroup A and B viruses to the development of malignancy and other FeLV-associated diseases.

FeLV subgroups may also be distinguished by their host range in vitro (60, 63,64): FeLV-A isolates grow poorly or not at all in cells from heterologous species; FeLV-B has an intermediate host range which includes certain dog and human cell lines in addition to many cat cells; and FeLV-C infects all of the cells above and has the capability, unique among FeLVs, to infect guinea pig cells.

Evidence in favor of an FeLV subgroup- or strain-determined outcome of infection was first put forth for the development of aplastic anemia (65,66) induced by FeLV-C (67). Like FeLV-B, cloned FeLV-C isolates are generally inefficient in establishing viremia except in newborn cats and are found only in association with FeLV-A (and sometimes FeLV-B) in nature (59). Although severe nonregenerative anemia is one of the common outcomes of FeLV infection, FeLV-C is evidently rare, being isolated from as few as 1% of viremic cats and only from those with aplastic anemia (59,62,67) (Figure 1). The appearance of FeLV-C in the blood of cats infected with FeLV-A/C mixtures is reported to correlate with development of anemia (59,68). Furthermore, it has been shown that FeLV-C by itself is sufficient for the induction of anemia in SPF (specific-pathogen-free) kittens (67,69,70). Whether the expression of FeLV-C or an endogenous FeLV-C-like element (see below) is involved in the many cases of aplastic anemia from which no FeLV-C has been cultivated is not yet clear.

V. ORIGIN OF FeLV SUBGROUPS

DNA sequence analysis of feline leukemia viruses has in most cases focused on analysis of the envelope gene because it encodes the 70-kD extracellular glycoprotein or surface protein, referred to as gp70. gp70 is the immunodominant moiety which elicits neutralizing antibody and also specifies the subgroup-specific determinants of the virus and, therefore, sequences involved in receptor recognition. To date, envelope genes from one enFeLV and 11 FeLVs have been determined (Table 1).

Jarrett and Russell (60) postulated that, because of its rare occurrence and association with FeLV-A, FeLV-C may arise by recombination between the ecotropic subgroup A virus present in all viremic animals and endogenous FeLV-related sequences. Supporting this hypothesis is the finding that FeLV-C-related envelope determinants are expressed on feline lymphoma cells in the absence of transmissable subgroup C viruses (71,72), and that endogenous viral genes are transcriptionally activated in certain tumor cells and developing tissues (73–76).

Recombination between exogenous retroviruses and endogenous virus elements was first noted nearly two decades ago in studies of avian retroviruses

TABLE 1 DNA Sequence Divergence Between Natural Isolates of FeLV

| Virus isolate | | | | % Divergence of DNA sequence | | | |
| | | | | vs. F6A | | | vs. enFeLV-5 |
Name	Date	City	Subgroup	gp70	p15E	LTR	gp70
F6A (FeLV-61E)	1983	Ft. Collins	A	—	—	—	21.1
F3A (FeLV-3281-50)	1977	New York	A	2.0	0.7	1.9	—
FGA (FeLV-Glasgow-1)	1971	Glasgow	A	2.1	1.9	1.9	—
61C (FeLV-16C)	1983	Ft. Collins	?	1.3	0.2	0.6	—
GA-FeSV	1969	Los Angeles	?	2.1	0.8	3.1	—
SM-FeSV	1971	Philadelphia	?	7.0	3.7	6.0	3.5
FGB (FeLV-Gardner Arnstein-B)	1969	Los Angeles	B	1.3	0.8	3.3	0.6
FRB (FeLV-Rickard-B)	1967	Ithaca	B	0.4	1.0	2.3	0.9
FSB (FeLV-Snyder Theilen-B)	1969	Davis	B	7.7	4.8[a]	ND	1.2
F1B (FeLV-B1-B)	1969[b]	Davis[b]	B	7.7	5.3[c]	ND	1.2
FSC (FeLV-Sarma-C)	1967	Davis	C	7.9	4.2	7.3	0.6
enFeLV-5	—	—	—	21.7	15.3	>30	—

Percent divergence was calculated giving equal weight to mismatches and gaps (irrespective of length). In the gp70 comparison the numbers indicate divergence relative to FeLV-A in regions derived from FeLV-A-like virus and to enFeLV in regions derived from enFeLV. References for each virus isolation, molecular cloning and proviral DNA sequences are: F6A (25,83); F3A (25,120,121); FGA (79,116); 61C (83); GA-FeSV (90,122-125); SM-FeSV (90,126,127); FGB (78,84,122,128,129); FRB (37,86,115); FSB (125,129-131); F1B (FeLV-λ1B;80); FSC (69,132); enFeLV-5 (37).

[a]Over the first 208 base pairs of the gene.

[b]Indicates that the precise origin of the culture from which FeLV-λ1B/F1B was isolated is not clear, although it probably is derived from the Snyder-Theilen isolate (130) (J. H. Elder, personal communication).

[c]Number refers to regions outside a 120-base pair segment probably derived from enFeLV in which it has 0.8% divergence from enFeLV and 11.7% divergence from F6A.

(77). Furthermore, precedent for a recombinational origin of pathogenic retroviruses occurs in AKR mice, in which endogenous xenotropic or polytropic viral sequences are involved in the generation of tumorigenic-specific determinants of MCF viruses through recombination with endogenous ecotropic virus (77).

The first evidence that certain FeLVs might originate by recombination was uncovered by demonstration that the deduced protein encoded by the Gardner-Arnstein (GA)-FeLV subgroup B virus envelope gene has a unique sequence homology with the portion of the murine MCF gp70 protein contributed by the endogenous xenotropic-like parent (78). Later it was found that some FeLV-B specific sequences, including the portion with MCF sequence homology, but not the corresponding FeLV-A sequences, are present in uninfected cat cell DNA and in a cloned representative of the major family of enFeLV sequences (79). Recombination between FeLV-A and enFeLV sequences was experimentally demonstrated by transfection of feline embryo cells with FeLV-A, transmission of the resulting virus to heterologous cells, and identification of replicating virus with enFeLV-specific envelope sequences (36). These recombinant viruses were shown to have acquired the extended host range of FeLV-B, although none were found with the host range of FeLV-C. Finally, nucleotide sequence analysis of the gp70 gene of enFeLV and representatives of each FeLV subgroup reveal that all FeLV-B gp70 genes are characterized by blocks of sequence with greater than 98% homology to enFeLV, adjacent to regions with usually >98% homology to a prototype of FeLV-A (37) (Figure 5 and Table 1). The unique structure of each FeLV-B gp 70 gene can be attributed almost entirely to the position at which recombination occurred. The fact that generation of FeLV-B genomes occurs frequently during propagation of FeLV-A in some cat cells raises the possibility that the true incidence of FeLV-B in vivo may actually be lower than previous estimates (59,62), which were in some cases based on subgroup determinations made following propagation of viruses for a time in feline cells. Curiously, recombinant FeLV genomes are not observed at high levels or with high frequency in cats inoculated with cloned FeLV-A (36; Hoover, Overbaugh, and Mullins, unpublished results), and certain cell lines give rise to recombinant viruses at much higher frequency than others (Overbaugh and Kristal, unpublished observations; J. Neil, personal communication).

The single example of an FeLV-C sequence examined to date, FeLV-C-Sarma (FSC) (69,70), has two regions of homology to enFeLV within the 3' portion of its gp70 gene (Figure 5). This more complex origin might conceivably account for the failure to readily detect FeLV-C recombinants in tissue culture (36). However, mapping studies indicate that other regions of the FSC gp70 gene, regions of unknown origin, are involved in determining its unique host range properties (see below). And, it is currently not known what features of the FSC sequence are shared with other anemogenic FeLV-C isolates.

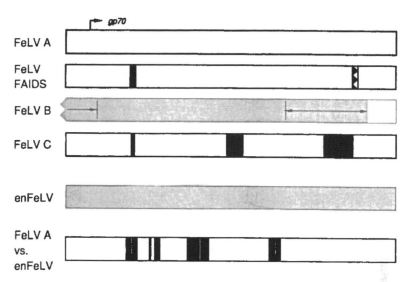

FIGURE 5 Origin of envelope gene diversity in feline leukemia viruses. Regions of 92–99% sequence conservation relative to FeLV-A (25) are shown by an open bar. Deletions introduced to maintain alignment are indicated by black boxes. The jagged box in the FeLV–FAIDS sequence (83) corresponds to an imperfect direct repeat in the DNA. Shaded regions indicate portions of gp70 derived by recombination with enFeLV (>98% sequence conservation). The arrows indicate the variety of points of recombination observed (37); for some viruses the 5′ point of recombination occurred upstream of the region shown.

gp70 genes are not the only portions of infectious FeLV found to have acquired enFeLV sequences. The FeLV-λ1B clone (80) (referred to as F1B in Table 1), likely derived from the original Snyder-Theilen isolate, has a block of enFeLV-derived sequence approximately 120 base pairs in length within the region of the p15E transmembrane protein gene. In contrast, the other Snyder-Theilen-derived clone does not (J. H. Elder, personal communication), whereas each clone has the same enFeLV-derived sequences within gp70, suggesting that the final FeLV-λ1B sequence was derived from an additional recombinational event with an enFeLV sequence (37,80).

VI. VIRAL GENETIC DETERMINANTS OF FeLV HOST RANGE IN VITRO

Host range properties of retroviruses are thought to be determined primarily by cell receptor recognition of the envelope surface protein (57), and studies with avian retroviruses suggest that several variable regions within gp85 are involved in determining cell receptor interaction (81,82). Recent results also indicate that FeLV gp70 domains may also interact in determining receptor binding properties (70).

Host range determinants of FeLV have been studied using chimeric proviruses generated between cloned prototypes of FeLV-A (F6A) (25,83) and FeLV-C (FSC) (69). Chimeras containing the 5' portion of the FSC gp70 gene, like FSC, replicated in all three cell lines tested—feline embryo fibroblasts, canine osteosarcoma cells, and guinea pig fibroblasts—demonstrating that this unique host range determinant can be mapped to the N-terminal portion of gp70 (70) (G. Viglianti, R. Khiroya, and J. I. Mullins, unpublished observations). However, chimeras containing the N-terminal portion of gp70 from F6A and the remainder of gp70 from FSC, including the two blocks of enFeLV sequence mentioned above, displayed an intermediate, FeLV-B-like host range and replicated in feline and canine cells but not in guinea pig cells. Thus, sequences in the N-terminal portion of the FSC gp70 are unique in their capacity to specify growth competency in guinea pig cells, yet either the C-terminal region from FSC alone (including two blocks of sequence derived from enFeLV) or the combination of N-terminal FeLV-A and C-terminal FeLV-C confers an intermediate host range and permits growth on dog cells (70).

VII. INFECTIVITY AND PATHOGENICITY OF FeLV SUBGROUPS

The distribution of FeLV subgroups in nature to a large extent is reflected in the ability of purified viruses to infect cats under experimental conditions. FeLV-A is found in all natural infections, and each of two molecularly cloned isolates has been shown to be capable of infecting a high proportion of inoculated cats regardless of their age at the time of inoculation (25). FeLV-B and FeLV-C have never been detected in the absence of FeLV-A in nature and cats are usually only susceptible to infection with purified B and C viruses during the first week of life or following induction of viremia with FeLV-A (59–62). Thus FeLV-A appears to act as an in vivo helper for infection with other FeLVs.

As with other naturally occurring retroviruses, the molecular basis of feline leukemia virus cell tropism and specific pathogenicity is understood only to a limited degree. The pathogenicity of representatives of each of the subgroups has been studied by inoculation of cats with molecular clone-derived virus. Two isolates of FeLV-A, F6A and F3A-clone 2, were inoculated into and induced viremia in a total of 23 cats ranging in age from newborn to 8 weeks old (weanling age). Only one developed disease, a T-cell lymphoma, 21 months after inoculation (25,83).

Studies of the pathogenicity of molecularly cloned FeLV-B isolates from the Gardner-Arnstein strain (84,85), referred to as FGB, and the Rickard strain (37,85,86), referred to as FRB, have thus far yielded divergent results. FGB is

capable of infecting newborns and a percentage of corticosteroid-treated weanling age cats. Among a total of six viremic cats, FGB has induced myelodysplas-
tic anemia in three (all newborns), T-cell lymphoma in one, fibrosarcoma in one,
and aplastic anemia in one (85). FRB failed to induce viremia in any of nine
newborn or four weanling-age cats with or without corticosteroid treatment to
enhance susceptibility to infection (87,88).

Molecularly cloned FeLV-C isolate FSC fails to induce viremia in cats
beyond 1 week of age, with or without corticosteroid treatment (69). However,
each of 13 cats inoculated as newborns developed viremia and fatal aplastic
anemia within 3 months of inoculation. Cats viremic with FeLV-A can be
infected with biologically or molecularly cloned FeLV-C and will also develop
fatal aplastic anemia (68,85). These findings are consistent with the original
studies demonstrating that consistent induction of anemia is associated with the
subgroup C phenotype (66,67). Together they confirm that the FeLV-C genome
contains dominant acting determinants for induction of fatal nonregenerative
anemia and indicate that FeLV-A can serve as an in vivo helper for the establishment of FeLV-C infection.

In summary, studies conducted to date indicate that FeLV-A isolates are
generally minimally pathogenic, FeLV-B isolates are strongly age-restricted and
are inconsistent in their pathogenicities, and FeLV-C isolates are also strongly
age-restricted but consistently induce fatal aplastic anemia.

The genetic determinants for induction of aplastic anemia have been
mapped by generating chimeras between cloned FeLV-A isolate F6A and FSC
in studies that parallel those mentioned earlier for identifying host range determinants (70). In these experiments, virus producer cells were inoculated directly
into the bone marrow of weanling age cats to circumvent the age-related host
resistance to infection (89). Ability to induce aplastic anemia colocalizes, along
with the guinea pig host range determinants, to a 723-base pair region of the
genome which encodes the N-terminal 241 amino acids of the mature gp70
protein (70) (E. Hoover, G. Viglianti, R. Dornsife, N. Riedel, R. Khiroya, and
J. Mullins, unpublished observations). Colocalization of the pathogenic and
unique host range determinants of FSC sugests the possibility that receptor
interaction plays a role in the pathogenicity exerted by this virus.

VIII. VARIABLE REGIONS AND CONSERVATION OF BACKBONE
STRUCTURE IN DEDUCED FeLV AND MURINE gp70 PROTEINS

Examination of the gp70 genes of F6A, F3A clone 50 (derived from the same
cell line as clone 2, the latter of which was used for in vivo studies) of the
Glasgow-1 subgroup A isolate (79), and the gp70 region of GA-FeSV (90)
reveals that they all share 98% DNA and deduced protein homology despite
having been isolated from naturally infected cats 6-13 years apart and from

widely different geographical locations (Table 1). These viruses are not acutely pathogenic but ultimately result in either proliferative (lymphoma) or antiproliferative (hemolymphosuppressive) disease, possibly via generation of more pathogenic viral mutants or recombinants. These subgroup A viruses are considered prototypes of the horizontally transmissable FeLV ubiquitous in domestic cat populations (25). The only exceptions to the uniformly high degree of conservation noted in FeLV gp70 genes are for the gp70 region of the McDonough isolate of FeSV (SM-FeSV), FSC, and the two Snyder-Theilen isolates (FSB and F1B) of FeLV-B. The latter sequences have 7–8% divergence outside of regions in which they are most closely related to enFeLV (Table 1).

Comparison of all the deduced FeLV gp70 proteins reveals that most nucleotide deletions, insertions, and point mutations leading to amino acid sequence variability cluster within five variable regions (Figure 6). Similar alignments can be generated for murine retrovirus sequences. In fact, when FeLV

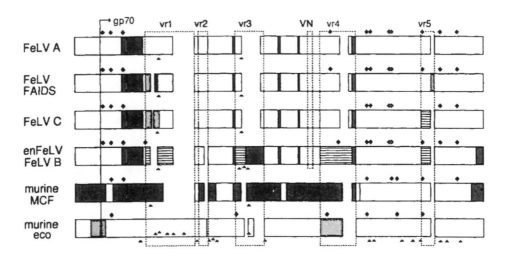

FIGURE 6 Variable regions in deduced feline and murine gp70 proteins. vr1–5 correspond to regions of maximal sequence divergence and deletions or insertions whose introduction was required to maintain alignment. Black diamonds correspond to potential N-linked glycosylation signals. Black triangles correspond to cysteine residues, triangles below the murine ecotropic virus sequence (144) are conserved between each of the sequences shown, triangles below individual bars on vr1 and vr3 are uniquely positioned in each sequence. Black boxes correspond to regions of homology uniquely shared with the murine mink cell focus–forming virus (MCF) sequence (145) and not with the murine ecotropic virus sequence. Striped boxes correspond to regions uniquely shared between enFeLV and FeLV-B and FeLV-C. Shaded areas correspond to virus-specific sequences. VN is a binding site for neutralizing antibody.

gp70 proteins are aligned with those of their murine counterparts, the same five variable regions are evident (Figure 6). The position of cysteine residues, which, because of their ability to form disulfide linkages, is an important determinant of the overall structure of proteins, is highly conserved between FeLV and MuLV (Figure 6). These findings indicate that feline and murine retrovirus glycoproteins have a conserved backbone structure on which is superimposed a regular array of relatively constant and variable regions.

FeLV gp70s have more consensus sites for N-linked glycosylation than their murine counterparts. The positions of these sites are conserved in several instances, even in regions of lesser homology (Figure 6). Both sets of proteins are predicted to be glycosylated near the N-terminus and especially in the conserved C-terminal third of the molecule.

Finally, comparison of FeLV and murine gp70 genes and of avian (82) and human and simian retroviruses (91-93) indicates a general pattern of extra-cellular glycoprotein structure. In all cases, regularly spaced regions of apparent sequence substitution are embedded within a highly conserved framework marked by retention of cysteine residue positions. In the case of divergent viruses, regions of sequence substitution may be domains that determine receptor recognition, binding, and possibly disease and cell type specificity.

IX. IDENTIFICATION OF FeLV NEUTRALIZATION EPITOPES

Because of the direct impact of FeLV as a widely disseminated pathogen in the domestic cat environment, and because of the relative rapidity of development of FeLV-induced disease and its association with induction of both proliferative and degenerative diseases, the feline retroviral disease model is one in which therapeutic regimens are being evaluated and vaccines developed. The importance of this model has increased in the last decade in which human T-cell leukemia and immunodeficiency viruses have been discovered and the AIDS epidemic has been identified and growing (12). Thus, it is particularly timely that the problem of host recognition and response to FeLV antigens has started to be dissected over the past few years (80,94-96). Three laboratories have independently developed monoclonal antibodies that neutralize infection by FeLVs of all subgroups in vitro and have mapped their site of interaction to the same conserved sequence within the gp70 protein (94,96) (Figure 6). Only one isolate, FeLV-λ1B, is not neutralized by these murine antisera, despite the fact that the minimal binding epitope is conserved in it's sequence (80). Rather a single amino acid change (pro -> leu) four residues upstream of the antibody binding site evidently alters the structure of the protein such that the antibody can no longer neutralize the virus. Thus, FeLV neutralization by heterologous sera, like that of HIV-1 (J. Goudsmit, personal communication), may involve a single dominant viral epitope. FeLV gp70 domains recognized by feline neutralizing antibodies, however, may be more complex and have not yet been defined.

X. IMMUNODEFICIENCY DISEASE-INDUCING FeLV VARIANT

Human AIDS has a long-standing precedent in a naturally occurring FeLV-induced disease of cats—the feline leukemia virus-associated acquired immunodeficiency syndrome (also referred to as feline AIDS)—a disease recognized for over 15 years in some cats experimentally or naturally infected with some isolates of FeLV (8-10,99-102). It has recently been shown that a specific FeLV variant (FeLV-FAIDS) will induce fatal immunodeficiency disease in all viremic SPF cats after an incubation period as short as 60 days, an interval that varies with, and can be manipulated by, age at time of inoculation (51,103,104). Experimental feline immunodeficiency, like the natural disease, is characterized by persistent viremia, progressive weight loss, and severe lymphoid depletion—which appears to be preceded by lymphoid hyperplasia, intractable diarrhea, and opportunistic infections (103). The course of disease is characterized by an age-dependent prodromal period during which a non-disease-specific, "common" form of proviral DNA is detected in bone marrow. Clinical onset is preceded by production of high levels of pathogenic variant genomes, primarily as unintegrated viral DNA in bone marrow (51,103,105), and by a drastic reduction of circulating T-lymphocyte progenitor cells, as measured in a colony-forming assay (106). Two clinical courses have been identified following inoculation with the original natural isolate (103,105): *Acute immunodeficiency syndrome*, which has a survival period of approximately 3 months, is associated with the appearance of a characteristic variant genome (designated originally as variant A) which persists as high levels of full-length, unintegrated viral DNA in bone marrow. *Chronic immunodeficiency syndrome*, characterized by survival for more than 1 year, is marked by a more gradual onset of severe clinical immunosuppression, is associated with a predominance of other variant genomes, and may culminate in the development of extranodal lymphoma.

Both common-form and variant genomes have been molecularly cloned and their DNA sequence and biological activity in vitro and in vivo studied (25, 83,106). Each of three examples of common-form genome is replication competent, and one, clone F6A described above, was shown to be noncytopathic for T cells (107) and minimally pathogenic in vivo (83). However, each of 24 variant genomes analyzed, including 10 that are apparently full-length, is replication defective in feline fibroblasts and in a feline T-cell line (83,107). Nonetheless, each of four variants, when rescued by cotransfection using the common form clone as helper virus, induced fatal immunodeficiency disease in vivo (83) and killed feline T cells in vitro (108). This surprising result suggests that the majority of pathogenic genomes in this strain are replication defective. This, in turn, implies that reliance on cultured virus for analysis, as is currently done for studies involving HIV and simian immunodeficiency virus, might select against detection of the more acute pathogens (83) (all FeLV-FAIDS clones were derived directly from tissue NDA).

In addition to causing profound lymphocyte depletion, cats inoculated with the F6A/variant A (clone 61C) mixture and a replication-competent chimera derived from these two cloned viruses developed enteritis and histological evidence of damage to intestinal epithelium, along with preferential localization of high levels of viral antigen and viral DNA in that tissue (83). The latter findings suggest that the virus may play a direct role in cytopathic damage to the intestine.

The subgroup of the pathogenic immunodeficiency variant has not been determined, but it is structurally most closely related to FeLV-A (83). Curiously, the causative immunodeficiency variant is distinguished from the minimally pathogenic virus found in the same tissue (F6A) by the same type and degree of variability (point mutational change, insertion of direct repeats and deletions) found among human AIDS viruses (91,92), rather than by evidence of the type of recombinational events that characterize the other known FeLV (subgroup) variants. As in the case of the aplastic anemia variant discussed previously, the pathogenic determinants of the prototype immunodeficiency variant of FeLV have been localized by chimera construction and testing in vivo and in vitro. The T-cell cytopathic determinants also map to changes within the gp70 gene, although to the C-terminal region (83,108) rather than to the N-terminal region as in the case of the anemogenic determinants (70). Thus, as is the case with the feline aplastic anemia virus, the specificity for induction of fatal immunodeficiency in cats resides within an FeLV-variant genome. Since only one immunodeficiency-inducing FeLV strain has been identified, further studies will be needed to determine whether most cases of FeLV-associated immunodeficiency in cats involve the generation of similar viruses to the FeLV-FAIDS variant, or to other factors or pathogens.

XI. GENETIC EVENTS ASSOCIATED WITH FeLV-INDUCED PROLIFERATIVE DISEASE

Although not the most frequent outcome of FeLV infection, leukemia and lymphoma in cats occur at the highest rate known in outbred species (109). Availability of a large number of naturally occurring tumors has allowed study of the genetic events leading to their development and permitted comparison to the genetic mechanisms involved in other models of experimental oncogenesis (see below). Furthermore, naturally occurring fibrosarcomas and lymphomas have served as a rich source for the identification of feline sarcoma viruses (FeSV) and FeLV-*myc* viruses containing cell-derived, viral oncogenes (Figure 1).

Virus-positive T-cell tumors account for the majority of FeLV-induced tumors, and these are most often clonal cell outgrowths, containing signature-fixed sites of virus integration (48). Approximately 10–30% of the field cases of FeLV-positive tumors are associated with activation of the *myc* onco-gene either by FeLV transduction (50,110–112), by provirus insertion in the vicinity of c-*myc* (7,110–112; Mullins and Hoover, unpublished results), or by events not directly linked to FeLV integration (112). These results were the first to challenge the prevailing view that transduction was a rare event in viral leukemogenesis (113). In contrast, a majority of the cases of T-cell lymphoma experimentally induced by FeLV are associated with provirus insertion near c-*myc* (7,110,114; Mullins and Hoover, unpublished results), consistent with the findings in the experimental avaian system (113). In chickens inoculated with avian leukosis virus, provirus insertion often occurs within or upstream of c-*myc* and in the same transcriptional orientation, such that the c-*myc* gene may be "promoter insertion" activated and expressed as part of a viral–c-*myc* fusion transcript (113). In most cases of feline lymphomas and MuLV-induced thymic lymphomas in which c-*myc* is rearranged, the provirus is integrated upstream and in the opposite transcriptional orientation relative to c-*myc*. Thus, the feline and murine tumors are consistent with the "enhancer insertion" model observed only occasionally in the avian system. It is presently unclear whether the predominance of provirus insertions near c-*myc* in experimental feline tumors is a consequence of the virus strain involved, since these studies have involved only two uncloned strains, FeLV-Rickard (115) and FeLV-Glasgow-1 (116). Furthermore, in none of four tumors induced by FeLV–FAIDS were proviruses detected in the vicinity of c-*myc* (105), although these tumors may have resulted from indirect effects of the virus associated with profound immunosuppression. Similarly, no *myc* involvement was detected in the two cases of lymphoma induced by cloned viruses F6A or FGB (85).

There is a disparity in the frequency of detection of c-*myc* rearrange-ments observed following experimental induction of feline tumors versus those observed in natural populations (7). Transduction of c-*myc* has been observed in nine field cases (15% of the total) and in none of 12 experimental cases, whereas provirus insertion near c-*myc* has been observed in four field cases (7%) and nine experimental cases (75%). It is not clear why this bias occurs; however, it might be explained by the selection process employed in experi-mental studies. These studies employ high doses of virus selected for high leukemogenicity and younger, less immunocompetent cats than would be expected to be exposed under natural conditions. These factors may lead to wider, more rapid dissemination of virus in the individual and a higher likelihood for *cis*-activation of oncogenes rather than the expected slower process of transduction.

Some FeLV-associated tumors are not clonal cell populations with respect to virus integration sites (48), suggesting that in these cases the clonal tumor cell outgrowth occurred prior to infection of that cell population with virus. In addition, approximately one-third of all FeLV-associated lymphomas are completely virus negative; that is, they express no viral antigens (20,117) and are devoid of complete exogenous proviruses, as evidenced by the lack of the U3 portion of the exogenous viral LTR (48). Deletions of portions of the avian leukosis virus (ALV) genome are commonplace in ALV-induced bursal lymphomas of chickens, although at least one LTR always remains at that locus (113). Despite a strong epidemiological association (117), FeLV is possibly not the direct cause of these tumors. Alternatively, proviruses could have been completely eliminated from the genome of the tumor cell following oncogenic activation, thereby indicating that the continued presence of viral genetic information is not necessary for tumor growth. As an additional alternative, endogenous FeLV-related sequences could play an interactive role in the induction or maintenance of the transformed state in some of these tumors, perhaps through the formation of subgroup B viruses. However, a consistent or direct role for any FeLV subgroup or cloned variant in the development of lymphoproliferative disease remains to be established. The possibility that host-specific factors play a determining role is also under investigation, following the identification of a c-*myc* allele which may be associated with an increased risk for development of lymphoma (118, 119).

XII. CONCLUSIONS

FeLV has provided substantial contributions to the study of retrovirus-induced pathogenesis. These include the first evidence that horizontally transmitted retroviruses were an important cause of naturally occurring leukemia, identification of a large number of viral oncogenes, and early evidence that retrovirus infection can result in antiproliferative diseases of the lymphoreticular system. More recent studies have shown that FeLV genome variants can confer disease specifically, due, in the cases that have been examined, to changes within an otherwise highly conserved extracellular glycoprotein (gp70) gene. Each of the three FeLV subgroups has a consistent pattern of disease specificity (FeLV–C induction of fatal anemia), or structure (FeLV–B origin via recombination between FeLV-A and endogenous FeLV-related sequences), or both (FeLV–A gp70 sequences are highly conserved and these viruses display minimal pathogenicity). The widespread prevalence of FeLV infection in the domestic cat population and the resulting importance of FeLV as a pathogen, as well as the recent emergence of an FeLV model for AIDS, will continue to focus research on this system directed toward understanding mechanisms of disease induction and the development of antiviral strategies and vaccine development.

ACKNOWLEDGMENTS

This work was supported by U.S. Public Health Service Grants CA43216, CA40646, CA01058, AI25273, and by grants from the AIDS Research Council of the state of Masschusetts. We thank John H. Elder, James C. Neil, Jack H. Nunberg, and Pradip Roy-Burman, for their comments on the manuscript.

REFERENCES

1. Hardy, W. D., Geering, G., Old, L. J., and deHarven, E. (1969). Feline leukemia virus: Occurrence of viral antigen in the tissues of cats with lymphosarcoma and other diseases. *Science* 166:1019–1021.
2. Jarrett, O. (1970). Evidence for the viral etiology of leukemia in the domestic mammals. *Adv. Cancer Res.* 13:39–62.
3. Hardy, W. D., Old, L. J., Hess, P. W., Essex, M., and Cotter, S. (1973). Horizontal transmission of feline leukaemia virus. *Nature* 27:266–269.
4. Jarrett, W., Jarrett, O., Mackey, L., Laird, H. M., Hardy, W., Jr., and Essex, M. (1973). Horizontal transmission of leukemia virus and leukemia in the cat. *J. Natl. Cancer Inst.* 51:833–841.
5. Hoover, E. A., Olsen, R. G., Hardy, W. D., Jr., and Schaller, J. P. (1977). Horizontal transmission of feline leukemia virus under experimental conditions. *J. Natl. Cancer Inst.* 58:443–444.
6. Besmer, P. (1983). Acute transforming feline retroviruses. *Curr. Topics Microbiol. Immunol.* 107:1–27.
7. Neil, J. C., Forrest, D., Doggett, D. L., and Mullins, J. I. (1987). The role of feline leukaemia virus in naturally-occurring leukaemias. *Cancer Surveys* 6:117–137.
8. Anderson, L. J., Jarrett, W. F., Jarrett, O., and Laird, H. M. (1971). Feline leukemia-virus infection of kittens: Mortality associated with atrophy of the thymus and lymphoid depletion. *J. Natl. Cancer Inst.* 47:807–817.
9. Perryman, L. E., Hoover, E. A., and Yohn, D. S. (1972). Immunologic reactivity of the cat: Immunosuppression in experimental feline leukemia. *J. Natl. Cancer Inst.* 49:1357–1365.
10. Hardy, W. D., Jr. (1982). Immunopathology induced by the feline leukemia virus. *Springer Semin. Immunopathol.* 5:75–106.
11. Hardy, W. D., Jr. (1984). Feline leukemia virus as an animal retrovirus model for the human T-cell leukemia virus. In: *Human T-Cell Leukemia/ Lymphoma Virus* (R. C. Gallo, M. E. Essex, and L. Gross, eds.), Cold Spring Harbor Laboratory, New York, pp. 35–43.
12. Wong-Staal, F., and Gallo, R. C. (1985). Human T-lymphotropic retroviruses. *Nature* 317:395–403.
13. Essex, M., McLane, M. F., Lee, T. H., Howe, C. W. S., Mullins, J. I., Cabradilla, C., and Francis, D. P. (1983). Antibodies to cell membrane antigens associated with human T-cell leukemia virus in patients with AIDS. *Science* 220:859–862.

14. Rogerson, P., Jarrett, W., and Mackey, L. (1975). Epidemiological studies on feline leukaemia virus infection. I. A serological survey in urban cats. *Int. J. Cancer* 15:781–785.

15. Rojko, J. L., Hoover, E. A., Mathes, L. E., Olsen, R. G., and Schaller, J. P. (1979). Pathogenesis of experimental feline leukemia virus infection. *J. Natl. Cancer Inst.* 63:759–768.

16. Hoover, E. A., Rojko, J. L., and Olsen, R. G. (1980). Host-virus interactions in progressive vs regressive feline leukemia virus infection in cats. In: *Viruses in Naturally Occurring Cancers.* Cold Spring Harbor Conference on Cell Proliferation. Cold Spring Harbor Laboratory, New York, pp. 86–95.

17. Hoover, E. A., Rojko, J. L., and Olsen, R. G. (1980). Pathogenesis of feline leukemia virus infection. In: *Feline Leukemia* (R. G. Olsen, ed.) CRC Press, Boca Raton, FL, pp. 32–51.

18. Hoover, E. A., Rojko, J. L., Wilson, P. L., and Olsen, R. G. (1981). Determinants of susceptibility and resistance to feline leukemia virus: I. Role of macrophages. *J. Natl. Cancer Inst.* 67:889–898.

19. Hoover, E. A., Olsen, R. G., Hardy, W. D., Jr., Schaller, J. P., and Mathes, L. E. (1976). Feline leukemia virus infection: Age-related variation in response of cats to experimental infection. *J. Natl. Cancer Inst.* 57:365–369.

20. Hardy, W. D., Jr., (1980). Feline leukemia virus diseases. In: *Feline Leukemia Virus* (W. D. Hardy, Jr., M. Essex, and A. J. McClelland, eds.) Elsevier/North Holland, New York, pp. 3–31.

21. McClelland, A. J., Hardy, W. D., and Zuckerman, E. E. (1980). Prognosis of healthy feline leukemia virus infected cats. In: *Feline Leukemia Virus* (W. D. Hardy, M. Essex, and A. J. McClelland, eds.) Elsevier/North Holland, New york, pp. 121–126.

22. Francis, D. P., and Essex, M. (1980). Epidemiology of feline leukemia. In: *Feline Leukemia Virus* (W. D. Hardy, Jr., A. J. McClelland, and M. E. Essex, eds.) Elsevier/Borth Holland, New York, p. 127.

23. Rojko, J. L., Hoover, E. A., Quackenbush, S. L., and Olsen, R. G. (1982). Reaction of latent feline leukemia virus infection. *Nature* 298:385–388.

24. Francis, D. P., Cotter, S. M., Hardy, W. D., Jr., and Essex, M. (1979). Comparison of virus-positive and virus-negative cases of feline leukemia and lymphoma. *Cancer Res.* 39:3866–3870.

25. Donahue, P. R., Hoover, E. A., Beltz, G. A., Riedel, N., Hirsch, V. M., Overbaugh, J., and Mullins, J. I. (1988). Strong sequence conservation among horizontally transmissable, minimally pathogenic feline leukemia viruses. *J. Virol.* 62:722–731.

26. Yoshinaka, Y., Katoh, I., Copeland, T. D., and Oroszlan, S. (1985). Translational readthrough of an amber termination codon during synthesis of feline leukemia virus protease. *J. Virol.* 55:870–873.

27. Laprevotte, I., Hampe, A., Sherr, C., and Galibert, F. (1984). Nucleotide sequence of the *gag* gene and *gag-pol* junction of feline leukemia virus. *J. Virol.* 50:884–894.

28. Copeland, T. D., Morgan, M. A., and Oroszlan, S. (1984). Complete amino acid sequence of the basic nucleic acid binding protein of feline leukemia virus. *Virology* 133:137–145.

29. Stephenson, J. R., Tronick, S. R., and Aaronson, S. A. (1974). Analysis of type specific antigenic determinants of two structural polypeptides of mouse RNA C-type viruses. *Virology* 58:1–8.

30. Copeland, T. D., Henderson, L. E., Vanlaningham-Miller, E. S., Stephenson, J. R., Smythers, G. W., and Oroszlan, S. (1981). Amino- and carboxyl-terminal sequences of proteins coded by *gag* gene of endogenous baboon and cat type C viruses. *Virology* 109:13–24.

31. Oroszlan, S., Henderson, L. E., Stephenson, J. R., Copeland, T. D., Long, C. W., Ihle, J. N., and Gilden, R. V. (1978). Amino- and carboxyl-terminal amino acid sequences of proteins coded by *gag* gene of murine leukemia virus. *Proc. Natl. Acad. Sci. USA* 75:1404–1408.

32. Dickson, C., Eisenman, R., and Fan, H. (1985). Protein biosynthesis and assembly. In: *RNA Tumor Viruses* (R. Weiss, N. Teich, H. Varmus, and J. Coffin, eds.) Cold Spring Harbor Laboratory, New York: pp. 135–146.

33. Okabe, H., Twiddy, E., Gilden, R. V., Hatanaka, M., Hoover, E. A., and Olsen, R. G. (1976). FeLV-related sequences in DNA from a FeLV-free cat colony. *Virology* 69:798–801.

34. Koshy, R., Gallo, R. C., and Wong-Staal, F. (1980). Characterization of the endogenous feline leukemia virus-related DNA sequences in cats and attempts to identify exogenous viral sequences in tissues of virus-negative leukemic animals. *Virology* 103:434–445.

35. Niman, H. L., Akhavi, M., Gardner, M. B., Stephenson, J. R., and Roy-Burman, P. (1980). Differential expression of two distinct retrovirus genomes in developing tissues of the domestic cat. *J. Natl. Cancer Inst.* 64:587–594.

36. Overbaugh, J., Riedel, N., Hoover, E. A., and Mullins, J. I. (1988). Transduction of endogenous envelope genes by feline leukaemia virus in vitro. *Nature* 332:731–734.

37. Mullins, J. I., Elder, J. H., Binari, R. C., Jr., Riedel, N., and Hoover, E. A. (1989). Envelope gene diversity and conserved backbone glycoprotein structure in feline leukemia viruses, in preparation.

38. Benveniste, R. E., Sherr, C. J., and Todaro, G. J. (1975). Evolution of type C viral genes: Origin of feline leukemia virus. *Science* 190:886–888.

39. Baluda, M. A., and Roy-Burman, P. (1973). Partial characterization of RD-114 virus by DNA–RNA hybridization studies. *Nature New Biol.* 244:59–62.

40. Spodick, D. A., Soe, L. H., and Roy-Burman, P. (1984). Genetic analysis of the feline RD-114 retrovirus-related endogenous elements. *Virus Res.* 1:543–555.

41. Spokick, D. A., Ghosh, A. K., Parimoo, S., and Roy-Burman, P. (1988). Long terminal repeat of feline endogenous RD-114 retroviral DNAs: Analysis of transcription regulatory activity and nucleotide sequence. *Virus Res.* 9:263–283.

42. Okabe, H., DuBuy, J., Hatanaka, M., and Gilden, R. V. (1978). Reiteration frequency of feline type C genomes in homologous and heterologous host cell DNA. *Intervirology* 9:253–260.
43. Mullins, J. I., Casey, J. W., Nicolson, M. O., Burck, K. B., and Davidson, N. (1980). Integration and expression of FeLV proviruses. In: *Feline Leukemia Virus* (W. D. Hardy, M. Essex, and A. J. McClelland, eds.) Elsevier/ North Holland, New York, pp. 373–380.
44. Okabe, H., DuBuy, J., Gilden, R. V., and Gardner, M. B. (1978). A portion of the feline leukaemia virus genome is not endogenous in cat cells. *Int. J. Cancer* 22:70–78.
45. Nicolson, M. O., Hariri, F., Krempin, H. M., McAllister, R. M., and Gilden, R. V. (1976). Infectious proviral DNA in human cells infected with transformation-defective type C viruses. *Virology* 70:301–312.
46. Soe, L. H., Devi, B. G., Mullins, J. I., and Roy-Burman, P. (1983). Molecular cloning and characterization of endogenous feline leukemia virus sequences from a cat genomic library. *J. Virol.* 46:829–840.
47. Soe, L. H., Shimizu, R. W., Landolph, J. R., and Roy-Burman, P. (1985). Molecular analysis of several classes of endogenous feline leukemia virus elements. *J. Virol.* 56:701–710.
48. Casey, J. W., Roach, A., Mullins, J. I., Burck, K. B., Nicolson, M. O., Gardner, M. B., and Davidson, N. (1981). The U3 portion of feline leukemia virus DNA identifies horizontally acquired proviruses in leukemic cats. *Proc. Natl. Acad. Sci. USA* 77:7778–7782.
49. Berry B. T., Ghosh, A. K., Kumar, D. V., Spodick, D. A., and Roy-Burman, P. (1989). Structure and function of endogenous feline leukemia virus long terminal repeats and adjoining regions, submitted.
50. Mullins, J. I., Brody, D. S., Binari, R. C., Jr., and Cotter, S. M. (1984). Viral transduction of c-myc gene in naturally occurring feline leukaemias. *Nature* 308:856–858.
51. Mullins, J. I., Chen, C. S., and Hoover, E. A. (1986). Disease-specific and tissue-specific production of unintegrated feline leukaemia virus variant DNA in feline AIDS. *Nature* 319:333–336.
52. Coffin, J. (1985). Genome structure. In: *RNA Tumor Viruses* (R. Weiss, N. Teich, H. Varmus, and J. Coffin, eds.). Cold Spring Harbor Laboratory, New York, pp. 17–74.
53. Laimins, L. A., Kessel, M., Rosenthal, N., and Khoury, G. (1983). Viral and cellular enhancer elements. In: *Enhancers and Eucaryotic Gene Expression* (Y. Gluzman, ed.) Cold Spring Harbor Laboratory, New York, pp. 28–37.
54. Jolly, D. J., Esty, A. C., Subramani, S., Friedman, T., and Verma, I. M. (1983). Elements in the long terminal repeat of murine retroviruses enhance stable transformation by thymidine kinase gene. *Nucleic Acid Res.* 11:1855–1872.
55. Sarma, P. S., and Log, T. (1971). Viral interference in feline leukemia-sarcoma complex. *Virology* 44:352–358.
56. Sarma, P. S., and Log. T. (1973). Subgroup classification of feline leukemia

and sarcoma viruses by viral interference and neutralization tests. *Virology* 54:160–169.

57. Weiss, R. (1984). Experimental biology and assay of RNA tumor viruses. In: *RNA Tumor Viruses* (R. Weiss, N. Teich, H. Varmus, and J. Coffin, eds.) Cold Spring Harbor Laboratory, New York, pp. 209–260.

58. Leis, J., Baltimore, D., Bishop, J. M., Coffin, J., Fleissner, E., Goff, S. P., Oroszlan, S., Robinson, H., Skalka, A. M., Temin, H. M., and Vogt, V. (1988). Standardized and simplified nomenclature for proteins common to all retroviruses. *J. Virol.* 62:1808–1809.

59. Jarrett, O., Hardy, W. D. J.r, Golder, M. C., and Hay, D. (1978). The frequency of occurrence of feline leukemia virus subgroups in cats. *Int. J. Cancer* 21:334–337.

60. Jarrett, O., and Russell, P. H. (1978). Differential growth and transmission in cats of feline leukaemia viruses of subgroups A and B. *Int. J. Cancer* 21: 466–472.

61. Sarma, P. S., Log, T., Skuntz, S., Krishnan, S., and Burkley, K. (1978). Experimental horizontal transmission of feline leukemia viruses of subgroups A, B, and C. *J. Natl. Cancer Inst.* 60:871–874.

62. Jarrett, O. (1980). Feline leukemia virus subgroups. In: *Feline Leukemia Viruses* (W. D. Hardy, Jr., M. Essex, and A. J. McClelland, eds.) Elsevier/ North Holland, New York, pp. 473–479.

63. Jarrett, O., Laird, H. M., and Hay, D. (1973). Determinants of the host range of feline leukaemia viruses. *J. Gen. Virol.* 20:169–175.

64. Sarma, P. S., Log, T., Jain, D., Hill, P. R., and Huebner, R. J. (1975). Differential host range of viruses of feline leukemia-sarcoma complex. *Virology* 64:438–446.

65. Hoover, E. A., Kociba, G. J., Hardy, W. D., Jr., and Yohn, D. S. (1974). Erythroid hypoplasia in cats inoculated with feline leukemia virus. *J. Natl. Cancer Inst.* 53:1271–1276.

66. Mackey, L., Jarrett, W., Jarrett, O., and Laird, H. M. (1975). Anemia associated with feline leukemia virus infection in cats. *J. Natl. Cancer Inst.* 54:209–217.

67. Onions, D., Jarrett, O., Testa, N., Frassoni, F., and Toth, S. (1982). Selective effect of feline leukaemia virus on early erythroid precursors. *Nature* 296:156–158.

68. Jarrett, O., Golder, M. C., Toth, S., Onions, D. E., and Stewart, M. F. (1984). Interaction between feline leukaemia virus subgroups in the pathogenesis of erythroid hypoplasia. *Int. J. Cancer* 34:283–288.

69. Riedel, N., Hoover, E. A., Gasper, P. W., Nicolson, M. O., and Mullins, J. I. (1986). Molecular analysis and pathogenesis of the feline aplastic anemia retrovirus, FeLV–C–Sarma. *J. Virol.* 60:242–260.

70. Riedel, N., Hoover, E. A., Dornsife, R. E., and Mullins, J. I. (1988). Pathogenic and host range determinants of the feline aplastic anemia retrovirus. *Proc. Natl. Acad. Sci. USA* 85:2758–2762.

71. Vedbrat, S. S., Rasheed, S., Lutz, H., Gonda, M. A., Ruscetti, S., Gardner, M. B., and Prensky, W. (1983). Feline oncornavirus-associated cell membrane antigen: A viral and not a cellularly coded transformation-specific antigen of cat lymphomas. *Virology* 124:445–461.

72. Snyder, H. W. Jr., Singhal, M. C., Zuckerman, E. E., Jones, F. R., and Hardy, W. D., Jr. (1983). The feline oncornavirus-associated cell membrane antigen (FOCMA) is related to, but distinguishable from, FeLV-C gp70. *Virology* 131:315–327.

73. Busch, M. P., Devi, B. G., Soe, L. H., Perbal, B., Baluda, M. A., and Roy-Burman, P. (1983). Characterization of expression of cellular retrovirus genes and oncogenes in feline cells. *Hematol. Oncol.* 1:61–75.

74. Roy-Burman, P., Busch, M. P., Rasheed, S., Gardner, M. B., and Lai, M. M. C. (1980). Oncodevelopmental gene expression in feline leukemia. In: *Feline Leukemia Virus* (W. D. Hardy, M. Essex, and A. J. McClelland, eds.). Elsevier/North-Holland, New York, pp. 361–372.

75. Niman, H. L., Stephenson, J. R., Gardner, , M. B., and Roy-Burman, P. (1977). RD-114 and feline leukaemia virus genome expression in natural lymphomas of domestic cats. *Nature* 266:357–360.

76. Niman, H. L., Gardner, M. B., Stephenson, J. R., and Roy-Burman, P. (1977). Endogenous RD-114 virus genome expression in malignant tissues of domestic cats. *J. Virol.* 23:578–586.

77. Linial, M., and Blair, D. (1985). Genetics of retroviruses. In: *RNA Tumor Viruses* (R. Weiss, N. Teich, H. Varmus, and J. Coffin, eds.) Cold Spring Harbor Laboratory, New York, pp. 147–186.

78. Elder, J. H., and Mullins, J. I. (1983). Nucleotide sequence of the envelope gene of Gardner-Arnstein feline leukemia virus B reveals unique sequence homologies with a murine mink cell focus-forming virus. *J. Vriol.* 46:871–880.

79. Stewart, M. A., Warnock, M., Wheeler, A., Wilkie, N., Mullins, J. I., Onions, D. E., and Neil, J. C. (1986). Nucleotide sequences of a feline leukemia virus subgroup A envelope gene and long terminal repeat and evidence for the recombinational origin of subgroup B viruses. *J. Virol.* 58:825–834.

80. Nicolaisen-Strouss, K., Kumar, H. P. M., Fitting, T., Grant, C. K., and Elder, J. H. (1987). Natural feline leukemia virus variant escapes neutralization by a monoclonal antibody via an amino acid change outside the antibody-binding epitope. *J. Virol.* 61:3410–3415.

81. Dorner, A. J., Stoye, J. P., and Coffin, J. M. (1985). Molecular basis of host range variation in avian retroviruses. *J. Virol.* 53:32–39.

82. Dorner, A. J., and Coffin, J. M. (1986). Determinants for receptor interaction and cell killing on the avian retrovirus glycoprotein gp85. *Cell* 45:365–374.

83. Overbaugh, J., Donahue, P. R., Quackenbush, S. L., Hoover, E. A., and Mullins, J. I. (1988). Molecular cloning of a feline leukemia virus that induces fatal immunodeficiency disease in cats. *Science* 239:906–910.

84. Mullins, J. I., Casey, J. W., Nicolson, M. O., Burck, K. B., and Davidson, N. (1981). Sequence arrangement and biological activity of cloned feline leukemia virus proviruses from a virus-productive human cell line. *J. Virol.* 38:688-703.

85. Hoover, E. A., Dornsife, R. E., and Mullins, J. I. (1989). Biological activity of molecularly cloned feline leukemia viruses, in preparation.

86. Elder, J. H., and Mullins, J. I. (1985). Nucleotide sequence of the envelope genes and LTR of the subgroup B, Rickard strain of feline leukemia virus. In: *RNA Tumor Viruses* (R. Weiss, N. Teich, H. Varmus, and J. Coffin, eds.) Cold Spring Harbor Laboratory, New York, pp. 1005-1010.

87. Rojko, J. L., Hoover, E. A., Krakowka, S., Olsen, R. G., and Mathes, L. E. (1979). Influence of adrenal corticosteroids on the susceptibility of the cats to feline leukemia virus. *Cancer Res.* 39:3789-3791.

88. Hoover, E. A., Rojko, J. L., and Olsen, R. G. (1980). Factors influencing host resistance to feline leukemia virus. In: *Feline Leukemia* (R. G. Olsen, ed.) CRC Press, Boca Raton, FL, pp. 69-76.

89. Dornsife, R. E., Gasper, P. W., Mullins, J. I., and Hoover, E. A., (1989). Rapid in vivo induction of aplastic anemia by intra bone marrow inoculation of a molecularly cloned feline leukemia virus. *Blood*, submitted.

90. Guilhot, S., Hampe, A., D'Auriol, L., and Galibert, F. (1987). Nucleotide sequence analysis of the LTR's and *env* genes of SM-FeSV and GA-FeSV. *Virology* 161:252-258.

91. Hahn, B. H., Gonda, M. A., Shaw, G. M., Taylor, M. E., Redfield, R. R., Markham, P. D., Salahuddin, S. Z., Wong-Staal, F., Gallo, R. C., Parks, E. C., and Parks, W. P. (1985). Genomic diversity of the acquired immune deficiency syndrome virus HTLV-III: Different viruses exhibit greatest divergence in their envelope genes. *Proc. Natl. Acad. Sci. USA* 82:4813-4817.

92. Coffin, J. M. (1986). Genetic variation in AIDS viruses. *Cell* 46:1-4.

93. Hirsch, V. H., Riedel, N., and Mullins, J. I. (1987). The genome organization of STLV-3 is similar to that of the AIDS virus except for a truncated transmembrane protein. *Cell* 49:309-317.

94. Nunberg, J. H., Rogers, G., Gilbert, J. H., and Snead, R. M. (1984). Method to map antigenic determinants recognized by monoclonal antibodies: Localization of a determinant of virus neutralization on the feline leukemia virus envelope protein gp70. *Proc. Natl. Acad. Sci. USA* 81:3675-3679.

95. Gilbert, J. J., Pedersoen, N. C., and Nunberg, J. H. (1986). Feline leukemia virus envelope protein expression encoded by a recombinant vaccinia virus: Apparent lack of immunogenicity in vaccinated animals. *Virus Res.* 7:49-67.

96. Elder, J. H., McGee, J. S., Munson, M., Houghten, R. A., Kloetzer, W., Bittle, J. L., and Grant, C. K. (1987). Localization of neutralizing regions of the envelope gene of feline leukemia virus using anti-synthetic peptide antibodies. *J. Virol.* 61:8-15.

97. Russell, P. H., and Jarrett, O. (1978). The specificity of neutralizing antibodies to feline leukaemia vies. *Int. J. Cancer* 21:768-778.

98. Russell, P. H., and Jarrett, O. (1978). The occurrence of feline leukaemia virus neutralizing antibodies in cats. *Int. J. Cancer* 22:351-357.

99. Hoover, E. A., McCullough, C. B., and Griesemer, R. A. (1972). Intranasal transmission of feline leukemia. *J. Natl. Cancer Inst.* 48:973-983.

100. Cockerell, G. L., Krakowka, S., Hoover, E. A., Olsen, R. G., and Yohn, D. S. (1976). Characterization of feline T- and B-lymphocytes and identification of an experimentally induced T-cell neoplasm in the cat. *J. Natl. Cancer Inst.* 57:907-913.

101. Cockerell, G. L., Krakowka, S., Hoover, E. A., Olsen, R. G., and Yohn, D. S. (1975). Lymphocyte mitogen reactivity in feline leukemia virus-infected cats and identification of feline T and B lymphocytes. *Bibl. Haematol.* 43:81-83.

102. Hoover, E. A., Perryman, L. E., and Kociba, G. J. (1973). Early lesions in cats inoculated with feline leukemia virus. *Cancer Res.* 33:145-152.

103. Hoover, E. A., Mullins, J. I., Quackenbush, S. L., and Gasper, P. W. (1987). Experimental transmission and pathogenesis of feline AIDS. *Blood* 70:1880-1892.

104. Hoover, E. A., Mullins, J. I., Quackenbush, S. L., and Gasper, P. W. (1986). Pathogenesis of feline retrovirus-induced cytopathic diseases: Acquired immune deficiency syndrome and aplastic anemia. In: *Animal Models of Retrovirus Infection and Their Relationship to AIDS* (L. A. Salzman, ed.). Academic Press, Orlando, FL pp. 59-74.

105. Mullins, J. I., Hoover, E. A., and Quackenbush, S. L. (1989). Disease progression and viral genome diversity in feline leukemia virus induced immunodeficiency syndrome, in preparation.

106. Quackenbush, S. L., Mullins, J. I., and Hoover, E. A. (1989). Colony forming T lymphocyte deficit in the development of feline retrovirus induced immunodeficiency syndrome. *Blood,* 73:509-516.

107. Hoover, E. A., Overbaugh, J., Quackenbush, S. L., Donahue, P. R., Wooley, D., and Mullins, J. I. Unpublished observations.

108. Donahue, P. R., Hoover, E. A., Overbaugh, J., Quackenbush, S. L., and Mullins, J. I. Manuscript in preparation.

109. Dorn, R. C., Taylor, D. O. N., and Hibbard, H. H. (1967). Epizootiologic characteristics of canine and feline leukemia and lymphoma. *Am. J. Vet. Res.* 28:993-1001.

110. Neil, J. C., Hughes, D., McFarlane, R., Wilkie, N. M., Onions, D. E., Lees, G., and Jarrett, O. (1984). Transduction and rearrangement of the *myc* gene by feline leukaemia virus in naturally occurring T-cell leukaemias. *Nature* 308:814-820.

111. Levy, L. S., Gardner, M. B., and Casey, J. W. (1984). Isolation of a feline leukaemia provirus containing the oncogene *myc* from a feline lymphosarcoma. *Nature* 308:853-856.

112. Miura, T., Tsujimoto, H., Fukasawa, M., Kodama, T., Shibuya, M., Hasegawa, A., and Hayami, M. (1987). Structural abnormality and overexpression of the *myc* gene in feline leukemias. *Int. J. Cancer* 40:564-569.

113. Teich, N., Wyke, J., and Kaplan, P. (1985). Pathogenesis of retrovirus-induced disease. In: *RNA Tumor Viruses* (R. Weiss, N. Teich, H. Varmus, and J. Coffin, eds.) Cold Spring Harbor Laboratory, New York, pp. 187–248.

114. Forrest, D., Onions, D., Lees, G., and Neil, J. C. (1987). Altered structure and expression of c-*myc* in feline t-cell tumours. *Virology* 158:194–205.

115. Rickard, D. G., Post, J. E., deNoronha, F., and Barry, L. M. (1969). A transmissible virus-induced lymphocytic leukemia of the cat. *J. Natl. Cancer Inst.* 42:987–1014.

116. Jarrett, O., Laird, H. M., and Hay, D. (1972). Restricted host range of a feline leukaemia virus. *Nature* 238:220–221.

117. Hardy, W. D., Jr., McClelland, A. J., Zuckerman, E. E., Snyder, H. W., MacEwen, E. G., Francis, D., and Essex, M. (1980). Development of virus non-producer lymphosarcomas in pet cats exposed to FeLV. *Nature* 288:90–92.

118. Soe, L. H., and Roy-Burman, P. (1984). Structure of the polymorphic feline c-*myc* oncogene locus. *Gene* 31:123–128.

119. Soe, L. H., Ghosh, A. K., Maxon, R. E., Hoover, E. A., Hardy, W. D., Jr., and Roy-Burman, P. (1986). Nucleotide sequences of the 1.2-kb 3'-region and genotype distribution of two common c-*myc* alleles of the domestic cat. *Gene* 47:185–192.

120. Snyder, H. W., Jr., Phillips, K. J., Hardy, W. D., Jr., Zuckerman, E. E., Essex, M., Sliski, A. H., and Rhim, J. (1980). Isolation and characterization of proteins carrying the feline oncornavirus-associated cell-membrane antigen. *Cold Spring Harbor Symp. Quant. Biol.* 44:787–799.

121. Grant, C. K., Weissman, D., Cotter, S., Zuckerman, E. E., Hardy, W. D., and Essex, M. (1980). Mechanisms of feline leukemia escape from antibody-mediated immunity. In: *Feline Leukemia Virus* (W. D. Hardy, M. Essex, and A. J. McClelland, eds.) Elsevier/North Holland, New York, pp. 171–179.

122. Gardner, M. B., Rongey, R. W., Arnstein, P., Estes, J. D., Sarma, P., Huebner, R. J., and Rickard, C. G. (1970). Experimental transmission of feline fibrosarcoma to cats and dogs. *Nature* 226:807–809.

123. Fedele, L. A., Even, J., Garon, C. F., Donner, L., and Sherr, C. J. (1981). Recombinant bacteriophages containing the integrated transforming provirus of Gardner-Arnstein feline sarcoma virus. *Proc. Natl. Acad. Sci. USA* 78:4036–4040.

124. Hampe, A., Laprevotte, I., and Galibert, F. (1982). Nucleotide sequences of feline retroviral oncogenes (v-*fes*) provide evidence for a family of tyrosine-specific protein kinase genes. *Cell* 30:775–785.

125. Hampe, A., Gobet, M., Even, J., Sherr, C. J., and Galibert, F. (1983). Nucleotide sequences of feline sarcoma virus long terminal repeats and 5' leaders show extensive homology to those of other mammalian retroviruses. *J. Virol.* 45:466–472.

126. McDonough, S. K., Larsen, S., Brodey, R. S., Stock, N. D., and Hardy, W. D., Jr., (1971). A transmissable feline fibrosarcoma of viral origin. *Cancer Res.* 31:953–956.

127. Donner, L., Fedele, L. A., Garon, C. F., Anderson, S. J., and Sherr, C. J. (1982). McDonough feline sarcoma virus: Characterization of the molecularly cloned provirus and its feline oncogene (v-*fms*). *J. Virol.* 41:489–500.

128. Wunsch, M., Schulz, A. S., Koch, W., Friedrich, R., and Hunsmann, G. (1983). Sequence analysis of Gardner-Arnstein feline leukaemia virus envelope gene reveals common structural properties of mammalian retroviral envelope genes. *EMBO J.* 2:2239–2246.

129. Nunberg, J. H., Williams, M. E., and Innis, M. A. (1984). Nucleotide sequences of the envelope genes of two isolates of feline leukemia virus subgroup B. *J. Virol.* 49:629–632.

130. Snyder, S. P., and Theilen, G. H. (1969). Transmissable feline fibrosarcoma. *Nature* 221:1074–1075.

131. Sherr, C. J., Fedele, L. A., Oskarsson, M., Maizel, J., and Woude, G. V. (1980). Molecular cloning of Snyder-Theilen feline leukemia and sarcoma viruses: Comparative studies of feline sarcoma virus with its natural helper virus and with Moloney murine sarcoma virus. *J. Virol.* 34:200–212.

132. Kawakami, T. G., Theilen, G. H., Dungworth, D. L., Munn, R. J., and Beall, S. G. (1967). C-type viral particles in plasma of cats with feline leukemia. *Science* 158:1049.

133. Snyder, H. W., Singhal, M. C., Zuckrman, E. E., and Hardy, W. D. (1984). Isolation of a new feline sarcoma virus (HZ1-FeSV): Biochemical and immunological characterization of its translation product. *Virology* 132:205–210.

134. Besmer, P., Murphy, J. E., George, P. C., Qui, F., Bergold, P. J., Lederman, L., Snyder, H. W., Brodeur, D., Zuckerman, E. E., and Hardy, W. D. (1986). A new acute transforming feline retrovirus and relationship of its oncogene v-*kit* with the protein kinase gene family. *Nature* 320:415–421.

135. Besmer, P., Hardy, W. D., Jr., Zuckerman, E. E., Bergold, P., Lederman, L., and Snyder, H. W., Jr. (1983). The Hardy-Zuckerman 2-FeSV, a new feline retrovirus with oncogene homology to Abelsen-MULV. *Nature* 303:825–828.

136. Irgens, K., Wyers, M., Moraillon, A., Parodi, A., and Fortuny, V. (1973). Isolement d'un virus sarcomatogene felin a partir d'un fibrosarcome spontane du chat: Etude du pouvoir sarcomatogene in vivo. *CR Acad. Sci.* 26:1783–1786.

137. Rasheed, S., Barbacid, M., Aaronson, S. A., and Gardner, M. B. (1982). Origin and biological properties of a new feline sarcoma virus. *Virology* 117:238–244.

138. Naharro, G., Robbins, K. C., and Reddy, E. P. (1984). Gene product of v-*fgr* onc: Hybrid protein containing a protein of actin and a tyrosine-specific protein kinase. *Science* 223:63–66.

139. Braun, M. J., Deininger, P. L., and Casey, J. W. (1985). Nucleotide sequence of a transduced *myc* gene from a defective feline leukemia provirus. *J. Virol.* 55:177–183.

140. Doggett, D. L., Drake, A. L., Rowe, M. E., Stallard, V., Hirsch, V., Neil, J. C., Elder, J. H., and Mullins, J. I. (1989). Structure, origin and oncogenic activity of the FeLV-*myc* recombinant provirus, FTT. J. Virol. in press.

141. Fulton, R., Forrest, D., McFarlane, R., Onions, D., and Neil, J. C. (1987). Retroviral transduction of T-cell antigen receptor B-chain and *myc* genes. *Nature* 326:190-194.

142. Pal, B. K., and Roy-Burman, P. (1975). Phosphoproteins: Structural components of oncorna-viruses. *J. Virol.* 15:540-549.

143. Pal, B. K., McAllister, R. M., Gardner, M. B., and Roy-Burman, P. (1975). Comparative studies on the structural phosphoproteins of mammalian type C viruses. *J. Virol.* 16:123-131.

144. Shinnick, T. M., Lerner, R. A., and Sutcliffe, J. G. (1981). Nucleotide sequence of Moloney murine leukaemia virus. *Nature* 293:543-548.

145. Bosselman, R. A., van Straaten, F., Van Beveren C., Verma, I. M., and Vogt, M. (1982). Analysis of the *env* gene of a molecularly cloned and biologically active Moloney mink cell focus-forming proviral DNA. *J. Virol.* 44:19-31.

5

Biology and Pathogenesis of Lentiviruses of Ruminant Animals

Opendra Narayan and Janice E. Clements *The Johns Hopkins University School of Medicine, Baltimore, Maryland*

I. INTRODUCTION

The concept that viruses, like certain intracellular bacteria, can cause slowly progressive disease after prolonged periods of subclinical infection was put forward originally by Bjorn Sigursson during his studies of two diseases in sheep in Iceland (1). The first was scrapie (rida), a slowly progressive degenerative disease of the brain and the prototype of a group of human diseases, including Creutzfeldt-Jakob disease, Gerstmann-Strausler disease, and kuru, that are caused by unconventional infectious agents (2). The second was maedi-visna, a slowly progressive pneumoencephalitic disease complex, caused by a conventional virus that is the biological prototype of a newly emerging taxonomic group of viruses that include the etiological viruses of progressive pneumonia of sheep, caprine arthritis encephalitis in goats, equine infectious anemia of horses, and ARC/ AIDS in humans (3). Both scrapie and maedi-visna agents are tropic for the lymphoreticular and nervous systems, where they replicate persistently but at a remarkably slow rate in their natural hosts (4,5). Clinical disease develops with gradual onset after a latent period of months to years and follows a chronic protracted course. Disease is manifest by cachexia and functional deficits in multiple organ systems. The etiological agent of scrapie is still uncharacterized in terms of genetic content. However, the causal agent of maedi-visna is a conventional RNA virus with the properties of a cytopathic, nononcogenic retrovirus (6). Experimental inoculation of sheep with virus cultivated in cell cultures reproduced the persistent infection and slowly progressive disease with its unusually long incubation periods (5). Some animals never became ill although the potential for late-onset disease was always present. Primarily because of this

117

slow pathogenic process, the etiological agent of the maedi-visna disease complex was named lentivirus.

II. THE DISEASE COMPLEX

The lentiviruses of ruminants cause diseases mainly in sheep and goats, and clinical signs are referable to lesions in any of, or a combination of, four organ systems—the central nervous system, the lungs, the joints, and the mammary gland (7). In nearly all cases disease occurs primarily among adult animals. The neurological disease, visna, appears initially as slight change in gait of the animal. This leads to ataxia and progressive paralysis during a period of weeks to months. The appetite of the animals and general body functions remain normal, but animals nevertheless develop a wasting syndrome. The pneumonic disease, maedi, is a dyspneic condition which becomes worse with time or after brief exercise. Severe weight loss also accompanies this disease. Disease in the joints is usually seen first as swelling in the carpal joints with excessive production of synovial fluid. This leads to arthritis with degeneration of cartilage and bone during the next several months. Mastitis is seen as hardening of the mammary glands and curtailment of milk production (8,9).

The basic lesion in all infected organs is chronic-active inflammation characterized at the cellular level by infiltration and proliferation of mononuclear cells. Normal tissue architecture is disrupted and degenerates in the face of invasion by inflammatory cell populations. This gives rise to functional deficits described above: progressive paralysis, dyspnea, synovitis/arthritis, and mastitis (7).

The main disease expression in infected sheep is dyspnea and is worldwide in dissemination except in Australia and New Zealand, where the disease has been kept out by restricted importation of sheep. The disease had occurred in epidemic form in Iceland following introduction of latently infected sheep from Europe in the 1930s. The Icelandic epidemic reached its peak in the 1950s, but the disease was eliminated from the island by slaughter of affected animals and prohibition of further importation of sheep. The disease occurs at a lower prevalence in most sheep-raising countries in the world and is known by several names—maedi in Iceland, Montana lung disease or, more recently, progressive pneumonia in the United States, zwoegerziekete in Holland, la bouhite in France, graaf reinet in South Africa, and so forth (1,10).

The neurological disease (visna) occurred mainly among Icelandic sheep during the maedi-visna epidemic of the 1950s and usually occurred as a complication of maedi (11). This disease occurs rarely among non-Icelandic sheep (12). The high prevalence of visna in Icelandic sheep has ben ascribed in part to greater genetic susceptibility of these unique animals in comparison to sheep of British breeds (13). Arthritis and mastitis also occur in sheep but in considerably lower incidence than the dyspneic condition (7).

The main disease in goats is synovitis and this occurs mainly in adult dairy animals (14,15). The gradual onset of synovitis and its slow progression to crippling arthritis are similar to the slow progression to the pneumonic disease (maedi) in sheep. Mastitis is also highly prevalent among goats (9). Newborn goats in endemically infected dairy herds frequently develop a rapidly progressive paralytic disease (7,16). First seen as weakness in hind legs at 3–6 weeks of age, the disease progresses to ataxia and paralysis during the following few weeks. The course of this disease is much faster than that of any of the diseases seen in adult goats or sheep and is the only break in the pattern of "slow" disease. However, a sporadic, slowly progressive neurological disease among adult goats has been observed rarely in Sweden and Germany (17). This disease is similar to visna of sheep. This brief overview thus illustrates that the spectrum of lentivirus-induced diseases not only varies between two closely related species of animals but also among different breeds and age groups of animals within the same species.

III. THE VIRUSES

A. Biological Properties

The etiological agent of maedi-visna in Icelandic sheep was the first member of the ruminant lentivirus group to be isolated and investigated. Since sheep with visna had inflammation in the choroid plexus—the lacelike structure in the ventricles of the brain that produces cerebrospinal fluid—it was assumed to be one of the target tissues of the virus. Cultures of choroid plexus were therefore prepared from normal sheep and inoculated with homogenates of brain from sheep with visna. The cultures developed multinucleated giant cell formation and produced large amounts of infectious particles which were found in the supernatant fluid (18). This culture system of visna virus is still used routinely for propagation of this agent. Early studies of this virus by Thormar in Iceland showed that the agent was enveloped and that its synthesis had a DNA-dependent phase (19). Lin and Thormar subsequently showed that the agent had reverse transcriptase activity and that its mechanism of replication was indistinguishable from that of retroviruses (20). Haase and Varmus (21) confirmed this replication scheme of the agent with the demonstration of proviral DNA in inoculated cell cultures. Further, they showed that proviral DNA was not present in normal sheep. This established that the agent was transmitted exogenously.

Studies on the morphogenesis of visna virus showed that the agent matured by budding off the cell membrane of the infected sheep choroid plexus (SCP) cells (22). Examination of the particles showed that the agent is enveloped and has a cylindrical core, a shape shared by the etiological agents of equine infectious anemia and ARC/AIDS (3). Host range studies of visna virus showed that it did not replicate in common laboratory animals and that it was not onco-genic and did not transform cells in culture (23). The virus replicated most effi-ciently in sheep cells. Inoculation of these cultures with virus at an MOI of 1 resulted in production of progeny virus between 16 and 20 hr later and develop-ment of multinucleated giant cell formation (cell fusion) 2–3 days later (18). Cell fusion was also produced by inoculating cell culture with high concentra-tions of virus. This was fusion "from without" and occurred within 2 hr after inoculation. Unlike fusion resulting from replication, fusion "from without" was not confined to ovine cells. Further, the reaction was produced using virus inac-tivated by irradiation with ultraviolet light. Thus, neither infectivity nor replica-tion was necessary for this effect (24). This type of fusion is analogous to that caused by inactivated paramyxoviruses (3.g., Sendai virus).

Prototype strains of carpine arthritis encephalitis virus have many biologi-cal properties similar to those of visna virus but also some unique properties (Table 1). The similarities include the retroviral type of replication and the ability to cause cell fusion during replication (25,26). These agents share nucleo-tide sequence homology in large regions of their genome with visna virus, and their major structural proteins cross-react antigenically with the latter virus (27, 28). Both viruses replicate in primary cultures of macrophages obtained from either sheep or goats (29). Despite these common properties, CAEV is biologi-cally distinct from maedi-visna virus. These two virus strains are, in effect, prototypes of two groups of ruminant lentiviruses whose common and variable

TABLE 1   Biological Properties of Ruminant Lentiviruses

	Group I	Group II
Constant properties		
1. Replication in terminally differentiated cells	+	+
2. Productive replication in vitro	+	+
3. Fusion from within	+	++
4. Restricted replication in vivo	+	+
5. Tropism for macrophages in vivo and in vitro	+	+
6. Virus mutation during persistent infection	+	+
7. Incubation of disease is weeks to months	+	+
Variable properties		
1. Fushion from without	++	–
2. Highly lytic in vitro	++	+/–
3. Transactivation	+	+/–
4. Neutralizing antibody	+	–
5. Antigenic variation	+	–

properties are outlined in Table 1. Unlike maedi-visna virus, CAEV replicates poorly in the fibroblastic sheep choroid plexus fibroblast cell cultures (25). Instead, in addition to primary macrophage cultures, CAEV replicates productively in cell lines developed from the synovial membrane of goats. These epithelial cells are very phagocytic but do not have Fc receptors. It is of interest that maedi-visna virus also replicates productively in these cultures (25). Both viruses cause multinucleated giant cell CPE in these goat synovial membrane cultures. However, CAEV does not cause fusion from without (30). Further, whereas maedi-visna viruses cause lysis in these cells, CAEV causes a less virulent infection and the culture remains persistently infected for several weeks. The reduced level of virulence of CAEV has been correlated with a lack of transactivation of transcription and postranscriptional events during replication in contrast to relatively high levels of transactivation of visna virus (31). (See Section III.C.) The two viruses also vary in their ability to induce neutralizing antibodies. Whereas visna virus induces both binding and neutralizing antibodies, CAEV induces antibodies that bind to virus polypeptides but do not neutralize infectivity (32,33). These differences are discussed in greater detail in Section VII.

The biological differences between prototype strains of visna virus and CAEV are representative of a wide range of biological properties of different field viruses. Many strains of field viruses obtained from American sheep with progressive pneumonia are not lytic in SCP cells; they do not cause fusion "from without" and are poor inducers of neutralizing antibodies (34; Narayan, unpublished). These properties are more similar to those of CAEV than of visna virus. In contrast, some strains of CAEV are very lytic during replication and cause fusion from without (35). These agents are more similar to visna virus. Thus, rather than fitting into a classification of specific serotypes, ruminant lentiviruses circulating in nature have a high degree of individual heterogeneity in biological properties that tends to make each strain unique. Since these biological properties are expressed variably by the viruses, it seems doubtful that each factor would be crucial in pathogenesis because all of the viruses had been obtained from persistently infected animals with chronic disease.

B. Genetic Structure and Replication of the Lentivirus

The common and variable biological properties of the viruses are determined by their genetic organization and molecular mechanisms of replication. The genomes of both visna virus and CAEV are positive stranded polyadenylated RNAs of 9400 base pairs (Fig. 1) (27,36–39). The three viral structural genes (organized 5'-3') are *gag*, *pol*, and *env*. In addition, these lentiviruses have two additional open reading frames (ORFs) located between the *pol* and *env* genes (28,40). These small ORFs are unique to the lentivirus subfamily and are found

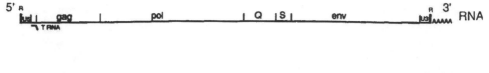

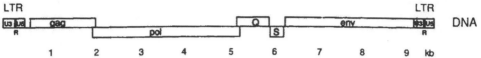

FIGURE 1   The genomic organization of ruminant lentiviruses. The genomic
RNA is contained in the virus; the viral-encoded, RNA-dependent DNA polymerase copies the RNA into DNA in the infected cell.

also in EIAV and HIV (41-44). Noncoding sequences of visna virus and CAEV
viruses are similar in nucleotide sequences and organization. At both the 5' and
3' ends of the RNA (R region) are sequences that contain the cap site, polyadenylation signal, and termination signal for viral RNA transcriptase (45).
Sequences at the 3' end of the viral RNA in the U3 region contain the enhancer-
promoter elements for the initiation of RNA transcription. The nucleotide
sequence immediately downstream of the U5 region in both visna virus and
CAEV is complementary to mammalian tRNA[1,2] (40,45). This serves as a
primer binding site for the synthesis of minus-strand viral DNA. The primer
binding sites for other lentiviruses (HIV and EIAV) are complementary to
tRNA[3] (41-44,46). Purine-rich sequences are located immediately upstream of
the U3 sequences and at the midpoint of both visna virus and CAEV genomes
(AAAAAAGAAAGGGTGG) and serve as the initiation site for synthesis of
plus-strand DNA. The nucleotide sequence of visna virus is A-T rich and the
partial nucleotide sequence of CAEV reflects a similar nucleotide content (40;
Clements, unpublished observations).

     The *gag* genes of the caprine and ovine lentiviruses are the most highly
conserved portion of their genomes, sharing 80% nucleotide sequence homology
(2,27,28). The *gag* gene of visna virus codes for a polyprotein of 55,000 daltons
which is cleaved into the three viral core proteins (5'-3' on the *gag* gene), p16,
p27, p14 (corresponding to their observed mw in Kd). The p27 is the major core
protein. This protein elicits a strong antibody response in the infected animal
and these antibodies cross-react between visna virus and CAEV as well as with
other related ovine caprine lentiviruses (27). The p14 is a highly basic protein
with a calculated molecular weight of 8.8 Kd. Within the protein is a repeated
motif of cysteines (Cys-X_2-Cys-X_9-Cys), characteristic of retroviral nucleic acid-
binding proteins (40). The amino acid composition of this protein is probably
responsible for its slow migration in SDS-PAGE gels and its large observed

molecular weight. A *gag-pol* precursor of 150 Kd has been observed on SDS-PAGE gels and this gives rise to the RNA-dependent DNA polymerase (RTase) and the endonuclease/integrase (40,47). A viral protease has been identified in the nucleotide sequence at the 5' end of the *pol* ORF (40). The *gag* and *pol* proteins of the ovine and caprine lentiviruses reflect the homology of their genes and have extensive serological cross-reactivity (27). A *gag* precursor of 55,000 has been observed and the major core protein is 27 Kd (47). The smaller core proteins of CAEV (p16 and p14 of visna virus) migrate differently than the visna proteins and may indicate some divergence in the amino acid sequence of these proteins (27).

The *env* gene of visna virus encodes a single large polypeptide of 115 Kd, which becomes glycosylated into the glycoprotein precursor of 150 Kd. This protein is processed into its final form of 135 Kd during assembly of the virion (40,47). A glycoprotein of the same size forms the envelope of CAEV (26,27). The predicted *env* protein of visna virus contains an amino acid sequence of Arg-X-Lys-Arg. This sequence is present also in the *env* gene of many retroviruses and represents the site of cleavage of the envelope glycoprotein into a hydrophilic outer membrane protein and a hydrophobic transmembrane portion (40). However, although the gp135 of both visna virus and CAEV is readily observed on SDS-PAGE gels, a transmembrane protein has only rarely been observed.

The homology in the *env* gene of visna and CAE viruses is reflected in extensive antigenic cross-reactivity in the glycoproteins (26,27). These cross-reactions are observed in sera from animals infected with virus and/or in sera from animals immunized with purified glycoproteins also from either virus. Further, several monoclonal antibodies prepared against the glycoprotein of visna virus recognize epitopes in the gp135 of CAEV with similar binding affinities (48). Thus, although the *env* genes of these lentiviruses are the least conserved of all the virus genes, they contain epitopes that are highly conserved.

The heterogeneity in the *env* genes of different lentiviruses is caused in part by a small number of point mutations which accumulate in the viral genomes. These mutations in the viral RNA occur during the persistent viral infection in immune animals. The mutant viruses are associated with escape from neutralization and are thus antigenic variants that have a strong selective advantage for replicating in the presence of neutralizing antibodies (49-51). This has been proven in studies of virus replication in cell culture in which the virus remains genetically and antigenetically stable during several passages (52-54). However, addition of neutralizing antibodies to infected cultures results in development of viruses that resist neutralization (55). Genetic analysis of these viruses showed mutations in the *env* gene (52,56).

Two mechanisms may be involved in the development of these variant viruses. First, they may arise as a result of the high intrinsic mutation rate (approximately $1:10^4$) of retroviruses whose RTases lack editing functions (57). As mentioned earlier, the mutations in the *env* gene probably confer selective advantage to the virus for replication in the immune host, hence the high prevalence of these mutants. In contrast, mutations in other genes probably confer no advantage to the virus and may even decrease the efficiency of replication of the virus. A second possibility is that the *env* gene contains highly mutable regions ("hot spots") and that some of these occur among sequences encoding neutralizing epitopes. These viruses are then selected by the appropriate neutralizing antibodies. This is supported by our earlier observations that mutations accumulate in viruses isolated from a single animal; further, the more the accumulated mutations, the greater the antigenic disparity from the parental virus (50-52,55). This evolution of variant viruses in a persistently infected animal leads to a widening spectrum of neutralizing antibodies. These antibodies develop late in infection (2-3 years after virus inoculation in the animal). They are less efficient in cell culture (55) (see Section VII).

C. Replication of Lentiviruses

Although lentiviruses are, by definition, retroviruses, their mechanism of replication differs from that of classical retroviruses in two major respects. The first relates to the physiological state of the host cell in which the viruses replicate. Retroviruses have a strong requirement for replicating in dividing cells (57). These cells provide optimal conditions for viral DNA synthesis and integration of the proviral DNA. In contrast, lentiviruses replicate efficiently in nondividing end-stage cells both in the host animal and in cell cultures. This allows for replication of visna virus and CAEV in the terminally differentiated macrophage populations, their natural host cells in the animal, and also for plaque formation in monolayer cultures of nondividing cells (22). Since these viruses must utilize the cellular DNA polymerase to synthesize viral DNA similar to the retroviruses, they either provide additional viral encoded replication functions or they activate the nondividing cells to synthesize proteins and other factors that are required for DNA replication. The effectors of these additional functions are probably encoded by the lentivirus genome and may be the products of any of three recently identified mRNA molecules whose functions at present are not known (58-60).

The second major difference in replication between lentiviruses and retroviruses is the cellular location of viral DNA synthesis. The synthesis of the proviral DNA of retroviruses takes place in the cytoplasm of the infected cell (57). In contrast, visna virus replicates its proviral DNA almost exclusively in the nucleus (61). This may be a common property of the lentiviruses and is probably linked to their ability to replicate in nondividing cells. Since the nucleus is the location of cellular DNA synthesis, the enzymes and cofactors necessary for this would be expected to be located there even when the cell is not actively dividing. Thus, by replicating in the nucleus the lentivirus may circumvent the need for the dividing cell or exert some function to stimulate the expression of the cellular enzymes required for complete viral DNA synthesis.

Synthesis of the (-) cDNA strand of the visna and CAE viral genomes via the RTase appears to be the same as that of other retroviruses. Studies suggest that the synthesis of the (+) DNA strand is initiated at the 3' end of the genome at a polypurine tract (AAAAAAGAAAGGGTGG) just upstream from the U3 region as well as from a polypurine tract with the identical sequence at the middle of the genome (nucleotide 4722) in the visna virus sequence (40,62). The nucleotide sequences of the polypurine tracts of visna virus and CAEV are identical and are located in analogous regions on their genome (40; Clements, unpublished observations). Evidence that initiation does occur at both these sites comes from the observation that linear unintegrated DNA of these viruses contains a single-stranded region that maps close to the internal polypurine tract (27,62).

The DNA replication schemes of retroviruses and lentiviruses are very similar to each other but they differ with respect to the ratio of linear to circular DNA that is synthesized during replication. During replication of retroviruses both types of DNA can be detected, but the circular form predominates. In contrast, only a small portion of the lentivirus DNA becomes circularized and only about five copies of these become integrated in each infected cell (37,38, 63). However, many more copies (100-200 copies/cell) are found as free, double-stranded linear DNA molecules in the nucleus of the infected cell (61). In other retroviral systems the integrated viral DNA is the most efficient template for viral RNA transcription (64). In the case of the lentiviruses, it is not currently known whether transcription of the linear unintegrated DNA also contributes to the large amount of viral RNA in the infected cell.

D. Gene Expression of Visna Virus and CAEV

Retroviruses and lentiviruses utilize the eukaryotic cell machinery to transcribe the viral DNA into genomic RNA and mRNAs. This is accomplished by the cellular RNA polymerase II (Pol II) whose function is to transcribe cellular mRNAs from cellular DNA. The region of the DNA to be transcribed by Pol II contains the nucleotide sequence "TATA" which is called a Goldberg-Hogness Box. It is located upstream (5') of the actual start site of transcription and is the recognition signal (promoter) for Pol II RNA transcription. The U3 regions of visna virus and CAEV contain elements commonly found in Pol II promoters (Fig. 2) (45). The nucleotide sequence TATAA is found a short distance upstream from the RNA initiation sites in both visna virus (16 bp) and CAEV (26 bp). In addition, both viruses contain exact tandem repeats 75 bp upstream from

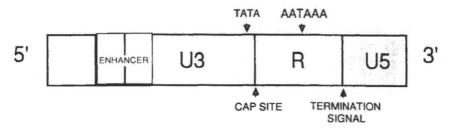

FIGURE 2   The structure of the lentivirus long terminal repeat (LTR).

the TATA box. The 41-bp tandem repeats of visna virus have been shown to act as enhancer elements; i.e. they stimulate transcription by the promoter in a position- and orientation-independent manner (31). The same location of the 71-bp repeats of CAEV in the U3 region and the homology of the repeats (at the 3′ end) make it likely that these repeats also act as enhancers in the CAEV LTR (Fig. 2) (45).

The overall nucleotide sequence homology between the LTRs of visna and CAE viruses LTR is 50%, although certain regions are more conserved than others (45). This suggests that conserved regions may have important functional roles in transcription since both viruses are efficiently transcribed in the same cell types (macrophages and goat synovial membrane cells). Regions with 100% homology are the 3′ ends of the tandem repeats, a 13-bp element upstream of the promoter and the TATA boxes. In the R region the polyadenylation signal contains an 11-bp exact nucleotide match, and the 3′ end of the U5 region is 82% homologous.

In order to study the transcriptional activity of the LTRs of visna and CAE viruses, the LTR sequences were attached to the bacterial gene chloramphenicol actyltransferase (CAT). The relative activity of the LTRs could be measured by transfecting the DNA into a variety of cell types and measuring the transient expression of CAT. These studies have demonstrated the presence of both enhancer and promoter elements in the visna virus U3 region and that the U3 region contains all the sequences required for the transcriptional activity of the visna virus LTR (31,45). Both the visna virus and CAEV U3 regions have high basal levels of activity in a variety of sheep and goat cells (sheep choroid plexus, goat synovial membrane, primary and transformed sheep macrophages). The highest basal level of activity for both viruses is in sheep macrophages (45). Both viral LTRs are active in sheep and goat cells, but the visna virus LTR is always substantially more active than the CAEV LTR (45). This correlates with the higher titer and lytic replication of visna virus in vitro (Table 1). The basal level of these promoters in sheep and goat cells is much higher than that of other viral promoters used as controls (i.e., SV40 and RSV LTR). The transcriptional activity of the visna virus and CAEV U3 regions in mouse L cells is comparable to the SV40 enhancer/promoter element (45).

Transcription of visna virus and CAEV begins in the 5' LTR. Pol II presumably binds to the viral DNA near the TATA box in the U3 region and the RNA is initiated at the "cap" site located at the U3-R junction (Fig. 2). transcription proceeds through the R–U5 region and through the viral genome to the polyadenylation site in the R region in the 3' LTR. The RNA termination signal is 19 bp or 22 bp, for visna virus or CAEV, respectively, beyond the polyadenylation site. The complete genomic RNA and the mRNAs contain a 7-methyguanosine "cap" at the 5' end and a poly (A) tail at the 3' end. Five subgenomic viral mRNAs have been identified in visna virus infected cells in vitro (5.0 kg, a 4.3-kb doublet, 1.8-kb and 1.5-kb mRNAs) in addition to the genomic RNA (9.4 kb) (Fig. 3) (58). At least one of the 4.3-kb mRNAs is the *env* gene messenger RNA; the function of the other 4.3-kb mRNA is presently unknown. The 5.0-kb mRNA contains the entire Q region and probably produces a protein that regulates viral gene expression. The small mRNAs (1.8

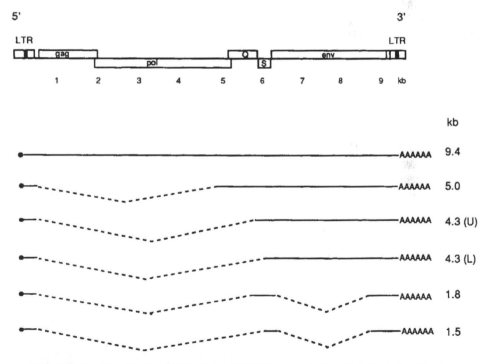

FIGURE 3 Structures of the viral mRNA transcripts in cells infected with visna virus.

and 1.5 kb) are produced after at least two splicing events and contain nucleotide sequences derived from the small ORFs Q and S as well as sequences from the 3' end of the genome (Fig. 3).

In addition to the complex pattern of transcription of visna virus, there is temporal regulation of transcription (58). Thus, the smallest mRNA molecule in infected cells is not synthesized until 2–4 hr after the other mRNA species have been transcribed. However, at 24 hr. after infection this mRNA is present at the same level as the other viral mRNAs. The function of the protein product of this small mRNA is not known. However, it may have a regulatory function not needed early in the replication cycle. One possible function is suppression of splicing in order to increase the level of genomic RNA for progeny virus. Another possible function is facilitation of efficient translation of viral proteins by the infected cell.

The visna virus–infected cell contains *trans*-acting factors which increase expression of any gene that is attached to the visna virus enhancer/promoter element and sequences that include the virus cap site (31,45). These *trans*-acting factors enhance the transcription of LTRs of both the visna virus and CAE viruses. CAEV, however, does not induce *trans*-acting factors (45), although its LTR is partially responsive to those produced by visna virus.

The regions responsive to *trans*-acting factors in the LTRs of visna virus and CAEV are apparently different (45). The U3 region of the CAEV LTR apparently contains the sequences that are responsive to the visna virus *trans*-acting factors. This result was obtained by comparing the levels of CAT activity in GSM cells infected with visna virus and transfected with either pCAEV–LTR–CAT or pCAEV–U3-CAT. The CAT activity of both plasmids was increased 16-fold over the activity measured in uninfected cells. Sequences in the R region of the visna virus LTR contribute to the level of transactivation by visna virus. Thus, the amount of CAT activity from the pvisna-LTR–CAT in visna virus–infected, compared to uninfected, GSM cells is 90-fold while the pvisna-U3-Cat is 45-fold. Thus, not only are the two viruses different in their ability to produce the *trans*-acting factors in vitro, but the presence of target sequences for these factors in the LTRs is variable.

The *trans*-activation of virus gene expression may be responsible for rapid mobilization of the cell transcriptional machinery for viral transcription which results in rapid accumulation of viral RNA in the infected cell. In visna virus–infected cells a posttranscriptional effect may account for some of the transactivation, and this posttranscriptional effect may result in the more efficient translation of viral protein products. These effects suggest that *trans*-acting factors are important in the replication of visna virus in vitro. However, these factors are not required for replication of all lentiviruses, since CAEV replicates as well as visna virus in vitro and both viruses transcribe equivalent levels of viral RNA at times of peak CPE (65; Pyper, unpublished observations). Despite this, visna virus is highly cytolytic in vitro whereas CAEV is not. Thus, it is tempting to speculate that these *trans*-acting factors may be involved in this highly cytolytic type of replication in vitro. Further, the importance of these *trans*-acting factors in vivo is questionable since all ruminant lentiviruses, including CAEV, cause persistent infection and disease in their natural hosts, irrespective of their ability to produce *trans*-acting factors in cell culture.

The small mRNAs of visna virus are analogous to those of HIV (58–60). One of these mRNAs in HIV produces proteins that regulate the gene expression of the virus in *trans* at a posttranscriptional level (tat III gene) (66-70). They are also thought to have regulatory functions in splicing of viral RNAs (rev gene) (71) and in the translation of viral structural genes (rev gene) (72). Thus, these small mRNAs appear to be important for the complex gene expression of lentiviuses and may play a role in the restricted gene expression observed for the lentiviruses in vivo.

Permissive tissue culture systems have been used to quantitate viral nucleic acids during replication of CAE and visna viruses. The levels of viral DNA and RNA of visna virus in permissively infected cell cultures and in cells from infected animals have been quantitated more thoroughly than those of CAEV. Viral RNA in cultures was measured by dot blot analysis and in tissue by in situ hybridization using ^{35}S-labeled cloned DNA probes. In the permissive SCP cell cultures there were 300-500 copies of viral DNA per cell at peak virus production (61). The viral RNA at this time was approximately 20,000 copies per cell (65). Much lower levels of virus gene expression were seen in the infected animal. In paraffin sections of choroid plexus from an infected lamb, 18% of nuclei contained viral DNA, but only 0.1% of cells expressed p27, the major core protein of the virus (73). In another experiment only 1-3% of cells contained proviral DNA or viral RNA; the numbers in positive cells were approximately 65 copies of proviral DNA and 140 copies of viral RNA (74).

Virus-specific RNA was harvested at sequential intervals from GSM cultures inoculated with CAEV at an MOI of 1 and quantitated by RNA dot blot analysis. This showed that viral RNA began to accumulate in the cells by 24 hr after inoculation. The RNA levels reached peak values of approximately 16,000 genomic copies per cell (Pyper and Clements, unpublished observations) by day 4, and remained at this level until day 7. The pattern of viral-specific DNA accumulation in CAEV-infected GSMs paralleled RNA accumulation (Pyper and Clements, unpublished observations).

IV. EPIDEMIOLOGY OF THE LENTIVIRUSES

The ability of lentiviruses to integrate their proviral DNA into host cell DNA, in particular the DNA of cells of the macrophage lineage, automatically makes these agents refractory to host defense mechansims. Further, the propensity of the viruses to undergo antigenic variation and to sequester their neutralization epitopes adds to their resilience against potential humoral immune defenses. Infection with these viruses therefore predicts indefinite persistence of the agent in the infected host despite production of large amounts of specific binding antibodies. Presence of antibodies to the virus in serum is an indicator of active infection in the animal, which may or may not develop disease at some point during its life.

The "Achilles heel" of the lentiviruses lies in their poor efficiency in transmission between hosts. Such transmission depends almost exclusively on exchange of body fluids. Among farm animals, agricultural management practices that increase the chances of such exchange have created epidemics from normally sporadic infections (7). Another potential factor in transmission of these viruses is that other types of infections may cause inflammation in the target organ and the inflammatory cells may become hosts for replication by the lentiviruses. Both factors have been identified in the epidemics of ruminant lentivirus infections. The epidemic of maedi-visna in Iceland may have been due in part to the genetically determined high susceptibility of these animals to disease. However, conditions were ripe for a lentivirus epidemic because infection with the maedi-visna virus followed on the heels of another respiratory disease—pulmonary adenomatosis (jaagsiekte) an oncogenic disease in the lung that results in production of voluminous amounts of respiratory exudate (75). It is thought that this disease provided an efficient means for nose-to-nose transmission of body fluids containing maedi-visna virus–infected cells. Furthermore, sheep housed under crowded and poorly ventilated barn conditions during the long Icelandic winter provided ideal conditions for such exchanges.

Colostrum and milk are another major source of virus for dissemination from mother to offspring. Macrophages in the mammary gland are host cells for virus replication (see below), and since colostrum and milk are high in macrophage content, these body fluids are highly infectious when they are produced by a latently infected animal. This is a major source of virus transmission among sheep in Europe and elimination of the virus from these sheep was accomplished mainly by quarantine of infected ewes and their offspring (76).

The epidemic of carpine arthritis-encephalitis among dairy goats in the United States is directly related to milk-borne infection (77,78). This vector system was magnified by the dairy management practice of pooling milk from all lactating does and feeding kids artificially from this common stock. CAE is a major problem only in the dairy goat populations. Free-ranging animals rarely show infection. This correlation was illustrated with dramatic effect by finding a very low incidence of infection among goats in developing countries where the dairy industry is minimal, in contrast to a high prevalence of infection in North America and Western Europe where the dairy industry is well established (79).

Unlike the dairy goat industry, the sheep industry in the United States is relatively small, but large sheep flocks in the western part of the country are nevertheless plagued with a high prevalence of infection and continuous attrition by disease in a small number of these animals. While the virus is probably perpetuated by the infectious milk vector, the development of disease may be caused in part by coinfection with the agent of pulmonary adenomatosis (J. DiMartini, University of Colorado, personal communication). Both of these are primary pathogens and the lesions that they induce in the lung consist in part of infiltration of inflammatory cells. Some of these cells are natural hosts for the lentiviruses.

V. PATHOGENESIS OF INFECTION

Replication of lentiviruses in vivo is maintained at a restricted level throughout the infection. This restricted life-cycle of the virus is the hallmark of the interaction between this virus and its host. Restriction in virus gene expression in tissues contrasts vividly with the highly productive type of virus replication, with cytopathic end results, in cultured host cells. These two extremes in virus replication are typical of all ruminant lentiviruses. One of the basic mechanisms for restricted replication in tissues is that host cells for virus replication in vivo are confined to a narrow lineage of cells and virus gene expression in these is limited by host factors that control maturation of these cells. The theme of this mechanism of restriction in virus replication is that immature cells are incapable of supporting the complete life-cycle of virus. Infection in immature cells is followed by integration of the proviral DNA and a minimal level of transcription from these templates. Further progress in the virus life-cycle is halted temporarily until the cells mature to a further stage. The infection in these precursor cells therefore is one of transient latency. As the cells mature under physiological or pathophysiological conditions, they gradually become more permissive for virus replication. Transcription of the viral RNA proceeds at an amplified level in mature cells, but this is followed by a restricted level of translation of virus proteins. The virus life-cycle slows down again in these healthy macrophages because the relatively high number of copies of viral RNA in these cells in not reflected by equivalent amounts of virus particles. The mechanisms regulating virus gene expression during different stages of maturation of the host cells hold one of the keys to understanding the pathogenesis of this disease complex. Some of the factors that restrict virus replication in vivo are described below.

Cells of the monocyte-macrophage series are the major cell type that support virus replication in the animal (29,55,80,81). This was determined by application of various immunological and nucleic acid probes to sections of tissues from experimentally infected sheep. Specifically, antibodies prepared to cells of the macrophage lineage and to viral antigens were used to visualize specific substrates with use of the avidin-biotin-peroxidase system (81,82). Other sections of tissues, or those that had been stained immunochemically were then hybridized with ^{35}S-labeled cloned viral DNA to ascertain the approximate number of copies of viral RNA in cell populations from different tissues (65,81, 82). Further, biological markers of macrophages, including adherence to plastic, phagocytosis via Fc receptors, and lysosomal enzyme content, were exploited to determine the limiting stages of virus gene expression in populations of macrophages at different stages of maturation (65). These experiments showed that small numbers of monocyte precursors in the bone marrow had limited numbers of copies of viral RNA, whereas macrophages in tissues had almost 10 times more.

During the normal ontogeny of tissue macrophages, promonocytes in bone marrow mature into monocytes and then leave the bone marrow and circulate in peripheral blood for several hours before leaving the blood to become tissue macrophages (83). This last stage involves a final maturation and differentiation step by which the monocytes, which are morphologically indistinguishable from one another in blood, become morphologically and functionally distinct macrophages that are specific for particular tissues. Typical examples are alveolar, synovial, splenic macrophages, Kupffer cells in the liver, microglial macrophages in brain, and so forth. All tissue macrophages are thought to be derived from a common granulocyting precursor cell in the bone marrow.

Infected monocytes were detected in peripheral blood of naturally and experimentally infected animals and were the only cell in which viral RNA was detected (55). Nonadherent T lymphocytes were not infected. Lysates of mononuclear cells from peripheral blood did not have infectivity. However, after these cells were cultivated for a few days, they matured into large phagocytic macrophages which then produced infectious virus particles. The suggestion that completion of the virus life-cycle required maturation of the cells was proven by inoculation of mononuclear cells from peripheral blood of normal sheep with visna virus and maintenance of the cells in culture under conditions that delayed maturation (low serum in the medium) or speeding up of maturation (high serum) of monocytes to macrophages (81). Infected stunted monocytes had considerably less numbers of copies of viral RNA 24 hr after inoculation than their counterparts that had matured to macrophages. Further, whereas the immature macrophages did not have viral proteins, the more mature cells did. Although the incipient macrophages were considerably more permissive for virus replication than the more immature monocytes, they did not produce virus particles until several more days had elapsed. By this time they had become large phagocytic cells in which virus particles could be seen in intracytoplasmic viscle vesicles as well as budding from the cell membrane. Whether this late permissive stage represented a further stage of maturation of the macrophages is not known.

The higher level of virus gene expression in mature macrophages compared with immature macrophages was also confirmed in studies of the tissue sections from infected sheep (65). Whereas promonocytes in the bone marrow had between 50 and 200 copies of viral RNA per cell, macrophages in spleen and lung had as many as 5000 copies of RNA per cell. Only a small amount of viral antigen was detected in these cells and no virions could be visualized. These observations in infected tissues thus were a close parallel to those of the earlier events following cultivation of infected monocytes. Whether the fully permissive macrophages seen in aging cultures have their counterparts in tissues is not known.

Histological examination of tissue sections that had been evaluated for virus gene expression showed that despite their restriction in completion of the virus life-cycle in tissue macrophages, large numbers of inflammatory cells had migrated into such tissues. This was particularly evident in the lung and in the mediastinal lymph nodes that drain the lung. These nodes had become enlarged as a result of such infiltrations. Previous studies had shown that cell-free homogenates from such tissues lack infectivity. This is in keeping with probing experiments of fixed tissues described above. Nevertheless, the coexistence of inflammatory cells with only partial expression of the virus genome in macrophages suggested that pathological responses did not require completion of the virus life-cycle. Since macrophages were the only infected cells in tissues, some early product of the virus replication cycle in the macrophages was probably responsible for triggering the inflammatory response. This product was probably produced only by the mature macrophages and not by latently infected precursor cells in the bone marrow because the inflammatory changes seen in the lung and other target organs was not and is almost never seen in the bone marrow (65).

Virus infection was confined to certain populations of macrophages, and these were tissue specific (82). Permissive macrophages included those in the spleen, lungs, synovia, and mammary glands but not histiocytic cells in connective tissues or Kupffer cells in the liver. Macrophages in the hyperplastic mediastinal lymph nodes draining the inflamed lungs also had viral RNA, but this was probably only a local response because lymph nodes draining unaffected tissues such as the gastrointestinal tract had minimal numbers of infected macrophages. These nodes were morphologically normal. This regional distribution of infected macrophages was paralleled by an identical distribution of inflammatory cells. These infected macrophages therefore provided the organ-specific basis of inflammation and clinical disease.

The minimal infection in Kupffer macrophages in the liver was a sharp contrast to the high degree of infection in alveolar macrophages (82). This provided a rational explanation for lack of hepatic lesions in the disease complex. Since most monocytes in the blood are destined to differentiate into

Kupffer macrophages (82), it was surprising that at least some of the macro-phages in this organ did not have viral RNA. This negative result introduces the possibility of yet another form of restriction of virus gene expression: Whereas maturation of infected monocytes to alveolar macrophages was associated with increased virus gene expression, maturation of the monocytes to Kupffer macro-phages may be associated with further down-regulation of the virus life-cycle, possibly by methylation of the viral DNA or even elimination of the virus genome from these cells. An alternative but less compelling explanation for lack of infection in Kupffer macrophages is that these cells are derived from bone marrow precursors that are resistant to infection and different from those that give rise to susceptible alveolar and splenic macrophages.

In contrast to the minimal susceptibility of liver cells to infection, suscep-tibility of cells in the central nervous system seems dependent on genetic factors in the host. Intracerebral inoculation of visna virus into Icelandic sheep resulted in infection of a large number of cells that expressed an equally large number of copies of viral RNA (84). The level of viral RNA transcription in brain cells was similar to that seen in macrophages in the lung, spleen, and lymph nodes of American sheep inoculated with this virus (Narayan, unpublished observations). Further, the inflammatory cell response in the brain of these intracerebrally inoculated Icelandic sheep was intense and similar in severity to that seen in the lungs of infected American sheep. Attempts to identify these cells in the brain have only been partially successful. A few oligodendrocytes with viral RNA have been identified, but most of the infected cells are still unidentified (84). Macro-phages in the inflammatory exudates did not have viral RNA. Whether the viral RNA-bearing cells in the brain are microglia is still not clear because of the lack of specific immunological markers for these cells.

Central nervous system cells of sheep belonging to U.S. breeds can be infected with visna virus but very little viral RNA transcription occurs from the proviral DNA templates in these cells. After intracerebral inoculation of these sheep, virus can be recovered from the brain after explantation of brain tissue (Narayan, unpublished observations). This tissue culture procedure is analogous to cultivation of infected monocytes which produce virus only after they have matured in culture. Presumably latently infected cells mature under conditions of explantation and begin to produce virus. However, hybridization of fixed or frozen sections from the brain of these infected animals with ^{35}S-labeled viral DNA did not show viral RNA transcripts. In contrast, sections of the lungs and sections of the cervical lymph nodes draining the brain of these inoculated animals had viral RNA in macrophages (Narayan, unpublished observations). Thus, a block in viral RNA transcription was confined to cells in the brain. These results fit with the clinical disease picture because whereas Icelandic sheep develop both maedi and visna, U.S. breeds of sheep develop mainly maedi.

VI. EFFECTS OF MACROPHAGE INFECTION ON IMMUNE RESPONSES IN SHEEP

The pathogenic effects of the lentiviruses in their animal hosts clearly differ from the cytopathic effects of these agents in cell culture. The fusion of cell cultures at the peak of virus production probably results from excessive production of viral glycoprotein and cell death from exhaustion of cellular metabolic pools that are diverted to production of virions. In the infected animal, however, virus gene expression is restricted with no evidence of virus-induced lysis. Disease results from lymphoproliferation, characterized histologically as inflammation, in particular organ systems in which viral RNA transcription occurred in macrophages but virus production was curtailed or restricted. This partial replication cycle in macrophage populations forms the basis for disease. An immunopathological mechanism for visna had been advanced earlier during experiments in which intracerebrally inoculated Icelandic sheep treated with antithymocyte serum and cyclophosphamide failed to develop the inflammatory lesions that were produced routinely in untreated virus-inoculated sheep (85). Studies at the cellular level support the immunopathological basis for disease.

Cultured ovine macrophages infected with any of the lentiviruses produce infectious virus particles. The virus-producing cells also process certain viral antigens and present them on the cell membrane in close association with their class II antigens of the major histocompatibility complex. This dual antigen is presented to T lymphocytes in an immunologically specific manner. One of the known by-products of this interaction is an interferon. Under tissue culture conditions this interferon develops in supernatant fluids within 24 hr after lymphocytes are added to infected macrophages. Production of this interferon was then used as an index to identify the specificity of the antigens required for its induction and the specificity of the cell types required for the reaction. These experiments showed that the interferon was produced only if the infected cells were macrophages (infected fibroblasts did not work) and only after T lymphocytes were added to the infected macrophages (86). One of the central properties of the macrophage was its class II antigens, because addition of antibodies to the class II antigen complex (monoclonal antibodies to sheep Ia antigen) prior to addition of lymphocytes prevented production of the interferon (87). Macrophages needed to be infected with the virus for production of the viral antigen. Addition of concentrated inactivated virus preparations to macrophages for "processing" did not result in production of the antigen. To prove viral particles were internalized we adopted a strategy used in previous studies on virus neutralization in macrophages. These experiments had shown that isotopically labeled infectious virus bound to neutralizing antibodies was internalized rapidly into macrophages and subsequently broken down into acid-soluble components (9).

Addition of complexes of virus and neutralizing antibodies to macrophages was performed to determine whether the processed antigen would be produced after internalization and digestion of the particles. However, the antigen was not produced under these conditions. Thus, infection in the macrophage with lentiviruses was an absolute requirement for production of the antigen. Although the reaction could be aborted by addition of anti-Ia antibodies to the infected macrophages, addition of antiviral antibodies to previously infected macrophages (after the antigen had been produced by the macrophage) had no effect on interferon production. This suggested that if the antigen were a structural protein, it was not in its native state. This antigen has not yet been identified. Whether it is a "processed" structural protein of the virus (similar to the processed nucleoprotein antigen of influenza virus) or a nonstructural protein encoded by one of the small viral RNA species is not known. However, production of virus was not essential for synthesis of this antigen (86,87).

The supernatant fluid containing the interferon has biological properties that indirectly down-regulate virus gene expression (86). When added to cultures of mononuclear cells from peripheral blood the fluid causes retarded maturation of the monocytes. Since the maturation of these cells is essential for completion of the virus life-cycle, the inhibition of the maturational process indirectly restricts virus replication. Second, the supernatant fluid containing the interferon is a potent inducer of expression of Ia antigens in uninfected and lentivirus-infected macrophages in culture (86). Infection of macrophages with the virus does not induce expression of Ia antigens. Since expression of Ia antigens by macrophages is as important as viral antigens for triggering T lymphocytes for the inflammatory reaction, the presence of the Ia antigen-inducing factor, possibly the interferon, is crucial for completion of the immunological reaction. The interaction of lymphocytes with infected macrophage provides all the essentials for this reaction. Since these two cell types are the main constituents of the inflammatory cell response in tissues, sections of fixed inflamed tissues were therefore examined by combined immunochemical and in situ hybridization to determine whether macrophages with viral RNA were also expressing Ia antigens. As reported earlier, combination of antimacrophage antibodies with in situ hybridization showed that nearly all of the cells in the spleen, the lung, and the mediastinal lymph node that had viral RNA also had the macrophage marker. Substitution of the anti-Ia antibodies for the antimacrophage antibodies in this combined test showed that more than 30% of the cells with virus RNA in the lung also had expressed Ia antigens (87). Since lentivirus infection in cultured macrophages does not result in expression of Ia, we presumed that expression of this antigen in tissue macrophages must have been induced by another factor. Such an inducer could be similar to the interferon found in the active supernatant fluids described earlier. The restriction in expression of the virus life-cycle in macrophages may also have been caused by soluble products of the interacting inflammatory cells and may be immunologically specific. Since these fluids can restrict maturation of macrophages in culture, it is possible that similar mechanisms may be at work in tissues to produce the known end result: restricted replication of virus in the macrophages.

The lymphocytes participating in the inflammatory responses have not been identified as yet in functional tests. It is evident, however, that these cells had not effect on macrophages expressing viral RNA because the lymphocytes and infected macrophages coexist in inflamed tissue. However, the failure to detect macrophages supporting the late stages of virus replication may be due to either inhibition of such maturational processes by lymphokines produced by the inflammatory cells or elimination by cytotoxic T lymphocytes. Such lymphocytes would obviously have no means of recognizing cells expressing only the early stages of the virus life-cycle.

VII. THE ANTIBODY RESPONSE TO LENTIVIRUSES

The structural and nonstructural proteins encoded by lentivirus genes become specific antigens when they are produced in vivo during the infection. Each antigen is comprised of multiple epitopes and some of these are shared with related proteins on other strains of ovine and caprine lentiviruses, depending on the degree of homology between the genes encoding these proteins. Since lentivirus infections are persistent, the antibodies to proteins encoded by genes with the greatest amount of shared homology become important diagnostic reagents. The *gag* proteins and certain epitopes in the virus envelope are of particular importance in this respect.

Most of the lentivirus antigens induce antibodies that are of minimal biological importance. However, the envelope glycoprotein of the virus contains the epitopes important for neutralization of the virus. Many of these epitopes are shared among various strains of lentiviruses, but their arrangement on the surface of the virion is apparently such that only few are exposed to the immune system in any particular strain of virus (48). This limitation in the number of neutralization epitopes presented on the infected cells may be the result of the restricted type of virus replication that occurs in macrophages. The limited repertoire of neutralization epitopes on the surface of the virus is genetically determined in each strain of virus and remains constant during multiple passage of the virus in cell culture (50,52). In the persistently infected animal, however, mutations occur in the *env* gene and these accumulate in the viral genome. These mutations are associated with rearrangement of the neutralization epitopes on the surface of the virus (48).

In order to evaluate the biology of the interaction of neutralizing antibodies and the virus during infection, sera were collected at sequential intervals during a 3-year study on sheep that had been inoculated experimentally with plaque-purified visna virus. Approximately 2 years after inoculation of plaque-purified virus into sheep, multiple viruses were isolated from peripheral blood mononuclear cells of two of these animals. Each virus was plaque purified, and viral RNA was extracted. Oligonucleotide analysis of RNA from these agents showed a series of cumulative mutations in the *env* gene (52,56). The virus with the greatest number of mutations was least recognized by the neutralizing antibodies to the parental virus, and conversely, the virus with the least mutations was serologically indistinguishable from the parental virus (52). Whereas the earliest neutralizing antibodies (produced during the first 2 months after inoculation) were specific for the inoculum parental virus, the sera collected 3 years later ("late" antibodies) had antibodies that recognized a wider range of viruses (51). Thus, the neutralizing antibody spectrum of the animals' sera had become wider with time.

Since neutralizing antibodies to the parental virus did not eliminate this agent from the animals, these antibodies may be of little beneficial value to the host. However, whether such antibodies are important for selection of mutants in the animal is not known. This question was examined in cell culture. Preliminary experiments had shown that a cell culture preparation of visna virus containing 1×10^7 pfu of virus did not have a single mutant that failed to be neutralized by early neutralizing antibodies from a sheep inoculated with this virus (52). However, addition of these early antibodies to tissue culture dishes that had been inoculated 1 day previously with portions of a suspension of a single plaque of visna virus resulted in development of mutant viruses in the cultures 3 weeks later (51). The mutants selected by these "early" antibodies were similar both serologically and genetically to the mutants obtained from sheep 2 years after inoculation. Further, whereas "early" immune sera with their narrow neutralizing spectrum were very efficient in selecting mutant viruses in culture, the "late" sera with wide spectrum of neutralizing antibodies were inefficient in selecting mutants (51). Thus, although neutralizing antibodies may not be able to rid the animal of virus, these antibodies may be important in selecting virulent, nonneutralizable mutant viruses.

The widening of the neutralizing antibody spectrum in sera obtained late in the infection may have been caused by antigenic stimulation by new epitopes on the mutant viruses that had developed late in the infection. To test whether these epitopic configurations on the mutant viruses were really new, we disrupted the parental virus with detergent and injected sheep repeatedly with this material in a hyperimmunizing protocol. Examination of this hyperimmune serum showed that it contained antibodies that neutralized not only the parental virus used for immunization but also the mutant viruses. Thus, the parental virus contained all the neutralization epitopes of the mutant viruses (53,54). Interpretation of these data in light of the known genetically determined narrow antigenic repertoire of each virus strain suggests that mutations in the *env* gene did *not* result in creation of new epitopes, but rather, they resulted in rearrangement of the epitopic mosaic into new configurations in which unique subsets of epitopes were represented on the surface of different virus strains.

Neutralizing antibodies, antigenic variation, and selection of mutants with antibodies are typical of lentiviruses in Group 1, of which visna virus is the protoype. Viruses belonging to Group 2 are exemplified by carpine arthritis encephalitis virus. During the infection with these viruses, the animals produce binding antibodies to the antigenic determinants of the virus in a similar manner as sheep infected with visna virus. Since CAEV has homology with visna virus through its genome, there is extensive cross-reactivity in *gag* and *env* proteins in these binding tests (27,88,89). However, viruses in Group 2 do not normally induce neutralizing antibodies during natural infection (53,54). Nevertheless, neutralizing antibodies have been induced experimentally under extraordinary conditions of immunization (53,54). Serum from one such animal had neutralizing properties similar to those of "early" immune sera to visna virus. Thus, anti-CAEV neutralizing antibodies neutralized only the virus used for immunization and not other field strains of viruses. This suggests that mutations may occur at random without the pressure of selecting neutralizing antibodies. Field viruses belonging to Group 2 may thus be as antigenically distinct a Group 1 viruses, despite difficulty of demonstrating this by neutralization. Mutations are reflected by the great variability in the restriction maps of cloned DNA from three strains of CAEV. Continuous mutation of the virus during persistent infection therefore seems to be the normal course of events. Neutralizing antibodies may be only one of several factors that select for variant viruses.

VIII. SUMMARY

The complex biology and pathogenesis of the lentiviruses described in this review clearly establish that these are unique and insidious pathogens. To compensate for their narrow host range and inefficient mechanism of transmission the lentiviruses have evolved with sophisticated strategies for continuous replication in immunocompetent hosts. These strategies include the ability of the viruses to integrate their genome into host DNA, select cells of the monocyte/macrophage lineage as target cells, and activate gene expression only upon terminal differentiation of the target cell. This ensures a reservoir for virus. These factors present a formidable combination of virulence factors against which the antimicrobial systems of the host have very little defense. The ineffectiveness of the cellular arm of the immune response is evident by lack of obvious specific cellular cytotoxic capabilities of the infected animal. The ability of the virus to either sequester its neutralization epitopes or change them rapidly enough that it keeps ahead of potential neutralizing capabilities of antibodies similarly nullifies the humoral arm of the immune response. The immunopathological responses that result from persistent presentation of viral antigens by infected macrophages and the elaboration of LV-interferon is a two-edged sword. On one hand, it perpetuates the lymphoproliferative pathological response, and on the other, it keeps the expression of the viral genome in check at all times. This highly regulated and restricted expression of viral genome with its attendant slowly progressive pathological processes is the hallmark of the infection and disease for which these etiological agents are so aptly named.

The *env* genes of antigenically distinct, but closely related, strains of visna virus have small numbers of unique sequences. These have been shown to be point mutations that occur in viruses recovered from infected immune sheep or from infected cell cultures maintained in medium containing neutralizing antibodies. Further, the mutations are associated with escape of the virus from neutralizing antibodies and thus are selected by these antibodies (see Section VII). Two mechanism may be involved in the development of these mutants. First, they may arise as a result of the high intrinsic mutation rate (approximately 1×10^4) of retroviruses whose RTases lack editing functions. The higher prevalence of mutations in *env* in comparison to other viral genes may reflect the compatibility of mutations in the former regions with survival of the virus in the host in contrast to mutations in other genes that may result in nonviable viuses. A second possibility is that the *env* gene contains highly mutable regions ("hot spots") and that some of these occur among sequences encoding neutralization epitopes. These viruses are then selected by appropriate neutralizing antibodies. This is supported by the accumulation of mutations in viruses that become progressively more antigenically distinct during evolution in a persistently infected animal whose neutralizing antibody spectrum becomes wider with time. Random, noncumulative mutations occur in the *env* gene of antigenic variant viruses. These mutations were not associated with neutralization and were identifiable only by monoclonal antibodies to epitopes in the glycoproteins.

ACKNOWLEDGMENTS

These studies have been supported by grants NS12127, NS16145, and NS21916 and gift funds from the Hamilton Roddis Foundation. We thank Linda Kelly for preparing this manuscript.

REFERENCES

1. Sigurdsson, B. (1954) Observations on three slow infections of sheep. Maedi, paratuberculosis, rida, a slow encephalitis of sheep with general remarks on infections which develop slowly, and some of their special characterizations. *Br. Vet.* 110:255–270, 307–333, 341–354.
2. Masters, C. L., Gajdusek, D. C., and Gibbs, C. J. (1981). Creutzfeldt-Jakob disease isolations from the Gerstmann-Straussler syndrome with an analysis

of the various of amyloid plaque deposition in the virus-induced spongiform encephalopathies. *Brain* 104:559–588.

3. Gonda, M. A., Wong-Staal, F., Gallo, R. C., Clements, J. E., Narayan, O., and Gilden, R. (1985). Sequence homology and morphologic similarity of HTLV III and visna virus, a pathogenic lentivirus. *Science* 227:173–177.

4. Hadlow, W. J., Race, R. E., Kennedy, R. C., and Eklund, C. M. (1979). Natural infection of sheep with scrapie virus. In: *Slow Transmissable Diseases of the Nervous System*, Vol. 2 (S. B., Prusiner and W. J. Hadlow, eds.) pp. 3–12. New York, Academic Press.

5. Gudnadottir, M. (1974). Visna-maedi in sheep. *Prog. Med. Virol.* 18:336–349.

6. Haase, A. T. (1975). The slow infection caused by visna virus. *Curr. Topics Microbiol. Immunol.* 72:101–156.

7. Narayan, O., and Cork, L. C. (1985). Lentiviral diseases of sheep and goats: Chronic pneumonia leukoencephalomyelitis and arthritis. *Rev. Infect. Dis.* 7:89–98.

8. Deng, P., Cutlip, R. C., Lehmkuhl, H. O., and Brogden, K. A. (1986). Ultrastructure and frequency of mastitis caused by ovine progressive pneumonia virus in sheep. *Vet. Pathol.* 23:184–189.

9. Kennedy-Stoskopf, S., Narayan, O., and Strandberg, J. D. (1985). The mammary gland as a target organ for infection with caprine arthritis encephalitis virus. *J. Comp. Pathol.* 95:609–617.

10. Dawson, M. (1980). Maedi/visna: A review. *Vet. Rec.* 106:212–216.

11. Sigurdson, B., Palsson, P. A., and Grissom, H. (1957). Visna, a demyelinating transmissable disease of sheep. *J. Neuropathol. Exp. Neurol.* 16:389–403.

12. Sheffield, D. W., Narayan, O., Strandberg, J. D., and Adams, R. J. (1980). Visna-maedi like disease in a Corriedale sheep associated with an ovine retrovirus infection. *Vet. Pathol.* 17:544–552.

13. Narayan, O., Griffin, D. E., and Silverstein, A. (1977). Slow virus infection: Replication and mechanism of persistence of visna virus in sheep. *J. Infect. Dis.* 135:800–806.

14. Crawford, T. B., Adams, D. S., Cheevers, W. P., and Cork, L. C. (1980). Chronic arthritis in goats caused by a retrovirus. *Science* 207:997–999.

15. Crawford, T. B., Adams, D. S., Sande, R. D., Gorham, J. R., and Hanson, J. B. (1980). The connective tissue component of the caprine arthritis encephalitis syndrome. *Am. J. Pathol.* 100:443–450.

16. Cork, L. C., Hadlow, W. J., Crawford, T. B., Gorham, J. R., and Piper, R. C. (1974). Infectious leukoencephalomyelitis of young goats. *J. Infect. Dis.* 129:123–141.

17. Sundquist, B., Jonsson, L., Jacobson, S. O., and Hammerberg, K. E. (1981). Visna virus meningoencephalomyelitis in goats. *Acta Vet. Scand.* 22:315–330.

18. Sigurdsson, B., Thormar, H., and Palsson, P. A. (1960). Cultivation of visna virus in tissue culture. *Arch. Virusforsch.* 10:368–381.

19. Thormar, H. (1965). Effect of 5-bromodeoxyuridine and actinomycin D on the growth of visna virus in cell cultures. *Virology* 26:36–43.

20. Lin, F. H., and Thormar, H. (1971). Characterization of ribonucleic acid from visna virus. *J. Virol.* 7:582–587.

21. Haase, A. T., and Varmus, H. E. (1973). Demonstration of DNA provirus in the lytic growth of visna virus. *Nature New Biol.* 245:237–239.

22. Thormar, H. (1961). Electron microscope study of tissue cultures infected with visna virus. *Virology* 14:463–475.

23. Thormar, H. (1969). Physical, chemical and biological properties of visna virus and its relationship to other animal viruses. Slow, latent and temperature virus infections. *Natl. Inst. Neurol. Dis. Blindness Mono.*, pp. 235–340.

24. Harter, D. H., and Choppin, P. W. (1967). Cell fusing activity of visna virus particles. *Virology* 31:279–288.

25. Narayan, O., Clements, J. E., Strandberg, J. D., Cork, L. C., and Griffin, D. E. (1980). Biological characterization of the virus causing leukoencephalitis and arthritis in goats. *J. Gen. Virol.* 41:343–352.

26. Clements, J. E., Narayan, O., and Cork, L. C. (1980). Biochemical characterization of the virus causing leukoencephalitis and arthritis in goats. *J. Gen. Virol.* 50:423–427.

27. Pyper, J. M., Clements, J. E., Molineaux, S. M., and Narayan, O. (1984). Genetic variation among ovine caprine lentiviruses: Homology between visna virus and caprine arthritis encephalitis virus is confined to the 5' GAG-POL region and a small portion of the ENV gene. *J. Virol.* 51:713–721.

28. Pyper, J. M., Clements, J. E., Gonda, M. A., and Narayan, O. (1986). Sequence homology between cloned caprine arthritis-encephalitis virus and visna virus, two neurotropic lentiviruses. *J. Virol.* 58:665–670.

29. Narayan, O., Wolinsky, J. S., Clements, J. E., Strandberg, J. D., Griffin, D. E., and Cork, L. C. (1982). Slow virus replication: The role of macrophages in the persistence and expression of visna viruses of sheep and goats. *J. Gen. Virol.* 59:345–356.

30. Klevjer-Anderson, P., and Cheevers, W. P. (1981). Characterization of the infection of caprine synovial membrane cells by the retrovirus caprine arthritis encephalitis virus. *Virology* 110:113–119.

31. Hess, J. L., Clements, J. E., and Narayan, O. (1985). *Cis* and *trans*-acting transcriptional regulation of visna virus. *Science* 229:482–485.

32. Klevjer-Anderson, P., and McGuire, T. C. (1982). Neutralizing antibody response of rabbits and goats to caprine arthritis encephalitis virus. *Infect. Immun.* 38:455–461.

33. Narayan, O., Sheffer, D., Griffin, D. E., Clements, J. E., and Hess, J. (1984). Lack of neutralizing antibody to caprine arthritis encephalitis lentivirus in persistently infected goats can be overcome by immunization with inactivated mycobacterium tuberculosis. *J. Virol.* 49:349–355.

34. Oliver, R. E., Gorham, J. R., Parish, S. F., Hadlow, W. J., and Narayan, O. (1981). Studies on ovine progressive pneumonia. I. Pathologic and virology studies on the naturally occurring disease. *Am. J. Vet. Res.* 42:1554–1559.

35. Ellis, T. M., Wilcox, G. E., and Robinson, W. F. (1985). Characteristics of cell fusion induced by a caprine retrovirus. *Arch. Virol.* 86:263-273.
36. Clements, J. E., and Narayan, O. (1981). A physical map of the linear unintegrated DNA of visna virus. *Virology* 113:412-415.
37. Harris, J. D., Scott, J. V., Traynor, B., Brahic, M., Stowring, L., Ventura, P., Haase, A. T., and Peluso, R. (1981). Visna virus DNA: Discovery of a novel gaped structure. *Virology* 113:573-583.
38. Yaniv, A., Dahlberg, J. E., Tronick, S. R., Chiu, I-M., and Aaronson, S. A. (1985). Molecular cloning of integrated caprine arthritis-encephalitis virus. *Virology* 145:340-345.
39. Molineaux, S., and Clements, J. E. (1983). Molecular cloning of unintegrated visna viral DNA and characterization of frequent deletions in the 3' terminus. *Gene* 23:137-148.
40. Sonigo, P., Alizon, M., Staskus, K., Klatzman, D., Cole, S., Danos, O., Retzel, E., Tiollais, P., Haase, A., and Wain-Hobson, S. (1985). Nucleotide sequence of the visna lentivirus: Relationship to the AIDS virus. *Cell* 42: 369-382.
41. Ratner, L., Haseltine, W., Patarca, R., Livak, K. J., Starcich, B., Josephs, S. F., Doran, E. R., Rafalski, J. A., Whitehorn, E. A., Baumeister, K., Ivanoff, L., Petteway, S. R., Jr., Pearson, M. L., Lavtenberger, J. A., Papas, T. S., Ghrayeb, J., Chang, N. T., Gallo, R. C., and Wong-Staal, F. (1985). Complete nucleotide sequence of the AIDS virus, HTLV-III. *Nature (London)* 313:277-284.
42. Sanchez-Pescador, R., Power, M. D., Barr, P. J., Steimer, K. S., Stempien, M. M., Brown-Shimer, S. L.,Gee, W. W., Renard, A., Randolph, A., Levy, J. A., and Luciw, P. A. (1985). Nucleotide sequence and expression of an AIDS-associated retrovirus (ARV-2). *Science* 227:484-492.
43. Wain-Hobson, S., Sonigo, P., Danos, O., Cole, S., and Alizon, M. (1985). Nucleotide sequence of the AID virus, LAV. *Cell* 40:9-17.
44. Rushlow, K., Olsen, K., Steigler, G., Payne, S. L., Montelaro, R. C., and Issel, C. J. (1986). Lentivirus genomic organization: The complete nucleotide sequence of the env region of equine infectious anemia virus. *Virology* 155:309-321.
45. Hess, J. L., Pyper, J. M., and Clements, J. E. (1986). Nucleotide sequence and transcriptional activity of the CAEV long terminal repeat. *J. Virol.* 60: 385-393.
46. Derse, D., Dorn, P. L., Levy, L., Stephens, R. M., Rice, N. R., and Casey, J. W. (1987). Characterization of equine infectious anemia virus long terminal repeat. *J. Virol.* 61:743-747.
47. Vigne, R., Filippi, P., Querat, G., Sauze, N., Vitu, C., Russo, P., and DeLori, P. (1982). Precursor polypeptides to structural proteins of visna virus. *J. Virol.* 42:1046-1056.
48. Stanley, J., Bhaduri, L. M., Narayan, O., and Clements, J. E. (1987). Topographical rearrangements of visna virus envelope glycoprotein during antigenic drift. *J. Virol.* 61:1019-1028.
49. Narayan, O., Griffin, D. E., and Chase, J. (1977). Antigenic shift of visna virus in persistently infected sheep. *Science* 197:376-378.

50. Narayan, O., Griffin, D. E., and Clements, J. E. (1978). Virus mutation during "slow infection." Temporal development and characterization of mutants of visna virus recovered from sheep. *J. Gen. Virol.* 41:343–352.

51. Narayan, O., Clements, J. E., Griffin, D. E., and Wolinsky, J. S. (1981). Neutralizing antibody spectrum determines the antigenic profiles of emerging mutants of visna virus. *Infect. Immun.* 32:1045–1050.

52. Clements, J. E., D'Antonio, N., and Narayan, O. (1982). Genomic changes associated with antigenic variation of visna virus. II. Common nucleotide changes detected in variants from independent isolations. *J. Mol. Biol.* 158: 415–434.

53. Narayan, O., Clements, J. E., Kennedy-Stoskopf, S., and Royal, W. (1987). In: *Antigenic Variation: Molecular and Genetic Mechanisms of Relapsing Disease* (J. M. Cruze, ed.). Basel, S. Karger Publ., in press.

54. Narayan, O., Clements, J. E., Kennedy-Stoskopf, S., and Royal, W. (1986). In: *Antigenic Variation in Infectious Diseases* (T. H. Birkbeck and C. W. Penn, eds.). Oxford, IRL Press.

55. Narayan, O., Kennedy-Stoskopf, S., Sheffer, D., Griffin, D. E., and Clements, J. E. (1983). Activation of caprine arthritis-encephalitis virus expression during maturation of monocytes to macrophages. *Infect. Immun.* 41:67–73.

56. Clements, J. E., Petersen, F. S., Narayan, O., and Haseltine, W. S. (1980). Genomic changes associated with antigenic variation of visna virus during persistent infection. *Proc. Natl. Acad. Sci. USA* 77:4454–4457.

57. Weiss, R., Teich, N., Varmus, H., and Coffin, J. (1982). *RNA Tumor Viruses: Molecular Biology of Tumor Viruses*, 2nd ed. New York, Cold Spring Harbor Laboratory.

58. Davis, J. L., Molineaux, S., and Clements, J. E. (1987). Visna virus exhibits a complex transcriptional pattern: One aspect of gene expression shared with the acquired immunodeficiency syndrome retrovirus. *J. Virol.* 61: 1325–1331.

59. Muesing, M. A., Smith, D. H., Cabradilla, C. D., Benton, C. V., Lasky, L. A., and Capon, D. J. (1985). Nucleic acid structure and expression of the human AIDS/lymphadenopathy retrovirus. *Nature* 313:450–463.

60. Rabson, A. B., Daugherty, D. F., Venkatesan, S., Baulukos, K. E., Benn, S. I., Folks, T. M., Feorino, P., and Martin, M. A. (1985). Transcription of novel open reading frames of AIDS retrovirus during infection of lymphocytes. *Science* 229:1388–1390.

61. Haase, A. T., Stowring, L., Harris, J. D., Traynor, B., Ventura, P., Peluso, R., and Brahic, M. (1982). Visna virus DNA synthesis and the tempo of infection in vitro. *Virology* 119:399–410.

62. Harris, J. D., Scott, J. V., Traynor, B., Stowring, L., Ventura, P., Haase, A. T., and Peluso, R. (1981). Visna virus DNA: Discovery of a naval gaped structure. *Virology* 113:573–583.

63. Clements, J. E., Narayan, O., Griffin, D. E., and Johnson, R. T., (1979). The synthesis and structure of visna virus DNA. *Virology* 93:377–386.

64. Panganiban, A. T., and Temin, H. M. (1983). The terminal nucleotides of retrovirus DNA are required for integration but not virus production. *Nature* 306:155-160.
65. Gendelman, H. E., Narayan, O., Molineaux, S., Clements, J. E., and Ghotbi, Z. (1985). Slow persistent replication of lentiviruses: Role of tissue macrophages and macrophage-precursors in bone marrow. *Proc. Natl. Acad. Sci. USA* 82:7086-7090.
66. Arya, S. K., Guo, C., Josephs, S. F., and Wong-Staal, F. (1985). *Trans*-activator gene of human T-lymphotropic virus type III (HTLV-III). *Science* 229:69-73.
67. Cullen, B. R. (1986). *Trans*-activation of human immunodeficiency virus occurs via a bimodal mechanism. *Cell* 46:973-982.
68. Dayton, A. I., Sodroski, J. G., Rosen, C. A., Goh, W. C., and Haseltine, W. A. (1986). The *trans*-activator gene of the human T cell lymphotrophic virus type III is required for replication. *Cell* 44:941-947.
69. Fisher, A. G., Feinberg, M. B., Josephs, S. F., Harper, M. E., Marselle, L. M., Reyes, G., Gonda, M. A., Aldovini, A., Debouk, C., Gallo, R. C., and Wong-Staal, F. (1986). The *trans*-activator gene of HTLV-III is essential for virus replication. *Nature* 320:367-371.
70. Sodroski, J., Patarca, R., Rosen, C., Wong-Staal, F., and Haseltine, W. (1985). Location of the *trans*-activating region on the genome of human T-cell lymphotropic virus type III. *Science* 74-77.
71. Feinberg, M. B., Jarrett, R. F., Aldovini, A., Gallo, R. C., and Wong-Staal, F. (1986). HTLV-III expression and production involve complex regulation at the levels of splicing and translation of viral RNA. *Cell* 46:807-817.
72. Sodroski, J., Goh, W. C., Rosen, C., Dayton, A., Terwilliger, E., and Haseltine, W. (1986). A second post-translational *trans*-activator gene required for HTLV-III replication. *Nature* 321:412-417.
73. Haase, A. T., Stowring, L., Narayan, O., Griffin, D., and Price, D. (1977). Slow persistent infection caused by visna virus: Role of host restriction. *Science* 195:175-177.
74. Brahic, M., Stowring, L., Ventura, P., and Haase, A. T. (1981). Gene expression in visna virus infection in sheep. *Nature* 292:240-242.
75. Palsson, P. A. (1976). Maedi and visna in sheep. In: *Slow Viral Diseases of Animals and Man* (R. H. Kimberlin, ed)., pp. 17-43. Amsterdam, North-Holland.
76. Houwers, D. J. (1985). Experimental maedi/visna control in the Netherlands. In: *Slow Viruses in Sheep and Goats and Cattle* (J. M. Shart and R. Hoff-Jorgensen, eds), pp. 115-21. Luxembourg, Commission of the European Communities.
77. Crawford, T. B., and Adams, D. S. (1981). Caprine arthritis-encephalitis: Clinical features and presence of antibody in selected goat populations. *J. Am. Vet. Med. Assoc.* 178:713-719.
78. Adams, D. S., Klevjer-Anderson, P., Carlson, J. L., McGuire, T. C., and Gorham, J. R. (1983). Transmission and control of caprine arthritis-encephalitis virus. *Am. J. Vet. Res.* 44:1670-1675.

79. Adams, D. S., Oliver, R. E., Ameghino, E., DeMartini, J. C., Verwoerd, D. W., Houwers, D. J., Waghela, S., Gorham, J. R., Hyllseth, B., Dawson, M., Trigo, F., and McGuire, T. C. (1984). Global survey of serologic evidence of caprine arthritis-encephalitis virus infections. *Vet. Rec.* 115:494–495.

80. Anderson, L. W., Klevjer-Anderson, P., and Liggitt, H. D. (1983). Susceptibility of blood-derived monocytes and macrophages to caprine arthritis-encephalitis virus. *Infect. Immun.* 41:837–840.

81. Gendelman, H. E., Narayan, O., Kennedy-Stoskopf, S., Kennedy, P. G. E., Ghotbi, Z., Clements, J. E., Stanley, J., and Pezeshkpour, G. (1986). Tropism of sheep lentiviruses for monocytes: Susceptibility to infection and virus gene expression increase during maturation of monocytes to macrophages. *J. Virol.* 58:67–74.

82. Gendelman, H. E., Narayan, O., Kennedy-Stoskopf, S., Clements, J. E., and Pezeshkpour, G. H. (1984). Slow virus-macrophage interactions: Characterization of a transformed cell line of sheep alveolar macrophages that express a marker for susceptibility to ovine-caprine lentivirus infection. *Lab. Invest.* 547–555.

83. vanFurth, R. (1980). Development of mononuclear phagocytes. In: *Heterogeneity of Mononuclear Phagocytes* (O. Forster and M. Landy, eds.), pp. 3–10. New York, Academic Press.

84. Stowring, L., Haase, A. T., Petursson, G., Georgsson, G., Palsson, P. A., Lutley, R., Roos, R., and Szuchet, S. (1985). Detection of visna virus antigens and RNA in glial cells in foci of demyelination. *Virology* 141: 311–318.

85. Petursson, G., Nathanson, N., Georgsson, G., Panitch, H., and Palsson, P. A. (1976). Pathogenesis of visna. I. Sequential virologic, serologic and pathologic studies. *Lab. Invest.* 35:402–412.

86. Narayan, O., Sheffer, D., Clements, J. E., and Tennekoon, G. (1985). Restricted replication of lentiviruses: Visna viruses induce a unique interferon during interaction between lymphocytes and infected macrophages. *J. Exp. Med.* 162:1954–1969.

87. Kennedy, P. G. E., Narayan, O., Ghotbi, Z., Hopkins, J., Gendelman, H. E., and Clements, J. E. (1985). Persistent expression of Ia antigen and viral genome in visna-maedi virus-induced inflammatory cells: Possible role of lentivirus-induced interferon. *J. Exp. Med.* 162:1970–1982.

88. Dahlberg, J. E., Gaskin, J. M., and Perk, K. (1981). Morphological and immunological comparison of caprine arthritis encephalitis and ovine progressive pneumonia viruses. *J. Virol.* 39:914–919.

89. Gogolewski, R. P., Adams, D. S., McGuire, T. C., Banks, K. L., and Cheevers, W. P. (1985). Antigenic cross-reactivity between caprine arthritis-encephalitis, visna and progressive pneumonia viruses involves all virion-associated proteins and glycoproteins. *J. Gen. Virol.* 66:1233–1240.

6

Adult T-Cell Leukemia/Lymphoma

Kiyoshi Takatsuki, Kazunari Yamaguchi, and Toshio Hattori *Kumamoto University Medical School, Kumamoto, Japan*

I. INTRODUCTION

Adult T-cell leukemia/lymphoma (ATL) was established as a distinct clinical entity in Japan. In about 1973 we came to recognize the presence of ATL, previously an unknown disease. This disease was internationally acknowledged in 1977 (1,2).

ATL shares some features of Sézary syndrome, but is a distinct entity. The disease is characterized as follows: (a) onset in adulthood; (b) ATL cells, derived from mature helper T cells, showing typical morphological features with deep indented nuclei; (c) frequent association of skin involvement, lymphadenopathy, and hepatosplenomegaly; (d) high cell count in peripheral blood without anemia and with little involvement of bone marrow; (e) hypercalcemia; and (f) clustering of birthplaces of patients. The most unique of these features is the clustering of birthplaces of ATL patients in southwestern Japan. Successive isolations of human T-cell lymphotropic virus type I (HTLV-I) (3-5) from cell lines established from ATL patients revealed that ATL is caused by HTLV-I infection and that southwestern Japan is the most endemic area in the world. ATL is the first and only human malignancy that has been proved to be caused by retrovirus infection.

The etiological association of HTLV-I with ATL is based on the following observations (6,7): (a) the areas of high incidence of ATL corresponded closely with those of high prevalence of HTLV-I infection in Japan; (b) all individuals with ATL had antibodies against HTLV-I; (c) HTLV-I proviral DNA was demonstrated in ATL neoplastic cells; and (d) HTLV-I immortalized human T cells. HTLV-I was thus proved to be directly associated with human malignancy.

Immediately after the discovery of ATL, acquired immune deficiency syndrome (AIDS) began to be recognized, and a retrovirus was also found to be a pathogen in this disease. The advances made in the study of AIDS owe much to the knowledge gained about the relationship between ATL and HTLV-I. For example, almost identical methods to those for isolation of HTLV-I have been used for isolation of human immunodeficiency virus (HIV) (8). HIV is tropic not only for helper T cells, but also for neurological tissues. Recently, HTLV-I was also proved to be etiologically associated with neurological diseases called "tropical spastic paraparesis" or "HTLV-I-associated myelopathy" (9,10). Thus, these two human retroviruses are tropic both to helper T cells and to neurological tissues.

II. CLINICAL FEATURES OF ATL

We studied 187 patients with ATL, 113 males and 74 females (1.5:1), in Kyushu, Japan. Their ages at onset ranged from 27 to 82 years (average 55 years).

The predominant physical findings were as follows: peripheral lymph node enlargement (72%), hepatomegaly (47%), splenomegaly (26%), and skin lesions (53%). Hypercalcemia (28%) was frequently found in association (11). Other symptoms at onset were abdominal pain, diarrhea, pleural effusion, ascites, cough, sputum, and abnormal shadow on chest X-ray films. White blood cell counts ranged from normal to 500×10^9/liter. Leukemic cells resembled Sézary cells, having indented or lobulated nuclei. The surface phenotype of ATL cells characterized by monoclonal antibodies was CD3+, CD4+, CD8-, and CD25 (Tac)+. Anemia and thrombocytopenia were rare. The survival time in acute and lymphoma-type ATL ranged from 2 weeks to more than 1 year. The causes of death were pulmonary complications including *Pneumocystis carinii* pneumonia (12), hypercalcemia, *Cryptococcus* meningitis, disseminated herpes zoster, and disseminated intravascular coagulopathies. All patients were positive for anti-HTLV-I antibodies and HTLV-I proviral DNA in the leukemic/lymphoma cells. Smoldering ATL is characterized by the long duration of a few ATL cells in the peripheral blood (13).

We also compared the clinical, hemotological, immunological, cytogenetic, and virological features of Japanese ATL patients with those of 24 Caribbean and African ATL patients diagnosed in the United Kingdom. These features and the clinical course, ATL subtype, frequency of hepercalcemia and opportunistic infections, cell morphology, phenotypic profile, and response to treatment were the same in both groups of patients. The only difference was the age of onset: the average for the Caribbean and African ATL patients was 43 years (ranging from 19 to 62), while that for the Japanese patients was higher (14). Blattner et al. (15) and Gibbs et al. (16) reported that the mean ages of ATL patients in the Unites States and Jamaica were 43 and 40, respectively.

III. CLASSIFICATION OF ATL

ATL patients are classified into five subtypes according to the clinical picture: acute, chronic, smoldering, crisis, and lymphoma types (17), as illustrated in Figure 1. There are intermediated states which cannot be definitely allocated to any of those five types.

The acute type is so-called prototypic ATL, which progresses acutely or subacutely. Patients with this type exhibit increased ATL cells, skin lesions, systemic lymphadenopathy, and hepatosplenomegaly. Most of them are resistant to combination chemotherapy using, for example, vincristine, cyclophosphamide, prednisolone, doxorubicin, and sometimes methotrexate. In general, a poor prognosis is indicated by the elevation of serum lactic dehydrogenase (LDH), calcium, and bilirubin, as well as by high white blood cell counts.

Chronic-type ATL patients have an increased white cell count. In a few of these patients, slight lymphadenopathy and hepatosplenomegaly are observed; elevation in serum LDH is also noted, but this is not associated with hypercalcemia or hyperbilirubinimia.

Smoldering ATL is characterized by the presence of a few ATL cells (0.5–3%) in the peripheral blood for a long period (13). Patients frequently have skin lesions as premonitory symptoms. The serum LDH values are within normal range and not associated with hypercalcemia. Lymphadenopathy, hepatosplenomegaly, and bone marrow infiltration are very slight. The HTLV-I proviral DNA is often integrated monclonally in patients with this type of ATL. However, it is difficult to draw a clear border between HTLV-I carriers.

Smoldering ATL and chronic-type ATL often progress to acute-type ATL after a long duration. This progress is termed "crisis" in ATL.

Lymphoma-type ATL is characterized by prominent lymphadenopathy. This type has been diagnosed as nonleukemic malignant lymphoma (18).

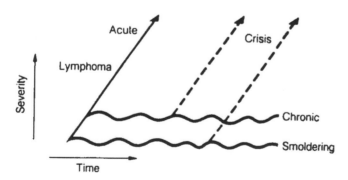

FIGURE 1   Heterogeneity in the clinical course of ATL.

IV. DIAGNOSIS OF ATL

A. Morphology of ATL Cells

Numerous abnormal lymphocytes, which vary considerably in size and cytoplasmic basophility, are seen in acute-type ATL. Cells from 30% of patients of this type possess small vacuoles, but not azurophilic granules. Most of these cells characteristically exhibit lobular division of their nuclei, most of which are bi- or multifoliate and marked by deep indentation. Cells with such a nuclear configuration are known as "flower cells." Leukemic cells with relatively coarsely clumped nuclear chromatin are not morphologically similar.

Cells from chronic ATL, relatively uniform in size and nuclear configuration, are smaller than those from acute- or smoldering-type ATL. Cells rarely possess small vacuoles and have no azurophilic granules. Cells in this type also exhibit lobular division of their nuclei; however, most of the lobulated neclei are bi- or trifoliate. Many cells exhibit deep indentation rather than lobulation. The nuclear chromatin is in coarse strands and deeply stained. The nucleocytoplasmic ratio is larger than in normal lymphocytes.

The proportion of abnormal cells in the peripheral blood varies from 0.5 to 3% in smoldering-type ATL. Cells from this type are realtively large and have no cytoplasmic granules or vacuoles. The lobulated nuclei are bi- or trifoliate; some of the nuclei exhibit indentation or clefts, which appear as a ridge formation. The nucleocytoplasmic ratio is large. The nuclear chromatin is in coarse strands and deeply stained.

B. Serology of HTLV-I

Serum specimens from all patients with ATL have anti-HTLV-I antibodies (19). The pattern of serum antibodies is not different in ATL patients and HTLV-I carriers. The presence of serum antibodies to HTLV-I was demonstrated by indirect immunofluorescence test, enzyme-linked immunosorbent assay, passive hemaggultination assay, and Western blot method.

C. Phenotypic Markers of ATL Cells

Immunological studies of ATL cells have revealed phenotypic heterogeneity of T-cell subset markers (10). However, most ATL cells have the CD4 antigen. There are few reports of different marker leukemic cells, e.g., CD4+ and CD8+, CD4- and CD8- phenotypes. Despite a helper/inducer phenotype, ATL cells suppress pokeweed mitogen-induced immunoglobulin synthesis by normal lymphocytes.

D. HTLV-I Provirus Genome in Leukemic Cells of ATL

The presence of HTLV-I provirus DNA was examined by a modified Southern blot method and was confirmed in peripheral blood mononuclear cells and/or lymph node cells from all patients with ATL and some T-cell malignant lymphomas (21,22). Without exception, the tumor cells were monoclonal with respect to the site of provirus integration. However, no provirus DNA was detected in HTLV-I carriers, or in anti-HTLV-I antibody-positive hematological patients, not exhibiting ATL, who had been infected by multiple blood transfusions. These results strongly suggest that HTLV-I is associated with malignant transformation of the infected cells, although the infection does not necessarily induce the leukemia. The detection of HTLV-I proviral DNA may therefore be useful for the diagnosis of ATL in an endemic area and provide a powerful tool for the classification of T-cell malignancies.

V. OTHER DISEASES ACCOMPANYING HTLV-I INFECTION

Infection with HTLV-I is a direct cause of ATL. However, infection with this virus can indirectly cause many other diseases, presumably due to the induction of immunodeficiency, such as chronic lung diseases, opportunistic lung infections, cancer of other organs (23), monoclonal gammopathy (24), chronic renal failure (25), strongyloidiasis (26), nonspecific intractable dermatomycosis, and nonspecific lymp node swelling.

Soon after we began to study immunological abnormalities of HTLV-I carriers, we started to notice the frequent complication of ATL, particularly the smoldering type, involving cancer of other organs. This led us to examine the HTLV-I seroprevalence in cancer patients in cooperation with many hospitals in Kumamoto Prefecture (23). Among the 394 patients with malignancies who had not had blood transfusions, 61 (15.5%) tested positive for HTLV-I antibody. The prevalence was significantly higher in men older than age 40 years and women of all ages compared to age- and sex-matched healthy individuals. There was no significant correlation between the site of malignancy and antibody prevalence. These results suggest that HTLI infection may contribute to the risk of other malignancies.

The recently highlighted disease HTLV-I-associated myelopathy (HAM) (9) is considered to be synonymous with tropical spastic paraparesis (TSP) (10). Elucidation of the mechanism responsible for its onset is now an important subject.

VI. EPIDEMIOLOGY OF ATL

Both HTLV-I and ATL have been shown to be endemic in some regions of the world, especially in southwest Japan (27,28), the Caribbean islands, the countries surrounding the Caribbean basin, and, probably, parts of Central Africa (15,29, 30). Many of the ATL patients studied in the United States and the United Kingdom came from Caribbean Sea countries. HTLV-I antibody-positive patients with T-cell malignancy have been reported in the United States, Italy, and Germany, although the number of such patients is small. HTLV-I is said to have its roots in Africa. However, no comprehensive data on HTLV-I carriers in Africa are available. Recently, HTLV-I carriers were found in some of the natives in Papua, New Guinea, and the Philippines. In China, almost no ATL patients have been reported, but considering its vast land and population, a large-scale survey is desirable. In Taiwan, the HTLV-I antibody-positive rate is slightly high in the northeast region, and a small number of ATL patients have been registered (31). In Korea, the HTLV-I antibody-positive rate is as low as 0.03%; HTLV-I carriers have been found in Pusan and Cheju Island, even though both are close to the Kyushu district of Japan (32).

We estimate that antibodies against HTLV-I are found in over one million persons in Japan, and approximately 500 new patients with ATL are diagnosed in a year.

Seroepidemiological studies in Japan indicated that the incidence of HTLV-I infection was variable among cities within the endemic region (33). This clustering is thought to be due to the limited transmission of virus between socially isolated populations. Anti-HTLV-I antibodies were found in the sera from 6–37% of healthy adults aged over 40 years in the endemic areas of ATL in Japan.

HTLV-I transmission occurs through three different modes. HTLV-I-infected mothers can transmit the virus to newborns via milk. Kinoshita et al. (34) demonstrated HTLV-I antigen in mononuclear cells from breast milk of HTLV-I-positive mothers after delivery. Nakano et al. (35) reported evidence for primary HTLV-I infection of infants from their mothers by follow-up studies. Of 16 infants born to seropositive mothers, 13 were found to have HTLV-I-infected cells in peripheral blood, whereas four of them were seroconverted from HTLV-I-negative to -positive 9–19 months after birth.

This virus can also be transmitted from male to female by sexual intercourse (36). The third route is through blood transfusion (37). The data pertaining to HTLV-I infection via transfusion indicate that infection does not occur unless live lymphocytes are transfused. With regard to vertical transmission from mother to infants, the lymphocytes contained in mother's milk are thought to carry HTLV-I. In the case of sexual transmission, lymphocytes in semen are thought to carry the virus. Some investigators claim that the virus can be transmitted by mosquitoes, but we do not agree with this view.

VII. FAMILIAL OCCURRENCE OF ATL

The familial occurrence of ATL and the prevalence of antibodies to HTLV-I in family members, which have already been reported by Ichimaru et al. (38), provide important information on the infectious nature of HTLV-I and its association with T-cell malignancies. We have examined the pedigree of a family including two ATL patients for anti-HTLV-I antibodies (36). Twenty-six relatives of the two ATL patients were available for examination. They were all healthy and hematologically normal. The results of this family study were as follows: (a) Two siblings suffered from ATL. (b) Among 26 healthy persons, seven were found to have anti-HTLV-I antibodies. (c) Eleven children and their spouses were negative, but one daughter of one ATL patient was positive. (d) In two of nine married couples, both partners were positive. Husband (-)/wife (+) was found in three couples; husband (+)/wife (-) was not found; and the wives of the two ATL patients were both positive. These results suggest two modes of transmission of HTLV-I: One is from parents to children, and the other is horizontal transmission between spouses, especially from husband to wife.

Three sisters, aged from 56 to 59 years, who developed lymphoma-type ATL during a 19-month period have been reported (39). They were born in Kumamoto Prefecture, an area where the incidence of malignant lymphoma is high. The histological diagnosis was diffuse medium-sized cell type in the elder sister and mixed cell type in the middle and younger sisters. These patients and their elder brother had serum antibodies against HTLV-I. As their lives and environments in adulthood were clearly different, HTLV-I infection may have occurred in their childhood. These findings suggest that the disease developed after a long latent period following the first viral infection.

VIII. PATHOPHYSIOLOGY OF ATL

Although ATL cells were etiologically associated with HTLV-I, HTLV-I-gene transcript or expression of HTLV-I-related antigens could not be detected in fresh ATL cells. Expression of HTLV-I gene was readily detectable immediately after the cells were transferred into culture systems, suggesting that here are unknown mechanisms by which expression of HTLV-I-related is suppressed in ATL cells (40). It is also possible that only ATL cells whose expression of these antigens is suppressed could escape the immunological surveillance of the infected host. Characterization of ATL cells by monoclonal antibodies revealed that most ATL cells were positive for CD4 and CD3 antigens and negative for CD8 and CD1, suggesting that they were derived from mature helper T cells. In addition, ATL cells expressed activated T-cell markers such as IL-2 receptor and HLA-DR antigen on their surface (20,41). These activated T-cell markers have been infrequently observed on T4-chronic lymphocytic leukemia cells or Sezary cells, which do not carry the proviral DNA of HTLV-I. The mechanisms of surface expression of these markers remain unclear. One responsible factor might be the "TAT" gene of HTLV-I, but negative expression of HTLV-I gene in fresh ATL cells makes this possibility less likely. Inolvement of CD3 and T-cell receptor complexes in proliferation of ATL cells could also be considered, because expression of these two antigens was reduced in fresh ATL cells (42). Progress in T-cell immunobiology has also given deep insight into the mechanism of proliferation of ATL cells. Autocrine proliferation of ATL cells by IL-2 was found in rare cases, although IL-2 did not support proliferation of ATL cells from most of the cases despite the expression of abundant IL-2-receptor gene (43). IL-1β-gene expression in the acute but not the chronic form of ATL cells was recently described, but again IL-1 did not support proliferation or IL-2 receptor expression of ATL cells (44). Activation of specific oncogenes or translocations of oncogenes to sites near the T-cell receptor locus were not found in ATL. Specific chromocomal abnormalities in ATL cells were also not found. The most common abnormalities in karyotypes of ATL were trisomy, trisomy 7, and loss of X chromosomes in the female. Frequencies of translocations involving chromosome 14 were controversial (45,46). In situ hybridization techniques disclosed that integrated sites of HTLV-I provirus were different in each patient. The random pattern of Southern blotting analysis of the integrated provirus also did not support the idea that HTLV-I is integrated to a specific site, activating cellular (onco)genes (47). Taking these findings together, it is speculated that expression of activated T-cell antigens on ATL cells may be important in interpreting leukemogenesis in ATL; however, it is difficult to determine the role of HTLV-I infection in producing ATL cells in activated T-cell states. HTLV-I might be one of the initiating factors for development of ATL.

Nevertheless, it is easy to imagine that ATL cells elaborate many cytokines, because they originate from activated T cells. These cytokines are presumed to play an important role in the pathophysiology of ATL. For example, elaborated IL-1 may promote bone resorption and act as one of the causative agents of hypercalcemia in ATL patients. IL-1 is well known to have multiple biological activities, such as may cause neutrophilia and fever. A high fever without overt infection and absolute neutrophilia were frequently found in ATL patients (48).

IX. TREATMENT OF ATL AND PREVENTION OF HTLV-I INFECTION

Patients with acute and lymphoma-type ATL should be treated with combination chemotherapy directed toward achieving a cure. However, patients showing hypercalcemia, high LDH levels, and an abnormal increase in white blood cells have a 50% survival time of less than 6 months (49). Even if aggressive combination chemotherapy is given, the prognosis does not improve. Patients often die of severe respiratory infection or hypercalcemia. On the other hand, regardless of treatment, chronic-type and smoldering-type ATL have a longer course. Aggressive chemotherapy may induce severe respiratory infection. Therefore, an independent treatment protocol for chronic and smoldering-type ATL, different from that for acute and lymphoma-type ATL, must be established.

There are now some reports of new treatment protocols. Waldmann et al. (50) administered toxin-labeled anti-Tac monoclonal antibodies to one patient and achieved a remission of six months. Deoxycoformycin (DCF) was successfully used as a single agent to treat a patient with acute ATL, resulting in an apparent long-lasting remission (41). DCF, a nucleoside analog produced by *Streptomyces antibioticus* or *Aspergillus nidulans*, is a potent tight-binding inhibitor of the enzyme adenosine deaminase (ADA). A therapeutic selectivity of DCF for lymphoid malignancies was inferred from the preferential lymphoid impairment in congenital ADA deficiency. We too have used this to treat seven patients with ATL (52). Prolonged complete remissions were documented in two patients. Patients treated with DCF developed profound lymphopenia but without myelosuppression. Serial measurements of serum ADA in the successfully treated patients showed good correlation with the levels of ATL cells during DCF treatment. The adverse effects of DCF therapy were gastrointestinal toxicity, nausea, and vomiting (52).

Although experiments have demonstrated that vaccination against HTLV-I is possible, there seems to be no necessity for vaccination. Screening of blood donors for anti-HTLV-I antibodies is necessary to prevent extension of HTLV-I from endemic areas and infection by an unnatural route, transfusion. The Japan Red Cross Blood Center started nation-wide screening for HTLV-I antibody testing by gelatin particle agglutination in November 1986. If HTLV-I is transmitted via breast milk, infants of HTLV-I carrier mothers should not become infected. HTLV-I carrier mothers are now instructed to refrain from breast feeding in some local communities within the framework of a pilot study.

REFERENCES

1. Takatsuki, K., Uchiyama, T., Sagawa, K., and Yodoi, J. (1977). Adult T-cell leukemia in Japan. In: *Topics in Hematology* (S. Seno, F. Takaku, and S. Irino, eds.), pp. 73-77. Amsterdam, Excerpta Medica.
2. Uchiyama, T., Yodoi, J., Sagawa, K., Takatsuki, K., and Uchino, H. (1977). Adult T-cell leukemia: Clinical and hematologic features of 16 cases. *Blood* 50:481-492.
3. Poiesz, B. J., Ruscetti, F. W., Gazdar, A. F., Bunn, P. A., Minna, J. D., and Gallo, R. C. (1980). Detection and isolation of type C retrovirus particles from fresh and cultured lymphocytes of a patient with cutaneous T-cell lymphoma. *Proc. Natl. Acad. Sci. USA* 77:7415-7419.
4. Hinuma, Y., Nagata, K., Hanaoka, M., Nakai, M., Matsumoto, T., Kinoshita, K., Shirakawa, S., and Miyoshi, I. (1981). Adult T-cell leukemia: Antigen in an adult T-cell leukemia cell line and detection of antibodies to the antigen in human sera. *Proc. Natl. Acad. Sci. USA* 78:6476-6480.
5. Miyoshi, I., Kubonishi, I., Sumida, M., Hiraki, S., Tsubota, T., Kimura, I., Miyamoto, K., and Sato, J. (1980). A novel T-cell line derived from adult T-cell leukemia. *Jpn. J. Cancer Res. (Gann)* 71:155-156.

6. Yoshida, M., Miyoshi, I., and Hinuma, Y. (1982). Isolation and character-
    ization of retrovirus from cell lines of human adult T-cell leukemia and its
    implication in the disease. *Proc. Natl. Acad. Sci. USA* 79:2031-2035.
7. Wong-Staal, F., and Gallo, R. C. (1985). Human T-lymphotropic retro-
    viruses. *Nature* 317:395-403.
8. Barre-Sinoussi, F., Chermann, J. C., Rey, F., Nugeyre, M. T., Chamaret, S.,
    Grust, J., Dauguet, C., Azler-Blin, C., Vé zinst-Brun, F., Rouzioux, C.,
    Rozenbaum, W., and Montagnier, L. (1983). Isolation of a T-lymphotropic
    retrovirus from a patient at risk for acquired immune deficiency syndrome
    (AIDS). *Science* 220:868-871.
9. Gessain, A., Barin, F., Vernant, J. C., Gout, O., Maurs, L., Calender, A., and
    de Thé, G. (1985). Antibodies to human T-lymphotropic virus type-I in
    patients with tropical spastic paraparesis. *Lancet* 2:407-410.
10. Osame, M., Usuku, K., Izumo, S., Ihichi, N., Amitani, H., Igata, A., Matsu-
    moto, M., and Tara, M. (1986). HTLV-I associated myelopathy, a new
    clinical entity. *Lancet* 1:1031-1032.
11. Kiyokawa, T., Yamaguchi, K., Takeya, M., Takahashi, K., Watanabe, T.,
    Matsumoto, T., Lee, S. Y., and Takatsuki, K. (1987). Hypercalcemia and
    osteoclast proliferation in adult T-cell leukemia. *Cancer* 59:1187-1191.
12. Yoshioka, R., Yamaguchi, K., Yoshinaga,T., and Takatsuki, K. (1985).
    Pulmonary complications in patients with adult T-cell leukemia. *Cancer*
    55:2491-2494.
13. Yamaguchi, K., Nishimura, H., Kohrogi, H., Jono, M., Miyamoto, Y., and
    Takatsuki, K. (1983). A proposal for smoldering adult T-cell leukemia: A
    clinicopathologic study of five cases. *Blood* 62:758-766,
14. Yamaguchi, K., Matutes, E., Catovsky, D., Galton, D. A. G., Nakada, K.,
    and Takatsuki, K. (1987). Strongyloides stercoralis as candidate co-factor
    for HTLV-I induced leulaemogenesis. *Lancet* 2:94-95.
15. Blattner, W. A., Kalyanaraman, V. S., Robert-Guroff, M., Lister, T. A.,
    Galton, D. A., Sarin, P. S., Crawfor, M. H., Catovsky, D., Greaves, M., and
    Gallo, R. C. (1982). The human type-C retrovirus, HTLV, in blacks from
    the Caribbean region, and relationship to adult T-cell leukemia/lymphoma.
    Int. J. Cancer 30:257-264.
16. Gibbs, W. N., Lofters, W. S., Campbell, M., Hanchard, B., La Grenade, L.,
    Clark, J., Cranston, B., Saxinger, C., Gallo, R., and Blattner, W. A. (1985).
    Adult T-cell leukemia/lymphoma in Jamaica and its relationship to human
    to T-cell leukemia/lymphoma virus type I associated lymphoproliferative
    disease. In: *Retrovirus in Human Lymphoma/Leukemia* (M. Miwa et al.
    eds.), p. 77. Tokyo, Japan Scientific Societies Press.
17. Kawano, F., Yamaguchi, K., Nishimura, H., Tsuda, H., and Taktsuki, K.
    (1985). Viriation in the clinical courses of adult T-cell leukemia. *Cancer*
    55:851-856.
18. Yamaguchi, K., Yoshioka, R., Kiyokawa, T., Seiki, M., Yoshida, M., and
    Takatsuki, K. (1986). Lymphoma type adult T-cell leukemia—A clinico-
    pathologic study of HTLV related T-cell type malignant lymphoma.
    Hematol. Oncol. 4:59-65.

19. Schneider, J., Yamamoto, N., Hinuma, Y., and Hunsmann, G. (1984). Sera from adult T-cell leukemia patients react with envelope and core polypeptides of dult T-cell leukemia virus. *Virology* 132:1-11.

20. Hattori, T., Uchiyama, T., Toibana, T., Takatsuki, K., and Uchino, H. (1981). Surface phenotype of Japanese adult T-cell leukemia cells characterized by monoclonal antibodies. *Blood* 58:645-647.

21. Yamaguchi, K. Seiki, M., Yoshida, M., Nishimura, H., Kawano, F., and Takatsuki, K. (1984). The detection of human adult T cell leukemia virus proviral DNA and its application for classification and diagnosis of T-cell malignancy. *Blood* 63:1235-1240.

22. Yoshida, M., Seiki, M., Yamaguchi, K., and Takatsuki, K. (1984). Monoclonal integration of human T-cell leukemia provirus in all primary tumors of adult T-cell leukemia suggests causative role of human T-cell leukemia virus in the disease. *Proc. Natl. Acad. Sci. USA* 81:2534-2537.

23. Asou, N., Kuamgai, T., Uekihara, S., Ishii, M., Sato, M., Sakai, K., Nishimura, H., Yamaguchi, K., and Takatsuki, K. (1986). HTLV–I seroprevalence in patients with malignancy. *Cancer* 58:903-907.

24. Matsuzaki, H., Yamaguchi, K., Kagimoto, T., Nakai, R., Takatsuki, K., and Oyama, W. (1985). Monoclonal gammopathies in T-cell leukemia. *Cancer* 56:1380-1383.

25. Lee, S.Y., Matsushita, K., Machida, J., Tajiri, M., Yamaguchi, K., and Takatsuki, K. (1987). Human T-cell leukemia virus type I infection in hemodialysis patients. *Cancer* 60:1474-1478.

26. Nakada, K., Yamaguchi, K., Furugen, S., Nakasone, T., Nakasone, K., Oshiro, Y., Kohakura, M., Hinuma, Y., Seiki, M., Yoshida, M., Matutes, E., Actovsky, D., Ishii, T., and Takatsuki, K. (1987). Monoclonal integration of HTLV–I proviral DNA in patients with strongyloidiasis. *Int. J. Cancer* 40:145-148.

27. Tajima, K., Tominaga, S., Kuroishi, T., Shimizu, H., and Suchi, T. (1979). Geographical features and epidemiological approach to endemic T-cell leukemia/lymphoma in Japan. *Jpn. J. Clin. Oncol.* 9(Suppl.):495.

28. The T- and B-cell Malignancy Study Group (1985). Statistical analyses of clinico-pathological, virological and epidemiological data on lymphoid malignancies with special reference to adult T-cell leukemia/lymphoma: A report of the second nationwide study of Japan. *Jpn. J. Clin. Oncol.* 15:517-535.

29. Catovsky, D., Greaves, M. F., Rose, M., Galton, D. A. G., Goolden, A. W. G., McCluskey, D. R., White, J. M., Lampert, I., Bourikas, G., Ireland, R., Brownell, A. I., Bridges, J. M., Blattner, W. A., and Gallo, R. C. (1982). Adult T-cell lymphoma/leukemia in blacks from the West Indies. *Lancet* 1:639-643.

30. Fleming, A. F., Yamamoto, N., Bhusnurmath, S. R., Maharajan, R., Schneider, J., and Hunsmann, G. (1983). Antibodies to ATLV (HTLV) in Nigerian blood donors and patients with chronic lymphatic leukaemia or lymphoma. *Lancet* 2:334-335.

31. Pan, I. H., Lin, C. Y., Komoda, H., Iami, J., and Hinuma, Y. (1986). Seropedemiology of adult T-cell leukemia virus infection and analysis of sero-positive cases in Taiwan. *Proc. Natl. Sci. Counc. B. ROC* 10:254–262.

32. Lee, S. Y., Yamaguchi, K., Takatsuki, K., Kim, B. K., Park, S., and Lee, M. (1986). Seroepidemiology of human T-cell leukemia virus Type-I in the Republic of Korea. *Jpn. J. Cancer Res. (Gann)* 77:250–254.

33. Yamaguchi, K., Nishimura, H., Seiki, M., Yoshida, M., and Takatsuki, K. (1983). Clinical diversity in adult T-cell leukemia-antibodies to adult T-cell leukemia-antibodies to adult T-cell leukemia virus-associated antigen (ATLA) in sera from patients with ATL and controls. *Recent Adv. RES Res.* 23:179.

34. Kinoshita, K., Hino, S., Amagasaki, T., Ikeda, S., Yamada, Y., Suzuyama, J., Momita, S., Toriya, K., Kamihira, S., and Ichimaru, M. (1984). Demonstration of adult T-cell leukemia virus antigen in milk from three seropositive mothers. *Jpn. J. Cancer Res. (Gann)* 75:103–105.

35. Nakano, S., Ando, Y., Saito, K., Moryama, I., Ichijo, M., Toyama, T., Sugamura, K., Imai, J., and Hinuma, Y. (1986). Primary infection of Japanese infants with adult T-cell leukemia-associated retrovirus (ATLV): Evidence for viral transmission from mothers to children. *J. Infect.* 12: 205–212.

36. Miyamoto, Y., Yamaguchi, K., Nishimura, H., Takatsuki, K., Motoori, T., Morimatsu, M., Yasaka, T., Ohya, I., and Koga, T. (1985). Familial adult T-cell leukemia. *Cancer* 55:181–185.

37. Okochi, K., Sato, H., and Hinuma, Y. (1984). A retrospective study on transmission of adult T-cell leukemia virus by blood transfusion: Seroconversion in recipients. *Vox Sang.* 46:245–253.

38. Ichimaru, M., Kinoshita, K., Kamihira, S., Ikeda, S., Yamada, Y., and Amagasaki, T. (1979). T-cell malignant lymphoma in Nagasaki district and its problems. *Jpn. J. Clin. Oncol.* 9:337–346.

39. Yamaguchi, K., Lee, S. Y., Shimizu, T., Nozawa, F., Takeya, M., Takahashi, K., and Takatsuki, K. (1985). Concurrence of lymphoma type adult T-cell leukemia in three sisters. *Cancer* 56:1688–1690.

40. Franchini, G., Wong-Staal, F., and Gallo, R. C.: Human T-cell leukemia virus (HTLV–I) transcripts in fresh and cultured cells of patients with adult T-cell leukemia.

41. Waldmann, T. A., Greene, W. C., Sarin, P. S., Saxinger, G., Blayney, D. W., Blattner, W. A., Goldman, G. K., Bongiovanni, K., Sharrow, S., Depper, J. M., Lepnald, W., Uchiyama, T., and Gallo, R. C. (1984). Functional and phenotypic comparison of human T-cell leukemia/lymphoma virus positive adult T-cell leukemia with human T-cell leukemia/lymphoma virus negative Sézary leukemia, and their distincion using anti-Tac monoclonal antibody identifying the human receptor for T-cell growth factor. *J. Clin. Invest.* 73: 1711–1718.

42. Matsuoka, M., Hattori, T., Chosa, T., Tsuda, H., Kuwata, S., Yoshida, M., Uchiyama, T., and Takatsuki, K. (1986). T3 surface molecules on adult T-cell leukemia are modulated in vivo. *Blood* 67:1070–1076.

43. Arima, N., Daitoku, Y., Ohgaki, S., Tanaka, H., Yamamoto, Y., Fujimoto, K., and Onoue, K. (1986). autocrine growth of interleukin 2-producing leukemic cells in a patient with adult T-cell leukemia. *Blood* 68:779–782.
44. Wano, Y., Hattori, T., Matsuoka, M., Takatsuki, K., Chua, A. O., Gubler, U., and Greene, W. C. (1987). Interleukin-I gene expression in adult T-cell leukemia. *J. Clin. Invest.* 80:911–916.
45. Ueshima, Y., Fukugara, S., Hattori, T., Uchiyama, T., Takatsuki, K., and Uchino, H. (1981). Chromosome studies in adult T-cell leukemia in Japan: Significance of trisomy 7. *Blood* 58:420–425.
46. Sanada, I., Tanaka, R., Kumagai, E., Tsuda, H., Nishimura, H., Yamaguchi, K., Kawano, F., Fujiwara, H., and Takatsuki, K. (1985). Chromosomal aberrations in adult T-cell leukemia: Relationship to the clinical severity. *Blood* 65:649–654.
47. Seiki, M., Eddy, R., Shows, T. B., and Yoshida, M. (1984). Nonspecific integration of the HTLV provirus genome into adult T-cell leukemia cells. *Nature* 309:640–642.
48. Yamamoto, S., Hattori, T., Asou, N., Nishimura, H., Kawano, F., Yodoi, J., and Takatsuki, K. (1986). Absolute neutrophilia in adult T-cell leukemia. *Jpn. J. Cancer Res. (Gann)* 77:858–861.
49. Shimoyama, M., Yunoki, K., Ichimaru, M., Ohta, K., and Ogawa, M. (1979). Combination chemotherapy with vincristine, cyclophosphamide, prednisolone and adreamycin (VEPA) in advanced adult non-Hodgkin's lymphoid malignancies: Relation between T-cell or non-T-cell phenotype and response. *Jpn. J. Clin. Oncol.* 9:397.
50. Waldmann, T. A., Davis, M. M., Bangiovanni, K. F., and Korsmeyer, S. J. (1985). Rearrangements of genes for the antigen receptor on T-cell as markers of lineage and clonality in human lyphoid neoplasms. *N. Engl. J. Med.* 313:776–783.
51. Daenen, S., Rojor, R. A., Smith, J. W., Halie, M. R., and Nieweg, H. O. (1984). Successful chemotherapy with deoxycoformycin in adult T-cell lymphoma-leukemia. *Br. J. Haematol.* 58:723–727.
52. Yamaguchi, K., Takatsuki, K., Dearden, G., Matutes, E., and Catovsky, D. (1989). Chemotherapy with deoxycoformycin in mature T-cell malignancies. In: *Cancer Chemotherapy: Challenges for the Future, Vol. 3.* (K. Kimura et al, eds.), pp. 216–220. Tokyo, Excerpta Medica.

7

Molecular Biology of HTLV-I
Biological Significance of Viral Genes in
Its Replication and Leukemogenesis

Mitsuaki Yoshida and Motoharu Seiki* *Cancer Institute, Kami-Ikebukuro, Toshima-ku, Tokyo, Japan*

I. INTRODUCTION

Since human T-cell leukemia virus type 1 (HTLV-I) was isolated in 1980 by R. C. Gallo and his colleagues (1), studies on HTLV-I and related viruses have progressed rapidly, mainly because a great number of scientists in a variety of fields have been keenly interested in HTLV-I. This virus has attracted their interest for the following reasons: (a) HTLV-I is the first human retrovirus to be well established (1,2). For many years numerous attempts to identify human retroviruses had been unsuccessful. There have, in fact, been many reports of isolation or detection of human retroviruses, but either these reports were too preliminary to be of value or the findings could not be confirmed by other groups. (b) Shortly after HTLV-I was discovered it was found to be associated with a unique, recently discovered human leukemia, adult T-cell leukemia (ATL) (2-4). Studies on animal retroviruses have proved that retroviruses are very useful in cancer research and have led to the discovery of oncogenes and their cellular precursors (protooncogenes) and to an understanding of the mechanism of activation of protooncogenes. In fact, protooncogenes are now a central concern in cancer research. Because so much information has been obtained in studies on animal retroviruses, it was expected that HTLV-I and ATL would be useful for direct study of the mechanism of tumorigenesis in human cancer. In

---

Present affiliation: Kanazawa University, Takara-cho, Kanazawa, Japan this respect, HTLV-I attracted attention because it was a retrovirus, rather than because it was the first. (c) Clinically, since HTLV-I was proposed to be associated with ATL, it was expected that, by studies on this virus, effective strategies could be developed for diagnosis, treatment, and prevention of ATL. Today, some of these expectations are becoming realities. In this chapter, we will review our own studies on the molecular biology of HTLV-I, focusing on the functions of pX and *env* genes and their significance in ATL development and its prevention.

II. ETIOLOGY OF ATL

ATL was first described by K. Takatsuki et al. (5) in 1977 in Japan. Today, ATL is known to be localized in three separate areas, southwestern Japan (3,5,6), the West Indies (7), and central Africa (8). HTLV-I is also found to be endemic in these areas. The prevalence of HTLV-I was determined by screening for serum antibodies against HTLV-I antigens. The identical geographical distributions of HTLV-I and ATL suggested the association of ATL with HTLV-I infection. Furthermore, almost all ATL patients were found to have antibodies to the HTLV-I proteins, indicating HTLV-I infection (3,4). Thus, HTLV-I infection and ATL overlap at the levels of both geographical areas and individuals. These findings strongly indicated a close association of ATL with HTLV-I infection. The overlapping of HTLV-I infection and disease at the cellular level was also demonstrated by the finding of proviral genomes integrated into primary leukemic cells (2,9,10). The concept behind these studies at a cellular level was that if HTLV-I is the etiological agent directly associated with leukemogenesis, HTLV-I should infect target cells that eventually become tumor cells, and therefore the final tumor cells should always be infected with HTLV-I. Since, on retrovirus infections, the provirus genome is integrated into the host chromosomal DNA, fresh ATL cells contain integrated HTLV-I proviral DNA. In fact, in surveys on more than 400 ATL patients, the proviral genome was consistently found to be integrated into the leukemic cells (2,9). Furthermore, in the majority of Japanese ATL patients, leukemic cells were found to be monoclonal with respect to the integration site of the proviral genome. Thus it was concluded that HTLV-I directly infects target cells that eventually become malignant cells. Since HTLV-I infects only a small population of T cells, accidental transformation of HTLV-I-infected T cells could be excluded as a main mechanism of ATL development.

With respect to HTLV-I association with ATL, Shimoyama et al. (11) recently reported six exceptional cases giving negative reactions for antibodies to HTLV-I but diagnosed as having ATL. Furthermore, the HTLV-I proviral genome was not integrated into tumor cells from these patients. These cases clearly demonstrated that a few cases of ATL, or at least an ATL-like disease, are not due to HTLV-I infection. This conclusion is not surprising, because some cases of Sézary syndrome or chronic lymphocytic leukemia of the T-cell type are very similar to ATL, but are not associated with HTLV-I infection (2). Nevertheless, almost all ATL cases are concluded to be associated with HTLV-I.

In the endemic area in Japan, 10-15% of adults are infected with HTLV-I. Since a small fraction of infected people (estimated nearly 1%) develop leukemia during life, some environmental factor(s) such as a plant tumor promoter was suspected to affect disease (12). However, the fact that the incidence of ATL among the HTLV-I-infected population was similar between the Japanese endemic and nonendemic areas (T. Osato, personal communication) does not support the idea that an environmental factor is involved in leukemogenesis. Thus at present, HTLV-I is the only known exogenous etiological agent of ATL. Possibly, host factors such as different immunological responses to HTLV-I infection or some genetic defects in a certain population might affect the incidence of ATL development. In connection with this possibility, HLA phenotypes in the Japanese population are now being analyzed and the preliminary data suggest some particular HLA haplotype might be associated with ATL development.

III. *TRANS*-ACTING VIRAL FUNCTION FOR ATL DEVELOPMENT

As discussed in Section II, HTLV-I is involved in leukemogenesis at the level of a single infected cell. Two mechanisms for such situations have been demonstrated for tumor induction by animal retroviruses: One is that the viruses have their own viral oncogenes, and the other that the provirus genome is integrated into a specific locus on the chromosomal DNA and then activates an adjacent cellular oncogene (insertional mutagenesis). The former mechanism has been demonstrated in many acute leukemia and sarcoma viruses (13,14), and the latter in chicken lymphoma (15) and erythroblastosis (16) induced by avian leukosis viruses, and also in mammary tumors (17) induced by mouse mammary tumor viruses.

Since HTLV-I has no typical oncogene in its genome (18), the first model cold not be applied. The second model, insertional mutagenesis, was examined by study of the integration sites of proviral genomes. For this purpose, a patient whose leukemic cells had a single copy of the HTLV-I provirus was selected and cellular DNA sequences flanking the integrated provirus were isolated. Using these cellular flanking regions as probes, the DNAs of other ATL patients were surveyed to determine the rearrangements of cellular DNA sequences induced by provirus integration. By use of combinations of other probes and different restriction enzymes, a region of about 25 kb was examined, which corresponded to the provirus integration site in the DNA of the original patient. However, no rearrangement was detected in the DNAs of 34 ATL patients (19). These results were confirmed using another set of probes isolated from a different patient (19). Thus, it was concluded that there was no common region for provirus integration in leukemic cells. These findings suggested that a *cis*-acting function of the integrated proviruses was not involved in the case of HTLV-I. This conclusion was further supported by the finding that the provirus was integrated into different chromosomes in different ATL patients (19).

Eventually, a *trans*-acting viral function, possibly mediated by viral proteins, was suspected. Infection with HTLV-I frequently immortalizes helper T cells in vitro, but no such immortalization or transformation has been observed with chronic leukemia virus. Therefore, a unique factor that acts in *trans* was predicted to be associated with in vitro immortalization.

IV. GENOMIC STRUCTURE AND GENE PRODUCTS

A provirus clone, λATK-1, of HTLV-I was isolated from cellular DNA of primary leukemic cells from an ATL patient and its total nucleotide sequence was determined (18). The genome contained the *gag*, *pol*, and *env* genes, indicating replication competence. In addition to these viral genes it had an extra sequence of about 1.6 kb, termed "pX" or "X," between the *env* and 3' LTR (Figure 1). The pX sequence appeared to be a virus-specific sequence, not a typical oncogene derived from a cellular sequence. Because of the presence of this unique sequence, HTLV-I was classified into a distinct retroviral subgroup; the other members of which are HTLV-II, simian T-cell leukemia virus type 1 (STLV-1), and bovine leukemia virus (BLV). The functions of this region in these retroviruses seem to be very similar, as discussed in Section VII, even though their sequences are only distantly related.

The viral proteins encoded by these viral genes are summarized in Figure 1. For identification of these gene products we used antisera against synthetic peptides predicted from the DNA sequence of the genomic clone (20-23). The *env* gene product is important in an initial step in viral infection and also as a target of host immune systems. The extra sequence pX was expected to have some unique function in viral replication. These two genes are the subject of this chapter.

V. THE ENVELOPE GENE PRODUCT

A. Characterization of the Product and Its Functions

In animals, continuous replication and spread of leukemia virus within individuals are requirements for development of viral leukemia (24). Thus an understanding of the mechanisms of infection and replication of HTLV-I is important for prevention of ATL. The glycoproteins encoded by the *env* gene of animal retroviruses are exposed on the surface of the viral particles and are known to be essential for infection, interacting with receptors on the surface of target cells. To establish a system for the study of the *env* gene function, we constructed expression plasmids containing DNA segments on the *env* gene to obtain large quantities of polypeptides (25).

The HTLV *env* gene was divided into two fragments, as illustrated in Figure 2, and the two fragments were inserted into the expression vectors

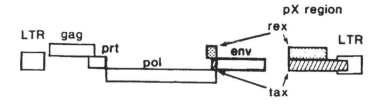

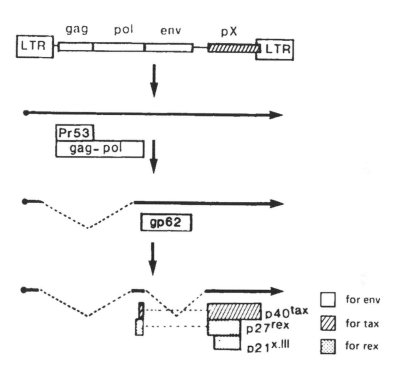

FIGURE 1 Proviral structure of HTLV-I, its gene products, and mechanism of expression of the pX region. Open bars represent open reading frames in genomic and mRNA sequences.

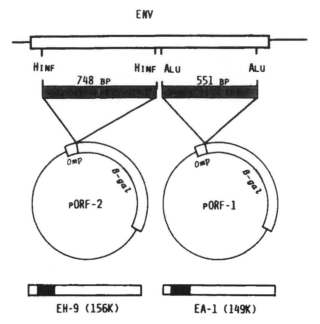

FIGURE 2 Schematic illustration of plasmids for expression of the HTLV–I *env* gene in *E. coli*. *Env* coding sequences are inserted at the upstream of the β-galactosidase gene in frame to produce the fusion protein.

pORF1 and pORF2, respectively (26). The resulting plasmids pEH9 and pEA1 expressed fusion protein containing the following three sequences: a short peptide from the *Escherichia coli* outer membrane protein, half the *env* polypeptide, and β-galactosidase. By transformation with these plasmids, new proteins with molecular weights of about 150 kD were produced at levels of as much as 10–20% of the total protein (25). The sizes of these proteins were the same as those deduced from the DNA sequences. Like the native *env* gene products, the hybrid proteins produced in this way, EH9 and EA1, both cross-reacted with sera from ATL patients, although they were not glycosylated.

Antiserum against the hybird protein produced in *E. coli* was raised in rabbits and used to characterize the gene products. Antibodies against the polypeptide of the N-terminus (EH9) detected gp62 and gp46. On the other hand, monospecific antiserum against the synthetic peptide predicted from the sequence of the 3' terminus of the putative *env* gene identified gp62 and p20 (E). Tunicamycin treatment and a pulse-chase experiment indicated that the translational product of the *env* gene is p46 (21), which is then glycosylated into gp62. The gp62 is then further processed to gp46 and p20. However, the arrangements of gp46 and p20 (E) on the cell surface are still not clear.

The antibodies to the *env* gene products were also used to study the function of *env* gene products, and two properties were identified: (a) *env* gene products induce cell fusion of certain types of cell lines; and (b) *env* gene products are the target for cytotoxic antibodies.

HTLV-producing cells can induce syncytia in certain types of cell lines (27,28). Antisera against hybrid proteins produced in *E. coli* markedly inhibited syncytium formation of cat S^+L^- cells induced with HTLV-I infection (25). This observation indicated that the *env* gene products have the capacity to induce cell fusion in certain cell lines, as those of murine leukemia virus do. As reported previously, sera from ATL patients also inhibited syncytium formation (27,28). These findings are consistent with the presence of antibodies against the *env* proteins in sera of ATL patients (20,21).

Although cell-free viruses are infectious, their infectious titers are very low. For more efficient transmission, cells are usually cocultivated with HTLV-I-producing cells. In this way, HTLV-I can be transmitted to not only T cells, but also B cells, fibroblasts, epithelial cells, and other cells even of other species (31–33). During cocultivation, viral transmission could be mediated either by binding to putative receptors or by cell fusion of recipient cells with HTLV-I-producing cells. Since antibodies against *env* protein can inhibit *env*-mediated cell fusion, they may be able to block viral infection even in case of cell-to-cell transmission. In the case of receptor-mediated transmission, the antibodies to fusion proteins could neutralize vesicular stomatitis virus (VSV)-pseudo-type bearing envelopes of HTLV-I (34). In both cases, antibodies against envelop proteins will inhibit the viral transmission. The vaccinelike activity of the *env-β-gal* fusion proteins supports this conclusion (discussed in Section B, below).

Antisera against *env* polypeptides were found to be cytotoxic to cell lines producing HTLV-I in the presence of complement (25). Since these toxicities were specific to HTLV-I-producing cells, presumably the *env*-gene products of HTLV-I, gp62 or gp46 or both, were exposed on the cell surface and served as targets in this system. Some sera from ATL patients were cytotoxic to HTLV-I-producing cells (25). The presence of cytotoxic antibodies in the sera of patients may explain the absence of viral antigen expression in primary ATL cells. In most patients with ATL, the tumor cells in the peripheral blood do not express HTLV antigens in amounts sufficient to detect by any assay, but the viral genomes are activated after short-term (1–2 days) culture (23,35,36). These findings provide a main basis for the model of ATL development proposed later.

B. Usages of the *env* Protein

1. Diagnostic Reagent

Fusion proteins between the *env* polypeptides and β-galactosidase were found to react with the sera of patients in blotting assay. As a diagnostic reagent, the fusion proteins produced in *E. coli* were coated on a plastic surface and an

ELISA system was established. In a survey of sera from HTLV–I-negative persons, healthy carriers, and ATL patients, all sera that gave a negative reaction in the indirect-immunofluorescence assay (IF) gave negative signals, and most of the seropositive sera gave positive signals, as expected. But about 7–8% of the seropositive sera gave a negative results. In a separate experiment with blot assay, these samples that were seronegative in our ELISA assay were shown to have either very low or scarcely detectable antibodies titers against *env* proteins (M. Yoshida, unpublished observations). Thus we concluded that the *env*-fusion protein alone might not be sufficient as a diagnostic reagent in survey of bloods for transfusion. Addition of *gag* proteins to this system would increase the sensitivity.

2. Vaccinelike Activity

As discussed in the previous section, HTLV–I is an etiological agent of ATL and its propagation in individuals is also a prerequisite for ATL development. In some experimental systems with animal retroviruses, antibodies against *env* proteins can block both infection and propagation, thus preventing new infection of other cells. Therefore, it was expected that an *env*-fusion protein produced in large quantity in *E. coli* would induce antibodies that prevented HTLV–I infection. In fact, six cynomolgus monkeys that were immunized with two *env-β-gal* fusion proteins, EH9 and EA1, developed antibodies to the *env* proteins gp62 and gp46 (37). When these monkeys were then inoculated with living MT-2 cells, which produce HTLV–I, none of them was infected with HTLV–I, whereas four nonimmunized monkeys were all infected (29). The establishment of infection in these monkeys was judged by the appearance of HTLV–I-antigen positive cells during cultivation of their peripheral blood lymphocytes for a few weeks. This test was carried out every week after inoculation of MT-2 cells. These observations clearly demonstrated that *env-β-gal* fusion proteins protected the monkeys from HTLV–I infection and thus could be useful as vaccine. This is direct evidence that the *env* protein is useful as a vaccine for HTLV–I infection. However, for clinical use, some improvements are necessary to produce protein of higher antigenicity (29).

VI. pX GENE PRODUCTS AND MECHANISM OF THEIR EXPRESSION

The unique sequence pX was first found by sequence analysis of HTLV–I and was proposed to have the capacity to code for four proteins (18). The arrangements of these open reading frames (ORF) are summarized in Figure 1. The pX genes are expressed by double splicing (38): the sequence within the R to 3′ region of the *pol* and the 5′ region of the *env* to the pX sequence (Figure 1). As a result of the second splicing, ORF III (*rex*) and IV (*tax*) are reconstructed in the mRNA: The AUG codon used for *env* protein translation is fused to ORF IV for *tax* translation, while *rex* translation can be initiated by the AUG codon that is 56 nucleotides upstream of the *tax* initiation site. Thus, most of *rex* overlaps *tax*.

Using peptide sera, cDNA construction, and site-directed mutations, *tax* was shown to code for p40^{tax} previously termed p40^{tax}) (22), and *rex* to code for p27^{rex} and p21^{x-III} from the first and fourth AUG codon in the mRNA, respectively (23,39). The first methionine of p40^{tax} and 29 amino acids of the N-terminus of p27^{rex} are encoded by the second exon that is derived from the 3' region of the pol gene. A unique point in these expressions is that a single mRNA codes for three pX proteins using independent AUG codons (39). Generally, a cellular mRNA codes for one protein, and the eukaryotic translational machinery can translate the gene furthest to the 5' end of mRNA. Therefore, there may be some biological significance in this unique mechanism: For example, the expressions of these proteins in a constant ratio may be crucial, or these proteins may function in association in some serial process. Two of the three proteins, p40^{tax} and p27^{rex}, were found to be localized in the nuclear fraction. p40^{tax} is reported to have a relatively short half-life and p27^{rex} is known to be phosphorylated at serine and threonine residues; so these proteins may have regulatory functions.

VII. FUNCTIONS OF pX PROTEINS

A. Transcriptional Activation of the LTR By p40^{tax}

The function of *tax* was first suggested from a project in which the tissue specificity of the LTR of HTLV-I was tested. Sodroski et al. (40) first tested this specificity using a plasmid pLTR-CAT containing a gene for (chloramphenicol acetyltransferase) under control of the LTR of HTLV-I and HTLV-II. After transfection of the pLTR-CAT into various cell lines, they assayed CATase activity by measuring the acetylation of chloramphenicol. The activity of the pLTR-CAT was found to be almost equal in epithelial, fibroblastic, B-, and T-cell lines, clearly demonstrating no significant T-cell specificity of the LTRs.

More surprisingly, Sodroski et al. (40) and we (41) found that the LTR function in HTLV-I-infected cell lines was more than 100 times that with SV40 promoter, whereas such high LTR activity was not observed in uninfected cell lines. Therefore, the activation was supposed to be mediated by a viral factor that acts in *trans*.

The pX protein p40^{tax} was proposed to be a *trans*-acting factor activating the LTR (40,41), since activation of the LTR was also detected in a rat cell line, in which only p40^{tax} was thought to be expressed (41). Direct evidence that pX proteins are in fact transcriptional *trans*-activators was obtained independently in several laboratories (42–45) using LTR-CAT and pX expression plasmids. The requirement for p40^{tax} was demonstrated with a mutant plasmid that can express *tax* only (42,43).

B. Sequence Responsible for *tax* Activation

p40^{tax} can activate not only its own LTR, but also HTLV–II LTR and adenovirus promoter (46), although it is rather specific for some promoters including its own LTR. For identification of the sequence responsible for this activation, a series of deletion mutants were prepared (47). Deletion of the sequence upstream of TATA in the LTR completely abolished in activation, while reinsertion of this deleted fragment in an antisense orientation restored the activity. Therefore, the sequence upstream of the TATA contains a conditional enhancer responsible for the *trans*-activation. Further deletion muations suggested that direct repeats of 21 nucleotides in this region are responsible for the activation. This conclusion was confirmed in two ways: First, a short fragment covering the two repeats of 21 nucleotides was cut out of the LTR and inserted into an enhancer-less SV40 promoter. This fragment rendered the plasmid responsive to activation by *tax*. The site and orientation of the insert did not affect the activity (47). Second, as a more direct method, the oligonucleotide of 21 bases was synthesized chemically and the hexamer of the oligonucleotide was inserted into the enhancer-less promoter of the LTR. The enhancer-less promoter showed no response to *tax* activation, but the construction containing the synthetic hexamer of the oligonucleotide was strongly activated by the *tax* (Nagashima et al., unpublished data). These results clearly demonstrated that the transcriptional enhancer responsible for *tax* activation is the sequence consisting of at least two repeats of the 21 nucleotides. Similar observations were also reported by two other groups (48,49).

To test the direct binding of the p40^{tax} to the enhancer, we used the synthetic hexamer of 21 nucleotides as a probe in tests by gel retardation assay. The nuclear extract from HTLV–I-infected cells gave a specific band that was retarded by protein binding (Nagashima et al., unpublished results). The specificity of the binding was confirmed by the efficient competition by the enhancer sequence with the 21 nucleotides and the LTR, but very inefficient competition by other enhancers. The same band was also obtained with various uninfected cells, indicating that enhancer binding proteins are coded by the cells. These observations strongly suggested that the *tax* function activates the enhancer indirectly through the cellular enhancer binding protein. Therefore, if the *tax* protein can interact with other DNA binding proteins that have different sequence specificity, it may have the capacity to activate various other enhancers also.

C. Posttranscriptional Regulation of *gag, pol,* and *env* Expression by p27*rex*

As described in the previous sections, the *tax* function is sufficient for transcriptional activation and this activation does not require either p27*rex* or p21x-III. Since transcriptional activation was mainly studied with a plasmid pLTR-CAT, we tested the effects of these proteins on expression of the viral protein using provirus constructions (50,51). For simplicity, we initially used a defective construction of the provirus genome. The defective provirus had two large deletions encompassing the *pol, env,* and pX sequences but had an intact *gag* gene (Figure 3). Transfection of this provirus alone did not result in expression of the *gag* protein, but its cotransfection with the wild-type pX construct induced *gag* protein expression. Here, the defect in pX expression of the provirus was complemented by the plasmid in *trans*. Plasmids expressing *tax* or *rex* only were unable to support *gag* protein expression, but cotransfection of both plasmids again induced the *gag* protein. These findings are clear direct evidence that both *tax* and *rex* are required for expression of the *gag* protein (50). These results were surprising, because *rex* was not required for activation of exprsssion of the CAT gene.

The effect of p27*rex* was also demonstrated at the level of mRNA. As shown in Figure 3 (51), in the presence of p40*tax* alone, the provirus was actively transcribed, but only spliced mRNA was detected. By expression of the p27*rex* in addition to the *tax*, the level of unspliced mRNA, a *gag* mRNA, was dramatically increased. Since *tax* alone is sufficient to activate transcription from the LTR, p27*rex* must modulate some posttranscriptional process, eventually controlling the level of unspliced mRNA. Expression of the gag protein was parallel with accumulation of the unspliced mRNA, so it was concluded that the *rex* operates at the level of mRNA, not at the translational level (51).

These observations on the defective proviruses seem to be applied to a whole provirus clone carrying a mutation in *rex* expression. Such mutant produced only spliced pX mRNA and could not express any *gag* protein unless active p27*rex* is complemented. Expression of the envelop protein was also found to be regulated by the *rex* function in exactly the same fashion as the *gag* protein expression; that is, unspliced *env* mRNA was accumulated by complementation of the *rex* function (52).

The simplest mechanism of the *rex* function is suppression of the splicing of viral transcripts. However, another defective construction that had a deletion at the 3' splice site also showed *rex* dependency for *gag* protein expression. These results suggest that the *rex* affects the level of unprocessed viral RNA, irrespective of whether it is eventually spliced or not; thus a direct target of the *rex* function is not the splicing process itself.

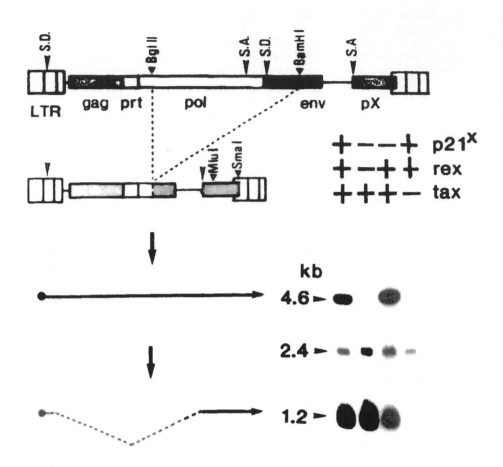

FIGURE 3 Effect of the *rex* on the viral RNA level. Defective proviral construction used for *gag* expression and the subgenomic viral RNAs are illustrated. Two RNA species are marked at positions corresponding to the bands in the blot. The probe was the exon 1. The bands of 2.4 kb represent RNA from pX expression plasmids.

Increasing doses of the *rex* expression plasmid increased the expression of unspliced *gag* and *env* mRNAs, but conversely decreased the level of spliced pX mRNA, which should result in decreased *tax* expression (51). Decreased expression of *tax* should reduce the *trans*-activation of transcription. Therefore, this function of *rex* consists of a feedback control of HTLV-I expression.

These functions of *rex*, increasing unspliced RNA levels and decreasing spliced form of RNA, should be specific to viral RNA. Otherwise, cellular gene expression would be completely destroyed by HTLV-I infection. Then, which elements are responsible for viral-specific regulation? To analyze the mechanism of this posttranscriptional regulation, the *gag* sequence in a defective provirus construction LTR-*gag*-LTR (in Figure 3) was replaced with a bacterial gene CAT constructing LTR-CAT-LTR (Figure 4a). As shown in Figure 4b, expression of CAT gene in presence of *tax* was further activated by increasing the dose of the *rex* plasmid (53). Therefore, it is evident that one or both LTRs are required for the *rex* regulation. Either replacement of the 3' LTR with a transcriptional terminator of thymidine kinase or of the 5' LTR with SV40 promoter impaired the *rex* regulation on CAT expression (Figure 4b). These results clearly indicated that both the 5' and 3' LTRs are required for the maximum response to the *rex* regulation, although the 3' LTR alone can confer a partial response. Further deletion analysis on the 3' and 5' LTRs showed that the essential element in the 3' LTR is the R sequence and in the 5' LTR it is the splice donor signal (see Ref. 53, data not shown).

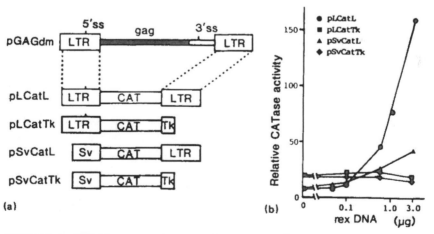

FIGURE 4 Identification of both LTR as responsible sequence for *rex* regulation. (a) Substitution constructions of defective provirus clone. (b) Effect of *rex* on CAT gene expression with various constructions.

The requirement for two *cis*-acting elements at the 3' and the 5' region of the RNA indicates a unique mechanism of the *rex* regulation: RNA transcripts containing a 5' splicing signal are either spliced or degraded when the RNA has no 3' splicing signal; thus they cannot be accumulated in the cytoplasm. When these RNA species have a specific element derived from the 3' LTR, the *rex* protein interacts with this element and then modulates the process. The modulation could be (a) direct suppression of splicing or (b) acceleration of trasport of the RNA into cytoplasm: Transport of nonspliced RNA could be accomplished by interfering with spliceosome formation, by opening another pathway of transportation, or by stabilizing the target molecules. The second possibility seems most likely, because the 3' splicing signal is dispensable for regulation by *rex*.

The p21^{x-III}, which consists of part of the amino acid sequence of p27^{rex}, has been detected in the cytoplasm or membranes of various T-cell lines infected with HTLV-I. However, its function is totally unknown. It is translated from the same mRNA as *tax* and *rex* but with a different initiation codon, and therefore, it is unlikely to be a breakdown product of p27^{rex}.

D. Significance of Two *Trans*-Acting Regulations in Viral Gene Expression and Replication

As discussed in the previous sections, viral gene products are required for expression of the other viral structural proteins. This situation leads us to postulate an early and late stage in viral replication (51; see also Figure 5): HTLV-I has two *trans*-acting regulations by *tax* and *rex*, and in the early stage, the *tax* activates the viral gene transcription in responding to the enhancer. Without *tax* function, the LTR in the integrated proviral genome probably can support low levels of transcription, resulting in low levels of pX mRNA, and thus producing low levels of pX proteins. At this stage, all viral RNA would be spliced into pX mRNA; thus viral structural proteins would not be synthesized. This low level of p40^{tax} would activate the LTR function further, in turn activating the transcription to produce high levels of pX mRNA (52,54). Once viral transcription is activated ready for high expression of the proteins, the accumulated *rex* protein p27^{rex} would start modulating RNA processing, eventually accumulating unspliced *gag* and *env* mRNAs, which would synthesize the viral structural proteins. Therefore, the early gene product p27^{rex} is a trigger of late gene expression. Without *rex* function in the early stage, viral gene transcription can be fully activated prior to viral antigen expression, that is, prior to host immune response to the infected T cells. This could be a significant factor in *rex* regulation.

Increasing the unspliced RNA level by the *rex* function consequently reduces the level of spliced pX mRNA resulting in reduction in the level of *tax*

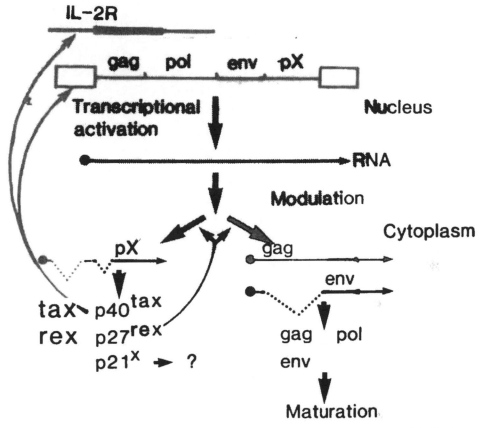

FIGURE 5 Schematic illustration of the functions of *tax* and *rex* in HLTV–I gene expression and activation of cellular IL-2R gene.

expression (52,53; Figure 5). The reduced *tax* function leads, in turn, to further reduction of viral gene transcription. Therefore, the *rex* mediates a feedback control of HTLV-I gene expression. By this feedback control, HTLV-I gene expression and replication become transient and are kept under the restriction; thus infected cells can escape from host immune rejection. This seems another aspect of *rex* regulation. The poor replicative capacity of HTLV-I and tendency to be latent in vivo might also be explained by the *rex* function.

　　To carry out efficient regulation by these two *trans*-acting factors operating at two serial processes, transcription and RNA processing, a coordinated expression of *tax* and *rex* is essential. In this respect, the expression mechanism that these two pX proteins are encoded independently by a single mRNA (39) seems significant. The fact that all viruses in the HTLV subgroup have a second pX gene corresponding to the *rex* strongly suggests that all members of this group have a similar mechanism of regulation by the *tax* and *rex*.

The regulation by *rex* is also important for retroviral replication from a different point of view: In the case of cellular mRNA, all splicing signals are processed very efficiently and no unspliced mRNA is accumulated in the cytoplasm, thus avoiding mistakes in RNA processing. However, in retroviral replication, the *gag* gene is an intron for *env* gene expression, and the *gag* and *env* genes are both introns for pX gene expression in HTLV-I. Therefore, unspliced mRNA should accumulate in the cytoplasm in a regulated fashion. *rex* explains how the viral unspliced mRNAs are protected from the cellular machinery and how the unspliced mRNA levels are regulated. However, it is still unknown how the replication is regulated in other oncoretroviruses that do not have pX equivalent gene.

Interestingly, another replicative retrovirus isolated from humans, human immunodeficiency virus (HIV), also has similar regulations with two *trans*-acting regulatory genes, *tat* (34) and *rev* (previously termed *trs/art*) (35,36), although they have no nucleotide sequence homologies with HTLVs. The mechanisms of these two functions of HIV seem to be different from those of the HTLV regulators (35-37), although they are similar to each other in some aspects: overlapping genes, *trans*-acting functions, operating at different levels, and signals for both activation and suppression. Therefore, the significance of *tax* and *rex* regulation discussed earlier might be extended to *tat* and *rev* of HIV also.

VII. *TRANS*-ACTIVATION OF INTERLEUKIN-2 RECEPTOR GENE

Immediately after the discovery of transcriptional *trans*-activation by the *tax*, this function was suspected to activate the expression of some cellular gene and eventually to induce proliferation of infected T cells. The interleukin-2α(IL-2Rα) gene was used to test this hypothesis, because primary leukemic cells in most ATL cases examined were found to express IL-2Rα on their cell surface (55-57). Furthermore, during in vitro cultivation of tumor cells, the number of IL-2Rα increased concomitantly with expression of viral proteins including pX gene products (23,57). In order to test this hypothesis, the *tax* expression plasmid was transfected into Jurkat or HSB-2 cells and expression of a subunit of the IL-2Rα (Tac antigen) was detected by indirect immunostaining. Transfection of the *tax* expression plasmid, but not its defective mutant, induced transient expression of IL-2Rα (58). Accumulation of the mRNA was also observed. Therefore, transcriptional activation of the IL-2Rα gene was concluded to be mediated by p40^{tax}. Activation of the IL-2 gene was also demonstrated in Jurkat cells, although it was very inefficient (58).

Transient induction of cellular IL-2Rα gene by the *tax* was found to show remarkable cell-type specificity (58): Expression of the *tax* could induce IL-2Rα in Jurkat and HSB-2 cells, but not in two other human T-cell lines, CEM and Molt, or in human B-cell lines such as Raji and Ball-1. This cell-line specificity is in sharp contrast to the cell-type nonspecific activation of the LTR by the *tax*, which demonstrated that the *tax* can function in any type of cells. Interestingly, identical cell specificity was observed in the induction of IL-2R by phytohemag-glutinin (PHA) and 12-O-tetradecanoyl phorbol 13-acetate (TPA), which are known to activate IL-2R in primary T cells. The specificities for these cell lines could be explained in three ways: (a) Some cellular factors may be involved which are stage specific in T-cell maturation and thus present only in some cell lines, not in others. (b) The gene configuration in the chromatin structure may keep the gene available for induction only in these cell lines. (c) A combination of explanations (a) and (b).

To obtain insight into this activation of the cellular IL-2Rα gene, we constructed CAT plasmids containing the 5'-flanking sequence of the IL-2 and IL-2Rα genes and tested their activations by the *tax* function (59). As in cellular gene activation, strong activation of the IL-2R gene and weak activation of the IL-2 gene were observed in transient activation of the exogenous CAT constructs. A similar observation in IL-2R–CAT expression was reported by another group (60). Therefore, it was proposed that differentiation-specific cellular factors are involved in activation of the IL-2R gene. Of course, other effects, including the gene configuration, are also possible. Recent experiments using exogenously introduced genes including the CAT construct suggested that besides specific factors, the gene configuration and the site of insertion are also important for regulation. During these experiments, we observed synergistic activation of the IL-2 gene by the *tax* and a mitogen (concanavalin A). This observation suggests unusually high expression of IL-2 by antigen stimulation of HTLV–I-infected T cells in vivo.

Very recently, it was reported that p40^{tax} activation of IL-2Rα gene is mediated by specific binding of NF-κB-like factor which had been originally identified as transcription factor for immunoglobulin κ gene (61,62). The mechanism by which p40^{tax} induces NF-κB-like factor would be interesting to understand. On the other hand, *tax* also activates the LTR, IL-2 (58), SV40 enhancer (63,64), and GM–CSF (65), and some of them have a sequence with little homology to the NF-κB binding site. These observation may suggest that p40^{tax} may induce or interact with multiple cellular factors involved in transcription (also see Section VII.B).

IX. SIGNIFICANCE OF IL-2R GENE ACTIVATION BY *tax*

As discussed in the previous sections, studies on provirus integration in primary tumor cells indicated an important role of a *trans*-acting viral function in ATL development. Independent research on the molecular biology of HTLV-I demonstrated the *trans*-acting function of the *tax*, which can activate not only the viral genome but also some cellular genes. Therefore, it is reasonable to postulate that the *trans*-acting function of the *tax* may represent a *trans*-acting viral function associated with leukemogenesis. The cellular genes activated by the *tax* are IL-2 and IL-2R, which are required for T-cell growth. Functional IL-2R with high affinity for IL-2 is known to consist of two subunits, α and β (66–68). We have demonstrated activation of the α subunit (Tac antigen). We have no information about activation of the β subunit, but this subunit is thought to be expressed on the surface of even resting T cells (69). Thus, induction of the α subunit by the *tax* may initiate abnormal proliferation of infected T cells. These cells would then be likely to grow continuously by their endogenous *tax* function, irrespective of antigen stimulation. This abnormal growth would be polyclonal but restricted to T cells at a certain stage of differentiation. This polyclonal expansion of certain infected T cells may increase the target size for malignant transformation that results in monoclonal expansion of ATL cells. In fact, polyclonal integration of infected T cells was observed in some HTLV-I carriers who were thought to be in the early stage of ATL development (M. Yoshida, unpublished observations). Therefore, it is likely that activation of the IL-2 and IL-2R genes by the *tax* function of HTLV-I is an operative mechanism in the early stage of ATL development.

Apparently, these cells directed by the *tax* function are not malignant cells in ATL. Fresh ATL cells express the IL-2R, but do not express the *tax* gene at either the protein level or the mRNA level. Thus, continuous expression of IL-2R could not be explained by the *tax* function. There seem to be three explanations for this discrepancy between the expressions of IL-2R and *tax*: (a) The *tax* function is no longer required for constitutive expression of IL-2R on ATL cells in vivo, although it is required in an early stage of leukemogenesis, as discussed earlier. (b) Leukemic cells in peripheral blood are all nonproliferating, and thus the absence of the *tax* function does not exclude their function in leukemic cells. (c) A trace level of *tax* expression is sufficient for maintenance of constitutive expression of IL-2R or the malignant state. It is difficult to determine which of these explanations is correct, but the second seems unlikely, since no *tax* expression was detected by immunostaining or RNA blotting assays in enlarged lymph nodes in which malignant cells are thought to proliferate (M. Yoshida, unpublished observations).

X. MULTISTEP MODEL OF ATL PROGRESSION

As discussed in the previous section, abnormal T-cell growth would be initiated by induction of IL-2R by the *tax*. However, ATL cells in vivo express the IL-2R without any detectable level of *tax* suppression. Other viral genes are also suppressed in infected T cells in vivo. A multistep mechanism for progression of ATL was proposed taking these observations into consideration (70,71; see also Figure 6).

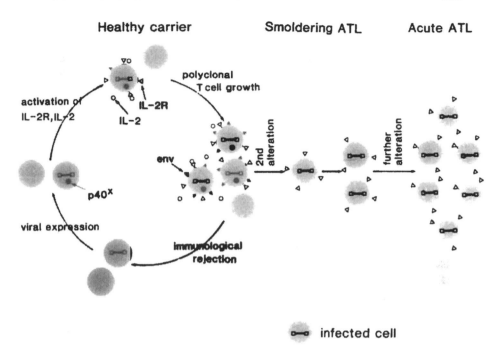

FIGURE 6 A model of the mechanism of ATL progression induced by HTLV-I infection. The scheme summarizes uncontrolled polyclonal growth of infected T cells in healthy carriers, monoclonal expansion of an infected cell, and malignant progression of ATL.

T cells infected with HTLV-I would abnormally proliferate, responding to exogenous or endogenous IL-2 without antigen stimulation, because the expressions of the genes for IL-2 and IL-2R are activated by the viral function *tax*. However, T cells activated by endogenous $p40^{tax}$ would also express viral antigens including the envelope glycoproteins, which are exposed on the cell surface. These glycoproteins are targets of host immune surveillance, as evidenced by the cytotoxic effects of antisera against envelope proteins or sera of patients (25). Thus, all cells expressing the viral antigens, that is, all cells driven by the *tax* function, would be rejected by the host, and only cells that did not express the viral antigens would survive. Some time later, these antigen-negative infected cells would enter a second cycle of viral gene expression and cell propagation. Similar cycles could be repeated again and again for 20 or 30 years or even longer in so-called healthy virus carriers.

During these repeating cycles of abnormal cellular expansion, a second event may take place in a cell during *tax* activation so that this cell can grow and express IL-2R independently of the *tax*. In the absence of the *tax* function, viral antigens would not be expressed, and thus the cell could escape from the host immune surveillance and establish smoldering ATL, which is the first clinical stage of ATL progression. This putative second event would explain monoclonal expansion of an HTLV-I-infected cell in a polyclonal population. In fact, in patients with smoldering ATL, only a few percent of the peripheral blood lymphocytes are morphologically abnormal, but these cells are infected with HTLV-I and are already monoclonally expanded (M. Yoshida, unpublished observations). The putative second event for malignant transformation may or may not be associated with viral function and is probably a rare event, since less than 1% of all HTLV-I carriers develop ATL.

This model explains why *tax* function is required for activation of cellular IL-2R in an early stage, but no longer in the acute stage. In this model, the *tax* function increases the population of target cells available for real malignant transformation. However, some viral function may be associated with the second or further alterations of the target cells. This model also suggests a possible method for preventing ATL development. Prevention of viral infection is, of course, a primary consideration in prevention of ATL, but suppression of the cycle of events in our model by inhibiting viral replication could also prevent ATL development. Therefore, for cancer prevention, studies on viral replication itself are important.

It is interesting that studies on oncogenes have led to the finding that some oncogenes are truncated versions of receptor genes for growth factors. Meanwhile studies on HTLV-I, which has no typical oncogene but is associated with a human tumor, have led to demonstration of activation of the IL-2R gene, which is the receptor gene for T-cell growth factor, IL-2.

XI. SUMMARY

Molecular biological studies on HTLV-I have clarified the regulatory mechanism of the viral gene expression and replication and have also indicated the possible mechanism involved in the early stage of ATL development. The main findings described here are as follows:

1. The envelop gene of HTLV-I was expressed in bacteria as fusion proteins and produced in large quantity. These fusion proteins were demonstrated to be useful as diagnostic reagents and vaccine to prevent HTLV-I infection.

2. The extra sequence "pX" was found to be a self-regulatory sequence that exerts positive and negative controls on viral gene expression. Two genes, *tax* and *rex*, were identified as involved in these regulations and their functions were characterized.

3. The *tax* gene codes for p40^{tax}, which is a *trans*-activator of transcription of the HTLV-I genome activating an enhancer in the LTR.

4. The *tax* also activates the cellular IL-2 receptor a (Tac antigen) in certain T-cell lines responding to the $5'$ regulatory sequences. This activation of the IL-2R gene may account for early events in ATL development.

5. The *rex* gene codes for p27^{rex}, which is a posttranscriptional regulator required for expression of the *gag, pol,* and *env* proteins. The *rex* increases the levels of unspliced mRNAs.

6. The *rex* mediates a feedback control of viral gene transcription. Thus, *rex* makes the viral gene expression transient and the infected cell can escape from host immune response.

7. Two *trans*-acting regulatory systems of HTLV–I are similar in regulating viral replication to those in HIV, although their functional mechanisms seem to be different.

REFERENCES

1. Poiesz, B. J., Ruscetti, F. W., Gazdar, A. F., Bunn, P. A., Minna, J. D., and Gallo, R. C. (1980). Detection and isolation of type C retrovirus particles from fresh and cultured lymphocytes of a patient with cutaneous T cell lymphoma. *Proc. Natl. Acad. Sci. USA* 77:7415–7419.
2. Yoshida, M., Miyoshi, I., and Hinuma, Y. (1982). Isolation and characterization of retrovirus from cell lines of human adult T-cell leukemia and its implication in the disease. *Proc. Natl. Acad. Sci. USA* 79:2031–2035.
3. Hinuma, Y., Nagata, K., Hanaoka, M., Nakai, M. Matsumoto, T., Kinoshita, K., Shirakawa, S., and Miyoshi, I. (1981). Adult T cell leukemia: antigen in an ATL cell line and detection of antibodies to the antigen in human sera. *Proc. Natl. Acad. Sci. USA* 78:6476–6480.
4. Kalyanaraman, V. S., Sarngadharan, M. G., Nakao, Y., Ito, Y., Aoki, T., and Gallo, R. C. (1982). Natural antibodies to the structural core protein (p24) of the human T cell leukemia (lymphoma) retrovirus found in sera of leukemia patients in Japan. *Proc. Natl. Acad. Sci. USA* 79:1653–1657.
5. Uchiyama, T., Yodoi, J., Sagawa, K., Takatsuki, K., and Uchino, H. (1977). Adult T cell leukemia: Clinical and hematological features of 16 cases. *Blood* 50:481–491.
6. Hinuma, Y., Komada, H., Chosa, T., Kondo, T., Kohakura, M., Takenaka, T., Kikuchi, M., Ichimaru, M., Yunoki, K., Sato, M., Matuo, R., Takiuchi, Y., Uchino, H., and Hanaoka, M. (1982). Antibodies to adult T cell leukemia virus associated antigen (ATLA) in sera from patients with ATL and controls in Japan: A nation-wide sero epidemiologic study. *Int. J. Cancer* 29:631–635.
7. Blattner, W. A., Kalyanaraman, V. S., Robert-Guroff, M., Galton, D. A. G., Sarin, P. S., Crawford, M. H., Catovski, D., Greaves, M., and Gallo, R. C. (1982). The human type C retrovirus, HTLV, in blacks from the Caribbean region, and relationship to adult T cell leukemia/lymphoma. *Int. J. Cancer* 30:257–264.

8. Hunsmann, G., Schneider, J., Schmitt, J., and Yamamoto, N. (1983). Detection of serum antibodies to adult T-cell leukemia virus in non-human primates and in people from Africa. *Int. J. Cancer* 32:329–332.

9. Yoshida, M., Seiki, M., Yamaguchi, K., and Takatsuki, K. (1984). Monoclonal integration of human T-cell leukemia provirus in all primary tumors of adult T-cell leukemia suggests causative role of human T-cell leukemia virus in disease. *Proc. Natl. Acad. Sci. USA* 81:2534–2537.

10. Wong-Staal, F., Hahn, H., Manzari, V., Colonbini, S., Franchini, G., Gelman, E. P., and Gallo, R. C. (1983). A survey of human leukemia for sequences of a human retrovirus. *Nature* 302:626–628.

11. Shimoyama, M., Kagami, Y., Shimotohno, K., Miwa, M., Minato, K., Tobinai, K., Suemasu, K., and Sugimura, T. (1986). Adult T-cell leukemia/lymphoma not associated with human T-cell leukemia virus type I. *Proc. Natl Acad. Sci. USA* 83:4524–4528.

12. Ito, Y. (1985). The Epidemiology of human T cell leukemia/lymphoma virus. *Curr. Topics Microbiol. Immunol.* 115:99–112.

13. Bishopn, J. M. (1982). Retroviruses and cancer genes. In: *Advances in Cancer Research*, Vol. 37 (G. Klein and S. Weinhouse, eds.), pp. 1–32. New York, Academic Press.

14. Toyoshima, K., Yamamoto, T., Kawai, S., and Yoshida, M. (1987). Viral oncogenes, v-yes and v-erb and their cellular counterparts. In: *Advances in Virus Research*, Vol. 32 (K. Maramorosch, F. A. Murphy, and A. J. Shatkin, eds.), pp. 97–128. Orlando, FL, Academic Press.

15. Hayward, W. S., Neel, B. G., and Astrin, S. M. (1981). Activation of a cellular onc gene by promoter insertion in ALV-induced lymphoid leukosis. *Nature (London)* 290:475–480.

16. Lewis, W. G., Crittenden, L. B., Kung, H-J. (1983). Activation of the cellular oncogene c-erbB by LTR insertion: Molecular basis for induction of erythroblastosis by avian leukosis virus. *Cell* 33:357–368.

17. Nusse, R., and Varmus, H. E. (1982). Many tumors induced by the mouse mammary tumor virus contain a provirus integrated in the same region of the host genome. *Cell* 31:99–109.

18. Seiki, M., Hattori, S., Hirayama, Y., and Yoshida, M. (1984). Human adult T cell leukemia virus: Complete nucleotide sequence of the provirus genome integrated in leukemia cell DNA. *Proc. Natl. Acad. Sci. USA* 80: 3618–3622.

19. Seiki, M., Eddy, R., Shows, T. B., and Yoshida, M. (1984). Nonspecific integration of the HTLV provirus genome into adult T-cell leukemia cells. *Nature (London)* 309:640–642.

20. Hattori, S., Imagawa, K., Shimizu, F., Hashimura, E., Seiki, M., and Yoshida, M. (1983). Identification of envelope glycoprotein encoded by *env* gene of human T cell leukemia virus. *Gann* 74:790–793.

21. Hattori, S., Kiyokawa, T., Imagawa, K., Shimizu, F., Hashimura, E., Seiki, M., and Yoshida, M. (1984). Identification of *gag* and *env* gene products of human T cell leukemia virus (HTLV). *Virology* 136:338–347.

22. Kiyokawa, T., Seiki, M., Imagawa, K., Shimizu, F., and Yoshida, M. (1984). Identification of a protein (p40^X) encoded by a unique sequence pX of human T-cell leukemia virus type 1. *Gann* 75:747-751.

23. Kiyokawa, T., Seiki, M., Iwashita, S., Imagawa, K., Shimizu, F., and Yoshida, M. (1985). p27^{X-III} and p21^{X-III}, proteins encoded by the pX sequence of human T-cell leukemia virus type 1. *Proc. Natl. Acad. Sci. USA* 82:8359-8363.

24. Vogt, P. K. (1977). Genetics of RNA tumor virus. *Compr. Virol.* 10:341-455.

25. Kiyokawa, T., Yoshikura, H., Hattori, S., Seiki, M., and Yoshida, M. (1984). Envelope proteins of human T cell leukemia virus: Expression in *E. coli* and its application to studies of *env* gene functions. *Proc. Natl. Acad. Sci. USA* 81:6202-6207.

26. Weinstock, G. M., Ap Phys, C., Berman, M. L. Hampar, B., Jackson, D., Silhavy, T. J., Weisemann, J., and Zweig, M. (1983). Open reading frame expression vectors: A general method for antigen production in *Escherichia coli* using proteins fusions to β-galactonsidase. *Proc. Natl. Acad. Sci. USA* 80:4432-4436.

27. Nagy, K., Clapham, P., Sheingsong-Popov, R., and Weiss, R. A. (1983). Human T-cell leukemia virus type 1: Induction of syncytia and inhibition by patients sera. *Int. J. Cancer* 32:321-328.

28. Hosino, H., Shimoyama, M., Miwa, M., and Sugimura, T. (1983). Detection lymhocytes producing a human retrovirus associated with adult T cell leukemia by syncytia induction assay. *Proc. Natl. Acad. Sci. USA* 80:7377-7341.

29. de Rossi, A., Aldovini, A., Franchini, G., Mann, D., Gallo, R. C., and Wong-Staal, F. (1985). Clonal selection of T-lymphocytes infected by cell-free human T-cell leukemia/lymphoma virus type 1: Parameters of virus integration and expression. *Virology* 143:640-645.

30. Hahn, B., Gallo, R. C., Franchini, G., Popovic, M., Aoki, T., Salahuddin, S. Z., Markham, P. D., and Wong-Staal, F. (1984). Clonal selection of human T-cell leukemia virus-infected cells in vivo and in vitro. *Mol. Biol. Med.* 2:29-36.

31. Tateno, M., Kondo, N., Itoh, T., Chubachi, T., Togashi, T., and Yoshiki, T. (1984). Rat lymphoid lines with human T-cell leukemia virus production. I. Biological and serological characterization. *J. Exp. Med.* 159:1105-1116.

32. Miyoshi, I., Yoshimoto, S., Taguchi, H., Kubonishi, I., Fujishita, M., Ohtsuki, Y., Shiraishi, Y., and Adagi, T. (1983). Transformation of rabbit lymphocytes with T-cell leukemia virus. *Gann* 74:1-4.

33. Clapham, K., Nagy, P., and Weiss, R. A. (1984). Pseudotypes of human T-cell leukemia virus types 1 and 2: Neutralization of patients' sera. *Proc. Natl. Acad. Sci. USA* 81:3083-3086.

34. Hoshino, H., Clapham, K., Weiss, R. A., Miyoshi, I., Yoshida, M., and Miwa, M. (1985). Human T cell leukemia virus type 1: Pseudotype neutralization

of Japanese and American isolates with human and rabbit sera. *Int. J. Cancer* 36:671–675.

35. Hinuma, Y., Gotoh, Y., Sugamura, K., Natgata, K., Goto, T., Nakai, M., Kamada, N., Matsumoto, T., and Kinoshita, K. (1982). A retrovirus associated with human adult T cell leukemia: In vitro activation. *Gann* 73:341–344.

36. Clarke, M. F., Trainor, C. D., Mann, D. L., Gallo, R. C., and Reitz, M. S. (1984). Methylation of human T-cell leukemia virus proviral DNA and RNA expression in short- and long-term cultures of infected cells. *Virology* 135:97–104.

37. Nakamura, H., Hayami, M., Ohta, Y., Tsujimoto, H., Kiyokawa, T., Yoshida, M., Sasagawa, A., and Honjo, S. (1987). Protection of cynomolgus monkeys against infection by human T-cell leukemia virus type-1 by their immunization with viral env gene products produced in *Escherichia coli*. *Int. J. Cancer*, 40:403–407.

38. Seiki, M., Hikikoshi, A., Taniguchi, T., and Yoshida, M. (1985). Expression of the pX gene of HTLV-1: General splicing mechanism in the HTLV family. *Science* 228:1532–1534.

39. Nagashima, K., Yoshida, M., and Seiki, M. (1986). A single species of pX mRNA of HTLV-1 encodes *trans*-activator $p40^X$ and two other phosphoproteins. *J. Virol.* 60:394–399.

40. Sodroski, J. G., Rosen, C. A., and Haseltine, W. A. (1984). Transacting transcriptional activation of the long terminal repeat of human T lymphotropic viruses in infected cells. *Science* 225:381–385.

41. Fujisawa, J., Seiki, M., Kiyokawa, T., and Yoshida, M. (1985). Functional activation of long terminal repeat of human T-cell leukemia virus type I by *trans*-acting factor. *Proc. Natl. Acad. Sci. USA* 82:2277–2281.

42. Seiki, M., Inoue, J., Takeda, T., Hikikoshi, A., Sato, M., and Yoshida, M. (1985). The $p40^X$ of human T-cell leukemia virus type 1 is a *trans*-acting activator of viral gene transcription. *Gann* 76:1127–1131.

43. Seiki, M., Inoue, J., Takeda, T., and Yoshida, M. (1986). Direct evidence that $p40^X$ of human T-cell leukemia virus type is a *trans*-acting transcriptional activator. *EMBO J.* 5:561–565.

44. Sodroski, J., Rosen, C., Goh, W., and Haseltine, W. A. (1985). A transcriptional activator protein encoded by the x-lor region of the human T-cell leukemia virus. *Science* 228:1430–1434.

45. Felber, B. K., Paskalis, H., Kleinman-Ewing, C., Wong-Staal, F., and Pavlakis, G. N. (1985). The pX protein of HTLV-1 is a transcriptional activator of its long terminal repeats. *Science* 229:675–679.

46. Chen, I. S. Y., Cann, A. J., Shah, N. P., and Gaynor, R. B. (1985). Functional relation between HTLV-2 *x* and adenovirus E1A protein in transcriptional activation. *Science* 230:570–573.

47. Fujisawa, J., Seiki, M., Sato, M., and Yoshida, M. (1986). A transcriptional enhancer sequence of HTLV-1 is responsible for trans-activation mediated by $p40^X$ of HTLV-1. *EMBO J.* 5:713–718.

48. Paskalis, H., Felber, B. K., and Pavlakis, G. N. (1986). *Cis*-acting sequences responsible for the transcriptional activation of human T cell leukemia

virus type 1 constitute a conditional enhancer. *Proc. Natl. Acad. Sci. USA* 83:6558–6562.

49. Shimotohno, K., Takano, M., Teruuchi, T., and Miwa, M. (1986). Requirement of multiple copies of a 21-nucleotide sequence in the U3 regions of human T cell leukemia virus type 1 and type 2 long terminal repeats for *trans*-acting activation of transcription. *Proc. Natl. Acad. Sci. USA* 83: 8112–8116.

50. Inoue, J., Seiki, M., and Yoshida, M. (1986). The second pX product p27^{x-III} of HTLV-1 is required for *gag* gene expression. *FEBS Lett.* 209: 187–190.

51. Inoue, J., Yoshida, M., and Seiki, M. (1987). Transcriptional (p40^{x}) and post-transcriptional (p27^{x-III}) regulators are required for the expression and replication of human T cell leukemia virus type 1 genes. *Proc. Natl. Acad. Sci. USA* 84:3653–3657.

52. Hidaka, M., Inoue, M., Yoshida, M., and Seiki, M. (1988). Posttranscriptional regulator (*rex*) of HTLV-1 initiates expression of viral structural proteins but suppresses expression of regulatory proteins. *EMBO J.* 7:519–523.

53. Seiki, M., Inoue, J., Hidaka, M., and Yoshida, M. (1988). Two *cis*-acting elements responsible for post-transcriptional trans-regulation of gene expression of human T cell leukemia virus type 1. *Proc. Natl. Acad. Sci. USA* 85:7124–7128.

54. Chen, I. S. Y., Slamon, D. J., Rosenblatt, J. D., Shah, N. P., Quan, S. G., and Wachsman, W. (1985). The x gene is essential for HTLV replication. *Science* 229:54–58.

55. Depper, J. M., Leonard, W. J., Kronke, M., Waldmann, T. A., and Greene, W. C. (1984). Augmented T-cell growth factor receptor expression in HTLV-1-infected human leukemic T-cells. *J. Immunol.* 133:1691–1695.

56. Popovic, M., Lange-Wantzin, G., Sarin, P. S., Mann, D., and Gallo, R. C. (1983). Transformation of human umbilical cord blood T cells by human T cell leukemia/lymphoma virus. *Proc. Natl. Acad. Sci. USA* 80:5402–5406.

57. Yodoi, J., Uchiyama, T., and Maeda, M. (1983). T-cell growth factor receptor in adult T-cell leukemia. *Blood* 62:509–511.

58. Inoue, J., Seiki, M., Taniguchi, T., Tsuru, S., and Yoshida, M. (1986). Induction of interleukin 2 receptor gene expression by p40 encoded by human T-cell leukemia virus type 1. *EMBO J.* 5:2883–2888.

59. Maruyama, M., Shibuya, H., Harada, H., Hatakeyama, M., Seiki, M., Fujita, T., Inoue, J., Yoshida, M., and Taniguchi, T. (1987). Evidence for aberrant activation of the interleukin-2 autocrine loop by HTLV-1-encoded p40^{x} and T3/Ti complex triggering. *Cell* 48:343–350.

60. Cross, S. L., Feinberg, M. B., Wolf, J. B., Holbrook, N. J., Wong-Staal, F., and Leonard, W. J. (1987). Regulation of the human interleukin-2 receptor *a* chain promoter: Activation of a nonfunctional promoter by the *trans*-activator gene of HTLV-1. *Cell* 49:47–56.

61. Leung, K., and Nabel, G. J. (1988). HTLV-1 transactivator induces interleukin-2 receptor expression through an NF-KB-like factor. *Nature* 333:776–778.

62. Böhnlein, E., Lowenthal, J. W., Siekevitz, M., Ballard, D. W., Franza B. R., and Greene, W. C. (1988). The same inducible nuclear proteins regulates mitogen activation of both the interleukin-2 receptor alpha gene and type 1 HIV. *Cell* 53:827-836.

63. Fujisawa, J., Seiki, M., Tokta, M., Miyatake, S., Arai, K., and Yoshida, M. (1988). Cell line specific activation of SV40 transcriptional enhancer by p40^{tax} of HTLV-1. *Jpn. J. Cancer Res.* 79:800-804.

64. Saito, S., Nakamura, M., Ohtani, K., Ichijo, M., Sugamura, K., and Hinuma, Y. (1988). *Trans*-activation of the SV40 enhancer by a pX product of human T-cell leukemia virus type I. *J. Virol.* 62:644-648.

65. Miyatake, S., Seiki, M., Malefijt, R. D., Heike, T., Fujisawa, J., Takebe, Y., Nishida, J., Shlomai, J., Yokota, T., Yoshida, M., Arai, K., and Arai, N. (1988). Activation of T-cell-derived lymphokine genes in T-cells and fibroblasts: Effects of human T-cell leukemia virus type I, p40^X protein and bovine papilloma virus encoded E2 protein. *Nucl. Acid. Res.*, 16: 6547-6566.

66. Sharon, M., Klausner, R. D., Cullen, B. R., Chizzonite, R., and Leonard, W. J. (1986). Novel interleukin-2 receptor subunit detected by cross-linking under high-affinity conditions. *Science* 234:859-863.

67. Teshigawara, K., Wang, H. M., Kato, K., and Smith, K. A. (1987). Interleukin 2 high-affinity receptor expression requires two distinct binding proteins. *J. Exp. Med.* 165:223-238.

68. Tsudo, M., Kozak, R. W., Goldman, C. K., and Waldmann, T. A. (1986). Demonstration of a non-Tac peptide that binds interleukin 2: A potential participant in a multichain interleukin 2 receptor complex. *Proc. Natl. Acad. Sci. USA* 83:9694-9698.

69. Dukovich, M., Wano, Y., Thuy, L. B., Katz, P., Cullen, B. R., Kehrl, J. H., and Greene, W. C. (1987). A second human interleukin 2 binding protein that may be a component of high-affinity interleukin 2 receptor. *Nature* 327:518-522.

70. Yoshida, M. (1987). Expression of the HTLV-1 genome and its association with a unique T-cell malignancy. *Biochim. Biophys. Acta* 970:145-161.

71. Yoshida, M., and Seiki, M. (1987). Recent advances in the molecular biology of HTLV-1: *trans*-activation of viral and cellular genes. *Annu. Rev. Immunol.* 5:541-559.

8

Molecular Biology and Pathogenesis of HTLV-II

Alan J. Cann,* Joesph D. Rosenblatt, William Wachsman, and Irvin S. Y. Chen
UCLA School of Medicine, Los Angeles, California

I. INTRODUCTION

Human T-cell leukemia virus type I (HTLV-I) is implicated as the causative agent of adult T-cell leukemia (ATL) (1–8). This disease is an aggressive T-lymphoid malignancy that is endemic in several areas of the world, including Japan, Africa, and the Caribbean Basin. In contrast, HTLV-II is reported to have been isolated from only a few patients. Of these, one patient had acquired immune deficiency syndrome (AIDS) (9), another, hemophilia (10). However, in two additional patients, isolation of HTLV-II was associated with the diagnosis of atypical hairy-cell leukemia (11–15). The isolation of HTLV-II in two patients with atypical hairy-cell leukemia suggests that the virus has an etiological role in this rare disease and implicates HTLV-II as the second known tumorigenic retrovirus in humans.

The first isolate of HTLV-II was identified in the Mo-T cell line, derived from the spleen of a patient diagnosed to have T-cell hairy-cell leukemia. The virus was detected by its antigenic cross-reactivity with the *gag*-encoded structural proteins, p19 and p24, of HTLV-I (11). However, competition radioimmunoassays indicated the p24 antigens of HTLV-I and HTLV-II showed only partial cross-reactivity, indicating that HTLV-II was a distinct new subtype of human T-cell leukemia virus. The complete nucleotide sequences of the provirus genomes of HTLV-I (16) and HTLV-II (17) have now been determined. They share approximately 60% overall nucleotide sequence homology. Both virus genomes of HTLV-I (16) and HTLV-II (17) have now been determined. They share approximately 60% overall nucleotide sequence homology. Both virus genomes are of approximately equal size and encode a similar set of proteins.

Present affiliation: Medical Research Council (MRC) Centre, Cambridge, United Kingdom

We have recently carried out a molecular investigation of the virus in the leukemia cells of a second patient with atypical hairy-cell leukemia (15). Our findings strongly suggest a role for the virus in the evolution of the malignancy. Also, the data we obtained reiterate a fundamental paradox in HTLV-associated malignancies. Although an oligoclonal pattern of provirus integration was detected in leukemic T cells, no viral RNA expression was detectable in the fresh leukemic cells. This property is common to all members of the HTLV/BLV (bovine leukemia virus) family of viruses. However, the difference in the relatively indolent course of HTLV-II-associated hairy-cell leukemia and the much more aggressive course of ATL suggest that subtle molecular variations between the closely related viruses HTLV-I and HTLV-II are responsible for the clinical differences between these syndromes. Therefore, we have carried out a detailed molecular analysis of the HTLV-II genome in order to explore the mechanism of cellular transformation of HTLV and the pathogenesis of HTLV-associated malignancy. Here, we review our findings and relate them to unanswered questions concerning the mechanism of cellular transformation by HTLV and the biology of HTLV infection.

II. STRUCTURE OF THE HTLV-II GENOME

While HTLV-II has been encountered only rarely, the availability of a closely related pair of viruses has allowed much of what is known about the HTLV-I genome to be derived from comparisons with HTLV-II. Early hybridization studies showed that neither HTLV-I nor HTLV-II contains a classical viral oncogene, homologous to normal cellular sequences. This was confirmed when the complete nucleotide sequences of the HTLV-I and -II proviruses were determined. However, both viruses do share a unique structural feature. In addition to the normal retroviral complement of *gag, pol,* and *env* genes, there is a fourth coding region which was termed the X region by Seiki et al. (16). This region, located at the 3' end of the genome, contains several open reading frames. The longest open reading frame, the *tax* gene, encodes proteins in both HTLV-I and HTLV-II, called p40^{taxI} and p34^{taxII} (previously referred to as *x, x-lor,* or *tat*). A second open reading frame within the X region encodes a pair of proteins called p27^{rexI}/p21^{rexI} in the case of HTLV-I (18), and p26^{rexII}/p24^{rexII} for HTLV-II (19). As the mechanism of cellular transformation by these viruses remains unknown, much interest has centered on the functions of these unique proteins.

Close inspection of the nucleotide sequence of HTLV-I and HTLV-II proviruses shows that they are similar in size and overall structural features. However, they nevertheless differ substantially from each other in certain regions of the genome. A comparative survey of the HTLV-I and -II genomes is given in Table 1. The complete nucleotide sequence of HTLV-II (17) was determined from a cloned provirus genome which produces infectious virus when transfected into susceptible cells (20). The availability of this infectious DNA clone has ensured the accuracy of the sequence data described below and summarized in Figure 1. The HTLV-II provirus is 8952 nucleotides in length and is bounded at both ends by long terminal repeats (LTRs). The LTR is 763 base pairs (bp) in length, of which the U3 region comprises 314 bp. The R region of the LTR is unusually long, at 248 bp (see below). The U5 region comprises 199 bp. This is comparable with HTLV-I, where U3, R, and U5 comprise 353, 229, and 173 bp, respectively. As with all retroviruses, the integrated virus DNA is terminated by the sequence TG. . . .CA (21,22). Integration of HTLV-II results in the duplication of six bases of cellular DNA on either side of the provirus (21), as does HTLV-I (16).

In contrast to other regions of the genome, there is no detectable nucleotide sequence homology between the LTRs of HTLV-I and -II, outside of certain conserved sequences which are described below. As with all retroviruses, the U3 region contains promoter and enhancerlike elements responsible for the transcription of the virus genome (21). These transcriptional control elements are conserved between HTLV-I and HTLV-II, suggesting that they are of functional importance. Twenty-one of twenty-two bases surrounding the cap site (the site at which RNA transcription is initiated) are identical, there is a sequence of 21 nucleotides which is repeated three times in the U3 region of the LTR. Similar sequences are also present at comparable positions in the HTLV-I LTR. These sequences are thought to function as an enhancer element for transcription, and to confer responsiveness to the virus-encoded *tax* protein (see below). The unusually long R region (248 bp) is also common to HTLV-I. The polyadenylation signal, AAUAAA, is located an unusually large distance (290 nucleotides) from the 3' end of the viral mRNA. A unique secondary structure has been proposed for the 3' end of HTLV-I mRNA that would bring the polyadenylation signal into proximity with the plyadenylation site of the message (16). Similar sequences are present in the HTLV-II LTR, and a similar structure could theoretically be formed, although there is no direct evidence that such structures actually exist in the mRNA.

The *gag* gene of HTLV-II comprises 1299 nucleotides (807-2106) and, by comparison to HTLV-I, is believed to encode three proteins of 15 kd, 24 kd, and 9 kd (Table 1). In common with other mammalian retroviral *gag* proteins, the NH_2 terminus of the HTLV-I *gag* peptide is linked to myristic acid (23). Although no direct information is available for the HTLV-II *gag* proteins, it is probable, by analogy to HTLV-I, that they are also myristylated at the NH_2 terminus (Table 1).

At its 3' end, the HTLV-II *gag* gene overlaps an open reading frame which encodes a putative protease (Figure 1). This reading frame is 534 nucleotides in length, sufficient to encode a polypeptide of 178 amino acids (17). By analogy to other retroviruses, the location of this second open reading frame suggests that it encodes a viral protease. The predicted amino acid sequence of this

TABLE 1 Genomic Structure of HTLV–I and HTLV–II

	Provirus	LTR	*gag*	Protease	*pol*
HTLV-I	9032 nt	755 nt	802–2089	?	2497–5184
HTLV-II	8952 nt	763 nt	807–2106	2078–2611	2239–5184
Proteins encoded	*HTLV-I:*		14, 26, 9 kd	?	896 aa
	HTLV-II:		15, 24 9 kd	178 aa	982 aa
Sequence homology			55%, 85% 68% (aa)	?	61% (aa)

Data for HTLV–I taken from Seiki et al. (16) and for HTLV–II from Shimotohno et al. (17).

reading frame shows significant homology to known proteases of other retroviruses, notably Rous sarcoma virus and Moloney murine leukemia virus.

Unlike most retroviruses, where the protease is expressed from the same open reading frame as *gag* and is a cleavage product of the *gag* precursor, expression of the HTLV–II protease involves translation with a ribosomal shift in reading frames and subsequent cleavage of a fused *gag*-protease precursor. Initially, it was suggested that the presence of hexadeoxy adenylate, hexadeoxy guanylate, and hexadeoxy cytidylate clusters at the *gag* protease junction may play a role in RNA processing necessary for expression of the protease (17). However, the HTLV–II genome contains the sequence AAAAAAC (nucleotide positions 2080–2087) at the *gag*-protease junction. The same sequence is also present at the 3′ end of the *gag* gene in HTLV–I (nucleotide positions 2064–2070). In other retroviruses, this consensus sequence has been shown to be involved in ribosomal frameshifting responsible for expression of fusion proteins (T. Jacks and H. Varmus, personal communication; 24). A second ribosomal frameshifting event occurs in the 3′ part of the protease coding region in order to allow expression of the *pol* gene (see below). Although the mechanism of expression of the HTLV protease has not been formally proven, based on homology with the protease gene of other retroviruses such as mouse mammary tumor virus, it is probable that the protease and *pol* genes are expressed by a similar mechanism involving frameshifting.

env	X region	tax	rex
5180–6643	6644–8356	7302–8356 (X–IV)	7302–7811 (X–III)
5180–6637	6638–8202	7214–8202 (X-c)	7214–8202 (X-b)
GP68 / \ GP46 GP21	—	40 kd	27 kd/21 kd
GP62 / \ GP35 GP20	—	37 kd	26 kd/24 kd
63%, 73% (aa)	33% (nt, untranslated region)	82% (aa)	61% (aa)

The protease protein itself has not been characterized, and further studies of this region of the genome are necessary to fully understand the expression and role of the protease molecule. In some molecular clones of HTLV-I, this putative protease reading frame is closed by a termination codon (16). It has not yet proved possible to isolate an infectious HTLV-I provirus clone (M. Yoshida, personal communication, unpublished observations), and the nucleotide sequence data available for HTLV-I were derived from noninfectious clones. Therefore, the significance of sequence differences between HTLV-I and -II in the protease gene is not clear, and the inability to produce a functional protease might be related to the lack of infectivity of the available HTLV-I provirus clones.

The *pol* region extends from nucleotide 2239 to nucleotide 5184, overlapping the protease gene at its 5′ end and the *env* gene at its 3′ end. This reading frame has the potential to encode 982 amino acids, and by comparison to other retroviruses, the 5′ portion of the gene is believed to encode reverse transcriptase and sequences further downstream, the probable nuclease and RNase H functions.

The *env* gene comprises nucleotides 5180–6637 , sufficient to encode a protein of 486 amino acids. Based on the size and cleavage of *env* glycoproteins in HTLV-I, the cleavage site of these proteins in HTLV-II is probably located at amino acid 308 in the *env* precursor polyprotein, producing amino acid chains of

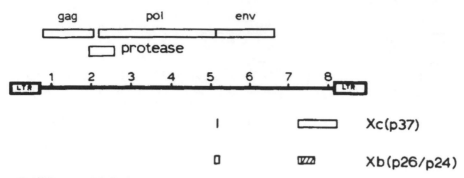

FIGURE 1 HTLV–II gene structure. Location of open reading frames in the HTLV–II genome known to encode protein products. Exons of genes in the X region are joined by RNA splicing (see text).

35 kd for the surface glycoprotein and 20 kd for the transmembrane protein. The HTLV *env* glycoprotein has been shown to be responsible for syncytium formation in a variety of nonlymphoid, adherent cell types (25). However, it is not clear whether this process plays any role in HTLV-associated disease in vivo. It is believed that the *env* protein is responsible for the interaction of the virus with the target cell membrane. The cellular receptor for HTLV has not been identified, although interference studies with VSV pseudotypes indicate that both HTLV-I and -II share the same receptor, and that this molecule is present on most cell types (25).

The X region is situated 3' of the *env* gene and comprises nucleotides 6638–8202. The 5' part of this region is a noncoding stretch of 572 nucleotides. In the 3' part of the X region, there are three open reading frames, which are believed to encode at least two proteins. Details of this region of the genome are shown in Figure 1. The *tax* protein of HTLV-II, p37^{taxII}, is encoded primarily by the largest open reading frame, X-c. In HTLV-I, the *tax* protein, p40^{taxI}, is encoded by the X-IV open reading frame. p37^{taxII} is present predominantly in the nucleus of HTLV-infected cells and has a relative short half-life of approximately 120 min (26). An additional pair of proteins, called p27^{rexI}/p21^{rexI}, was first identified in the nucleus of HTLV-I-infected cells (18). These proteins are both encoded by the X-III reading frame, and arise from initiation of translation at alternative ATG codons (27). All three X proteins are encoded by a single 2.1-kb RNA transcript, composed of three exons (27–29) (see below). The HTLV-II X region also encodes a pair of proteins from the X-b reading frame, p26^{rexII} and p24^{rexII} (19). These proteins, which probably have related primary structures but appear distinct owing to posttranscriptional modification, are analogous to p27^{rexI} of HTLV-I (Figure 1). Like p27^{rexI}, p26^{rexII} and p24^{rexII} are located in the nuclear fraction of infected cells. The HTLV-II p26^{rexII} protein shares 61% amino acid homology with p27^{rexI} of HTLV-I (Table 1).

III. EXPRESSION OF VIRAL GENES

Transcription of viral RNA is initiated at the cap site, which corresponds to the U3/R boundary in the 5' LTR of the provirus. At least three messenger RNAs are produced to encode the viral proteins (30). These messages are shown in Figure 2. The *gag* and *pol* gene products are translated from an unspliced genomic mRNA, as described previously. The protease gene product is probably also produced from this species of mRNA (see above). *env* gene products are produced from a singly spliced 4.5-kb messenger RNA (30).

It has been shown that p37^{taxII} is produced from a 2.1.-kb doubly spliced mRNA (29). A short second exon which includes the *env* methionine codon and one additional nucleotide at its 3' end are spliced to the third exon downstream, containing the major *tax* and *rex* open reading frame. HTLV-I *tax* mRNA is produced by a similar double-splicing mechanism (28,29). As described earlier, this message also encodes the minor X region products, p27^{rexI}/p21^{rexII} (27),

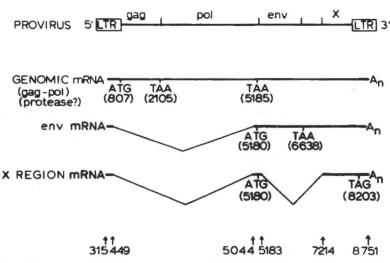

FIGURE 2 Expression of the HTLV-II genome. Top line shows HTLV-II provirus genome. Three species of messenger RNA which encode the gene products are shown beneath. Heavy lines represent exons of mature mRNAs; light lines, introns which are removed from mature RNA species by splicing. Nucleotide positions of splice donor and acceptor sites are indicated by arrows at bottom of figure.

and presumably encodes p26^{rex}II and p24^{rex}II in HTLV-II. The unusual exon structure mechanism of the *tax* gene splicing may be important in the regulation of expression of the *tax* gene (29). One report indicates that p27^{rex}I/p21^{rex}I of HTLV-I are involved in expression of the *gag* gene (31). Based on their amino acid sequence homologies, p26^{rex}II/p24^{rex}II of HTLV-II presumably act in an analogous way to regulate HTLV-II gene expression. It has recently been shown that HTLV-II *trans*-activation is regulated by both the *tax* and *rex* genes (31a). The function of the *tax* protein in *trans*-activation and its possible involvement in cellular transformation are discussed in the following sections.

IV. *TRANS*-ACTIVATION OF TRANSCRIPTION

DNA transfection experiments in which recombinant LTR-chloramphenicol acetyl transferase (CAT) constructions are introduced into HTLV-infected cells indicated that there is a factor(s) in infected cells that activates transcription from the LTR (*trans*-activation) (32,33). Using cotransfections of LTR-CAT constructions plus recombinant constructions that express the *tax* gene, it was shown that the tax protein was responsible for *trans*-activation (34-38). Later results demonstrated that p37^{rex}II alone was responsible for this phenomenon, and that no other viral genes were required (35-38). By making a series of recombinant *tax* gene constructions that efficiently express the *tax* protein from a nonspliced mRNA, we have performed a direct quantitative comparison of the function of p37^{tax}II and p40^{tax}I (35). These studies showed that, as had been indicated by earlier results (32,34), p37^{tax}II could *trans*-activate the HTLV-I LTR, in addition to the HTLV-II LTR. In contrast, p40^{tax}I activates the HTLV-I LTR, but not the heterologous HTLV-II LTR, indicating a potential difference in the mechanism of *trans*-activation by the two proteins (Figure 3). Our results also showed that both *tax* proteins were functional in all the mammalian cell lines tested, but that p40^{tax}I did not function in avian cells (35). This finding indicates that cellular factors are involved in *trans*-activation, which may also explain the different patterns of *trans*-activation seen with the two proteins. Recent data indicate that the *tax* protein does not bind directly to the LTR, and that its effect is probably mediated by one or more cellular factors (38a).

As stated earlier, the *tax* protein is an integral part of the HTLV genome, and not a classical oncogene with homology to a cellular sequence. The critical importance of the *tax* protein for HTLV replication was confirmed by the demonstration that p37^{tax}II is essential for the replication of HTLV-II (33). Using an infectious cloned HTLV-II provirus, we introduced deletions into the *tax* gene. The resulting mutant virus genomes were stably transfected into a B-cell line. Transcription of virus genes in these cell lines was very low (1% of wild-type), and infectious virus was not produced. By superinfecting these cell

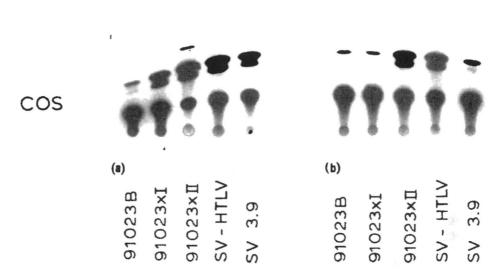

FIGURE 3 *Trans*-activation of transcription from the HTLV LTR. CAT activity in COS fibroblasts cotransfected with recombinant HTLV LTR-CAT constructions is demonstrated by conversion of ^{14}C-chloramphenicol (lower band) into acetylated forms. As described in the text, constructions that express both p40$^{tax\text{I}}$ (91023 xI and SV-3.9) and p34$^{tax\text{II}}$ (91023 xII and SV-HTLV) activate the HTLV–I LTR (LTR-I CAT) in *trans* (left panel). However, while p34$^{tax\text{II}}$ activates the HTLV–II LTR (LTR-II CAT), p40$^{tax\text{I}}$ does not activate above the background level seen with the 91023-B vector (right panel). (From Ref. 35.)

lines with wild-type replication-competent HTLV-II, transcription of the mutant genome was restored to normal levels seen in infected cells, and mutant virus genomes were rescued from the cells.

   In order to investigate the mechanism of *trans*-activation and possibly provide insights into the role of the *tax* gene in immortalization of HTLV-infected cells, we studied a number of *tax* gene mutants, with phenotypes that differ from those of the wild-type virus. The *tax* gene expression vectors we have developed facilitate the construction of mutant *tax* proteins (35). One of these mutations consists of a deletion of amino acids 2-16 at the amino-terminus of p37$^{tax\text{II}}$ and results in a protein that is functionally unable to *trans*-activate either the HTLV-I or -II LTR (39). The phenotype of this mutant demonstrates the importance of the NH$_2$-terminal region of the protein to its function. Accordingly, we have concentrated our studies on this region. Substitution of leucine for proline at amino acid position 5 of p37^{tax}II gives rise to a protein that is unable to *trans*-activate the HTLV-II LTR (39). However, unlike the deletion mutant, the mutant Leu^5 protein inhibits *trans*-activation of the HTLV-II LTR by wild-type p37^{tax}II. Possibly, the altered phenotype of this mutant unmasks a potential negative autoregulatory function of p37^{tax}II which may be important in the life-cycle of the virus (see Section VI, "Biology of HTLV-II Infection"). These two observations, together with other mutations in a 17-amino-acid domain at the amino-terminus of the *tax* protein, indicate that this region of the protein is functionally important in *trans*-activation.

The kinetics of virus replication following HTLV infection of cells are related to the level of *tax* gene expression. In a typical retrovirus infection, stable integration and RNA expression are dependent on passage of the infected cells through the S phase of the cell cycle (40). In contrast, the kinetics of HTLV infection in vitro are very slow (unpublished observations). We propose that following integration into the genome of the host cell, low levels of *tax* mRNA are transcribed. Production of small amounts of *tax* protein stimulates transcription from the LTR, in turn producing higher levels of *tax* protein, and so on, in a positive feedback cycle. A single amino acid substitution five residues from the NH$_2$-terminus of p37^{tax}II (see above) results in a mutant protein which is no longer capable of activating the LTR, but does block *trans*-activation by the wild-type protein (39). The altered phenotype of this mutant protein suggests that the wild-type *tax* protein might possess a potential negative regulatory function which may be important to control the positive regulation of expression that occurs upon infection of a target cell (see above). Such a scheme would enable HTLV to regulate gene expression more closely than other retroviruses and, under certain circumstances, might enable the virus to establish a latent state (see Section VII, "Biology of HTLV-II Infection"). However, the *rex* protein also has the potential to down-regulate transcription of virus genes and might be involved in establishment and/or maintenance of latency (unpublished observations).

The precise mechanism of *trans*-activation remains unknown. However, the *tax* protein does not appear to possess DNA binding activity (40a), and it appears that the *tax* protein must interact with other cellular factors and the *rex* protein in order to stimulate transcription. Subtle differences in the interaction of p37^{tax}II and p40^{tax}I with cellular transcription factors may explain the observed differences in the transcriptional properties of HTLV-I and -II.

V. CELLULAR TRANSFORMATION IN VITRO

Infection of peripheral blood T cells in vitro with HTLV-I or HTLV-II results in immortalization of clones of cells which continue to proliferate indefinitely without exogenous interleukin-2 (IL-2) (41–43). These cells resemble normal activated T cells by the criteria of surface marker phenotype and lymphokine production. Both CD-8- and CD-4-positive clones of cells may arise; these cells are Tac-antigen-positive (44; unpublished observations). Transformation is defined only by immortalization. The cells will not clone in soft agar without IL-2 and are initially density-dependent. Later passages of some HTLV-II-immortalized cells will clone in soft agar at limiting dilution. We hypothesize that in these cases, secondary changes have occurred to confer additional properties on these cells. Quantitative data are not available to compare the efficiency of transformation and properties of the cells transformed by HTLV-I or –II, but the properties and growth of cloned cells in vitro appear to be similar (42,45). The clinical conditions associated with the two viruses are distinct, and the relationship between cellular immortalization in vitro and tumorigenesis in vivo is unclear (see below). An understanding of the mechanism of action of the *tax* protein, which is central to the biology of HTLV, may explain the process of cellular transformation by the virus and account for some of the clinical differences between ATL and hairy-cell leukemia.

Since the *tax* gene plays a central role in the life-cycle of HTLV, we have conducted extensive investigations into the requirement for and function of the *tax* protein (see Section IV, "*trans*-Activation of Transcription"). We have also performed quantitative comparisons of the *trans*-activating capabilities of the HTLV-I and -II *tax* proteins. As discussed earlier, p37^{taxII} *trans*-activates both the HTLV-I and HTLV-II LTRs, whereas p40^{taxI} significantly *trans*-activates only its own LTR (34,35). With the discovery of the *tax* gene, it was suggested that this was the transforming gene of HTLV. Its function in *trans*-activation further suggested that the mechanism of cellular transformation might be aberrant transcriptional activation of cellular promoters in *trans*. However, it is possible that T-cell transformation results from a more complex interaction between the *tax* protein and cellular gene expression, rather than merely by inappropriate transcriptional activation.

p37^{taxII} activates, in addition to the HTLV-I and HTLV-II LTRs, a non-HTLV-related promoter, the adenovirus early region promoter, E3 (46). The basis for this functional homology between the adenovirus E1A protein, which normally activates E3, and the *tax* protein is not clear, since neither the *trans*-activating proteins nor the E3 or HTLV-II promoters share any significant amino acid or nucleotide sequence homology. In spite of the homology between p37^{taxII} and p40^{taxI} (Table 1), only p37^{taxII} is able to activate this heterologous promoter in *trans*. The discovery that p37^{taxII} is able to activate transcription from heterologous promoters that share no obvious homology to its own LTR supports the theory that promiscuous *trans*-activation might result in cellular transformation. It has been reported that p40^{taxI} activates expression of interleukin 2 and its receptor (IL-2R or Tac) in HTLV-I-infected T cells (47,48). p37^{taxII} was reported to activate IL-2R (49), although this study has recently been questioned (50). The significance of these observations is uncertain, especially since the HTLV-II leukemic cells are Tax antigen-negative (above) and IL-2 is not expressed by all cell lines transformed in vitro (51).

The transcriptional regulatory properties of the *tax* protein are more complex than was initially thought. Mutants of p37^{tax}II we have constructed may unmask a potential negative regulatory function present in the wild-type protein (39). Therefore, the *tax* protein may be involved in transformation by mechanisms other than by straightforward *trans*-activation of endogenous genes. In addition, the interaction of p40^{tax}I/p37^{tax}II and the additional gene products, p26^{rex}II/p24^{rex}II and p27^{rex}I/p21^{rex}I, has not been fully determined. It is not known whether these additional proteins are involved in transformation, although we have demonstrated through site-directed mutagenesis that the *rex* gene is essential for HTLV replication (unpublished observations). Thus, it is possible that cellular transformation results from the composite action of the products of the X region.

VI. BIOLOGY OF HTLV-II INFECTION

The original isolate of HTLV-II was derived from a patient (Mo) diagnosed with atypical T-cell variant hairy-cell leukemia at UCLA Medical Center in 1976 (14). We have recently described a second case of HTLV-II infection associated with atypical hairy-cell leukemia (15). There have also been a limited number of other reports of HTLV-II seropositivity. In 1984, one study revealed a cluster of HTLV-II-seropositive individuals among British intravenous (i.v.) drug users (52). Five percent of those sampled were positive for HTLV-II. Recently, a serological survey of i.v. drug abusers in New York showed 18% to be seropositive for HTLV-II (53). In both these reports, criteria for detection of HTLV-II were based on positive reactions in enzyme-linked immunosorbent assays, relying on antigen competition to differentiate between HTLV-I and HTLV-II infection. In neither case was the virus isolated. Two other HTLV-II-infected patients with T-lymphoid malignancies have been reported (54,55). These patients apparently suffered from an aggressive malignancy more similar to T-prolymphocytic leukemia than hairy-cell leukemia. Although tumor cells from one of these patients were shown to be infected with HTLV-II and able to transmit the virus to normal human T cells in cocultures (54), a firm etiological association between HTLV-II and malignancy in these cases was not established.

We have recently had the opportunity to carry out an extensive clinical and molecular examination of a second case of HTLV-II-associated atypical hairy-cell leukemia (15). The patient (NRA) was a 74-year-old white man, with a 6-year history of lymphoproliferative disease. His initial illness was characterized by mild leukocytosis, anemia, thrombocytopenia, splenomegaly, and recurrent cutaneous and pulmonary infections. The patent's peripheral blood contained many atypical lymphocytes which stained positively with Leu[4] (pan T-cell) monoclonal antibody. The majority of these cells were of T-suppressor phenotype, and a small minority contained tartrate-resistant acid phosphate (TRAP), a characteristic of hairy-cell leukemia. Therapeutic splenectomy was performed with a transient improvement in the patient's clinical condition.

Continuous cell lines were cultured from the patient's purified peripheral blood lymphocytes. These cells were shown to be infected with HTLV-II by expression of viral antigens, Southern hybridization, and passage of the virus to uninfected cells. Immortalized cell lines were also obtained by infection of peripheral blood lymphocytes from normal donors by cocultivating with lethally irradiated cell lines derived from the patient. Southern hybridization analysis of genomic DNA from the NRA cell lines using probes derived from the original HTLV-II isolate ($HTLV-II_{Mo}$) revealed hybridization to specific bands. Restriction mapping of the $HTLV-II_{NRA}$ provirus genome showed that it was closely related to, but not identical to, $HTLV-II_{Mo}$ (15).

Detection of the virus genome in peripheral blood mononuclear cells revealed an oligoclonal pattern of integration similar to that seen with HTLV-I in ATL (Figure 4). In addition, the patient's peripheral blood cells showed another unique characteristic similar to that of HTLV-I-infected ATL patients. While the provirus genome was readily detected in the tumor cells, no HTLV-II viral RNA expression could be detected under conditions where RNA was readily detectable in cell lines infected with the virus in vitro (55a). Therefore, HTLV-II-associated atypical hairy-cell leukemia, although clinically distinct from ATL, shares certain similarities to HTLV-I-associated disease at the molecular level. The pattern of virus integration, together with the repeat isolation of HTLV-II associated with atypical hairy-cell leukemia, strongly argues that the virus has a role in the etiology of the malignancy. However, it is clear that not all cases of hairy-cell leukemia or T-cell hairy-cell leukemia are associated with HTLV. Hairy-cell leukemia may therefore result from a spectrum of causes, some of which are of viral origin.

The patient eventually succumbed to extensive pleural and hepatic infiltrations of lymphoid cells. At autopsy, analysis of the patient's peripheral blood cells revealed the presence of two distinct malignant clones of cells. One of these was a clonal expansion of TRAP-positive B cells (demonstrated by an immunoglobulin gene rearrangement) associated with the lymphoid infiltrations. These cells did not contain the HTLV-II provirus. The other clone was a Tac⁻, OKTi⁺ (T-suppressor) T-cell fraction in which the HTLV-II genome had been detected. At present, the connection between these two malignancies is not clear.

This investigation of the role of HTLV-II in atypical hairy-cell leukemia, together with our studies on the kinetics of HTLV-II infection in vitro, and in particular, on the role of the *tax* gene in the virus life-cycle, have begun to build a picture of the natural history of HTLV-II infection. The virus displays strong

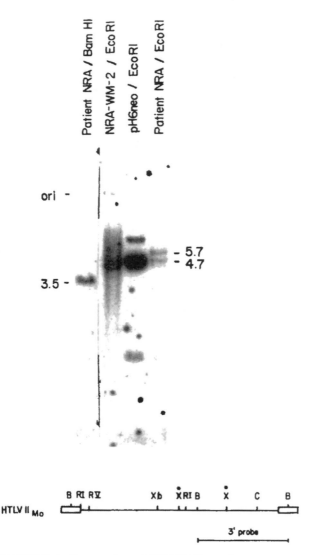

FIGURE 4 Integration of HTLV-II provirus genomes in infected cells. Southern hybridization analysis of DNA from cell lines infected with HTLV-II$_{NRA}$ (lane 2), HTLV-II$_{Mo}$ (lane 3), or fresh cells from patient NRA (lanes 1 and 4). In lane 1, the hybridization probe used was an internal 3.5-kb Bam HI fragment from HTLV-II$_{Mo}$. In lanes 2–4, the probe was a 5.5-kb Eco RI fragment of the HTLV-II$_{Mo}$ provirus, spanning the 3' end of the genome, plus some flanking chromosomal sequences. This probe detects multiple integration sites in the cellular DNA of cultured cell lines (lanes 2 and 3), but detects only two discrete fragments in fresh cells from the patient (lane 4), indicating an oligoclonal pattern of integration of HTLV-II provirus in these cells.

T-cell tropism; clones of immortalized cells growing out following transformation in vitro are invariably of T-cell phenotype. However, in the case of the patient NRA, the HTLV-II-infected tumor cells were Tac antigen (IL-2 receptor)-negative and of CD-8 (T-suppressor) phenotype. In ATL, the tumor cells are generally Tac-positive and of CD-4 (T-helper) phenotype, although there is one report of an HTLV-I-infected patient with a T-suppressor proliferation (56). However, this patient also had AIDS, and therefore, presumably, a depleted T-helper cell population.

We believe that our observations of HTLV-II infection in vitro and in vivo show that HTLV is capable of replication by two independent but complementary processes. Under the influence of the *tax* (and *rex*) proteins, high levels of transcription occur, and infectious virions are produced. In contrast, HTLV-induced T-cell proliferation would also result in replication of the proviral genome, as the virus-infected clone of cells expands. If this process results from products encoded by the X region, structural proteins may not be expressed, thus enabling the virus to escape host immune surveillance.

This HTLV-induced T-cell proliferation, while resembling immortalization of cells in vitro, is likely to be distinct from the end stages of tumorigenesis. This is suggested by the rarity of adult T-cell leukemia in HTLV-I-seropositive individuals (57), the long latent period preceding the appearance of a tumor (57), mono- or oligoclonal provirus insertion on the tumor cells, and lack of viral RNA expression, which contrast with productive HTLV infection associated with cellular transformation in vitro (above). Thus, multiple events in addition to viral infection appear to be necessary to generate adult T-cell leukemia. Other cofactors are probably necessary which are as yet unknown. Certain karyotypic abnormalities have been reported to occur with increased frequency in advanced ATL (58–60). While the degree of cytogenic aberration appears to be related to the severity of the disease, the precise molecular abnormalities that occur and their significance to the pathogenesis of the disease are unclear. Differences in the severity and prognosis of ATL compared with atypical hairy-cell leukemia possibly result from secondary events involved in tumorigenesis, subsequent to virus infection.

VII. SUMMARY

Comparative molecular biology of HTLV-I and HTLV-II affords the opportunity to investigate the pathogenesis of these two viruses. Both viruses have a similar genetic organization and share approximately 60% overall nucleotide sequence homology. However, in spite of their evident relatedness, the pathology of the diseases with which they are associated is distinct. The *tax* gene was thought to be involved in leukemogenesis when it was first identified, and the discovery that the *tax* protein is a transcriptional activator led to the hypothesis that it was involved in cellular transformation by causing aberrant transcriptional regulation in infected cells.

The transcriptional properties of the *tax* protein have now been characterized. Our studies, described here, have revealed that the HTLV-II *tax* protein is able to activate transcription from heterologous promoters. Moreover, the capabilities of *tax* protein in control of viral gene expression are complex and suggest mechanisms of cellular transformation other than aberrant transcriptional activation. $p37^{tax}I$ is required for HTLV-II replication and is central to the biology of the virus. However, direct proof of a role for the *tax* gene in the transformation of T cells is still lacking. The events involved in subsequent progression to tumorigenesis are almost completely unknown.

Our studies have shown that, as with ATL and its etiological agent, HTLV-I, HTLV-II is associated with a human T-lymphocyte malignancy. Details of the mechanism of tumorigenesis are yet to be explained. Differences in the severity and prognosis of HTLV-II-associated atypical hairy-cell leukemia compared with ATL likely results from late (nonviral) events in the formation and outgrowth of the tumor cells.

ACKNOWLEDGMENTS

We are grateful to our colleagues for helpful discussions, and to W. Aft for preparation of the manuscript. A. J. Cann and J. D. Rosenblatt are fellows of the Leukemia Society of America.

REFERENCES

1. Poiesz, B. J., Ruscetti, F. W., Reitz, M. S. Kalyanaraman, V. S., and Gallo, R. C. (1981). Isolation of a new type C retrovirus (HTLV) in primary uncultured cells of a patient with Sézary T-cell leukaemia. *Nature* 294:268–271.
2. Hinuma, Y., Nagata, K., Hanaoka, M., Nakai, M., Matsumoto, T., Kinoshita, K-I., Shirakawa, S., Miyoshi, I., (1981). Adult T-cell leukemia: Antigen in an ATL cell line and detection of antibodies to the antigen in human sera. *Proc. Natl. Acad. Sci. USA* 78:6476–6480.
3. Gallo, R. C., Kalyanaraman, V. S., Sarngadharan, M. G., Sliski, A., Vonderheid, E. C., Maeda, M., Nakao, Y., Yamada, K., Ito, Y., Gutensohn, N., Murphy, S., Bunn, P. A., Jr., Catovsky, D., Greaves, M. F., Blayney, D. W., Blattner, W., Jarrett, W. F. H., zur Hausen, H., Seligmann, M., Brouet, J. C., Haynes, B. F., Jegasothy, B. V., Jaffe, E., Cossman, J., Broder, S., Fisher, R. I., Golde, D. W., and Robert-Guroff, M. (1983). Association of the human type C retrovirus with a subset of adult T-cell cancers. *Cancer Res.* 43:3892–3899.
4. Yoshida, M., Miyoshi, I., and Hinuma, Y. (1982). Isolation and characterization of retrovirus from cell lines of human adult T-cell leukemia and its implication in the disease. *Proc. Natl. Acad. Sci. USA* 79:2031–2035.

5. Yoshida, M., Seiki, M., Yamaguchi, I., and Takatsuki, K. (1984). Monoclonal integration of human T-cell leukemia provirus in all primary tumors of adult T-cell leukemia suggests causative role of human T-cell leukemia virus in the disease. *Proc. Natl. Acad. Sci. USA* 81:2534–2537.
6. Blattner, W. A., Kalyanaraman, V. S., Robert-Guroff, M., Lister, T. A., Galton, D. A. G., Sarin, P. S., Crawford, M. H., Catovsky, D., Greaves, M., and Gallo, R. C. (1982). The human type-C retrovirus, HTLV, in blacks from the Caribbean region, and relationship to adult T-cell leukemia/lymphoma. *Int. J. Cancer* 30:257–264.
7. Blayney, D. W., Jaffe, E. S., Blattner, W. A., Cossman, J., Robert-Guroff, M., Longo, D. L., Bunn, P. A., Jr., and Gallo, R. C. (1983). The human T-cell leukemia/lymphoma virus associated with American adult T-cell leukemia/lymphoma. *Blood* 62:401–405.
8. Saxinger, W., Blattner, W. A., Levine, P. H., Clark, J., Biggar, R., Hoh, M., Moghissi, J., Jacobs, P., Wilson, L., Jacobson, R., Crookes, R., Strong, M., Ansari, A. A., Dean, A. G., Nkrumah, F. K., Mourali, N., and Gallo, R. C. (1984). Human T-cell leukemia virus (HTLV-I) antibodies in Africa. *Science* 225:1473–1476.
9. Hahn, B. H., Popovic, M., Kalyanaraman, V. S., Shaw, G. M., LoMonico, A., Weiss, S. H., Wong-Staal, F., and Gallo, R. C. (1984). Detection and characterization of an HTLV–II provirus in a patient with AIDS. In: *Acquired Immune Deficiency Syndrome* (M. S. Gottlieb and J. E. Groopman, eds.), pp. 73–81. New York, Alan R. Liss.
10. Kalyanaraman, V. S., Naryanan, P., Feorino, P. Ramsey, R. B., Palmer, E. L., Chorba, T., McDougal, S., Getchell, J. P., Holloway, B., Harrison, A. K., Cabradilla, C. D., Telfer, M., and Evatt, B. (1985). Isolation and characterization of a human T-cell leukemia virus type II from a hemophilia-A patient with pancytopenia. *EMBO J.* 4:1455–1460.
11. Kalyanaraman, V. S., Sarngadharan, M. G., Robert-Guroff, M., Miyoshi, I., Blayney, D., Golde, D., and Gallo, R. C. (1982). A new subtype of human T-cell leukemia virus (HTLV–II) associated with a T-cell variant of hairy cell leukemia. *Science* 218:571–573.
12. Chen, I. S. Y., McLaughlin, J., Gasson, J. C., Clark, S. C., and Golde, D. W. (1983). Molecular characterization of genome of a novel human T-cell leukaemia virus. *Nature* 305:502–505.
13. Gelmann, E. P., Franchini, G., Manzari, V., Wong-Staal, F., and Gallo, R. C. (1984). Molecular cloning of a unique human T-cell leukemia virus (HTLV–II$_{Mo}$). *Proc. Natl. Acad. Sci. USA* 81:993–997.
14. Saxon, A., Stevens, R. H., and Golde, D. W. (1978). T-lymphocyte variant hairy-cell leukemia. *Ann Intern. Med.* 88:323–326.
15. Rosenblatt, J. D., Golde, D. W., Wachsman, W., Jacobs, A., Schmidt, G., Quan, S., Gasson, J. C., and Chen, I. S. Y. (1986). A second HTLV–II isolate associated with atypical hairy-cell leukemia. *N. Engl. J. Med.* 315:372–375.
16. Seiki, M., Hattori, S., Hirayama, Y., and Yoshida, M. (1983). Human adult T-cell leukemia virus: Complete nucleotide sequence of the provirus

nome integrated in leukemia cell DNA. *Proc. Natl. Acad. Sci. USA* 80: 3618–3622.

17. Shimotohno, K., Takahashi, Y., Shimizu, N., Gojobori, T., Chen, I. S. Y., Golde, D. W., Miwa, M., and Sugimura, T. (1985). Complete nucleotide sequence of an infectious clone of human T-cell leukemia virus type II: A new open reading frame for the protease gene. *Proc. Natl. Acad. Sci. USA* 82:3101–3105.

18. Kiyokawa, T., Seiki, M., Iwashita, S., Imagawa, K., Shimizu, F., and Yoshida, M. (1985). p247^{xIII} and p21^{xIII}, proteins encoded by the pX sequence of human T-cell leukemia virus type I. *Proc. Natl. Acad. Sci. USA* 82:8359–8363.

19. Shima, H., Takano, M., Shimotohno, K., and Miwa, M. (1986). Identification of p26^{xb} and p24^{xb} of human T-cell leukemia virus type II. *FEBS Lett.* 209:289–294.

20. Chen, I. S. Y., McLaughlin, J., and Golde, D. W. (1984). Long terminal repeats of human T-cell leukaemia virus II genome determine target cell specificity. *Nature* 309:276–279.

21. Shimotohno, K., Golde, D. W., Miwa, M., Sugimura, T., and Chen, I. S. Y. (1984). Nucleotide sequence analysis of the long terminal repeat of human T-cell leukemia virus type II. *Proc. Natl. Acad. Sci. USA* 81:1079–1083.

22. Temin, H. M. (1981). Structure, variation and synthesis of retrovirus long terminal repeat. *Cell* 27:1–3.

23. Oroszlan, S., Copeland, T. D., Kalyanaraman, V. S. Sarngadharan, M. G., Schultz, A. M., and Gallo, R. C. (1984). Chemical analyses of human T-cell leukemia virus structural proteins. In: *Human T-Cell Leukemia/Lymphoma Virus* (R. C. Gallo, M. E. Essex, and L. Gross, eds.), New York, pp. 101–110. Cold Spring Harbor Laboratory,

24. Jacks, T., and Varmus, H. (1985). Expression of the Rous sarcoma virus *pol* gene by ribosomal frameshifting. *Science* 230:1237–1242.

25. Nagy, K., Clapham, P., Cheingsong-Popov, R., and Weiss, R. A. (1983). Human T-cell leukemia virus type I: Induction of synctia and inhibition by patients' sera. *Int. J. Cancer* 32:321–328.

26. Slamon, D. J., Press, M. F., Souza, L. M., Cline, M. J., Golde, D. W., Gasson, J. C., and Chen, I. S. Y. (1985). Studies of the putative transforming protein of the type I human T-cell leukemia virus. *Science* 228:1427–1430.

27. Nagashima, K., Yoshida, M., and Seiki, M. (1986). A single species of pX mRNA of human T-cell leukemia virus type I encodes *trans*-activator p40^x and two other phosphoproteins. *J. Vriol.* 60:394–399.

28. Seiki, M., Hikikoshi, A., Taniquchi, T., and Yoshida, M. (1985). Expression of the px gene of HTLV-I: General splicing mechanism in the HTLV family. *Science* 228:1532–1535.

29. Wachsman, W., Golde, D. W., Temple, P. A., Orr, E. C., Clark, S. C., and Chen, I. S. Y. (1985). HTLV x gene product: requirement for the *env* methionine initiation codon. *Science* 228:1534–1537.

30. Wachsman, W., Shimitohno, K., Clark, S. C., Golde, D. W., and Chen, I. S. Y. (1984). Expression of the 3' terminal region of human T-cell leukemia viruses. *Science* 226:177–179.

31. Inoue, J., Seiki, M., and Yoshida, M. (1986). The second pX product p27^{xIII} of HTLV–I is required for *gag* gene expression. *FEBS Lett.* 209: 187–190.

31a. Rosenblatt et al. (1988). *Science* 240:916–919.

32. Sodroski, J. G., Rosen, C. A., and Haseltine, W. A. (1984). *Trans*-acting transcriptional activation of the long terminal repeat of human T lymphotropic viruses in infected cells. *Science* 225:381–385.

33. Chen, I. S. Y., Slamon, D. J., Rosenblatt, J. D., Shah, N. P., Quan, S. G., and Wachsman, W. (1985). The x gene is essential for HTLV replication. *Science* 229:54–58.

34. Cann, A. J., Rosenblatt, J. D., Wachsman, W., Shah, N. P., and Chen, I. S. Y. (1985). Identification of the gene responsible for human T-cell leukemia virus transcriptional regulation. *Nature (London)* 318:571–574.

35. Shah, N. P., Wachsman, W., Souza, L., Cann, A. J., Slamon, D. J., and Chen, I. S. Y. (1986). Comparison of the *trans*-activation properties of the HTLV–I and HTLV–II x proteins. *Mol. Cell. Biol.* 6:3626–3631.

36. Sodroski, J., Rosen, C., Goh, W. C., and Haseltine, W. (1985). A transcriptional activator protein encoded by the x-lor region of the human T-cell leukemia virus. *Science* 228:1430–1434.

37. Felber, B. K., Paskalis, H., Kleinman-Ewing, C., Wong-Staal, F., and Pavlakis, G. N. (1985). The pX protein of HTLV–I is a transcriptional activator of its long terminal repeats. *Science* 229:675–679.

38. Fujisawa, J., Seiki, M., Kiyokawa, T., and Yoshida, M. (1985). Functional activation of the long terminal repeat of human T-cell leukemia virus type I by a *trans*-acting factor. *Proc. Natl. Acad. Sci. USA* 82:2277–2281.

38a. Nyborg et al. (1988). *Proc. Natl. Acad. Sci. USA* 85:1457–1461.

39. Wachsman, W., Cann, A., Williams, J., Slamon, D., Souza, L., Shah, N., and Chen, I. S. Y. (1987). HTLV x gene mutants exhibit novel transcriptional regulatory phenotypes. *Science* 235:674–677.

40. Chen, I. S. Y., and Temin, H. M. (1982). Establishment of infection by spleen necrosis virus: Inhibition in stationary cells and the role of secondary infection. *J. Virol.* 41:183–191.

40a. Nyborg et al. (1988). *Science* 85:1457–1461.

41. Miyoshi, I., Kubonishi, I., Yoshimoto, S., Akagi, T., Ohtsuki, Y., Shiraishi, Y., Nagata, K., Hinuma, Y. (1981). Type C virus particles in a cord T-cell line derived by co-cultivating normal human cord leukocytes and human leukaemic T cells. *Nature* 294:770–771.

42. Chen, I. S. Y., Quan, S. G., and Golde, D. W. (1983). Human T-cell leukemia virus type II transforms normal human lymphocytes. *Proc. Natl. Acad. Sci. USA* 80:7006–7009.

43. Popovic, M., Lange, Wantzin, G., Sarin, P. S., Mann, D., and Gallo, R. C. (1983). Transformation of human umbilical cord blood T cells by human T-cell leukemia/lymphoma virus. *Proc. Natl. Acad. Sci. USA* 80:5402–5406.

44. Koeffler, H. P., Chen, I. S. Y., and Golde, D. W. (1984). Characterization of a novel HTLV-infected cell line. *Blood* 64:482–490.

45. Popovic, M., Kalyanaraman, V. S., Mann, D. S., Richardson, E., Sarin, P., and Gallo, R. C. (1984). Infection and transformation of T cells by human T-cell leukemia/lymphoma virus of subgroups I and II. In: *Human T-Cell Leukemia/Lymphoma Virus*. (R. C. Gallo, M. Essex and L. Gross, eds.) pp. 217–228. New York, Cold Spring Harbor Laboratory.

46. Chen, I. S. Y., Cann, A. J., Shah, N. P., and Gaynor, R. B. (1985). Functional relationship of HTLV-II x and adenovirus E1A proteins in transcriptional activation. *Science* 230:570–573.

47. Inoue, J., Seiki, M., Taniguchi, T., Tsuru, S., and Yoshida, M. (1986). Induction of interleukin-2 receptor gene by $p40^{xI}$ encoded by human T-cell leukemia virus type I. *EMBO J.* 5:2883–2888.

48. Maruyama, M., Shibuya, H., Harada, H., Hatakeyama, M., Seiki, M., Fujita, T., Inoue, J., Yoshida, M., and Taniguchi, T. (1987). Evidence for aberrant activation of the interleukin-2 autocrine loop by HTLV-I-encoded $p40^{xI}$ T3/Ti complex triggering. *Cell* 48:343–350.

49. Greene, W. C., Leonard, W. J., Wano, Y., Svetlik, P. B., Peffer, N. J., Sodroski, J. G., Rosen, C. A., Goh, W. C., and Haseltine, W. A. (1986). *Trans*-activator gene of HTLV-II induces IL-2 receptor and IL-2 cellular gene expression. *Science* 232:877–880.

50. Greene, W. C., Leonard, W. J., Wano, Y., Svetlik, P. B., Peffer, N. J., Sodroski, J. G., Rosen, C. A., Goh, W. C., and Haseltine, W. A. (1987). *Trans*-activator gene of HTLV-II: Interpretation. *Science* 235:1073.

51. Saladhuddon, S. Z., Markham, P. D., Lindner, S. G., Gootenberg, J., Popovic, M., Hinumi, H., Sarin, P. S., and Gallo, R. C. (1984). Lymphokine production by cultured human T-cells transformed by human T-cell leukemia virus. *Science* 223:703–706.

52. Tedder, R. S., Shanson, D. C., Jeffries, D. J., Cheingsong-Popov, R., Dalgleish, A., Clapham, P., Nagy, K., and Weiss, R. A. (1984). Low prevalence in the UK of HTLV-I and HTLV-II infection in subjects with AIDS, with extended lymphadenopathy, and at risk of AIDS. *Lancet* 2: 125–128.

53. Robert-Guroff, M., Weiss, S. H., Giron, J. A., Jennings, A. M., Ginzburg, H. M., Margolis, I. B., Blattner, W. A., and Gallo, R. C. (1986). Prevalence of antibodies to HTLV-I, -II and -III in intravenous drug abusers from an AIDS endemic region. *JAMA* 255:3133–3157.

54. Cervantes, J., Hussain, S., Jensen, F., and Schwartz, J. M. (1986). T-prolymphocytic leukemia associated with human T-cell lymphotropic virus II (abstr.). *Clin. Res.* 34:454A.

55. Sohn, C. C., Blayney, D. W., Misset, J. L., Mathé, G., Flandrin, G., Moran, E. M., Jensen, F. C., Winberg, C. D., and Rappaport, H. (1986). Leukopenic chronic T-cell leukemia mimicking hairy cell leukemia: Association with human retroviruses. *Blood* 67:949–956.

55a. Rosenblatt et al. (1988) *Blood* 71:363–369.

56. Harper, M. E., Kaplan, M. H., Marselle, L. M., Pahwa, S. G., Chayt, K. J., Sarngadharan, M. G., Wong-Staal, F., and Gallo, R. C. (1986). Concomitant

infection with HTLV–I and HTLV–III in a patient with T$_8$ lymphoproliferative disease. *N. Engl. J. Med.* 315:1073–1078.

57. Kinoshita, K., Amagasaki, T., Ikeda, S., Suzuyama, J., Toriya, K., Nishino, K., Tagawa, M., Ichimaru, M., Kamihira, S., Yamada, Y., Momita, S., Kusano, M., Morikawa, T., Fujita, S., Ueda, Y., Ito, N., and Yoshida, M. (1985). Preleukemic state of adult T cell leukemia: Abnormal T lymphocytosis induced by human adult T cell leukemia-lymphoma virus. *Blood* 66:120–127.

58. Miyamoto, K., Sato, J., Kitajima, K.-I., Togawa, A., Suemaru, S., Sanada, H., and Tanaka, T. (1983). Adult T-cell leukemia. *Cancer* 52:471–478.

59. Sanada, I., Tanaka, R., Kumagai, E., Tsuda, H., Nishimura, H., Yamaguchi, K., Kawano, F., Fujiwara, H., and Takatsuki, K. (1985). Chromosomal aberrations in adult T cell leukemia: Relationship to the clinical severity. *Blood* 65:649–654.

60. Whang-Peng, J., Bunn, P. A., Knutsen, T., Kao-Shan, C. S., Broder, S., Jaffe, E. S., Gelmann, E., Blattner, W., Lofters, W., Young, R. C., and Gallo, R. C. (1985). Cytogenetic studies in human T-cell lymphoma virus (HTLV)-positive leukemia-lymphoma in the United States. *J. Natl. Cancer Inst.* 74:357–369.

9

Epidemiology of Adult T-Cell Leukemia/Lymphoma and the Acquired Immunodeficiency Syndrome

Angela Manns and William A. Blattner *National Cancer Institute, National Institutes of Health, Bethesda, Maryland*

I. INTRODUCTION

The discovery of the first human retrovirus, human T-lymphotropic virus (HTLV-I), was facilitated by advances in virological and cell culture techniques. In particular, highly sensitive assays for the virally encoded reverse transcriptase and availability of T-lymphocytic growth factors, specifically interleukin-2 (formerly T-cell growth factor), greatly facilitated this discovery (1). As a consequence, HTLV-I was discovered and linked with a specific clinical syndrome, adult T-cell lymphoma/leukemia (ATLL). This discovery set the stage for finding HTLV-II, and subsequently, the human immunodeficiency virus type 1 (HIV-1). HTLV-II was initially isolated from a patient with hairy-cell leukemia, but has not been consistently associated with any one disease. HIV-1 is associated with the clinical syndrome of acquired immunodeficiency (AIDS). The AIDS epidemic has developed worldwide and poses the major public health problem of the decade. Much remains to be learned from these viral-mediated disease processes which will serve as a template for understanding the etiology of human malignancy and immunodeficiency states with their subsequent sequelae.

II. EPIDEMIOLOGY OF ATLL

A. Historical Overview

ATLL is a clinicopathological entity that clusters in regions where HTLV-I is endemic but also occurs sporadically in other areas, often among migrants from viral endemic areas. The disease entity is a prototype for exemplifying how the interface of molecular biology and epidemiology can lead to new etiological insights.

The first cases of ATLL were described in 1977 in Japan (2). The clinical characteristics and hematologic findings were: (a) adult onset; (b) acute, smol-dering, or chronic leukemia with a rapidly progressive terminal course; (c) phenotypically mature T lymphocytes, often with the CD-4 phenotype; (d) frequent skin involvement; (e) lymphadenopathy and hepatosplenomegaly; (f) hypercalcemia; and (g) sometimes cytologically atypical pleomorphic cells in lymph node biopsies and in peripheral blood. An important epidemiological observation was the recognition that these cases tended to cluster, particularly in the southernmost islands of Kyushu, Shikoku, and Okinawa. An infectious cofactor was postulated, but the true etiology of HTLV-I was discovered half a world away by Gallo and co-workers.

In 1978, cells from a patient with mycosis fungoides, d'emblée variant, were placed into cell culture with the then newly discovered IL-2 growth factor, and the recognition of reverse transcriptase and morphologically characteristic retrovirus particles heralded the discovery of HTLV-I (3-7). Subsequently Japanese virologists also isolated HTLV-I, and collaborative studies between the United States and Japanese researchers led to serological investigations showing a high rate of seropositivity for HTLV-I among ATLL cases (8-10). In retro-spect, the U.S. patient with mycosis fungoides (MF) also had clinical features consistent with ATLL. This finding and the failure to show an association of HTLV-I to most classical cases of mycosis fungoides led to further understand-ing of the clinical and virological distinction between MF and ATLL.

With the establishment of this distinction, Catovsky et al. (18) described ATLL in Caribbean-born patients in the United Kingdom; and Blattner, working with Kalyanamaran, Robert-Guroff, and Gallo (19), established the seroepi-demiological link of these cases to HTLV-I. Once the endemic foci were identi-fied, an attempt was made to elaborate on the modes of virus transmission, understand the natural history of the infection, discover its relationship to human malignancy and potential for malignant transformation, and define other associated diseases.

B. Geographical Distribution

Major endemic areas for ATLL with HTLV-I seropositivity have been identified in Japan, the Caribbean, the southeastern United States, equatorial Africa, parts of Central America, and the northern coast of South America. Other areas are further defined in Table 1. Identification of endemic regions has generally been supported by case reports of ATLL, and by HTLV-I seropositivity in the major-ity of cases with ATLL-like features (11-14).

TABLE 1 Geographical Distribution of HTLV–I Infection

Endemic	Low viral prevalence	Nonendemic
Asia		
Kyushu, Japan	Taiwan	Vietnam
Shikoku, Japan	People's Republic of China	Philippines
Okinawa, Japan		Malaysia
New Guinea		
Africa and Mideast		
Nigeria	Tunisia	
Ghana	Egypt	
Zaire	South Africa	
Uganda		
Kenya		
Israel (Ethiopian Jews)		
Seychelle Islands		
Europe		
U.K. (Caribbean immigrants)		Denmark
Southern Italy		Spain
		Germany
		U.K. (Caucasians)
		France
Western Hemisphere		
Jamaica		Northern U.S.
Trinidad (blacks)		(Caucasians)
Martinique		
Guadeloupe		
Venezuela		
Panama		
Colombia		
Guyana		
Haiti		
Barbados		
Southern U.S. (blacks)		
Hawaii (Japanese immigrants)		

Source: Adapted from Ref. 41.

The most frequently studied areas in Japan have been the southern coastal/rural regions, which include Kyushu, parts of Shikoku, and Okinawa (15–17). Surprisingly, areas of close proximity, with as little as 40-km distance between them but with varying geographical topography, may have disparate prevalences of HTLV-I positivity. Highest rates are often observed in isolated coastal/rural areas, with lower rates generally observed in highly urban areas.

The recognition that the Caribbean basin was an HTLV-I endemic area came from the identification of HTLV-I-positive cases of ATLL in migrants from this region. Among the first cases identified were a cluster of West Indian-born patients in the United Kingdom (18,19). Cases appeared among blacks from the West Indies, where most had resided until adolescence or early adulthood. Classic clinicopathological features are evident in these cases.

A typical patient, a 47-year-old black man, was admitted with weight loss and weakness. He was lethargic and had a nondesquamative, erythematous rash and generalized lymphadenopathy. Blood counts were normal, but bizarre lymphoid cells were present in the films. Calcium level was raised and continued to rise over the next 2–3 weeks. The patient had numerous osteolytic lesions in the ribs. Films of aspirated bone marrow showed 34% of abnormal lymphoid cells. He was treated with i.v. saline, oral phosphate, prednisone, and two courses of CHOP. He improved considerably, but later his white blood count increased to $64 \times 10^*9/1$. (His HTLV-I serology was positive.) He died a month later of septicemia (18).

Recent surveys in Jamaica reveal relatively fewer cases of the chronic or smoldering leukemia variety of ATLL than in Japan. The majority of cases are non-Hodgkin's lymphomas with an aggressive clinical course. Topographically, as in Japan, there is a rural regional predominance of cases, although referral bias could contribute to this pattern (unpublished data and Ref. 20).

In Africa, particular attention was focused on the aggressive clinical course of non-Hodgkin's lymphoma patients. Initial studies were performed prospectively in a search for cases that were consistent with ATLL. The first cases reported were from Nigeria, with serological confirmation of HTLV-I positivity (21). However, compared to HTLV-I endemic areas in Japan and the West Indies, the proportion of adult non-Hodgkin's lymphoma cases associated with HTLV-I was low.

Since the initial isolation of HTLV-I from an American patient, few clusters of ATLL have been identified in the United States. Patients with this diagnosis have been born, for the most part, in the southeastern United States or immigrated from other endemic areas. The majority of cases occur among blacks, although cases do occur among Caucasians who have traveled to or emigrated from an endemic area. The clinicopathological features in the United States are similar to those previously described; however, hypercalcemia and bony abnormalities may occur more commonly in American cases (14,22–24).

Recently, several ATLL cases were reported among blacks in New York City (25). These patients were born in either the southeastern United States or the Caribbean basin, known endemic areas. These data plus the data from the United Kingdom emphasize that place of birth is an important variable, which indicates that viral exposure predates the onset of disease often by what would appear to be many years to decades. The extent of the worldwide distribution of HTLV-I is yet to be fully defined, but the apparent restricted distribution of the virus suggests that it is maintained in populations through inefficient modes of transmission (26). Further studies are needed to better define the distribution of virus infection.

C. Incidence/Prevalence of HTLV-I and ATLL

Because of the infrequent occurrence of ATLL cases worldwide, few systematic surveys of actual disease incidence have been done. Prevalence rates have been estimated in some affected areas. In Jamaica, the crude annual incidence of ATLL is estimated at 2 cases /100,000 population. The prevalence of HTLV-I seropositivity among ATLL cases is 80%. The prevalence rate among healthy individuals is 5-6% (unpublished data). Trinidad-Tobago, a country in the eastern Caribbean, reports an annual incidence of ATLL of 2.8 cases/100,000 population. This country has an ethnic composition of African, Asian, and Caucasian people. All cases of ATLL are of African descent with low socioeconomic status. The HTLV-I prevalence among healthy individuals is 2.2%. The seroprevalence by ethnic group is Africans, 3.4%; Asians 0.2%; and Caucasian, 0%. This racial predisposition is further emphasized when the island of Tobago, whose population is entirely of African descent, is evaluated. The seroprevalence is reported to be 12% among healthy individuals (27,28).

In Japan, the overall incidence of ATLL is 3.5 cases/100,000 population and 5.7 cases/100,000 population in persons greater than 40 years of age. The prevalence of HTLV-I seropositivity among these ATLL cases is over 95%, although a few classic cases that were HTLV-I negative have been reported. The prevalence rate among healthy individuals in the endemic area of Kyushu is 8% (29). Incidence rates are not available for ATLL case occurrence in the United States and Africa. The prevalence rate of HTLV-I seropositivity among healthy individuals is estimated as 2% and 2-8%, respectively, in endemic areas (23,26, 30-32).

Southern Italy has been identified as a possible endemic area for HTLV-I among a white population (11). Among T-cell leukemia cases, only 20% antibody seropositivity was found. A serosurvey among a normal population from southern Italy revealed a prevalence rate of 8%. However, recent molecular virological data suggest that the virus associated with these cases may be a new variant termed HTLV-V, and careful analysis of the cases suggests distinctive clinical and laboratory features, particularly an absence of T-cell receptor markers on the cell surface.

There have been instances where HTLV-I seropositivity has been reported but no documented positive cases of lymphoma or leukemia, and conversely, cases of ATLL have been reported that are antibody negative (26,33). This points to the fact that in the former instance, further epidemiological investigation is indicated to document cases; and in the latter case, an etiology other than HTLV-I is possible, or perhaps available technology fails to detect some cases. When malignancies other than lymphomas and leukemias have been surveyed for antibody prevalence in endemic areas, seroprevalence rates as high as 15–30% have been obtained (33–36). Although no correlation between site of disease and antibody prevalence can be established, one must consider the possibility that development of malignancy contributes to expression of latent HTLV-I reactivation or that HTLV-I infection may contribute to the risk of other malignancies.

The risk for development of ATLL appears to be similar among men and women (37,38). On the other hand, prevalence rates among healthy individuals for HTLV-I seropositivity in Jamaica and Japan have shown a female predominance after age 30–40 (37,39). The significance of this is not apparent, given the similar occurrence in cases of ATLL among both sexes.

There is a consistent age-dependent increase in seropositivity rates, with prevalence rates increasing significantly after 40 years of age. For example, in a Jamaica serosurvey, an endemic area, the prevalence rate was 5% for persons under 40 years of age and 12% for those greater than 40 years of age. The mean age for ATLL occurrence is 45 years of age, with a range from 19 to 84 years of age (20).

The reasons for this pattern of age-dependent increasing prevalence are not known although several hypotheses have been proposed (40,41). The simplest explanation is that the majority of infections occur later in life, possibly through a combination of sexual transmission, blood transfusion, possible vector transmission, and other yet undetermined modes of environmental exposure. Since the majority of individuals who are infected appear to remain infected for life, the increasing antibody prevalence with age could be due to the continual addition of newly infected individuals. An alternative hypothesis that has recently been raised is that HTLV-I can infect individuals early in life, but without sustained detectable level of antibodies. Also, the age-specific rise in seroprevalence, at least in part, could represent reactivation of dormant virus expression resulting from antigen stimulation of viral-infected T cells that have harbored the virus quiescently.

D. Modes of Virus Transmission

HTLV-I clusters in geographical locations where ATLL cases occur. From these endemic areas, migrants who move to nonendemic areas still have the propensity to develop disease (42,43). Migrant Japanese populations to Hawaii have been evaluated in comparison to lifetime residents of Japan. Hawaii is not an endemic area; however, healthy persons who immigrated from the endemic area of Okinawa were found to have high seroprevalence rates for HTLV-I. On the other hand, persons who immigrated from Niigata, a nonendemic region of Japan, demonstrated a low seroprevalence for HTLV-I. The exposures for developing seropositivity were acquired outside of the Hawaiian environment. Differences in seropositivity rates were noted between persons who migrated in their late teens versus those who emigrated before age 12. The former group had rates twice that of the latter group, again focusing on the enigmatic age-dependent rise in seroprevalence. Still unknown is the mechanism by which persons become infected.

Studies have been aimed at defining the mode of transmission by focusing on family members of ATLL cases. Antibodies were found to be more prevalent among healthy relatives of patients than among healthy normal donors (44). This indicated that close person-to-person contact was necessary for transmission. These observations have been further defined by demonstrating the pattern of infection within families which points to parent-to-child or vertical transmission and spouse-to-spouse or horizontal transmission.

As a model to demonstrate horizontal transmission, animal studies were performed with Japanese monkeys (*Macaca fuscata*) which were seropositive for HTLV-I-associated antigens that carry identical serological, morphological, and biological activity to HTLV-I (45). Infected and uninfected monkeys were allowed to mate. It was observed that previously negative monkeys seroconverted after 8 weeks. The suspected mode of transmission was sexual contact. To further demonstrate sexual transmission in humans, husband-wife units were studied (46). Semen lymphocytes from antibody-positive men, when cultured, were HTLV-I antibody positive. Wives whose husbands were HTLV-I antibody positive demonstrated seropositivity 100% of the time. An additional study has confirmed that the main direction of transmission is from male to female (47). Observed rates are 60.8% for male-to-female transmission and only 0.4% for female-to-male transmission over an estimated 10-year period. To further evaluate risk factors for sexual transmission, a cross-sectional study of sexually transmitted disease clinic patients in Jamaica, by Murphy et al., demonstrated independent risk factors among women for HTLV-I infection to be a greater number of lifetime sexual partners and a current diagnosis of syphilis. Risk factors for infected men were penile sores or ulcers and current diagnosis of syphilis (48).

Another route of transmission within the family explored the possible spread vertically from parent to child. Epidemiological observations have shown that children of antibody-positive fathers and mothers have the highest seropositivity rate, and those of antibody-positive mothers and antibody-negative fathers have the next highest seropositivity rate. Children of antibody-positive fathers and negative mothers, however, have no demonstrated antibodies. These data suggest that mother-to child transmission is a principal route (47,49,50). Cord blood samples of infants born from seropositive mothers were positive for antibody to HTLV-I of the IgG class, suggesting possible maternal transmission via germinal cells, transplacentally, or possible intrapartum infection. The IgG antibody disappeared by the time of follow-up in some infants, so persistent infection was not demonstrated, only passive IgG transfer from mother to child. On the other hand, some of the infants became seropositive at later follow-up. Further studies focused on the establishment of postnatal transfer of HTLV-I infection by evaluation of possible maternal milk-borne infection. It has been demonstrated that seropositive mothers contain HTLV-I-positive lymphocytes in breast milk (51). Postnatal mother-to-child transmission is thought to be the dominant route of infection in children (52). ATLL is probably more likely to develop in individuals infected early in life.

With this familial clustering, the question of possible genetic predisposition arises. Some families have been reported with numerous cases of ATLL among members (53,54). HLA typing was performed on antibody-positive and antibody-negative persons in one study (55). HLA-B7 was identified with discordant rates among similarly endemic areas, and therefore, no conclusive evidence to support or refute the possibility of an associated genetic susceptibility was established. Similar conclusions were reached in evaluation of HLA antigens of ATLL cases versus seropositive and seronegative familial controls (56).

Blood transfusions have been identified as a reservoir for transmission of HTLV-I. In one study, the rate of seroconversion after transfusion of 1-2 units of positive blood products resulted in a 63.4% seroconversion rate (57). The population remains persistently infected, which allows for further spread of infection. Only blood units containing cellular components are able to transmit the virus (58,59). A recent survey of 40,000 U.S. blood donors identified a seroprevalence rate of 0.025%, and screening of U.S. blood donors has been recommended (60). From infected individuals producing antibodies after transfusion, it has been possible to isolate clonal cell lines carrying viral antigens with the same surface markers as neoplastic cells of adult T-cell leukemia and HTLV-I (61). Infected individuals are at increased risk for development of ATLL and other HTLV-I-associated diseases.

Among intravenous drug abusers, a high prevalence of retroviral infection, including HTLV-I, has been demonstrated (62). Transmission of infection is related to the practice of using shared drug paraphernalia, including nonsterile needles and syringes. In the New York City area, the HTLV-I prevalence rate was 9% in this group, well above the level expected (less than 1%) for this non-endemic area. This suggests that once the virus is introduced into a population, its transmission is affected by practices of the group involved. Similarly, the background prevalence of virus in the group involved may be enhanced. This particular study had a large number of blacks under study, and although most were born in New York, pertinent collaborative data to suggest other sources (i.e., familial, transfusion, travel history) were not available.

E. Other Disease Associations

Until recently only ATLL and other hematological malignancies have been associated with increased HTLV-I seroprevalence. Tropical spastic paraparesis (TSP) is a term introduced in 1969 for a slowly progressive myelopathy affecting mainly the pyramidal tracts, and to some extent, the sensory system (63). The disease is prevalent in Jamaica, Colombia, southern India, Africa, and Martinique. HTLV-I has been isolated from sera and cerebrospinal fluid of patients with this disorder.

Initially, the distribution of this neurological disease raised the possibility that environmental factors such as poor nutrition, toxins, or an infectious agent were possible etiological factors. The finding of two HTLV-I-positive TSP cases in a retrospective serological survey triggered further investigation into this possible association (64). Fifty-nine percent of patients with TSP were positive for HTLV-I antibodies in a subsequent survey in Martinique, suggesting that HTLV-I is neurotropic or contributes to the pathogenesis of the disease. Jamaica and Colombia reported 67% and 73% HTLV-I-positive TSP cases, respectively (65). Japan and Trinidad have also reported seropositive cases (66,67).

The neurological symptoms of TSP are generally of gradual onset and progress over many years. It is not clear at what point HTLV-I infection occurs, but one would expect early infection during childhood or adolescence, with latency followed by viral initiation or promotion of a cofactor and subsequently disease manifestation. This hypothesis is supported by reports of TSP among West Indies migrants to the United Kingdom where a pattern resembling that for ATLL was present. Thus, in these cases, many years to decades transpired between migration and subsequent development of disease. Patients with neurological symptoms have been later identified as cases of ATLL demonstrating a continuum of disease secondary to infection with the virus.

Conversely, recipients of HTLV-I-infected blood transfusions in Japan have been observed to develop symptoms of TSP within several weeks to months after transfusion (68). This observed shorter latent period may help to explain the female excess of TSP cases observed in most parts of the world. Since HTLV-I is more prevalent in women than men, presumably owing to more efficient male-to-female transmission in older age, perhaps the female excess of TSP cases represents cases occurring among women who acquired infection through sexual contact with their infected male spouse.

Other disease associations have been sought. Of particular interest are diseases with T-cell abnormalities or similar clinical features to T-cell leukemia.

In one study, Boeck's sarcoid (sarcoidosis) patients were surveyed because of similar clinical features (hypercalcemia and skin lesions) to T-cell leukemia. Fifty-one patients were tested, and all were negative for serum antibodies to HTLV-I (69). Multiple sclerosis, which often demonstrates imbalances in proportions of T cells during disease exacerbations or remissions, has been studied for retrovirus associations (70,71). In one study, weak reactivity to the HTLV-I p24 was observed in cerebrospinal fluid but was not uniform, and an association could not be ascribed to HTLV-I. The data raise the possibility of a distantly related retrovirus. A second study failed to find antibodies against any retroviruses in the sera or cerebrospinal fluid of multiple sclerosis patients. The role of a retrovirus still remains a possibility in this disease and, hopefully, interest in further study has been stimulated.

Unanswered questions remain with regard to the implications and impact of coinfection with HTLV-I and HIV-I, which through more widespread testing is observed to be quite frequent. Drug addicts, leukemia patients, and other high-risk-group members are frequently affected (62,72,73).

F. Summary

What we have learned thus far regarding ATLL and HTLV-I infection is: (a) the disease and virus are endemic among the Japanese, blacks in the southeastern United States, Caribbean, and Africa, and some individuals exposed via travel or born in an endemic area; (b) there is an age-dependent increase in prevalence of infection, with disease incidence peaking in the mid-40s; (c) there are no M:F differences in occurrence of ATLL; (d) there is a worldwide geographical distribution, with rural and coastal areas showing increased prevalence; (e) transmission has been demonstrated to be sexual, specifically male-to-female, mother-to-child, through cellular blood products, and needle sticks; (f) there is generally a long incubation period between infection and disease onset; and (g) there is an apparently low rate of infectivity secondary to inefficient transmission, although the discovery of new disease associations may establish a more prominent role for this retroviral infection.

III. EPIDEMIOLOGY OF AIDS

A. Historical Overview

The occurrence of *Pneumocystis carinii* pneumonia and Kaposi's sarcoma in previously healthy homosexual males and drug abusers was first reported in 1981 (74,75). Hemophiliacs and blood transfusion recipients were the next groups to be identified with these clinical manifestations which were commonly associated with immunodeficient states. It became evident that these individuals had an acquired T-cell defect. Although initial reports of cases originated in the

United States, it soon became apparent that the disease was more widespread as cases were reported from Europe, Haiti, and Africa. It was evident that an epidemic of a new disease was evolving. The disease was subsequently called the acquired immunodeficiency syndrome (AIDS).

In 1982, a Centers for Disease Control (CDC) case definition for AIDS was established and subsequently revised, and surveillance mechanisms were implemented for the United States and abroad (Table 2). By December 1983, 3000 cases had been reported in the United States. As further clinical and epidemiological data were accumulated, the etiological factor was thought to be an infectious transmissable agent. It was not until 1984 that the causative agent was definitely identified. Human immunodeficiency virus type 1 antibody was detected repeatedly from sera in over 90% of patients with AIDS and pre-AIDS, or ARC, and virus isolation was achieved in over 80% of cases (76-81).

By 1985, an ELISA blood test was developed to detect antibodies to HIV-1, and in the United States nationwide screening of donated blood products was undertaken. Over the past 5 years, the case numbers have escalated, with a doubling of the rate each year in the United States. Thus far the cost in lives and impact on health care economics has been great. There is no cure, as yet. Clinical trials of an anti-HIV drug, azidothymidine, have shown therapeutic benefit (82, 83). Current projection of the course of the epidemic indicates no change in the escalating numbers of cases through the mid-1990s.

B. Demographics of AIDS

As of July 1988, 138 countries had reported over 100,000 cases of AIDS (84). The majority of cases were reported from the United States, western Europe, central Africa, Brazil, and Haiti. One hundred thousand cases is an underestimation of the actual number of cases. Potentially 250,000 cases are thought to exist worldwide, with 5-10 million persons with symptoms of HIV-1 infection or asymptomatic carrier state. The actual number of reported cases is underrepresented because of failure to recognize cases, failure to report cases, and lack of facility for testing for the virus in order to meet the CDC case definition, especially in developing countries. An additional 1 million AIDS cases are estimated to be reported over the next 5 years worldwide. In the United States, the CDC currently estimates that there are 1-1.5 million infected persons (85,86).

1. United States

The early cases of AIDS seemed to be confined to certain patient groups. Seventy-one percent of cases were identified among homosexual/bisexual males. Among the first 1000 U.S. cases reported between June 1981 and February 1983, patients presented with Kaposi's sarcoma (28%), PCP (50%), combined KS/PCP (8%), and other opportunistic infections (14%) (87,88). Eighty percent of case reports emanated from New York, California, Florida, and New

TABLE 2 Surveillance Definition of the Acquired Immunodeficiency Syndrome (AIDS) (Revised)

A. Presence of reliably diagnosed diseases at least moderately indicative of cellular immunodeficiency.

B. Absence of known causes of underlying reduced resistance to those diseases, other than due to HIV infection.

Indicative diseases of underlying immunodeficiency

1. Protozoan infection:
   Pneumocystis carinii pneumonia
   Toxoplasma gondii encephalitis or disseminated infection (excluding congenital infection)
   Chronic *Cryptosporidium* enteritis (duration > 1 month)

2. Fungal diseases:
   Candida esophagitis
   Cryptococcal meningitis or disseminated infection
   Disseminated histoplasmosis[a]
   Chronic enteric isosporiasis[a]

3. Bacterial infection:
   Salmonella septicemia[a]
   Disseminated *Mycobacterium avium* complex or *M. kansasii*
   Disseminated *Mycobacterium tuberculosis*[a]

4. Noncongenital virus infection:
   Chronic mucocutaneous herpes simplex (age > 1 month)
   Cytomegalovirus infection other than of liver or lymph node
   Progressive multifocal leukoencephalopathy

5. Cancers:
   Kaposi's sarcoma (age < 60 years)
   Primary brain lymphoma (age < 60 years)
   Non-Hodgkin's lymphoma (high-grade B-cell)[a]

6. Helminthic infections:
   Strongyloidiasis disseminated beyond gastrointestinal tract

7. Other:
   Chronic lymphoid interstitial pneumonitis (age > 13 years)[a]

---

[a]In association with HIV seropositivity.
Source: Refs. 156,157. Refer to Ref. 157 for complete definition.

Jersey. Ninety percent of affected individuals were between 20 and 49 years of age. Fifty-nine percent of cases occurred among whites, 26% among blacks, and 14% among persons of Hispanic origin. Seven percent of cases occurred among women.

Over the past several years, the number of cases has increased in all patient groups (Figure 1). The race, age, and sex distribution of adults with AIDS has remained relatively constant. *P. carinii* continues to be the most common opportunistic infection. The incidence of Kaposi's sarcoma, on the other hand, has begun to decline, accounting for only 15% of reported cases. Updated information reveals that 62,740 cases have been reported in 50 states as of June 1988. The cumulative rate for the United States is 9 per 100,000 population. New York and San Francisco have the highest rates, 72 per 100,000 and 68 per 100,000 population, respectively (89).

Other striking aspects of disease occurrence in the United States are reflected in the marked differences in disease occurrence among various ethnic groups. Most recent statistics reveal that U.S. blacks account for 22-26% of all AIDS cases and Hispanics for 14% (90-92). They comprise only 12% and 6%,

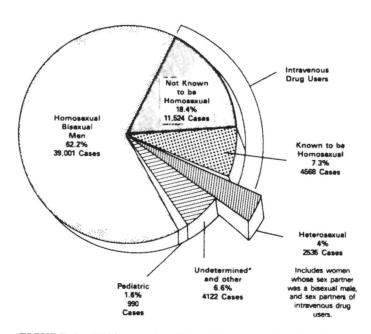

FIGURE 1   AIDS cases by risk groups since 1981 (total as of 5/23/88:62,740). *Includes blood transfusion recipients, hemophiliacs, and undetermined cases. (From Centers for Disease Control.)

respectively, of the U.S. population. Cumulative incidences are 2.5-10 times those for Caucasians. Drug abusers account for greater than 50% of these cases. Women are affected in larger numbers among blacks and Hispanics, 52% and 29%, respectively. The cumulative incidence rates are 13.3 and 11.1 times, respectively, the incidence for white women. As a result, heterosexual transmission of AIDS is significant, occurring in blacks (50%), Hispanics (25%), and Caucasians (25%). This heterosexual transmission patient group and increase in numbers of affected females have further impact on the occurrence of pediatric AIDS cases.

2. Pediatric AIDS

Pediatric AIDS cases in the United States are defined by the CDC case definition as AIDS cases occurring under age 13 years (93 and Table 2). Fifty-seven percent of these cases are black children and 22% are children of Hispanic origin. Seventy-nine percent of pediatric cases come from families in which parents are affected with AIDS or are at increased risk for developing AIDS. Thirteen percent have received transfusion of blood or blood components and 6% are hemophiliacs. Cases have been reported from 23 states, the District of Columbia, and Puerto Rico. Seventy-nine percent of cases reported are from New York, Florida, New Jersey, and California. Pediatric AIDS patients are diagnosed with *P. carinii* pneumonia (52%), other opportunistic infections (47%) and Kaposi's sarcoma (1%) (90). They usually suffer from an accelerated course, with a disease incubation period shorter than that seen in adult cases. Fortunately, the total number of pediatric cases is less than 2% of all reported cases. In order to reduce or eliminate these cases, widespread dissemination of information on disease prevention and education to involved risk groups is necessary.

3. Europe and Elsewhere

In Europe and Latin America, the epidemiology of AIDS is similar to the pattern seen in the United States (84,89,94-96). The majority of cases are homosexual/ bisexual males and intravenous drug abusers, although a lower proportion (11%) are drug abusers in Europe vs. 25% in the United States. Cases generally present with opportunistic infection (65%) or Kaposi's sarcoma (20%). Males account for 91% of cases. Greater than 80% of cases occur between 20 and 49 years of age. The European countries with the highest rates of disease occurrence are Switzerland (12 per million population), Denmark (11 per million population), and France (9 per million population), compared to a United States rate of 90 per million population. Approximately 11% of European cases are of non-European origin. Most of these cases are of African origin. These European cases with African ethnic origins helped to identify other potential areas to search for epidemic cases of AIDS. Significant numbers of cases were traced to central and eastern Africa. Eighty-nine percent of these cases were found to have no identifiable risk factor. Surprisingly, the male:female ratio was 2:1, indicating increased heterosexual disease transmission.

4. Africa

Similar to the European cases with African origins, the majority of AIDS cases in Africa originate in countries of central and eastern Africa. The first country cited with epidemic problems was Zaire (97). The estimated annual incidence is 550–1000 cases per million adults (98). Very few people with AIDS in these countries report homosexual or bisexual activity, and intravenous drug abuse is a rare occurrence in these countries. Most cases are attributed to bidirectional heterosexual transmission (99). Other patient groups identified are transfusion recipients, recipients of medical injections, and perinatally infected newborns. Thirty-five to fifty-seven percent of reported cases are women (89). The disease affects primarily young and middle-aged persons. The mean age of persons in Kinshasa, Zaire, for example, is 33. Pediatric cases account for approximately 7% of total cases versus a U.S. rate of less than 2%. This higher rate is partly attributable to the frequent transfusion of blood to children with severe malaria. The African AIDS cases are more likely to present with a syndrome character- ized by weight loss, diarrhea, opportunistic infection, and an interaction with endemic diseases such as tuberculosis and malaria. Other opportunistic infections such as cryptococcosis and toxoplasmosis present rather than *P. carinii* (97,100). These differences may be secondary to difficulty in diagnosis or simply differ- ences in patterns of exposure to potential opportunistic infections.

C. Clinical Spectrum of AIDS

1. Opportunistic Infections

P. carinii (PC) is the most common opportunistic pathogen in AIDS, associated with severe pulmonary infection (101). It has been suggested that PC represents reactivation of a latent infection. In support of this, studies have demonstrated antibody to PC in normal children by the age of 4 years, and lymphocytes from adults proliferate when exposed to PC antigen (102). In fact, a number of patho- gens that commonly affect AIDS patients are agents that are usually latent or chronic infections in normal hosts. These include cytomegalovirus, herpes simplex virus, *Toxoplasma gondii, Mycobacterium tuberculosis, Salmonella,* and *Cryptococcus neoformans.* Other agents, such as *Mycobacterium avium* and *Candida,* colonize patients readily. The basis for overt infection is, of course, a result of the underlying defect in cellular immunity associated with AIDS.

2. AIDS-Related Cancers

Cancer risk assessment in AIDS has been approached indirectly from SEER registry data or case reports. The most common neoplasms associated with AIDS are Kaposi's sarcoma and B-cell, non-Hodgkin's lymphomas (103). Other malignancies have been reported in the literature, including leukemia, Hodgkin's disease, oral cancer, rectal and anal cancers, non-small cell lung cancer, and renal cell carcinoma (104–112). Prospective cohort studies are needed to establish incidence estimates for these putative AIDS-associated cases. In fact, there is currently no way of knowing whether these AIDS-related cases occur simply by chance or whether they occur in excess of cases presenting in the general population. What has been observed with AIDS-related cancers is a tendency to be more aggressive, to present in advanced pathological stages, and to be resistant to conventional therapy.

Kaposi's Sarcoma Kaposi's sarcoma (KS) was the first identified malignancy associated with AIDS (74). The malignancy usually presents in elderly men of Mediterranean or Jewish extraction in the United States and Europe. It is generally localized to the lower extremities and has an indolent clinical course (113). An endemic form of KS has been identified in equatorial Africa. The disease accounts for as much as 9% of total cancers in some areas (114). This form of KS is also generally localized to the extremities, with an indolent course and subsequent dissemination only after several years. A more aggressive lymphadenopathic form exists which usually occurs in children.

In the United States, the first reported cases associated with AIDS were unusual in that the distribution of skin lesions was widespread rather than localized, with generalized lymphadenopathy, visceral involvement, and a rapid clinical course with deaths occurring less than 20 months after diagnosis (115). This pattern was also similar to KS seen in immunosuppressed renal transplant patients. Once these unusual cases were identified, further evaluation of African KS revealed emerging differences in the pattern of case occurrence in 1983. A report from Zambia revealed an increase in the more aggressive form of KS (114). Subsequent serological testing for HIV-1 confirmed that this atypical KS could be distinguished from the endemic form because cases of the aggressive disease were positively correlated with HIV-1 infection 90% of the time (116). The common manifestation in these varying forms of KS is that histopathological distinctions cannot be readily identified.

The etiology of this disease, characterized by endothelial cell proliferation, is not known. It has been suggested that the disease results from effects of oncogenic viruses, i.e., HIV-1 alone or in concert with other viruses such as cytomegalovirus (CMV), or that perhaps impaired tumor surveillance results because of the underlying immunosuppressed state associated with AIDS. Much attention has focused on CMV's role in the pathogenesis of KS (117). CMV has been considered a possible oncogenic virus with potential to stimulate synthesis of DNA and RNA and transform human embryo fibroblasts to sarcomatous tumors. Biopsies of KS lesions have also demonstrated CMV nucleic acids and antigens. Epidemiological data in immunosuppressed renal transplant patients corroborate that of AIDS patients with high rates of primary or reactivated

CMV, and also an associated increased incidence of KS. However, recent studies have shown that CMV is most likely an incidental pathogen rather than the underlying etiology, since CMV is only rarely found to be integrated in most KS tumors from Africa and the United States. Recent data from Salahuddin, Gallo, and co-workers suggest that retrovirus-induced growth factors may also play a role in stimulating endothelial cell proliferation (118).

When AIDS/KS patients were evaluated with case-control studies, an attempt was made to identify possible risk factors. Significant associations first identified included use of amyl nitrite and other recreational drugs, history of mononucleosis, and promiscuity measured by number of different sexual partners (119). Later studies did not completely concur with these early findings with respect to the use of nitrites, and this issue has not been totally resolved (120-122). Of interest now is the fact that KS has low incidence rates among risk groups other than homosexual/bisexual males, and even in this group a decline in the frequency of KS cases has occurred. Further research at this time needs to be implemented to establish what factor(s) have been altered to account for the current decreasing incidence and why KS is more prevalent among homosexuals.

Lymphomas In mid-1985, the CDC case definition was expanded to include cases of AIDS characterized by aggressive, B-cell non-Hodgkin's lymphoma. Numerous case reports clinically describe this disease association (123–126). Characteristically, the majority of patients are HIV antibody positive, have histologically high-grade lymphomas (small noncleaved cell, or immunoblastic), have extranodal disease (central nervous system, gastrointestinal, bone marrow) with advanced stage at presentation, and have an overall poor prognosis. The pathogenesis of these lymphomas is of particular interest because of their explosive onset and possible link to lymphadenopathy. Presence of HIV-1 infection is not sufficient to explain the development of this malignancy, so other hypotheses have been proposed (127). One study evaluated the roles of HIV-1, Epstein-Barr virus (EBV), and the c-*myc* oncogene (128). HIV-1 DNA sequences were not found in the tumor and there was no evidence of infection of malignant B lymphocytes. EBV DNA sequences and c-*myc* gene rearrangement were detected in tumor tissue. The authors concluded that EBV potentiates the development of B-cell neoplasia via polyclonal B-cell activation and HIV-1 only alters immunoregulatory mechanisms against neoplasms, rather than playing a direct role in neoplastic transformation.

4. Neurological Syndromes

HIV-1 is neurotropic (129). This predilection results in the frequent complications of infections and neoplasms of the central nervous system (CNS) associated with systemic HIV-1 infection, as well as primary brain infection with HIV-1 and its resulting sequelae. Neuroepidemiological evaluation of CDC data revealed that 7.46% of AIDS cases presented with CNS disease (130). The diseases most frequently occurring among adults included cryptococcosis (47%), toxoplasmosis (40%), progressive multifocal leukoencephalopathy (7%), and primary CNS lymphoma (6%). Not included in this survey are dementia (subacute encephalitis) and myelopathies, which are now most frequently attributed to HIV-1 infection (131).

D. Transmission of HIV Infection

1. Blood and Blood Products

Transfusion donor recipient pairs proved to be an ideal model for illustrating that AIDS was transmitted by an infectious agent. Once HIV-1 was isolated, serological testing of presumed high-risk donor sera and subsequent evaluation of recipients assisted in establishing HIV-1 as the etiological agent in AIDS (132). Unlike HTLV-I, blood products including plasma, whole blood, as well as cellular blood components are capable of transmitting HIV-1. Blood component products such as gamma globulin, hepatitis B vaccine, albumin, and plasma protein fractions have not been found to be infectious (133,134). Transfusion recipients represent 2% of total AIDS cases. Of reported pediatric cases, 13% received transfusions of contaminated blood prior to onset of AIDS (90). Individuals with coagulation disorders, such as hemophiliacs, are another group at increased risk from contaminated blood products. The prevalence of HIV-1 in hemophiliacs in 1985, in one serosurvey, revealed 92% of persons with hemophilia A, and 52% of those with hemophilia B in a U.S. cohort had antibodies to HIV-1 (135). The risk of infection seems to be related to the amount of factor concentrate used (136). With the institution of blood bank screening for HIV-1, it has been documented that there is a decline in blood and blood product–associated infection.

2. Sexual Transmission

HIV-1 transmitted via homosexual and heterosexual sex practices. It has been repeatedly shown that heterosexual contacts of high-risk-group members can readily lead to HIV-1 infection (137–139). The potential for HIV-1 infection among heterosexuals in the general population creates a route for dissemination of disease to catastrophic proportions. This, in fact, has been postulated to explain the current pattern of disease in Africa (98). The actual modes of HIV-1 infection involve close personal contact involving the exchange of bodily secretions, particularly semen. Other risk factors identified in an STD clinic population in Kenya further identified heightened risk for seropositivity among men who were uncircumcized or had genital ulcers (140). Bidirectional transmission has been documented from male to male, male to female, female to male, and possibly in a single case from female to female.

3. Intravenous Drug Use

Intravenous drug users overlap with other risk groups and serve as a conduit for other routes of transmission. Intravenous drug users may be homosexuals, prostitutes, or mothers. Intravenous drug use is a major route for transmission of HIV-1 infection among blacks and Hispanics. Behavioral patterns among drug users, such as sharing drug paraphernalia and reuse of previously used needles, facilitate HIV-1 transmission. In addition, it has been reported that drug users have immunological abnormalities in the absence of HIV-1 infection which may enhance their potential for acquiring infection and perhaps for developing AIDS (141,142).

4. Maternal-Infant Transmission

A majority of pediatric AIDS cases result from HIV-1 infection of the mother. Infants with AIDS developing disease in early infancy have onset of disease at approximately 4 months of age, suggesting perinatal transmission (143). Direct evidence supporting actual transplacental transmission has been reported (144). Postnatal transmission has also been demonstrated where a mother transfused HIV-1-contaminated blood postpartum transmitted HIV-1 presumably in breast milk or by close contact (145). The effects of HIV-1 on pregnant women are of great interest. In one ongoing study, 14 seropositive and 21 seronegative women were evaluated. Adverse pregnancy outcomes were seen for seropositive mothers and included babies with younger gestational ages at birth, lower birth weights, more common perinatal complications, greater number of pregnancy complications, and lower mean APGAR scores (146).

5. Occupational Exposure/Casual Contact

Health care workers are at the forefront caring for the victims of AIDS. They run the day-to-day risk of coming into contact with HIV-1-contaminated blood products and body fluids. Prospective surveillance of health care workers with documented parenteral or mucous membrane exposures to patients who have AIDS was initiated in 1983 (147). The majority of exposures occurred in direct patient care areas. Types of exposures included needle-stick injuries (65%), cuts with sharp instruments (16%), mucosal exposure (14%), and contamination of open skin lesions with potentially infective body fluids (6%). The risk of HIV-1 transmission from a single needle-stick accident is estimated to be approximately 0.35% (148). Among laboratory workers exposed to concentrated virus, a similar incidence rate of 0.48 per 100 person-years of exposure has been estimated (149). To date, the number of positive cases developing after exposure has remained small, with seroprevalence rates of less than 1% (150–152). Guidelines for protection against exposure in the workplace have been published (153). Risk reduction can be accomplished by strict adherence to these Public Health Service guidelines.

Further concern focuses on transmission of HIV infection to individuals with casual contact with AIDS patients at home, at school, and in the workplace. One study of nonsexual household contacts which included children, spouses, and other relatives evaluated 101 household contacts; 100 of the 101 contacts did not contract HIV infection (154,155). The only infected contact acquired infection via perinatal transmission. There is minimal risk of infection through casual contact.

IV. CONCLUSION

Retroviruses, as illustrated here, have been recognized to have an important new role in human disease. Epidemiology has contributed considerably to establishing the retrovirus association with human disease. The full impact of retrovirus biology is still evolving as new viruses and new disease associations continue to be discovered.

REFERENCES

1. Poiesz, B. J., Ruscetti, F. W., Mier, J. W., Woods, A. M., and Gallo, R. C. (1980). T cell lines established from human T-lymphocytic neoplasias by direct response to T-cell growth factor. *Proc. Natl. Acad. Sci. USA* 77: 6815–6819.
2. Uchiyama, T., Yodoi, J., Sagawa, K., Takatsuki, K., and Uchino, H. (1977). Adult T-cell leukemia: Clinical and hematological features of 16 cases. *Blood* 50:481–492.
3. Gallo, R. C., and Meyskens, F. L., Jr. (1978). Advances in viral etiology of leukemia and lymphoma. *Semin. Hematol.* 15:379–398.
4. Poiesz, B. J., Ruscetti, F. W., Reitz, M. S., Kalyanaraman, V. S., and Gallo, R. C. (1981). Isolation of a new type C retrovirus (HTLV) in primary uncultured cells of a patient with Sézary T-cell leukaemia. *Nature* 294:268–271.
5. Rho, H. M., Poiesz, B., Ruscetti, F. W., and Gallo, R. C. (1981). Characterization of the reverse transcriptase from a new retrovirus (HTLV) produced by a human cutaneous T-cell lymphoma cell line. *Virology* 112:355–360.
6. Reitz, M. S., Jr., Poiesz, B. J., Ruscetti, F. W., and Gallo, R. C. (1981). Characterization and distribution of nucleic acid sequences of a novel type C retrovirus isolated from neoplastic human T-lymphocytes. *Proc. Natl. Acad. Sci. USA* 78:1887–1891.
7. Kalyanaraman, V. S., Sarngadharan, M. G., Poiesz, B., Ruscetti, F. W., and Gallo, R. C. (1981). Immunological properties of a type C retrovirus isolated from cultured human T lymphoma cells and comparison to other mammalian retroviruses. *Virology* 38:906–915.
8. Robert-Guroff, M., Nakao, K., Ito, Y., Sliski, A., and Gallo, R. C. (1982). Natural antibodies to human retrovirus HTLV in a cluster of Japanese patients with adult T cell leukemia. *Science* 215:975–8.

9. Popovic, M., Sarin, P. S., Robert-Guroff, M., Kalyanaraman, V. S., Mann, D., Minowada, J., and Gallo, R. C. (1983). Isolation and transmission of human retrovirus (human T-cell leukemia virus). *Science* 219:856–859.
10. Watanabe, T., Seiki, M., and Yoshida, M. (1984). HTLV type 1 (U.S. isolate) and ATLV (Japanese isolate) are the same species of human retrovirus. *Virology* 133:238–41.
11. Manzari, V., Gradilone, A., Brillari, G., Zani, M., Collalti, E., Pandolfi, F., DeRossi, G., Liso, V., Babbo, P., Robert-Guroff, M., and Frati, L. (1985). HTLV-I is endemic in southern Italy: Detection of the first infectious cluster in a white population. *Int. J. Cancer* 36:557–559.
12. Su, I. J., Chan, H. L., Kuo, T. T., Eimoto, T., Maeda, Y., Kikuchi, M., Kuan, Y. Z., Shih, L. Y., Chen, M. J., and Takeshita, M. (1985). Adult T-cell leukemia/lymphoma in Taiwan. A clinicopathologic observation. *Cancer* 56:2217–2220.
13. Follezon, J. Y., Audorim, J., Reynes, M., Thulliez, M., Diebold, J., Perrot, J. Y., and Boucheix, C. (1986). Non-epidemic tropic cutaneous lymphoma ending in leukemia in case of ATL in France? *Pathol. Res. Pract.* 181:93–100.
14. Blayney, D. W., Jaffe, E. S., Fisher, R. I., Schechter, G. P., Cossman, J., Robert-Guroff, M., Kalyanaraman, V. S., Blattner, W. A., and Gallo, R. C. (1983). The human T-cell leukemia/lymphoma virus, lymphoma, lytic bone lesions, and hypercalcemia. *Ann. Intern. Med.* 98:144–151.
15. Hinuma, Y., Komoda, H., Chosa, T., et al. (1982). Antibodies to adult T-cell leukemia-viruses-associated antigen (ATLA) in sera from patients with ATL and controls in Japan: A nation-wide sero-epidemiologic study. *Int. J. Cancer* 29:631–635.
16. The T- and B-cell Malignancy Study Group. (1985). Statistical analyses of clinico-pathological, virological and epidemiological data on lymphoid malignancies with special reference to adult T-cell leukemia/lymphoma: A report of the second nationwide study of Japan. *Am. J. Clin. Oncol.* 15:517–535.
17. Clark, J. W., Robert-Guroff, M., Ikehara, O., Henzan, E., and Blattner, W. A. (1985). Human T-cell leukemia-lymphoma virus type 1 and adult T-cell leukemia-lymphoma in Okinawa. *Cancer Res.* 45:2849–2852.
18. Catovsky, D., Greaves, M. F., Rose, M., Goolden, A. W., White, J. M., Bourikas, G., Brownell, A. I., Blattner, W. A., Greaves, M. F., Galton, D. A., McCluskey, D. R., Lampert, I., Ireland, R., Bridges, J. M., and Gallo, R. C. (1982). Adult T-cell lymphoma-leukaemia in blacks from the West Indies. *Lancet* 1:639–643.
19. Blattner, W. A., Kalyanaraman, M., Robert-Guroff, M., Lister, T. A., Galton, D. A. G., Sarin, P. S., Crawford, M. H., Catovsky, D., Greaves, M., and Gallo, R. C. (1982). The human type-C retrovirus, HTLV in blacks from the Caribbean region and relationship to adult T-cell leukemia/lymphoma. *Int. J. Cancer* 30:257–264.
20. Gibbs, W. N., Lofters, W. S., Campbell, M., et al. (1987). Non-Hodgkin's lymphoma in Jamaica and its relation to adult T-cell leukemia/lymphoma. *Ann. Intern. Med.* 106:361–368.

21. Williams, C. K. O., Saxinger, W. C., Alabi, G. O., Junaid, T. A., Blayney, D. W., Greaves, M. F., Gallo, R. C., and Blattner, W. A. (1984). HTLV-associated lymphoproliferative disease: Report of 2 cases in Nigeria. *Br. Med. J.* 288:1495–1496.

22. Blayney, D. W., Jaffe, E. S., Blattner, W. A., Cossman, J., Robert-Guroff, M., Longo, D. L., Bunn, P. A., Jr., and Gallo, R. C. (1983). The human T-cell leukemia/lymphoma virus associated with American adult T-cell leukemia/lymphoma. *Blood* 62:401–405.

23. Blayney, D. W., Blattner, W. A., Robert-Guroff, M., Jaffe, E. S., Fisher, R. I., Bunn, P. A., Jr., Patton, M. G., Rarick, H. R., and Gallo, R. C. (1983). The human T-cell leukemia-lymphoma virus in the southeastern United States. *JAMA* 250:1048–1053.

24. Levine, P. H., Jaffe, E. S., Manns, A., et al. (1989). Human T-cell lymphotropic virus type I and adult T-cell leukemia/lymphoma outside of Japan and the Caribbean Basin. *Yale Univ. J. Biol. Med.*, 61; 215–222.

25. Dosik, H., Denic, S., Patel, N., Krishnamurthy, M., Levine, P. H., and Clark, J. W. (1988). Adult T-Cell leukemia/lymphoma in Brooklyn. *JAMA* 259: 2255–2257.

26. Levine, P. H., Blattner, W. A., and Biggar, R. J. (1986). Issues in the sero-epidemiology of human retroviruses. In: *Viruses and Human Cancer*, Vol. 43, Proceedings of UCLA Symposium, Park City, Utah, Feb. 2–9, 1986, Alan R. Liss, New York.

27. Bartholomew, C., Saxinger, C., Cleghorn, F., et al. (1989). A study of HTLV–I and its associated risk factors in Trinidad and Tobago, in preparation.

28. Miller, G. J., Pegram, S. M., and Kirkwood, B. (1986). Ethnic composition, age, sex, together with location and standard of housing as determinants of HTLV–I infection in an urban Trinidadian community. *Int. J. Cancer* 38:801–808.

29. Tajima, K., and Kuroishi, T. (1985). Estimation of rate of incidence of ATL among ATLV (HTLV–I) carriers in Kyushu, Japan. *Jpn. J. Clin. Oncol.* 15:423–430.

30. Williams, C. K. O., Johnson, A. O. K., and Blattner, W. A. (1984). Human T-cell leukemia virus in Africa: Possible roles in health and disease. *IARC Sci. Publ.* 63:713–726.

31. Biggar, R. J., Saxinger, C., Gardiner, C., Collins, W. E., Levine, P. H., Clark, J. W., Nkrumah, F. K., and Blattner, W. A. (1984). HTLV Type-I antibody in urban and rural Ghana, West Africa. *Int. J. Cancer* 34:215–219.

32. Fleming, A. F., Maharajan, R., and Abraham, M. (1986). Antibodies to HTLV–I in Nigerian blood donors, their relatives and patients with leukemia, lymphomas and other diseases. *Int. J. Cancer* 38:809–813.

33. Shimoyama, M., Kagami, Y., Shimotohno, K., Miwa, M., Minato, K., Tobinani, K., Suemasu, K., and Sugimura, T. (1986). Adult T-cell leukemia/lymphoma not associated with human T-cell leukemia virus type I. *Proc. Natl. Acad. Sci. USA* 83:4524–4528.

34. Clark, J., Saxinger, C., Gibbs, W. N., Lofters, W., Lagranade, L., De-
    ceulaer, K., Ensroth, A., Robert-Guroff, M., Gallo, R. C., and Blattner,
    W. A. (1985). Seropeidemiologic studies of human T-cell leukemia/lym-
    phoma virus type I in Jamaica. *Int. J. Cancer* 36:37–41.
35. Asou, N., Kumagai, T., Uekihara, S., Ishii, M.,Sato, M., Sakai, K., Nishi-
    mura, H., Yamaguchi, K., and Takatsuki, K. (1986). HTLV–I seropreva-
    lence in patients with malignancy. *Cancer* 58:903–907.
36. Blattner, W. A., Blayney, D. W., Jaffe, E. S., Robert-Guroff, M., Kalyanar-
    aman, V. S., and Gallo, R. C. (1983). Epidemiology of HTLV-associated
    leukemia. *Hematol. Blut* 28:148–155.
37. Murphy, E. L., Hanchard, B., Figueroa, J. P., et al. (1989). Modelling
    the risk of adult T-cell leukemia/lymphoma in persons infected with
    human T-lymphotropic virus type 1. *Int. J. Cancer*, 43:250–252.
38. Okamoto, T., Ohno, Y., Tsugane, S., et al. (1989). A multistep carcino-
    genesis model for adult T-cell leukemia. *Lancet*, submitted for publica-
    tion.
39. Tajima, K., Tominaga, S., Suchi, T., Kawagoe, T., Komoda, H., Hinuma,
    Y., Oda, T., and Fumita, K. (1982). Epidemiological analysis of the dis-
    tribution of antibody to adult T-cell leukemia-virus-associated antigen:
    Possible horizontal transmission of adult T-cell leukemia virus. *Gann* 73:
    893–897.
40. Blattner, W. A. and Gallo, R. C. (1985). HTLV: comparative epidemiology.
    In: *Proceedings: XIIth Symposium for Comparative Research on Leukemia
    and Related Diseases*, Hamburg, Germany 7–11 July 1985 (Froiecrech
    Reinhardt, ed.), pp. 361–382.
41. Clark, J. W., Blattner, W. A., and Gallo, R. C. Human T-cell leukemia
    viruses and T-cell lymphoid malignancies. In: *Harrison's Update of Internal
    Medicine*, Vol. VII (R. Q. Petersdorf, R. D. Adams, E. Braunwald, K. J.
    Isselbach, J. B. Martin, and J. D. Wilson, eds.), pp. 29–48. New York,
    McGraw-Hill.
42. Blattner, W. A., Nomura, A., Clark, J. W., Ho, G. Y., Nakao, Y., Gallo, R.,
    and Robert-Guroff, M. (1986). Modes of transmission and evidence for viral
    latency from studies of HTLV–I in Japanese migrant populations in Hawaii.
    Proc. Natl. Acad. Sci. USA 83:4895–4898.
43. Tajima, K., Tominaga, S., Suchi, T., Fukui, H., Komoda, H., and Hinuma,
    Y. (1986). HTLV–I carriers among migrants from an ATL-endemic area
    to ATL non-endemic metropolitan areas in Japan. *Int. J. Cancer* 37:383–
    387.
44. Robert-Guroff, M., Kalyanaraman, V. S., Blattner, W. A., Popovic, M.,
    Sarngadharan, M. G., Maeda, M., Blayney, D., Catovsky, D., Bunn, P. A.,
    Shibata, A., Nakao, Y., Ito, Y., Aoki, T., and Gallo, R. C. (1983). Evidence
    for human T-cell lymphoma-leukemia virus infection of family members of
    HTLV positive T cell leukemia-lymphoma patients. *J. Exp. Med.* 159:248–
    258.
45. Miyoshi, I., Fujishita, M., Tagrechi, H., Niiya, K., Kobayashi, M., Matsu-

bayashi, K., and Miwa, M. (1983). Horizontal transmission of adult T-cell leukaemia virus from male to female Japanese monkey. *Lancet* 1: 241.

46. Ando, Y., Nakano, S., Shimato, I., Ichijo, M., Sugamura, K., and Hinuma, Y. (1985). Possible route of ATLV transmission in husband-wife unit. *Jpn. J. Cancer Res.* 30:417.

47. Kajiyama, W., Kashwagi, S., and Ikematsu, H. (1986). Intrafamilial transmission of adult T-cell leukemia virus. *J. Infect. Dis.* 154(5):851.

48. Murphy, E. L., Figueroa, J. P., Gibbs, W. N., et al. (1989). Sexual transmission of HTLV-I. *Ann. Int. Med.*, submitted for publication.

49. Komuro, A., Hayami, M., Fujii, H., Miyahara, S., and Hirayama, M. (1983). Vertical transmission of adult T-cell leukemia virus. *Lancet* 1:240.

50. Hino, S., Yamaguchi, K., Katamine, S., et al. (1985). Mother to child transmission of human T-cell leukemia virus type-I. *Jpn. J. Cancer Res. (Gann)* 76:474–480.

51. Kinoshita, K., Hino, S., Amayaski, T., Ikeda, S., Yamada, Y., Suzuyama, J., Mimita, S., et al. (1984). Demonstration of ATL virus antigen in milk from three seropositive mothers. *Gann* 75(2):103–105.

52. Sugiyama, H., Doi, H., and Yamaguclu, K. (1986). Significance of postnatal mother-to-child transmission of human T-lymphotropic virus type I on the development of adult T-cell leukemia/lymphoma. *J. Med. Virol.* 20:253–260.

53. Miyamoto, Y., Yamaguchi, K., Nishimura, H., et al. (1985). Familial adult T-cell leukemia. *Cancer* 55:181–185.

54. Ichimaru, M., Kinoshita, K., Kamihara, S., Ikeda, S., Yamada, Y., Suzuyama, J., Momita, S., and Anagasaki, T. (1986). Familial disposition of adult T-cell leukemia and lymphoma. *Hematol. Oncol.* 4(1):21–29.

55. Tajima, K., Akaza, T., Koike, K., Hinuma, Y., Suchi, T., and Tominaga, S. (1984). HLA antigens and adult T-cell leukemia infection: A community-based study in the Goto Islands, Japan. *Jpn. J. Clin. Oncol.* 14: 347–352.

56. T- and B-cell Malignancy Study Group (1988). The third nation-wide study on adult T-cell leukemia/lymphoma in Japan: Characteristic patterns of HLA antigen and HTLV-I infection in ATL patients and their relatives. *Int. J. Cancer* 41:505–512.

57. Okochi, K., Sato, H., and Hinuma, Y. (1984). A retrospective study in transmission of adult T-cell leukemia virus by blood transfusions seroconversion in recipients. *Vox Sang* 46:245–253.

58. Jason, J. M., McDougal, J. S., Cabradilla, C., Kalyanaraman, V. S., and Evatt, B. L. (1985). Human T-cell leukemia virus (HTLV-I) p24 antibody in New York City blood product recipients. *Am. J. Hematol.* 20:129–137.

59. Miyamoto, K., Tomita, N., Ishii, A., Nishizaki, T., Kitajima, K., Tanaka, T., Nakamura, T., Watanabe, S., and Oda, T. (1984). Transformation of ATLA-negative leukocytes by blood components from anti-ATLA-positive donor in vitro. *Int. J. Cancer* 33:721–725.

60. Williams, A. E., Fang, C. T., Slamon, D. J., et al. (1988). Seroprevalence and epidemiological correlates of HTLV-I infection in U.S. blood donors. *Science* 240:643-646.
61. Sato, H., and Okochi, K. (1986). Transmission of human T-cell leukemia virus (HTLV-I) by blood transfusion: Demonstration of proviral DNA in recipients' blood lymphocytes. *Int. J. Cancer* 37:395-400.
62. Robert-Guroff, M., Weiss, S. H., Giron, J. A., Jennings, A. M., Ginzburg, H. M., Margolis, I. B., Blattner, W. A., and Gallo, R. C. (1986). Prevalence of antibodies to HTLV-I, -II, and -III in intravenous drug abusers from an AIDS endemic region. *JAMA* 255:3133-3137.
63. Mani, K. S., Mani, A. J., and Montgomery, R. D. (1969). A spastic paraplegic syndrome in South India. *J. Neurol. Sci.* 9:179-199.
64. Gessain, A., Barin, F., Vernant, J. C., Maurs, L., Barin, F., Gout, O., Calender, A., and de The G. (1985). Antibodies to human T-lymphotropic virus type-I in patients with tropical spastic paraparesis. *Lancet* 2:407-409.
65. Rodgers-Johnson, P., Gajdusek, D. C., Morgan, O. S., Zaninovic, V., Sarin, P. S., and Graham, D. S. (1985). HTLV-I and HTLV-III antibodies and tropical spastic paraparesis. *Lancet* 2:1247-1248.
66. Osame, M., Usuku, K., Izumo, S., Ihichi, N., Amitani, H., Igata, A., Masumoto, M., and Tara, M. (1986). HTLV-I associated myelopathy a new clinical entity. *Lancet* 1:1031-1032.
67. Bartholomew, C., Cleghorn, F., Charles, W., Ratan, P., Roberts, L., Maharaj, K., Jankey, N., Daisley, H., Hanchard, B., and Blattner, W. A. (1986). HTLV-I and tropical spastic paraparesis. *Lancet* 2:99-100.
68. Osame, M., Matsumoto, M., Usuku, K., et al. (1987). Chronic progressive myelopathy associated with elevated antibodies to human T-lymphotropic virus type I and adult T-cell leukemia-like cells. *Ann. Neurol.* 21(2):117-122.
69. Blayney, D. N., Rohatgi, P. K., Hines, W., Robert-Guroff, M., Saxinger, W. C., Blattner, W. A., and Gallo, R. C. (1983). Sarcoidosis and the human T-cell leukemia-lymphoma virus. *Ann. Intern. Med.* 99:409.
70. Koprowski, H., DeFreitas, E. C., Harper, M. E., Sandberg-Wollheim, M., Sheremata, W. A., Robert-Guroff, M., Saxinger, C. W., Feinberg, M. B., Wong-Staal, F., and Gallo, R. C. (1985). Multiple sclerosis and human T-cell lymphotropic retrovirus. *Nature* 318:154-160.
71. Karpas, A., Kampf, U., Siden, A., Koch, M., and Poser, S. (1986). Lack of evidence from the involvement of known human retrovirus in multiple sclerosis. *Nature* 322:177-178.
72. Gradilone, A., Zani, M., Barillari, G., et al. (1986). HTLV-I and HIV infection in drug addicts in Italy. *Lancet* 2:752-754.
73. Harper, M. E., Kaplan, M. H., and Manselle, L. (1986). Concomitant infection with HTLV-I and HTLV-III in a patient with T8 lymphoproliferative disease. *N. Engl. J. Med.* 315:1073-1078.
74. Gottlieb, M. S., Schroff, R., Schanker, H. M., Weisman, J. D., Fan, P. T., Wolf, R. A., and Saxon A. (1981). *Pneumocystis carinii* pneumonia and mucosal candidiasis in previously healthy homosexual men: Evidence

of a new acquired cellular immunodeficiency. *N. Engl. J. Med.* 305:1425–1431.

75. Masur, H., Michelis, M. A., Greene, J. B., Onorato, I., Vande Stouwe, R. A., Holzman, R. S., Wormer, G., Brettman, L., Lange, M., Murray, H. W., and Cunningham-Rundles, S. (1981). An outbreak of community-acquired *Pneumocystis carinii* pneumonia: Initial manifestation of cellular immune dysfunction. *N. Engl. J. Med.* 305:1431–1438.

76. Popovic, M., Sarngadharan, M. G., Read, E., and Gallo, R. C. (1984). Detection, isolation, and continuous production of cytopathic retrovirus (HTLV-III) from patients with AIDS and pre-AIDS. *Science* 224:497–500.

77. Gallo, R. C., Salahuddin, S. Z., Popovic, M., Shearer, G. M., Kaplan, M., Haynes, B. F., Palker, T. J., Redfield, R., Oleske, J., Safai, B., et al. (1984). Frequent detection and isolation of cytopathic retroviruses (HTLV-III) from patients with AIDS and at risk for AIDS. *Science* 224:500–503.

78. Schupbach, J., Popovic, M., Gilden, R. V., Gonda, M. A. Sarngadharan, M. G., and Gallo, R. C. (1984). Serological analysis of a subgroup of human T-lymphotropic retroviruses (HTLV-III) associated with AIDS. *Science* 224:503–505.

79. Sarngadharan, M. G., Popovic, M., Bruch, L., et al. (1984). Antibodies reactive with a human T-lymphotropic retrovirus (HTLV-III) in the serum of patients with AIDS. *Science* 224:506–508.

80. Levy, J. A., Hoffman, A. D., Kramer, S. M., Landis, J. A., Shimabukuro, J. M., and Oshiro, L. S. (1984). Isolation of lymphadenopathic retroviruses from San Francisco patients with AIDS. *Science* 225:840–842.

81. Shaw, G. M., Hahn, B. H., Arya, S. K., Groopman, J. E., Gallo, R. C., and Wong-Staal, F. (1984). Molecular characterization of human T-cell leukemia (lymphotropic) virus type III in the acquired immune deficiency syndrome. *Science* 226:1165–1171.

82. Mitsuya, H., Kent, J., Furma, P., et al. (1985). 3'-Azido-3' deoxythymidine (BW A50GW) an antiviral agent that inhibits the infectivity and cytopathic effect of human T-lymphotropic virus type III in vitro. *Proc. Natl. Acad. Sci. USA* 82:7096–7100.

83. Yarchoan, R., Weinheld, K. J., and Lyerly, H. (1986). Administration of 3' azido-3' deoxythymidine, an inhibitor of HTLV-III/LAV replication to patients with AIDS or ARC. *Lancet* 1:575.

84. Mann, J. M., and Chin, J. (1988). AIDS: A global perspective. *N. Engl. J. Med.* 319:302–303 (Editorial).

85. Centers for Disease Control (1988). Quarterly report to the Domestic Policy Council on the Prevalence and Rate of Spread of HIV and AIDS in the United States. *Morbid. Mortal. Wkly. Rep.* 37(14):223–226.

86. WHO Weekly Epidemiological Record (1986). AIDS: Global data. Nov. 21.

87. Jaffe, H. W., Bregman, D. J., and Selik, R. M. (1983). Acquired immune deficiency syndrome in the United States: The first 1,000 cases. *J. Infect. Dis.* 148:339.

88. Centers for Disease Control (1984). Update: Acquired immunodeficiency syndrome—United States. *Morbid. Mortal. Wkly. Rep.* 32(52):688–691.

89. Population Reports (1986). Issues in World Health. AIDS–A Public Health Crisis. Population Information Program, *Johns Hopkins Univ. Bull.*, Series L, 6, Jul-Aug, pp. 194-195.
90. Centers for Disease Control (1986). Update: Acquired immunodeficiency syndrome–United States. *Morbid. Mortal. Wkly. Rep.* 35(49):757-766.
91. Centers for Disease Control (1986). AIDS among blacks and Hispanics–United States. *Morbid. Mortal. Wkly. Rep.* 35(44):695.
92. Curran, J. W., Jaffe, H. W., Hardy, A. M., et al. (1988). Epidemiology of HIV infection and AIDS in the United States. *Science* 239:610-616.
93. Centers for Disease Control (1986). Update: Acquired immunodeficiency syndrome–United States. *Morbid. Mortal. Wkl. Rep.* 35(2):17-21.
94. Centers for Disease Control (1985). Update: Acquired immunodeficiency syndrome–Europe. *Morbid. Mortal. Wkly. Rep.*, 34:583-589.
95. Centers for Disease Control (1986). Update: Acquired immunodeficiency syndrome–Europe. *Morbid. Mortal. Wkly. Rep.* 35:35-46.
96. *WHO Weekly Epidemiological Record* (1986). AIDS: Sitution in the WHO European Region as of 30 June 1986. Oct. 3.
97. Biggar, R. J. (1986). The AIDS problem in Africa. *Lancet* 1:79-83.
98. Quinn, T. C., Mann, J. M., Curran, J. W., et al. (1986). AIDS in Africa: An epidemiological paradigm. *Science* 234:955-963.
99. Mann, J. (1986). Epidemiology of LAV/HTLV-III in Africa. *Proceedings of the International Conference on AIDS*, Paris, France, June 23-25, 1986. (Abstract 49:Slok), p.101.
100. Colebunders, R., and Frances, H. (1986). Clinical manifestations of African AIDS patients. *Proceedings of the International Conference on AIDS*, June 23-25, 1986 (Abstract 139:Szob), p. 36.
101. Gold, J. (1985). Clinical spectrum of infections in patients with HTLV-III-associated diseased. *Cancer Res.* 45:465-472(s).
102. Blaser, M. J., and Cohn, D. L. (1986). Opportunistic infections in patients with AIDS: Clues to the epidemiology of AIDS and the relative virulence of pathogens. *Rev. Infect. Dis.* 8:21-30.
103. Biggar, R. J., Horm, J., Lubin, H. H., Goedert, J. J., Greene, M. H., and Fraumeni, J. F. (1985). Cancer trends in a population at risk of AIDS. *J. Natl. Cancer Inst.* 74:793-797.
104. Ernberg, I., Bonkholm, M., et al. (1986). An EBV genome carrying pre-B cell leukemia in a homosexual man. *J. Clin. Oncol.* 4(10):1481-1488.
105. Arlin, Z. A., Mittelman, A., Gebhard, D., and Danieu, L. (1983). Chronic lymphocytic leukemia in a bisexual male. *Cancer Invest.* 1:549-550.
106. Unger, P. D., and Stranchen, J. A. (1986). Hodgkin's disease in AIDS complex patients. *Cancer* 58:821-825.
107. Robert, N. J., and Schneiderman, H. (1984). Hodgkin's disease and the acquired immunodeficiency syndrome. *Ann. Intern. Med.* 101:142-143.
108. Lozada, F., Silverman, S., and Conant, M. (1982). New outbreak of oral tumors, malignancies and infectious diseases strikes young male homosexuals. *Calif. Dent. J.* 10:39-42.

109. Leach, R. D., and Ellis, H. (1981). Carcinoma of the rectum in male homosexuals. *J. Roy. Soc. Med.* 74:490.

110. Li, P., Osborn, D., and Cronin, C. M. (1982). Anorectal squamous carcinoma in two homosexual men. *Lancet* 2:391.

111. Case Records of Massachusetts General Hospital (1986). Case 51-1985. *N. Engl. J. Med.* 315:1660.

112. Bleiweiss, I. J., Pervez, N. K., and Hammen, G. S. (1986). Cytomegalovirus induced adrenal insufficiency and associated renal cell carcinoma in AIDS. *Mt. Sinai J. Med.* 53(8):676–679.

113. Rosenberg, S. A., Suit, H. D., and Baker, L. H. (1982). Sarcomas of soft tissues. In: *Cancer–Principles and Practice of Oncology* (V. T. DeVita et al., eds.), pp. 1262. Philadelphia: Lippincott.

114. Bayley, A C. (1984). Aggressive Kaposi's sarcoma in Zambia, 1983. *Lancet* 1:1318–1320.

115. Hymes, K. B., Cheung, T., Greene, J. B., Prose, N. S., Marcus, A., Ballard, H., et al. (1981). Kaposi's sarcoma in homosexual men: A report of eight cases. *Lancet* 2:598–600.

116. Bayley, A. C., Cherng-Popov, R., and Dalgleish, A. G. (1986). HTLV–III distinguishes atypical and endemic Kaposi's sarcoma in Africa. *Lancet* 1: 359–361.

117. Drew, W. L. (1986). Is cytomegalovirus a cofactor in the pathogenesis of AIDS and Kaposi's sarcoma? *Mt. Sinai J. Med.* 53(8):622–626.

118. Ensoli, B., Bibenfeld, P., Nakamura, S., et al. (1988). Possible role of growth factors and cytokines in the pathogenesis of Kaposi's sarcoma. IV. International AIDS Conference, Stockholm, Sweden, 1988. 2:131 (Abstract 2647).

119. Marmor, M., Friedman-Kien, A. E., Laubenstein, L., et al. (1982). Risk factors for Kaposi's sarcoma in homosexual men. *Lancet* 1:1083–1087.

120. Jaffe, H. W., Choi, K., Thomas, P. A., Haverkos, H. W., Auerbach, D. M., Guinan, M. E., Rogers, M. F., et al. (1983). National case-control study of Kaposi's sarcoma and *Pneumocystis carinii* pneumonia in homosexual men; Part I. Epidemiologic results. *Ann. Intern. Med.* 99(2):145–151.

121. Polk, F., Fox, R., Brookmeyer, R., Kanchanaraksa, S., et al. (1987). Predictors of AIDS developing in a cohort of seropositive homosexual men. *N. Engl. J. Med.* 316:61–66.

122. Goedert, J. J., Biggar, R. J., Melbye, M., et al. (1987). Effect of T4 count and cofactors in incidence of AIDS in homosexual men infected with human immunodeficiency virus. *JAMA* 257(3):331–334.

123. Ziegler, J. L., Drew, W. L., Miner, R. C., Mintz, L., Rosenbaum, Gershow, J., Lennette, E. T., Greenspan, J., Shillitoe, E., Beckstead, J., Casavant, C., and Yamamoto, K. (1982). Outbreak of Burkitt's-like lymphoma in homosexual men. *Lancet* 2:631–633.

124. Ziegler, J. L., Beckstead, J. A., Volberding, P. A., et al. (1984). Non-Hodgkin's lymphoma in 90 homosexual men. Relation to generalized lymphadenopathy and the acquired immunodeficiency syndrome. *N. Engl. J. Med.* 311:565–570.

125. Kalter, S. P., Riggs, S. A., and Butler, J. J. (1985). Aggressive non-Hodgkin's lymphomas in immunocompromised homosexual males. *Blood* 66(3):655–659.

126. Levine, A. M., Gill, P. S., Meyer, P. R., Burkes, R. L., Ross, R., Dworsky, R. D., et al. (1985). Retrovirus and malignant lymphoma in homosexual men. *JAMA* 254:1921–1925.

127. Ciobanu, N., and Wiennik, P. (1986). Malignant lymphomas, AIDS, and the pathogenic role of Epstein-Barr virus. *Mt. Sinai J. Med.* 53(8):627–638.

128. Groopman, J. E., Sullivan, J. L., Mulder, C., Ginsburg, D., Orkin, S. H., O'Hara, C. J., Falchuk, K., Wong-Staal, F., and Gallo, R. C. (1986). Pathogenesis of B-cell lymphoma in a patient with AIDS. *Blood* 67(3): 612–615.

129. Ho, D. D., Rota, T. R., Schooley, R. T., et al. (1985). Isolation of HTLV–III from cerebrospinal fluid and neural tissues of patients with neurologic syndromes related to the acquired immunodeficiency syndrome. *N. Engl. J. Med.* 313:1493–1497.

130. Levy, R. M., Morgan, N. M., and Rosenblum, M. L. (1986). The neuro-epidemiology of AIDS in the United States. *Proceedings of the Second International Conference on AIDS*, Paris, France, June 23–25, 1986, p. 56 (Poster 312).

131. Herman, P. (1986). Neurologic effects of HTLV–III infection in adults: An overview. *Mt. Sinai J. Med.* 53(8);616–621.

132. Peterman, T. A., Jaffe, H., Feorino, P. M., Getchell, P., et al. (1985). Transfusion-associated acquired immunodeficiency syndrome in the United States. *JAMA* 254(2):2913.

133. Centers for Disease Control (1985). Provisional public health service inter-agency recommendations for screening donated blood and plasma for antibody to the virus causing AIDS. *Morbid. Mortal. Wkly. Rep.* 34:1–5.

134. Poiesz, B., Tomar, R., Lehr, B., and Moore, J. (1984). Hepatitis B vaccine; evidence confirming lack of AIDS transmission. *Morbid. Mortal. Wkly. Rep.* 33:685–687.

135. Centers for Disease Control (1986). Surveillance of hemophilia-associated acquired immunodeficiency syndrome. *Morbid. Mortal. Wkly. Rep.* 35 (43):669.

136. Ramsey, R. B., Palmer, E. L., McDougal, J. S., Kalyanaraman, V. S., Jackson, D. W., Chorba, T. L., et al. (1984). Antibody to lymphadeno-pathy-associated virus on hemophiliacs with an without AIDS. *Lancet* 2:397–398.

137. Redfield, R. R., Markham, P. D., Salahuddin, S. Z., Bodner, A. J., Bolks, T. M., Ballou, W. R., Wright, D. C., and Gallo, R. C. (1985). Frequent transmission of HTLV–III among spouses of patients with AIDS-related complex and AIDS. *JAMA* 253:1571–1573.

138. Kreiss, J. K., Koech, D., Plummer, F. A., et al. (1986). AIDS virus infection in Nairobi prostitutes: Spread of the epidemic in East Africa. *N. Engl. J. Med.* 314:414–418.

139. Redfield, R. R. (1986). Heterosexual transmission of human T-lympho-tropic type III: Syphilis revisited. *Mt. Sinai J. Med.* 53(8):592.

140. Simonsen, J. N., Cameron, D. W., Gakinya, M. N., et al. (1988). Human immunodeficiency virus infection among men with sexually transmitted diseases. *N. Engl. J. Med.* 319(5):274–277.

141. Brown, S. M., Stimmel, B., Taub, R. N., Kochwa, S., and Rosenfield, R. E. (1974). Immunologic dysfunction in heroin addicts. *Arch. Intern. Med.* 134:1001–1006.

142. Brown, L. S., Evans, R., and Murphy, D. (1986). Drug use patterns: Implications for the acquired immunodeficiency syndrome. *J. Natl. Med. Assoc.* 78(12):1145–1150.

143. Scott, G. B., Fischl, M. A., Klimas, N., Fletcher, M. A., Dickinson, G. M., Levine, R. S., and Parks, W. P. (1985). Mothers of infants with the acquired immunodeficiency syndrome. Evidence for both symptomatic and asymptomatic carriers. *JAMA* 253:363–366.

144. LaPointe, N., Michaud, J., Pekovic, D., Chausseau, J. P., and Dupuy, J. M. (1985). Transplacental transmission of HTLV-III virus. *N. Engl. J. Med.* 312(20):1325–1326.

145. Ziegler, J. B., Cooper, D. A., Johnson, R. O., and Gold, J. (1985). Post-natal transmission of AIDS-associated retrovirus from mother to infant. *Lancet* 1:896–898.

146. Willoughby, A. D., Mendez, H., and Minkoff, H. (1987). Human immuno-deficiency virus in pregnant women and their offspring. III International AIDS Conference, June 1–5, 1987, Washington, DC, p. 158, Abstract No. TH. 7.3.

147. Centers for Disease Control. (1984). Prospective evaluation of health-care workers exposed via parenteral or mucous-membrane routes to blood and body fluids of patients with acquired immunodeficiency syndrome. *Morbid. Mortal. Wkly. Rep.* 33:181–182.

148. Wormser, G. P., Rabkin, C. S., and Jolone, C. (1988). Frequency of nosocomial transmission of HIV infection among health care workers. *N. Engl. J. Med.* 319(5):307–308 (Letter).

149. Weiss, S. H., Goedert, J. J., Gartner, S., et al. (1988). Risk of human immunodeficiency virus (HIV-1) infection among laboratory workers. *Science* 239:68–71.

150. Centers for Disease Control (1985). Update: Evaluation of human T-lymphotropic virus type III/lymphadenopathy-associated virus infection in health-care personnel—United States. *Morbid. Mortal. Wkly. Rep.* 34:575–579.

151. Weiss, S. H., Saxinger, W. C., Rechtman, D., Grieco, M. H., Nadler, J., Holman, S., Ginzburg, H. M., Groopman, J. E., Goedert, J. J., Markham, P. D., et al. (1985). HTLV-III infection among health care workers. Association with needle-stick injuries. *JAMA* 254(15):2089–2093.

152. McCray, E. (1986). Occupational risk of the acquired immunodeficiency syndrome among health care workers. *N. Engl. J. Med.* 314:1127–1132.

153. Centers for Disease Control (1985). Summary: Recommendations for preventing transmission of infection with human T-lymphotropic virus type III lymphadenopathy-associated virus in the workplace. *Morbid. Mortal. Wkly. Rep.* 34:681–696.
154. Friedland, G. H., Saltzman, B. R., Rogers, M. F., et al. (1986). Lack of transmission of HTLV–III/LAV infection to household contacts of patients with AIDS or AIDS-related complex with oral candidiasis. *N. Engl. J. Med.* 314:344–349.
155. Sande, M. A. (1986). Transmission of AIDS: The case against casual contagion. *N. Engl. J. Med.* 314(6):380.
156. Centers for Disease Control (1985). Revision of the case definition of acquired immunodeficiency syndrome for retroviral reporting—United States. *Morbid. Mortal. Wkly. Rep.* 34:373–375.
157. Centers for Disease Control (1987). Revision of the CDC surveillance case definition for acquired immunodeficiency syndrome. *Morbid. Mortal. Wkly. Rep.* 36(Suppl ls).

10

Structure and Function of Human Pathogenic Retroviruses

William A. Haseltine and Ernest F. Terwilliger *Harvard Medical School, Boston, Massachusetts*

Craig A. Rosen *Roche Institute of Molecular Biology, Nutley, New Jersey*

Joseph G. Sodroski *Harvard Medical School, Boston, Massachusetts*

I. INTRODUCTION

The human retroviruses represent an emerging class of complex pathogens involved in a wide variety of maladies including leukemias and lymphomas, diseases of the central nervous system, and immune function impairment. Four different types of human retroviruses have been isolated: HTLV-I, the etiological agent of a malignant T-cell leukemia/lymphoma (ATLL); HTLV-II, a virus associated with a more benign form of T-cell leukemia; HIV-1 (previously called HTLV-III or LAV-1), the etiological agent of acquired immune deficiency syndrome and related disorders; and HIV-2 (also called LAV-2), a virus that has recently been isolated from people living in West Africa. Salient features of the pathogenesis and structure of these viruses are summarized in this chapter.

II. THE HUMAN LEUKEMIA RETROVIRUSES

A. Pathogenesis

The T-cell leukemias and lymphomas induced by HTLV-I and -II appear after a long incubation period measured in decades (1). Infection is generally marked by seroconversion. However, there is some evidence that seroconversion may occur only after very prolonged periods, as long as 15-20 years for neonatal infection.

Infected people are not viremic at any stage of the disease. There is also a notable lack of virus expression even in fresh tumor cell populations (2). Stimulation of infected patient T cells with mitogens results in the expression of high levels of viral RNA and protein and the budding of virus particles (3).

Cocultivation of mitogen-activated T cells from infected people with normal cells results in immortalization of peripheral CD4+ lymphocytes (4–7). Such transformation is generally accomplished by cocultivation and is very difficult to accomplish with cell-free virus. The cells transformed in vitro have the appearance of tumor cells, carrying a distinctive set of surface markers including the CD4 antigen. They have large, lobulated nuclei. The fresh tumor cells and cell lines immortalized by HTLV-I express abnormally high levels of the IL-2 surface receptor (8,9).

The absence of viremia in infected persons and the difficulty of inducing cell-free infection may help to explain the distribution of infection in populations. The virus is endemic in Japan (10) and Taiwan, the Caribbean (1,11), as well as in Africa (12,13), the Southeast United States (14), and in parts of South America (11). In these areas, transmission occurs predominantly in a family context (15). Transmission from mother to child and from infected male to female partner is documented (16–18). Transmission from infected female to male sex partners is rare. The virus is also transmitted by needle parenterally, either by blood transfusion or by sharing of contaminated needles by i.v. drug users. The latter route appears to be a significant factor in current transmission patterns of the virus, as large drug abuse populations have been found to be infected with either HTLV-I or HTLV-II (20–40% penetrance), in metropolitan centers in the United States and Europe (19).

1. Central Nervous System Disease

Infection with HTLV-I also presents a risk for neurological disorders (20,21). Originally described as tropical spastic paraparesis in African patients, this disorder in HTLV-I-seropositive individuals has also been detected in Japan, where it has been called HTLV-I-associated myelopathy (HAM), and in the Caribbean. This disorder is characterized by chronic progressive myelopathy. The virus present in patients with this disorder is similar to the HTLV-I associated with ATLL (22).

B. Genomic Organization

Like all other retroviruses, the human leukemia retroviruses contain genes that encode the virion internal capsid proteins (the *gag* gene proteins), genes that encode replication functions (the reverse transcriptase, integrase, and protease), and genes that specify the exterior envelope proteins that are embedded in the lipid layer that surrounds the virion (Figure 1) (23–29). The envelope protein is comprised of an exterior glycoprotein and an integral transmembrane protein.

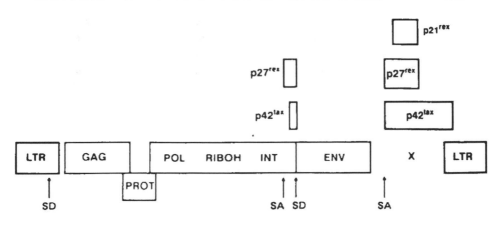

FIGURE 1 A schematic diagram of the genome of HTLV-I. The open reading frames that encode the *gag, pol,* and *env* genes are indicated. The open reading frames that encode the p42 and pp27 and pp21 proteins specified by the pX region are also shown.

The organization of the virion structural genes and replicative genes is similar to that of the simple avian murine and feline retroviruses.

The genomes of HTLV-I and HTLV-II differ from those of other retroviruses by the presence of about 1500 nucleotides located between the 3' end of the envelope glycoprotein and the 3' LTR (23,30,31). This region, called pX, has the capacity to encode multiple polypeptides of size 100 amino acids or greater (28,32).

It has been demonstrated that the pX region of HTLV-I specifies at least three polypeptides that are made in infected, activated T cells (31,33,34). The largest of these proteins, of size 42 kD for HTLV-I (35,36), 38 kD (also designated p40 or p41) for HTLV-II (37), and 36 kD for bovine leukemia virus (BLV) (38), respectively, is located primarily in the nucleus (36,39). The protein was initially called the *X-lor* protein for the product of the long open reading frame within the pX region (31). It is now often called the *tat* or *tax* gene product for *trans*-activator (40).

The *tax* product is synthesized from a doubly spliced messenger RNA species that includes transcripts of portions of the 5' LTR, a small sequence located immediately 5' to the envelope gene, and the distal two-thirds of the pX region through the end of the 3' LTR (40-43).

The same messenger RNA species encodes two other polypeptides from an overlapping reading frame (Figure 1) (44,45; Haseltine and Dokhelar, in press). The initiation codon for the larger of these two polypeptides is located 5' to the site of initiation of the *tax* gene product. This polypeptide is translated from a different reading frame within the same spliced mRNA species used to make the *tax* gene product. The product of this second initiation event is a 27-kD protein. The protein is phosphorylated and located predominantly in the nucleus (44). The protein is called pp27, denoting both its size and the fact that it is phosphorylated. A second polypeptide is also synthesized from the same reading frame as the pp27 protein. This third product of the pX region is thought to be initiated at an AUG codon within the second coding exon of the messenger RNA. This protein is also phosphorylated, has an apparent molecular weight of 21 kD, and is called pp21. It is also located primarily in the cytoplasm. The gene that encodes these proteins is called *rex*, for regulator of virion protein expression.

It is notable that the genomes of HTLV-II (46,47), STLV-I (48), and BLV (49–51) all have the capacity to encode similar alternative reading frame polypeptides. It is curious that these proteins do not raise antibodies in the infected host. The complete coding capacity of these viruses has not yet been fully explored. It is conceivable that other virally encoded proteins that are of low antigenicity in the infected host are present in virus-infected cells.

1. *Trans*-Activation: The *tax* Protein

The phenomenon of *trans*-activating retroviral gene products was first reported for HTLV-I and -II (52). It was observed that the LTRs of HTLV-I and -II function much more efficiently as promoter elements in infected as compared to uninfected cells (40). A positive *trans*-activating genetic regulatory system requires at least two elements, the *trans*-activator product and a *cis*-acting responsive element.

The *trans*-activator product of HTLV-II was initially identified as the product of the HTLV-II pX open reading frame. It has since been reported that the pX open reading frame of HTLV-I and of BLV also encodes a *trans*-activator (53,54). Gene expression directed by a plasmid that carries the *trans*-activator gene has been shown to stimulate the homologous LTRs of HTLV-II, and of BLV (52,53,55). Plasmids constructed to eliminate the possibility of producing the pp27 and pp21 gene products are also capable of *trans*-activation as measured in transient cotransfection assays (44; Dokhelar et al., in press). The LTR of HTLV-I can also be activated by the *tax* gene product of HTLV-II but not by the BLV *tax* product (52,53).

The increase in LTR-directed gene expression induced by the *tax* gene is accompanied by an increase in the steady-state level of corresponding messenger RNA species (40,56). The increase in the messenger RNA levels of heterologous genes directed by the LTR corresponds roughly to the level of increase observed for protein expression. However, precise correspondence is difficult to document, and posttranscriptional alterations in the efficiency of mRNA utilization cannot be ruled out.

The *cis*-regulatory sequences, called TAR (for *trans*-acting responsive region) were initially found to be located in the U3 region of the viral long terminal repeat (LTR), entirely 5′ to the site of initiation of viral RNA synthesis (57). It was noted early on that the U3 element of the HTLV-I and -II LTRs contained 21-nucleotide sequences repeated several times, and that the sequences of these repeat units were preserved between HTLV-I and -II (52). It was also observed that except for these repeated sequences and a short region near the site of RNA initiation, the sequences of the HTLV-I and -II LTRs are notably different as compared to the extent of conservation of other parts of the genomic sequences. Recently, synthetic oligonucleotides that correspond to these 21 nucleotide long sequences have been demonstrated to convey a response to the *trans*-activator upon heterologous promoters (58-60). The response to the *trans*-activator is observed when the 21-nucleotide repeat sequence is located proximal to the promoter and is irrespective of orientation of the 21-nucleotide sequence with respect to the promoter. In some experiments, a single repeat unit suffices to convey the *trans*-activation response (59). Others report that two or more 21-nucleotide sequences, tandemly repeated, are required for the *trans*-activation effect (58,60). These repeat units are called TAR-21 sequences to denote the observation that they convey a responsive phenotype (59).

A cautionary note is appropriate. Although the TAR-21 sequences do permit increased expression of both homologous and heterologous promoters in the presence of the *trans*-activators, the response is weak and the level of expression of heterologous genes is one or two orders of magnitude below that observed for promoters in their natural configuration—even promoters that contain 5′ deletions that preserve only the TAR-21 sequence located proximal to the promoter (57). This observation suggests that the promoter strength and inducibility depend on the sequences adjacent to TAR-21.

Two other features of the viral promoters are notable. Promoter strength of HTLV-I is dependent on sequences located 3′ to the site of RNA initiation, within the R and U5 regions of the LTR (54,59). A set of nested deletions originating within the U5 region of the HTLV-I LTR and extending to the site of RNA initiation results in a progressive weakening of promoter activity. Gene expression directed by such altered LTRs is inducible by the *trans*-activator genes, although the ultimate level of LTR-directed gene expression is progressively diminished in both the induced and uninduced states by these deletions. Evidently, the R and U3 region of the viral LTR encodes sequences important for high-level, LTR-directed gene expression. Location of these sequences 3′ to the site of RNA initiation raises the possibility that they may be involved in posttranscriptional regulatory events as well as contributing to the rate of RNA initiation. Sequences that have similar effect are reported to exist 3′ to the site of RNA initiation within the BLV LTR (54).

The second notable feature of the viral LTRs is an asymmetry in the function of the HTLV-I and -II sequences. The LTR of HTLV-I functions well as a promoter in a wide variety of cell types (57). The HTLV-II LTR is markedly limited and functions well in very few cell types (52). It is remarkable that the HTLV-II LTR does not function as a promoter in most human lymphoid cell lines, whether they be T cells or B cells. In fact, no promoter activity was observed in two human lymphoid cell lines that expressed a functional HTLV-II *tax* product (52). The HTLV-II *tax* product in these cell lines was found to be capable of stimulating the HTLV-I LTR. In the same cell lines, no HTLV-II promoter activity was observed. It can be concluded that the HTLV-II promoter is either extremely fastidious as regards its requirement for cell-specified expression factors or that viral gene products besides the *trans*-activator are required for activity of the HTLV-II LTR. The activity of the BLV promoter is also narrowly restricted regarding host range. It is a very poor promoter in most uninfected cell types.

2. *Trans*-Activation: The pp27 and pp21 Proteins

Recent reports indicate that the pp27 protein may play an important role in virus replication via a *trans*-acting mechanism (61,62). An integrated provirus deleted for the amino terminal portion of the *env* gene was found to be defective for RNA synthesis and for *gag* gene production. The deletion eliminated the 5' coding exons of the HTLV-I *tax* and pp27 proteins. Transfection of a cell line containing this defective provirus with plasmids capable of expression of the HTLV-I *tax* and/or pp27 proteins revealed that *gag* gene synthesis was dependent on both *tax* and pp27 gene expresson from the transfected plasmids. Moreover, no *gag* gene RNA was detected upon transfection with the *tax*-expressing plasmid alone. It was also reported that in transient cotransfection experiments both *tax* and pp27 are required for *gag* protein expression but that only *tax* is required for the expression of *gag* mRNA (62).

These findings suggest that both the *tax* and pp27 proteins are needed for the expression of viral genes. However, heterologous gene synthesis directed by the HTLV-I LTR located 5' to the gene is not dependent on pp27, nor does the expression of pp27 markedly affect the rate of expression of such constructs (Sodroski, Dokhelar, and Haseltine, unpublished observations).

The function of the pp27 gene resembles, in a formal sense, that of the *rev* gene of HIV-1 (see below). Neither pp27 nor *rev* is required for expression of heterologous genes under the control of the LTR. However, in the absence of a second gene product, the *trans*-activator genes (*tat* genes) of these viruses are insufficient to permit expression of viral *gag* proteins. It is notable that the *trans*-activator genes of both viruses can be synthesized in the absence of auxiliary proteins. For both viruses, the regulatory genes are controlled independently from the structural genes.

Although the *rev* and pp27 genes display a formal analogy in functional terms, such similarity does not imply that the mechanism of action is the same. However, functional cross-complementation of the *rev* function by the *rex* protein of HTLV-I has recently been demonstrated (191). The *trans*-activator gene of HTLV-I acts primarily as a transcriptional *trans*-activator of the viral LTR (40,56,63-65).

3. The Mechanism of Transformation

The process of in vitro formation of tumors by HTLV-I, -II, and BLV has not been fully characterized. Infection of T cells by the virus does not result in immediate tumor formation. Rather, tumors arise rarely at a frequency of about 1 in 100 HTLV-I-infected people (N. Mueller, personal communication). The role of viral genes in the transformation process is inferred from epidemiological studies that link seropositivity to disease, as well as the observation that T-cell tumors in infected people contain at least one integrated copy of the provirus (66). It is sometimes observed that tumors contian only the 3' portion of the genome (2,67). However, most of the tumors found in patients contain, as a minimum, the 5' LTR and the pX region and the 3' LTR. This genomic arrangement indicates that LTR-directed pX gene expression is required throughout tumor development.

The T-cell tumors in patients are clonal with respect to the site of integration of the provirus (66,68). The long latent period and clonal nature of the tumors indicate that events, in addition to infection of T cells with the virus, are likely to be required for the appearance of malignant tumors. Such events may represent either secondary changes occurring within the infected cell, such as somatic mutations, or changes in the immunological status of the host.

Two additional observations indicate that the viral genes play an important role in the initiation and maintenance of tumors. Tumorigenesis by the avian, murine, and feline retroviruses that contain only those genes required for virion formation and virus replication depend on activation of cellular growth regulatory genes. This conclusion is reached from the observation that independent virally induced tumors contain proviruses that are found integrated near the same cellular genes. Such is not the case for tumors induced by HTLV-I or BLV, for which no repeated chromosomal sites of integration have been observed in naturally occurring tumors (66). It is therefore inferred that viral genes themselves play a key role in the initiation and maintenance of the tumor phenotype.

The role of the viral genes in the transformation process is also inferred from in vitro transformation studies. Primary T cells can be immortalized by cocultivation with mitomycin C-treated HTLV-I- or -II-infected cells (4,6,7). In contrast to normal T cells, the transformed T cells continue to proliferate without continued antigen stimulation. Eventually, immortalized cells emerge from such cultures (4-7). The expanding population of T cells is initially polyclonal with respect to the site of provirus integration. Eventually, cell lines that are monoclonal with respect to the sites of viral integration emerge from the population and dominate the culture. Such cell lines may remain dependent on IL-2 for growth. However, IL-2-independent cell lines can also be obtained. Such immortalized primary cells are typically CD4+ as are most HTLV–I-induced tumors. CD8+ cell lines can be derived by cocultivation of infected mitomycin-treated infected cells with primary populations of lymphocytes enriched for cells that bear the CD8 antigen (69).

Events that occur between the initiation of infection and establishment of IL-2-dependent or -independent T-cell lines have not been well characterized. Selection of specific fast-growing clones may occur in infected patients as well as in vitro (70,71). It is possible that secondary changes occur within the infected cells that permit rapid growth. Alternatively, the clonality of the tumor cells may represent selection of a cell population that expresses high levels of viral proteins that promote cellular growth.

An additional series of experiments indicates that the pX region of HTLV–I contains genes with transforming potential. Transgenic mice that carry an HTLV–I LTR *tax* gene are found to develop soft tissue fibrosarcomas (72,73). Other transgenic mice that carry the same gene develop a lethal disorder characterized by acellularity of the thymus. The region of the genome that is introduced into these mice could produce both the 42-kD and the 27-kD viral-encoded proteins.

4. Induction of the IL-2 Receptor by the *Trans*-Activator Gene

The promoters of the IL-2 receptor and the IL-2 genes have been cloned (74–76). Cotransfection of the promoters placed 5′ to reporter genes, such as the chloramphenicol acetylase transferase gene with the *trans*-activator gene of HTLV–I, has been shown to increase the level of expression of the IL-2 receptor gene promoter (77). The level of expression of the genes under the control of the IL-2 promoter was found to be increased slightly in similar experiments (74). The *trans*-activator gene of HTLV–II has also been shown to increase the level of expression of the IL-2 receptor gene, albeit more weakly than that observed for the *tax* gene, at least in the particular experimental configuration used (192). It is noteworthy that *tax* activation of the IL-2 receptor promoter in transient transfection assays is restricted to a few Jurkat cell lines and such activation is not observed in many other human T-cell lines or in other mammalian cell lines. The *trans*-activation of the IL-2R promoter is much weaker than that observed for the HTLV–I LTR in similar experiments.

These observations suggest the *trans*-activator gene of HTLV–I and –II can contribute to the growth properties of the T cell by deregulation of genes that normally control T-cell proliferation in response to antigen stimulation. Such a model for T-cell transformation must include at least one more consideration, that the expression of the viral genes is dependent on T-cell activation. Thus, an infected resting T cell should not be transformed as the viral genes are not expressed.

Although simple, this explanation for transformation does not account for the clinical observations with ATLL patients. If the *tax* genes were sufficient to induce both IL-2 and IL-2 receptors, then infection should lead to transformation. However, malignant growth of T cells in infected patients is a rare event. It is also possible that the pp27 and pp21 proteins play a role in the activation of cellular genes.

C.   A Model for HTLV–I Replication and Tumorigenesis

The broad outlines of mechanisms of tumorigenesis by the HTLV-I family of viruses are beginning to emerge. The viruses encode at least three genes, in addition to the genes *gag*, *pol*, and *env* required for virus replication. These additional genes encoded by the pX region are likely to affect in a specific fashion the growth of lymphocytes. The *tax* gene appears to mimic at least part of the response of mature lymphocytes to recognition of the cognate antigen. That is, in T lymphocytes, the *tax* gene seems to induce the IL-2 and IL-2 receptor genes. The alternative reading frame proteins may contribute to the transformed phenotype in cooperation with the *tax* gene.

The expression of viral genes in infected lymphocytes, the *tax* gene, and the pp21 and pp27 proteins, and possibly other viral genes (the coding capacity of the pX region is not exhausted by the *tat* and pp21 and pp27 proteins), may be sufficient to account for the transformation of T cells in culture. A secondary change in the infected cells in culture is not required to explain the outgrowth of cells that are clonal with respect to the site of viral genomic integration, as selection of the most rapidly growing infected cell could account for this observation.

The situation in infected patients is more complex. Infection of T cells with the HTLV-I or –II virus is not sufficient to produce malignant disease. Failure of the virus to induce malignancy in all infected T cells may be attributed to diverse causes. It is possible that viral gene expression is suppressed in most infected T cells. Certainly, no viral RNA is detected in peripheral lymphocytes in infected patients including the tumor cells themselves (2). Transcriptional repression of viral genes in infected cells is a sufficient explanation for the failure of the virus to transform most T cells in patients.

It is also possible that T cells that do express viral antigens are eliminated by the immune system. The observation that many tumor cell lines derived from patients contain deletions of virus structural proteins is consistent with this notion (43). Patients infected with HTLV-I and –II do make good immune responses to virion structural proteins.

An additional explanation may lie in homeostatic regulatory mechanisms of the immune response itself. Lymphocytes are thought to possess regulatory mechanisms that limit their proliferation response to antigen recognition. The early proliferative response of T cells in response to the presence of the cognate antigen is followed by reestablishment of a resting phase. Stabilization of the stimulated population of T cells was thought to involve activation of an internal cellular program of a repressive nature. Interaction of the activated T cells with other components of the immune system may also contribute to reestablishment of the resting state. It is conceivable that the homeostatic mechanisms regulating T-cell proliferation also regulate HTLV-I and -II gene expression and thereby limit the growth of infected cells in patients. In this view, malignant transformation by HTLV-I and -II requires bypass of the normal homeostatic mechanisms of growth control of lymphocytes. Such bypass may occur by a secondary intracellular change that occurs in the infected cells. Alternatively, it may be due to a systemic failure of normal immunoregulatory mechanisms. Either explanation could give rise to a tumor cell population, the first by outgrowth of a cell that contains a secondary genetic lesion, and the second by overgrowth of the infected cell population by a fast-growing infected cell as is observed in cell culture.

The molecular biology and in vivo replication of the virus provide some insight into the mechanisms of transmission of the virus as well. This family of viruses seems to be either poorly infectious or altogether noninfectious for uninfected cells. For establishment of infection it is likely that viral gene products transferred from an infected cell by cell fusion are required. The infectious unit may well be an infected cell rather than a virion. In this context, the X genes of this family of viruses are required for replication and may be viewed as replicative genes. Tumorigenesis may be a by-product of the natural replicative cycle of this family of viruses.

III. THE HUMAN IMMUNODEFICIENCY VIRUS

The second family of human retroviruses includes the human immunodeficiency virus (here called HIV-1) and the recently described viruses here called HIV-2.

A. Pathogenesis

Infection with HIV-1 initiates a progressive degenerative disease of the immune and central nervous systems (78). Primary lesions include ablation of the T-cell population and atrophy of the central nervous system (79-84). The disease usually progresses slowly in adults. Many years may elapse between infection and the first appearance of symptoms of the progressive disease. The mean latent period in adults has not yet been determined but is likely to exceed 5 years and may be as long as 15 years (85). The incidence of fatal disease may exceed 35% in those infected for 5 years or more. A recent prospective study indicates that the majority of those infected will develop terminal illness (Redfield, personal communication).

An acute mononucleosis-like symptom accompanied by viremia has been documented within a few weeks of exposure (78). During the early stage of infection, virus can sometimes be isolated from cerebrospinal fluid as well as serum (78,80). A prolonged asymptomatic or mildly symptomatic state may ensue. Persistent mild to severe lymphadenopathy may be present in otherwise healthy adults. Immune function and T-cell count may remain normal for some time. A decline of the total T-cell population usually marks the onset of progressive immunological disease (86–88). After a variable asymptomatic period (1–10 years), total T-cell counts often decline steadily over a period of an additional 2-3 years, often reaching a very low to nondetectable level. The severity of the disease as defined by the incidence of opportunistic infections, weight loss, and other signs of serious illness is correlated with total T-cell number (89). Progressive neurological impairment is also manifest over a prolonged period (80,81). Progressive enlargement of macroscopic lesions in the central nervous system is also observed in some patients. In extreme cases the brain mass can be reduced to one-third normal size (79). Despite an extensive search, no cofactors for disease have yet been found. It should be assumed until otherwise demonstrated that progressive degeneration of the immune and central nervous systems is a sole consequence of HIV-1 infection.

Although the defects in the immune function in advanced cases are pleomorphic, including depletion of the helper-T-cell populations, macrophage and monocyte malfunction, and hypergammaglobulinemia, many of these effects may be explained by the specific destruction of the CD4+ T-helper subset of lymphocytes. The CD4+ subset of cells apparently coordinates much of the immune response. Destruction of this cell type has been proposed to lead to the multiple abnormalities observed.

Early studies of the virus indicated the CD4+ T cells were specifically infected and killed (90). More recent studies demonstrate that although the virus is specifically cytotoxic for the CD4+ T cells in culture, other cell types can be infected. Infected monocytes (91,92), macrophages (91,93), Langerhans cells (94), and B lymphocytes (94), as well as microglial cells (96,97) and possibly certain endothelial cells (98), have been observed to produce virus in infected people. There is some indication that bone marrow stem cells may also be infected (99,100). Many of these cell types can be productively infected in vitro. These studies show that many of these cell types express low levels of the CD4+ surface antigen which serves as a receptor for the virus. The virus is not markedly cytopathic for these cells and they may provide a reservoir of infection in infected people.

Proviral DNA can be detected in the T-cell population in infected people. However, the total amount of proviral DNA in the lymphocyte population is low. Less than 1 in 100 T cells harbors the provirus. Even so, it appears that most T cells are latently infected, as only about 1 in 10,000-100,000 T cells are found to express viral antigens (101). Proviral DNA and some viral antigens can also be detected in the brains of those infected (79,102). However, abundant virus expression is not observed in the central nervous system tissues, even in patients undergoing the most severe central nervous system destruction.

Infectious virus can be isolated from a variety of body fluids, including blood, semen, tears, and vaginal secretions (80,103-107). In this respect, HIV differs from HTLV-I and -II as the virus particle itself is highly infectious. Infection occurs by cell-free virus as well as by cell-associated virus.

The virus is transmitted venereally (bidirectionally), maternally (both pre- and postnatally), and parenterally, as well as by organ and blood donation (85). The relatively rapid spread of HIV in various populations as compared to that of HTLV-I and -II may be attributed, at least in part, to the infectivity of the cell-free virus.

The humoral immune response to infection is usually vigorous, high antiviral antibody levels being detectable within 2-3 weeks to 6 months postinfection (108). The envelope glycoprotein is the most antigenic of the viral proteins as antibodies to the protein are both the first to appear and the last to disappear during late stages of infection (Lee, personal communication). High-titer antibodies to other virus structural proteins as well as to some of the regulatory proteins encoded by the viruses are observed in some patients. A cell-mediated, cytotoxic immune response that is histocompatibility restricted has been reported in asymptomatic infected people as well as patients during the early stages of disease (109). Destruction of infected cells by the host immune response has been proposed to account for at least part of the observed pathogenesis. However, firm evidence of such a reaction is lacking. In particular, the central nervous system disease is neither focal nor characteristic of reactive hyperplasia.

1. The Pathogenesis of HIV-2 and Related Simian Viruses

Much less is known regarding the pathogenesis of the HIV-2 viruses. Seroepidemiology studies show that infection with a virus that serologically cross-reacts with the protoype simian virus (SIV) is present in West Africa. Rates of seropositivity for this virus range from 1 to 10% of the population in sub-Saharan West African countries in preliminary surveys (110). Despite high rates of infection in these areas, cases of HIV-2-related AIDS are extremely rare. No AIDS-like symptoms were found upon examination of several hundred seropositive individuals in five West African countries (111).

By contrast, in a separate study, about 20 patients with diseases characterized as immunosuppressive disorders were found to be seropositive for HIV-2 (112). Most of these patients either lived in West Africa or spent some time in the endemic region. In some cases, the clinical spectrum of disease closely resembled that of AIDS. Both prospective studies of people known to be seropositive for these viruses as well as detailed transmission studies are required to assess the diverse potential of HIV-2.

Viruses that have close immunological and nucleic acid similarities to HIV-2 have been isolated from healthy feral African green monkeys as well as several different primate species in captivity. Serological surveys of feral African green monkeys show that between 30 and 70% of the population, depending on the region, are infected with SIV (113).

Studies with the simian virus demonstrate that the pathogenicity of this family of viruses is not solely a property of the virus itself, but depends on the host. The SIV causes no diseases in its natural host but induces a severe immunosuppressive disease in macaques (114). The HIV-1 virus induces immunosuppressive and central nervous system disorders in humans but has not induced disease in chimpanzees. Although chimpanzees can be infected with the virus, HIV-1 does not cause disease from 1 to 5 years postinoculation in this species.

2. The Genetics of HIV

The genetic organization of HIV-1 differs from that of other retroviruses. Like all known retroviruses, the HIV-1 genome encodes genes for the capsid protein *gag* genes, the replicative genes (*pol*), and an envelope glycoprotein (*env*). The initial and surprising finding was that, unlike previously characterized retroviruses, there are about 1000 nucleotides between the 3' end of the *pol* gene and the 5' end of the *env* gene (115–118). Another sequence of about 500 nucleotides is present between the 3' end of the *env* gene and the LTR.

3. The LTR

The long terminal repeat (LTR) of HIV-1 contains regulatory sequences that govern the expression of the viral genes from the integrated provirus. These include the sequences required for transcription initiation. The LTR functions as a promoter in a wide variety of cell types, including human T and B cells, cells of epithelial origin, and rodent cells such as murine fibroblast and Chinese hamster ovary cells (63,65,119). No striking tissue specificity has been demonstrated for the activity of the HIV-1 promoter (63,119). In most cell types, the LTR functions much more poorly as a promoter than does the early region promoter of the SV40 virus. The exception is the function of the HIV LTR in lymphoid cells, a cell type in which both the SV40 and HIV promoters function poorly.

Sequences within the LTR required for promoter function in uninfected cells include a set of nucleotides located around the position +1 relative to the site of initiation. A TATAA sequence is 5' to the site of initiation. However, it is noteworthy that deletion of the TATAA sequence does not prevent correct initiation at +1 provided an enhancer is located 5' to the deleted HIV LTR (63). Evidently, sequences near +1 contain sufficient information to assure initiation in the absence of the TATAA sequence itself.

The LTR also contains sequences that stimulate the activity of heterologous promoters independent of orientation and distance (63,120). These sequences fulfill the definition of transcriptional enhancer sequences (121). The HIV-1 enhancer functions to stimulate heterologous promoters in a wide variety of human and rodent cells (63,65).

Deletion of sequences 5' to the enhancer increases the activity of the HIV-1 promoter (63). Sequences located 5' to the enhancer exert a negative regulatory effect not only on the HIV promoter, but also on heterologous promoters independent of orientation and distance. The negative regulatory effect is observed for multiple promoters in a wide variety of cell types. This region of the HIV-1 LTR has been termed NRE for negative regulatory region and may be similar to "silencer sequences" reported to be present in some other promoters.

It is curious that although the HIV-1 LTR does not function well as a promoter in uninfected cells, it is very active in in vitro experiments using nuclear extracts of HeLa cells (122,123). In such experiments there is no effect of deletion of the NRE sequences on the activity of the HIV promoter.

Several cellular proteins have been identified that interact with HIV-1 LTR sequences. These include the transcription factor SP1 (122). Three SP1 binding sites have been identified in the HIV LTR. The SP1 transcription factor has also been shown to be required for high-level in vitro activity of the HIV-1 promoter. A second protein, which has previously been recognized to bind to a sequence in the adenovirus latent promoter, was also shown to bind to the HIV LTR (Sawadago, Roeder, Patarca, and Haseltine, unpublished). The sequences recognized by this cellular transcription factor are identical in the HIV-1 and adenovirus promoters. The promoter activity of the HIV-1 LTR in a human T-cell line Jurkat is also shown to be increased by 5- to 10-fold by stimulation with PHA and phorbal esters (124; Luciw, personal communication). Such treatment of Jurkat cells induces a set of genes that are induced by the activation of normal T cells upon recognition of the appropriate antigen. Extracts from stimulated Jurkat cells have been shown to contain a binding activity for a sequence within the HIV-1 LTR (124). The factor that binds to the HIV-1 LTR that is present in stimulated but not unstimulated Jurkat cells is similar to a factor previously recognized to bind to the enhancer region of several promoter elements-NF$\kappa\beta$ protein. The significance of the binding of the factor in PHA-stimulated Jurkats is uncertain, as HIV-1 grows in and kills unstimulated Jurkat cells.

In addition to the sequences that have been shown to have a functional role in promoter activity, or to bind proteins and cellular extracts, comparative sequence analysis demonstrates that the HIV-1 LTR contains a number of sequences that have previously been recognized to play functional roles in other promoters.

The HIV LTR has been shown to respond to a variety of trans-acting factors encoded by DNA viruses. The activity of the HIV-1 promoter is increased by transcription factors of the herpes simplex viruses I and II, and by a transcription factor encoded by cytomegalovirus (125,126).

B. The Structural Genes

1. The *gag* Gene

The virion particle is comprised of a core surrounded by a lipid membrane that contains the envelope glycoprotein (Figure 2). The core particle contains a viral RNA, as well as reverse transcriptase and integrase proteins, and is surrounded by a protein shell or capsid. The capsid proteins are derived from the *gag* gene. The capsid contains a p17, p24–25, and p15 (p7, p9) proteins derived from the amino terminus, central region, and carboxy terminus of the *gag* gene precursor, respectively. The amino terminal protein is phosphorylated. The major capsid protein can be found in two forms that differ at the carboxy terminus. The p15 protein is further cleaved into p7 and p9 proteins.

Capsid proteins of HIV-1 share many features with other retrovirus capsid proteins. The amino-terminal protein is very rich in proline and is phosphorylated. The p17 protein is thought to comprise the outer surface of the capsid, possibly interacting with the lipid membrane surrounding the virion particle. The amino terminus of the p17 protein is myrisylated (193). The p24 protein is the major structural protein of the capsid. The carboxy-terminal protein contains repeated patterns of cysteine residues characteristic of metal binding sites. By analogy with other retroviruses, it is thought the p9 protein binds to viral RNA.

2. The *pol* Gene

The *pol* gene encodes several distinct activities, including the virus-specified protease, the reverse transcriptase, and the integrase. The *pol* gene is almost certainly synthesized as a *gag-pol* polyprotein precursor. However, the precursor half-life of this species must be short, as such a precursor is not detected in

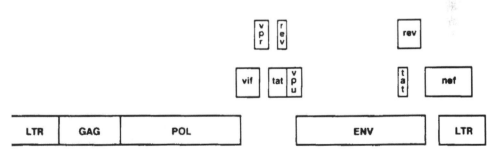

FIGURE 2 Schematic organization of the HIV-1 genome. The open reading frames that encode structural proteins are indicated as well as those that encode regulatory proteins.

infected cells. There is a long overlap in the open reading frames of the *gag* and *pol* genes. The *gag-pol* precursor arises from a translational frame shift between the *gag* and *pol* open reading frames such as that described for the avian leukosis virus (127; Salfeld, Gottiliger, and Haseltine, unpublished observation).

The amino terminus of the *pol* gene contains sequences that are conserved among virus-specific proteases. A protease activity was recently detected in a *gag-pol* gene fusion product synthesized in *Escherichia coli* (Rosenberg and Swanstrom, personal communication; 128).

The reverse transcriptase gene is present as both 65-kD and 53-kD proteins within the virion (129). These two proteins share a common amino terminus and differ at the carboxy terminus. The amino-terminal portion of the reverse transcriptase contains sequences conserved among all retroviruses. The enzyme prefers magnesium as a divalent cation. The polymerase also encodes a ribonuclease H function that also prefers magnesium.

A 34-kD protein is derived from the 3' end of the *pol* gene. By analogy with other retroviruses, the 34-kD protein is thought to provide the integrase function (130). This protein contains sequences conserved among other integrase proteins. Mutations that interrupt these regions yield replication defective virus (Dayton, Gottiliger, and Haseltine, unpublished results). HIV-1 is unusual as the virion contains a relatively large number of reverse transcriptase and integrase proteins as compared to other retroviruses. Moreover, these proteins are highly immunogenic, as most infected people make antibodies to the p65, p53, and p34 proteins.

3. The *env* Gene

The envelope gene encodes a bipartite envelope protein. The protein is synthesized from a spliced mRNA. The protein is synthesized as a precursor that is posttranslationally modified. Modifications of the *env* precursor include cleavage of the membrane signal sequence and cleavage of an internal site to yield an exterior protein and an integral membrane protein. The exterior protein is heavily glycosylated (131,132). There exist more than 25 potential sites for *N*-linked glycosylation in the region of the protein that is entirely exterior to the cell. The transmembrane protein is also glycosylated. Recent studies show that the protein does not contain *O*-linked sugars (194).

Precursor forms of the envelope protein are evident in infected cells. These include a 160-kD precursor form (gp160). The gp160 form may represent an immature form of the protein that is hyperglycosylated, as proteins of slightly faster electrophoretic mobility appear during pulse chase experiments (194). The precursor is cleaved to an exterior glycoprotein of apparent molecular weight 120 kD. Over half of the apparent molecular weight of this protein is attributable to sugar residues, as complete digestion of the gp120 protein with endoglycosylase F results in a reduction of the apparent size of the protein to 55 kD (131). The smaller carboxy-terminal cleavage product is also glycosylated and has an apparent molecular weight of 41 kD (gp41) (131,133). The calculated molecular weight of the transmembrane protein is 32 kD. This observation indicates that the protein probably contains two sites of glycosylation. Deletions that remove portions of the 3' terminus of the *env* gene alter the apparent electrophoretic mobility of the transmembrane protein (134). This result indicates that the entire length of the *env* gene is used in synthesis of *env*-specific proteins.

The exterior glycoprotein, gp120, does not appear to be firmly bound to the transmembrane protein, gp41, at least in the strains of HIV-1 most frequently studied, HTLV-III B and LAV-BRU. The amount of gp120 shed into the medium of infected cells is approximately equal to the amount that is associated with the infected cell (135,136). The gp120 is readily dissociated from virions. The amount of gp120 shed into the medium or released from the virions is not increased by reducing agents, indicating that the gp120 is not linked to the transmembrane protein by disulfyhdryl bonds (137).

The cellular CD4 molecule serves as a receptor for the virus via a specific reaction with the envelope glycoprotein. Vesicular stomatitis virus pseudotypes carrying HIV-1 envelope glycoprotein can infect CD4+ but not CD4- cell lines (138,139). A subset of monoclonal antibodies that recognize CD4 blocks virus infection of CD4+ cells (138,140-142). Cells that do not naturally express the CD4 molecule but do so after transfection with the CD4 gene can be infected by the virus (143).

Recognition by the virus of CD4 occurs via a high-affinity binding of the cell surface protein by the exterior portion of the envelope glycoprotein (gp120). The purified gp120 protein binds to CD4 cells with an affinity of about 4×10^{-9} M (Laskey, personal communication). Antibodies to either CD4 or to gp120 have been shown to precipitate a stable gp120-CD4 complex (141). Monoclonal anti-CD4 antibodies that block virus infection block gp120 binding to the CD4 molecule (135,137).

The specificity of CD4 as a receptor is shared by HIV-2 and SIV. A similar set of anti-CD4 monoclonal antibodies that block HIV-1 infection also block infection of these related viruses (112,144; Kowalski, Doyle, Haseltine, and Sodroski, unpublished observations).

The tissue specificity of infection can be largely accounted for by the distribution of the CD4 surface protein. The CD4 protein is found abundantly on the surface of the helper subset of T cells. The abundance of the CD4 molecule increases by three- to fourfold during the course of T-cell activation. The CD4 molecules are also found on other cell types, albeit at a much lower concentration. These cell types include monocytes, macrophages, and Langerhans cells, and some normal as well as Epstein-Barr transformed cells (143,145,146). Both CD4 messenger RNA and protein have been detected in these cells (92,143,145, 146). Infection of these cells can be demonstrated in vitro, and infection is blocked by the same set of monoclonal antibodies as block viral infection of CD4+ helper T cells (145–147). The CD4 mRNA and protein can also be found on certain glial cells, on some cell lines derived from colon cell carcinoma, and on some endothelial cell lines in culture (143,148).

The envelope glycoprotein has been shown to mediate a fusion reaction between cells that express the intact gp120-gp41 complex and CD4+ cells (139, 149,150). Introduction of the envelope gene into CD4+ cells results in the formation of giant multinucleated syncytia (139,149). This reaction can be blocked by anti-CD4 antibodies that block gp120 binding (139,149,150). The number and size of syncytia formed depend partly on the concentration of the CD4 receptor in the recipient cell. The syncitia that form are not viable over the long term. In one experiment, the half-life of the syncytia was found to be about 7 days (149).

Expression of the envelope glycoprotein in CD4– cell lines does not result in syncytium formation. However, such cell lines will initiate syncytium formation when mixed with CD4+ cell lines (149).

The observation that fusion events occur between cells that express the envelope glycoprotein and uninfected CD4+ cells indicated that fusion occurs between juxtaposed surfaces and does not require endocytotic events. This conclusion is further strengthened by the observation that the fusion reaction occurs at neutral pH in buffered culture media. The observation that high concentrations of UV-inactivated extracellular virus can induce cell-mediated fusion also supports the notion that the gp120-gp41 complex can initiate extracellular fusion events (151).

The reaction between the envelope glycoprotein and CD4 is postulated to play a key role in HIV-induced disease, both in early stages of infection as well as cell death (149,150). It is proposed that virus-to-cell transmission occurs when the virus membrane fuses with a CD4+ cell membrane and the core of the virus is introduced into the cell. Cell-to-cell transmission is postulated to occur by a similar mechanism whereby core particles that have assembled near the surface of the virus-producing cells are transmitted to uninfected cells after the cell-to-cell fusion event (152).

CD4 cell death is postulated to occur via two pathways. The first pathway involves the formation of giant multinucleated syncytia whereby one cell that produces the envelope glycoprotein serves as a nucleus for the formation of a multicellular complex which eventually dies. Careful examination of the thymus and spleen, as well as the brain and lymph nodes, of AIDS patients reveals the existence of large multinucleated CD4+-virus-infected cells (153,154). Such syncytia are not expected to be found in the circulation as they are too large to pass through capillaries. The second pathway proposed is the death of individual cells that express simultaneously high concentrations of the envelope glycoprotein and the CD4 surface molecules. The death of individual cells can be expressed by the following equation:

$$[\text{gp-120-gp41 surface complex}] \times [\text{CD4}]\ [\text{cell factor}] = \text{cell death}$$

The term [gp120-gp41] refers to a functional complex that is normal with respect to CD4 binding function, fusion function, gp120-gp41 self-association, and gp120-gp41 association with the membrane. The term [CD4] refers to a functional-unblocked CD4 surface glycoprotein. The specific mechanism proposed for the death of cells that simultaneously express high concentrations of both proteins is an autofusion mechanism whereby the gp120/gp41 complex embedded in the cell membrane reacts with the CD4 antigen on the same membrane. Multiple autofusion reactions are proposed to lead to loss of membrane integrity and cell death. The [cell factor] term is introduced to account for the inability of rodent cells that express CD4 to participate in the fusion reaction (143).

The requirements for the envelope glycoprotein–CD4 membrane fusion reaction have been investigated by the analysis of mutant envelope glycoproteins. Several different phenotypes result from introduction of in-frame point mutations, deletions or insertions of small groups of amino acids into the coding sequences of the envelope glycoprotein. These mutants were all analyzed for their ability to produce the gp160, gp120, and gp41 proteins, for the amount of protein produced on the cell surface, for the amount of protein released into the cell medium, for the ability of proteins to bind to the CD4+ antigen, and for the envelope glycoprotein to initiate fusion events when transfected into CD4+ cultures.

C. Fusion Reaction Mutants

Mutants for the fusion reaction can be classified into several independent groups: mutants that affect binding of gp120 to CD4; mutants that inhibit the fusion reaction but permit gp120 binding to CD4; mutants that inhibit anchorage of gp41 to the membrane; mutants that disrupt gp120-gp41 association; mutants that prevent cleavage of the gp160 precursor. Additionally, some mutants prevent the association of the envelope with the virion particles but permit association of the envelope glycoprotein with the cell membrane. For the most part, the mutants cluster into discrete areas of the coding sequence of the envelope glycoprotein and permit definition of functional regions in terms of specific amino acid sequences (135).

1. CD4 Binding Region

Mutants that affect CD4 binding are located in three separate regions near the carboxy terminus of gp120. The mutants in all three regions are highly conserved among HIV-1 strains and between HIV-1, HIV-2, and SIV. The conserved regions are separated by hypervariable regions and changes in these regions do not affect gp120 binding to CD4. Additional experiments also indicate that this region binds CD4. Antisera raised to a conserved peptide to one of these regions inhibit gp120 binding to CD4. It is likely that the three noncontiguous conserved sequences are brought together by the tertiary structure of the gp120 to form a CD4 binding pocket that is interdigitated with hypervariable regions that comprise the folds between the faces of the binding pocket. Deletion analysis of the *env* gene by Lasky and co-workers also indicates that the central of the three carboxy-terminal conserved regions of gp120 is required for CD4 binding (155). Additionally, monospecific and monoclonal antibodies that are raised to specific amino acid sequences located in the three conserved regions block gp120 binding to CD4.

2. Fusion

Mutants in the amino terminus of gp41 permit normal synthesis and processing of the *env* protein and permit gp120 binding to CD4 but block the fusion reaction. The hydrophobic region affected by these mutations is analogous to the hydrophobic region of other enveloped viruses that are involved in fusion reactions. It is likely that the amino terminus of gp41 is inserted into the opposing membrane and disrupts the lipid bilayer, thus initiating a fusion event.

3. gp120-gp41 Cleavage

Some mutants prevent the cleavage of gp160 (149,156). Such mutants do not bind or fuse to CD4+ cells. The wild-type gp160 also does not bind to CD4. It is likely that a cleavage reaction is necessary to release the carboxy terminus of gp120 from the amino terminus of gp41 so that they may assume a configuration required for the binding and fusions reactions, respectively.

4. gp120-gp41 Association

Some mutants weaken the association of gp120 with gp41. These mutants cluster at the amino terminus of both proteins. Such mutants produce normal levels of the envelope protein, but the gp120 is not attached to the cell surface and is released into the cell supernatant. Such mutants do not initiate fusion events. It is likely that the gp120 and gp41 proteins associate via a series of noncovalent interactions between the amino termini of both proteins and that the gp120 association with gp41 is critical for the binding and fusion reaction.

5. Association of gp120-gp41 to the Membrane

The gp41 protein is anchored to the membrane. The mutants that remove or destroy the hydrophobic character of the second hydrophobic sequence of the gp41 protein result in release of the gp120-gp41 complex into the medium. No fuison reaction occurs. It is likely that the second hyrophobic region of gp41 is the membrane-spanning region that serves as an anchor for the *env* protein to the membrane of infected cells and virions.

D. The Tail of gp41

One of the most unusual features of HIV is the 150-amino-acid-long carboxy terminus of gp41. The corresponding region of many other retroviruses is only 12 to 15 amino acids long. Deletion of much of this region does not destroy the ability of the *env* protein to bind CD4 and to initiate fusion reactions as measured by cell fusion, a reaction that occurs where the protein is located on the surface of an infected cell. However, transfection of cells with plasmid that are deleted for the region that encodes this region of gp41 results in decreased levels of production of infectious virus (157). The carboxy terminus of gp41 or other *cis*- or *trans*-acting functions encoded for by this region appears to be critical to the formation of infectious virions.

   The functional analysis described above permits a model for the binding and fusion reactions to be constructed. In this model, the carboxy terminus of gp120 binds to CD4. The binding of gp120 to CD4 brings the membranes bearing CD4 and the envelope glycoprotein into close juxtaposition. The weak association between gp120 and gp41 permits dissociation of the gp120 and even closer opposition of the membranes. Fusion is initiated by insertion of the amino-terminal region of gp41 into the membrane of the CD4+ cell.

E. Antibody Recognition

Antisera from HIV-infected patients often contain high-titer antibodies to the envelope glycoproteins, gp120 and gp41 (133,158). Antienvelope antibodies are the first to appear postinfection and are often the last to disappear in the late stages of disease. Such antibodies are made to highly conserved epitopes within the envelope glycoprotein as deduced from the observation that all patients infected with HIV recognize the envelope glycoprotein of a single strain of virus (159). These immunodominant epitopes account for greater than 90% of the total antibodies produced to the envelope glycoprotein in the course of normal infections (160). It is this antigenic reaction which is used in the standard serological tests for AIDS virus infection. Although the immunodominant epitopes of the envelope glycoprotein are conserved among all HIV-1 strains, such antigenic conservation does not generally extend to HIV-2 or to SIV.

   The small immunodominant subregions of the HIV-1 envelope glycoprotein are located near the carboxy terminus of gp120 and the amino-terminal portion of the gp41. These conserved epitopes permit serodiagnosis of most HIV infections (159,160).

   · Despite the presence of high-titer anti-*env* antibodies, patient sera inhibit only weakly envelope-glycoprotein-mediated fusion events (149,161). High concentrations of patient antisera can reduce the rate of envelope-glycoprotein-mediated CD4 cell-to-cell fusion reactions but do not stop the reaction. The failure of the humoral response to stop this key reaction may account, at least in part, for the progressive nature of HIV disease. Evidently, the immune response is not sufficient to stop the envelope-mediated virus-to-cell or cell-to-cell spread of infection.

F. Regulatory Genes and Other Proteins

1. The *vif* Gene

An open reading frame capable of encoding a 23-kD protein is located immediately 3' to the *pol* gene (115–118). Studies show that the region, now called *vif*, for virion infectivity factor, encodes a 23-kD protein that can be detected in infected cells (162). The *vif* protein is probably synthesized from a 5.5-kD mRNA. Cell fractionation studies show that the *vif* product protein is located in the cytoplasm of infected cells (162). Antibodies to the *vif* protein are detectable in the serum of some infected patients. A similar protein is encoded by SIV and HIV-2.

The role of the *vif* gene in virus replication has been examined by the introduction of mutants that disrupt the ability of an infectious provirus to produce the *vif* protein (162). Transfection of CD4+ cell cultures with such proviruses results in an infection that spreads very slowly through the culture as compared to cultures transfected with the intact provirus. However, the amount of virus-specific proteins produced in the initial transfection by *vif*-defective viruses is the same as that produced upon transfection with the intact provirus. Equal amounts of virions are formed in the initial round of transfection as well. This observation indicates that the defect in the mutant *vif* virus does not involve the synthesis of viral RNA or protein from provirus.

Recent studies suggest that the infectivity of the virions for CD4+ cells of the *vif*-defective virus is reduced as compared to wild-type virus (163,164; Sodroski, Dorfman, Dayton, and Haseltine, unpublished). The extent of the defect depends on the recipient cell line. However, cell-free *vif*-defective virus can infect CD4+ cells in culture. The *vif* gene product appears to increase the specific infectivity of virions without dramatically altering cell-to-cell transmission.

In this regard, the spread of infection of *vif*-defective viruses in CD4 cultures is of interest. Ultimately, such cultures become entirely positive for viral antigens and eventually die (162). The latter observation indicates that under these circumstances, cell-to-cell transmission does occur. This observation implies that the requirements for cell-to-cell transmission may differ from those for virus-to-cell transmission. As it is likely that cell-to-cell transmission of infection requires the formation of envelope glycoprotein and involves reverse transcription, the results favor a role of the *vif* gene product in determining the infectivity of the virion particle itself.

2. The *vpr* Region

A small open reading frame that partially overlaps both the *vif* and first coding exon of the *tat* gene is present in most strains of HIV-1. A similar region is also present in the genomes of HIV-2 and SIV. A protein encoded by the *vpr* region was produced in bacteria. This polypeptide was recognized by some patient antisera in a Western blot test (165). This observation suggests that a protein that contains all or part of the *vpr* sequences is produced upon infection. As yet, this protein has not been identified in infected cells.

There is no known function of the *vpr* open reading frame. Many infectious clones of HIV-1 provirus contain stop translation termination codons which would truncate the *vpr* product. Viruses that are mutated in the *vpr* are cytopathic for CD4+ cells (164).

3. The *nef* Gene

The *nef*, negative factor, gene encodes a 27-kD protein (157,166). The protein is located in the cytoplasm and is likely to be membrane associated as it is myrislated at the amino terminus (166). The *nef* protein is synthesized from an mRNA that is spliced to remove the *gag* and *env* genes. Antibodies to the protein are found in some infected people. A similar protein is encoded by the SIV and HIV-2 viruses. The ungulate lentiviruses do not have regions that correspond to the *nef* gene (167).

Viruses that are defective in the *nef* gene replicate very well and are cytopathic for CD4+ cells (157,168,169). The original infectious provirus derived from the HTLV-III B strain of HIV (pHXBc2) contains a point mutation in the *nef* gene that results in premature translation termination. Comparison of the rate of replication of this virus with one into which a complete *nef* gene has been inserted indicated that the *nef*-expressing virus replicated somewhat more slowly than the original. Deletion of portions of the *nef* gene from this recombinant provirus resulted in mutants that replicate more rapidly than parental virus that produces the *nef* product. Similar results have been reported from mutants introduced into the ARV provirus.

The results demonstrate that the *nef* gene is not required for either the replication or cytopathic effect of HIV-1. The observed function of this gene is to retard the replication of HIV-1.

Several interesting properties of the *nef* product have been described. The protein is myristylated (166) and presumably is associated with a lipid bilayer.

A fraction of the protein is also phosphorylated in infected cells. The product of *nef* has also been purified from *E. coli*. This purified protein has been shown to be a substrate for protein kinase C at a residue close to the N-terminus (residue 14) (170). The bacterial protein has also been demonstrated to possess GTPase, GTP-binding, and autophosphorylation activities (170). The autophosphorylation reaction uses GTP as a substrate and occurs at a site different from that which is phosphorylated by protein kinase C. Distant amino acid sequence similarities with other proteins possessing GTP-binding and GTPase activities—such as the *ras* oncogene product—have been noted. Homologies have also been reported between the amino terminus of the *nef* product and the amino termini of other proteins such as pp60-SRC and the EGF receptor, all of which contain a common site for phosphorylation by protein kinase C.

It was recently observed that high-level expression of *nef* inhibits transcription by the LTR of HIV-1 (171). Cotransfection with large amounts of a plasmid that expresses *nef* with a plasmid that expresses the CAT gene under the control of the HIV-1 LTR results in decreased CAT enzyme activity, as compared to levels observed in the absence of the *nef* product. Deletion of the 5' LTR sequences (−450 to −167) abolishes the inhibitory effect of the *nef* product on the HIV-1 LTR. This suggests that the target for *nef* activity is located in this region. It is likely that the *nef* product acts indirectly to alter the factors that affect transcription, as the *nef* product is located predominantly in the cytoplasm. It is possible that the *nef* product acts to increase the activity of cellular repressors (or decrease the activity of cellular activators) that act in uninfected cells to mildly alter HIV LTR transcription by interaction with the NRE sequences. A second group has recently confirmed the effect of the *nef* protein upon transcription from the HIV-1 LTR, however they mapped the cis-acting sequences to a region straddling the start site of transcription (195).

4. The *tat* Gene

The *tat* gene of HIV-1 is a bipartite gene encoded by two exons, one located between the *vpr* and *env* and a second located within the 3' portion of the *env* itself. The *tat* protein is synthesized from a doubly spliced mRNA (119,172). In all strains of virus sequenced, the 5' coding exon encodes 74 amino acids and terminates with a stop codon. The length of the second coding exon is variable. Most of the *tat* protein is located in the nucleus of the infected cells (173). Some infected people make antibodies to the *tat* protein (174). A gene corresponding to the *tat* protein exists in HIV-2 and SIV.

Viruses that contain mutants defective in the *tat* protein are replication defective (175,176). No virus-specified proteins can be detected in cells infected with *tat*-defective virus. Mutants in the amino-terminal portion of the *tat* protein can be complemented in cells that constitutively produce the *tat* gene product (175). Virions produced by such cells are morphologically normal but are noninfectious.

The *tat* gene was discovered not by analysis of the open reading frames of the provirus, as were the *vif* and *nef* genes, but rather because of the demonstrated function, that of *trans*-activation. The existence of the *trans*-activation gene was deduced from experiments that showed that the level of LTR-directed CAT activity was elevated, more than 1000-fold, in HIV-infected as compared to uninfected cells (119). Deletion analysis of an infectious provirus demonstrated that none of the recognized open reading frames (*gag, pol, env, vif,* or *nef*) encoded the *trans*-activation function (119). However, deletions within the 1000-nucleotide region located between the *pol* and *env* genes eliminated the *trans*-activation function of a provirus. A more detailed analysis of the open reading frames of this region revealed the possibility that a small protein was synthesized. Comparisons of the sequences with the structure of messenger RNAs suggested that the *trans*-activation function was encoded by a doubly spliced messenger RNA. Upon further investigation, this turned out to be the case (119,172).

The sequences within the LTR that respond to the *trans*-activator have been mapped to between nucleotides +1 and 46 (63,120,177,178,179). These sequences are designated TAR, for *tran*-acting *r*esponsive *r*egion.

The function of the TAR depends on its position relative to the site of transcription initiation. The TAR sequence is located at the 5' end of all viral messages. Promoters that contain TAR sequences located 5' to the site of RNA initiation do not respond to *tat*. Insertion of sequences between the site of transcription initiation and TAR also eliminates the response to *tat*.

The mechanism of *trans*-activation is not fully understood. Both transcriptional and posttranscriptional modes of regulation have been proposed. Transfection of cells with proviruses defective for the *tat* function results in the synthesis of some virus-specific RNA, but no detectable synthesis of virus-specified proteins. However, the size and abundance of RNAs synthesized in the absence of *tat* have not been carefully analyzed.

Most experiments in the literature analyze *tat* function by measuring the effects of the *tat* gene on heterologous genes under the control of the HIV LTR. Introduction of such plasmids into cells that constitutively express the *tat* function produces a very marked increase in protein expression, 1000–2000-fold (63,65,177). Increases in the RNA concentration in these experiments between 20- and 50-fold have also been measured (63,65,120,177,178).

These results suggest the *tat* protein acts posttranscriptionally to increase the efficiency of utilization of the messenger RNA. The experiments also indicated the *tat* protein increases the steady-state concentration of mRNA, either by increasing the stability of the RNA and or by increasing the transcription rate of LTR-directed genes. A dual role of the protein in both the activation of transcription and posttranscriptional regulation has been proposed (63,123). No direct interaction between the *tat* protein and the TAR sequences, in either the

DNA or RNA form, has yet been reported. Certain recently reported data favor a hypothesis that *tat* protein functions as an anti-terminator of transcription; i.e. it acts to overcome a premature termination of transcription induced by the presence of the TAR sequence (196).

The *tat* protein has an unusual structure. The amino terminus is rich in proline. There is a cysteine-rich region that is conserved in all strains of the virus and is also similar in HIV-2 as well as SIV. There are, in addition, two regions of clustered positive charges. It has been proposed that the *tat* protein has a configuration similar to other eukaryotic regulatory metal-binding proteins (178). The *tat* protein has been shown to bind both cadmium and zinc (180).

Cells have been constructed that express constitutively a functional *tat* protein. Constitutive expression of *tat* protein in CD4+ cells does not result in cell death (63). Moreover, there is no known marked change in the function of T-cell-specific genes in such cell lines. This observation demonstrates that the cytopathic effect of HIV infection in CD4+ cells is not a sole consequence of expression of the *tat* gene product itself.

5. The *rev* Gene

The *rev*, for regulator of expression of virion proteins gene (also called *art* and *trs*), is an essential replication gene. It is encoded in the same region as *tat* although in a different reading frame. The initiation codon for the *rev* product is located 3′ to that of the *tat* gene. There is recent evidence that the splice acceptors for the *tat* and *rev* genes are different (181). A splice acceptor site located 95 nucleotides from the initiation codon of *rev* has been detected in cells infected with the IIIB strain of HIV-1. A similar splice acceptor site has also been found in cDNAs derived from cells infected with SIV (Wong-Staal, personal communication.) The product of the *rev* gene is a 20-kD protein that is immunogenic in some patients (182). An open reading frame corresponding to the *rev* gene is present in the genomes of SIV as well as HIV-2, although the predicted amino acid sequences of the putative *rev* genes are considerably different from that which is encoded for by HIV-1.

Viruses that encode functional *tat* genes but defective *rev* genes are defective for replication (64,183,184). The defect can be complemented by cotransfection with a plasmid that expresses the *rev* gene product. Mutants in the *rev* gene fail to make virion-structural proteins, *gag* and *env*. However, the *tat* gene as well as the *nef* gene can be made in *rev*-defective provirus (64,183). Mutants that are defective for *rev* expression accumulate abundant amounts of fully spliced mRNAs. Little or no mRNA species corresponding to the *gag* and *env* mRNAs are observed (64,184).

It is deduced that the *rev* gene acts posttranscriptionally. Since all HIV-1 mRNAs are thought to be derived from the same primary transcript, the expression of *tat* and of *nef* by a *rev*-defective mutant implies that the primary transcript is made and that the *rev* gene product affects the expression of proteins encoded by different segments of the primary transcript differently.

This hypothesis is supported by the observation that some virus-specific RNAs are detectable in cells transfected with a *rev*-defective provirus (183,186). The simplest explanation for the difference in the expression of virus regulatory and structural genes is that the mRNAs that encode the *gag* and *env* proteins contain sequences that inhibit their expression, and that these sequences are absent from the mRNAs that encode the *tat* and *nef* proteins. A common intron, comprising part of the coding sequence of the *env* gene, is absent from the mRNAs that encode the *tat* and *nef* genes but is present in the mRNAs that encode the *gag* and *env* genes (118).

The hypothesis that this intron contains *rev*-responsive sequences was recently verified (187). Placement of the *env* coding sequences 3' to heterologous genes suppresses the expression of these genes by 50- to 100-fold. The suppression was relieved by the presence of the *rev* gene product. The response to *rev* was evident when heterologous promoters and poly-A signals were substituted for the HIV LTRs. The *env* coding sequences do not function as a constitutive or *rev*-inducible enhancer when located 5' to a promoter. These observations demonstrate that the segment contains sequences that negatively regulate the expression of mRNAs that contain such sequences. These results have more recently been confirmed and extended by others (197).

6. The *vpu* Gene

The most recently identified gene of HIV-1 is *vpu*. The apparent size of the gene product on polyacrylamide gels varies between 15 and 20 kD depending upon the strain of the virus (188,189,190). The coding sequences for the *vpu* product are contained within a single conserved open reading frame within the central coding region of the genome. The *vpu* coding sequence begins immediately following the splice donor site at the end of the first exons of *tat* and *rev*, and overlaps the 5' end of the *env* gene coding sequence. Unlike the other genes of HIV-1, no correlate to the *vpu* reading frame occurs in the published sequences of either HIV-2 or the different SIV strains.

A high proportion of serum samples from HIV-1 infected individuals contain antibodies against the *vpu* product (188,190). Yet *vpu* is not an essential gene function, as several of the sequenced proviral clones, including HXBc2, carry defects within this reading frame. The protein is phosphorylated (M. Martin, personal communication) and appears to be associated with a membrane within infected cells (M. Martin, personal communication; Cohen, Terwilliger and Haseltine, unpublished observations). The *vpu* product appears to play a role in virus export from infected cells. A *vpu*-expressing strain of virus has been shown to produce up to ten times more reverse transcriptase activity in the culture supernatant than a related *vpu*-defective virus (189). Study of cultures infected with isogenic *vpu*⁺ or *vpu*⁻ strains of virus has also revealed a dramatic shift in the apportionment of viral proteins between intracellular viral proteins and extracellular virions in favor of free virions in the presence of the *vpu* protein (Terwilliger et al., in press).

G. A Model for HIV Replication and Pathogenesis

It is likely that HIV-1 may establish three different states of infection: latent, controlled replication, and active proliferation.

1. Latent Infection

The most extreme form of latent infections are those in which no viral RNA is made. Such may be the case for infections of nonreplicating T cells.

Cell cultures can be established in vitro that contain HIV-1 proviral DNA but do not produce RNA. Virus can be induced to replicate from such cultures by treatment with azacytidine. It is not certain whether the nonproductive infections achieved in replicating cultures represent similar states of latency to those established in nonreplicating T cells.

2. Controlled Replication

Infection of replicating cultures of functional T-cell clones results in the production of viral antigens without cell death. Cell death occurs in such cultures upon antigen activation. Expression of viral proteins in replicating, but not activated, cultures of T cells may reflect controlled noncytopathic replication. Infection of monocytes and macrophages also may result in chronic infection that results in poor virus yield and little cytopathic effect. Such infections may also reflect controlled replicative states.

3. Active Proliferation

HIV-1 is produced in abundance and is lytic for rapidly replicating CD4+ T cells in culture. Rare CD4+ T cells can be observed in infected people that also express abundant envelope glycoprotein. These replication states are likely to be controlled by interaction of the *cis* and *trans* regulatory viral sequences with cellular factors. Determination of the transcription activity of the viral LTR must be directed in the first instance by cellular transcription factors. Such factors are widely available in replicating cells, as the LTR of HIV-1 functions well in a wide variety of cell types as well as in cells of rodent origin. The extent to which further increases in transcription activity in *replicating* cells are affected by activation of specific genetic programs by external stimuli remains to be investigated.

It is likely that the *rev* gene and the corresponding *cis*-acting regulatory sequences play a key role both in the establishment of latency and in the lytic cycle. The *rev* gene function permits differential synthesis of viral structural and regulatory proteins. *Tat, rev,* and *nef* can be synthesized in the absence of *rev*. Synthesis of large amounts of *gag, pol,* and *env* requires the *rev* gene functions.

In principle, differential expression of regulatory and replication genes for virus structural genes is similar to the early-late transitions of DNA viruses. Early-late switching is generally, but not exclusively, controlled in DNA viruses by regulation of mRNA transcription. HIV-1 could achieve similar discrimination and selective use of different coding regions via *rev* gene function.

In this regard, it is noteworthy that two of the *rev*-independent functions have antagonistic effects with respect to HIV-1 replication. Expression of the *tat* function accelerates viral protein synthesis, whereas expression of the *nef* gene retards viral replication. Control of the relative abundance of *tat* and *nef* could determine the level of HIV-1 replication.

The *tat* gene function appears to be the key to prolific replicative states. The expression of viral protein is accelerated several-hundred-fold by this gene product. A relatively weak promoter characteristic of HIV-1 is capable of directing large amounts of protein synthesis in the presence of the *tat* gene function. Simultaneous expression of both *rev* and *tat* permits rapid production of virion particles.

Cytopathic effects of HIV-1 infection are also controlled No cytopathic effect is observed in latent infections. Similarly, no cytopathic effect is observed in controlled infection states, even controlled infection of CD4+ cells. The cytopathic effect is restricted to the case in which HIV-1 replicates prolifically and then only in cells that express a high concentration of the CD4 protein. Prolific replication per se is *not* cytotoxic to T cells, as many T-cell lines have been derived that are capable of continually producing high titers of infectious HIV.

The controlled replication and selected cytotoxicity of the virus account for much of the clinical spectrum of HIV-1-associated disease. The long asymptomatic period and accompanying low level of persistent viremia can be accounted for by controlled replication of HIV-1 in monocytes, macrophages, and T cells. The specific ablation of CD4+ cells can be attributed to attrition of the total population as a result of direct infection and subsequent cytotoxic activity of activated T cells or via syncytia formation. Any cell that expresses the envelope glycoprotein on the surface, whether the cell is killed or not, should be able to form syncytia with CD4+ cells. Syncitia up to 100–500 cells large have been observed in culture and large syncytia have been observed in the thymus, spleen, and brain of HIV-infected patients. The failure of the immune response to prevent the continued depredations of HIV infection can be attributed, at least in part, to the failure of the antibodies produced to HIV to block functional components of HIV envelope glycoprotein, including the binding and fusion reactions.

ACKNOWLEDGMENT

The authors thank Mark Kowalski and Bruce Walker for helpful discussions.

REFERENCES

1. Catovsky, D. (1982). Adult T-cell lymphoma-leukemia in blacks from the West Indies. *Lancet* 1:639–643.

2. Franchini, G., Wong-Staal, F., and Gallo, R. C. (1984). Human T-cell leukeumia virus (HTLV–I) transcripts in fresh and cultured cells of patients with adult T-cell leukemia. *Proc. Natl. Acad. Sci. USA*, 81:6207–6211.

3. Poiesz, B., Ruscetti, R. W., Gadzar, A. F., Bynn, P. A., Minna, J. D., and Gallo, R. C. (1980). T-cell lines established from human T-lymphocytic neoplasias by direct response to T-cell growth factor. *Proc. Natl. Acad. Sci. USA* 77:7415–7419.

4. Miyoshi, I., Kubonishi, I., Yoshimoto, S., Akagi, T., Ohtsuki, Y., Slaronski, Y., Nagata, K., and Hinuma, Y. (1981). Type C virus particles in a cord T-cell line derived by co-cultivating normal human cord leukocytes and human leukaemic T cells. *Nature* 294:770–774.

5. Yamamoto, M., Okada, M., Koyanagi, Y., Kannagi, M., and Hinuma, Y. (1982). Transformation of human leukocytes by cocultivation with an adult T cell leukemia virus producer cell line. *Science* 217:737–739.

6. Popovic, M., Lange-Wantzin, G., Sarin, P. S., Mann, D., and Gallo, R. C. (1983). Transformation of human umbilical cord blood T cells by human T-cell leukemia/lymphoma virus. *Proc. Natl. Acad. Sci. USA* 80:5402–5406.

7. Chen, I. S., Quan, S. G., and Golde, D. W. (1983). Human T-cell leukemia virus type II transforms normal human lymphocytes. *Proc. Natl. Acad. Sci. USA* 80:7006–7009.

8. Poiesz, B. J., Ruscetti, F. W., Mier, J. W., Woods, A. M., and Gallo, R. C. (1980). T-cell lines established from human T-lymphocytic neoplasias by direct response to T-cell growth factor. *Proc. Natl. Acad. Sci. USA* 77:6815–6819.

9. Leonard, W. J., Depper, J. M., Uchiyama, T., Smith, K. A., Waldmann, T. A., and Greene, W. C. (1982). A monoclonal antibody that appears to recognize the receptor for human T-cell growth factor; partial characterization of the receptor. *Nature* 300:267–269.

10. Okochi, K., Sato, H., and Hinuma, Y. (1984). A retrospective study on transmission of adult T cell leukemia virus by blood transfusion: Seroconversion in recipients. *Vox Sang* 46:245–253.

11. Schupbach, J., Kalyanaraman, V. S., Sarngadharan, M. G., Blattner, W. A., and Gallo, R. C. (1983). Antibodies against three purified proteins of the human type C retrovirus, human T-cell leukemia-lymphoma virus, in adult T-cell leukemia-lymphoma patients and healthy Blacks from the Caribbean. *Cancer Res.* 43:886–891.

12. Hunsmann, G., Schneider, J., Schmitt, J., and Yamamoto, N. (1983). Detection of serum antibodies to adult T-cell leukemia virus in non-human primates and in people from Africa. *Int. J. Cancer* 32:329–332.

13. Hunsmann, G., Bayer, H., Schneider, J., Schmitz, H., Kern, P., Dietrich, M., Buttner, D. W., Goudeau, A. M., Kulkarni, G., and Fleming, A. F.

(1984). Antibodies to ATLV–HTLV-1 in Africa. *Med. Microbiol. Immunol.* *(Berl.)* 173:167–170.

14. Blayney, D. W., Blattner, W. A., Robert-Guroff, M., Jaffee, E. S., Fisher, R. I., Bunn, P. A., Jr., Patton, M. G., Rarick, H. R., and Gallo, R. C. (1983). The human T-cell leukemia-lymphoma virus in the southeastern United States. *JAMA* 250:1048–1052.

15. Blattner, W. A., Blayney, D. W., Robert-Guroff, M., Sarngadharan, M. G., Kalyanaraman, V. S., Sarin, P. S., Jaffee, E. S., and Gallo, R. C. (1983). Epidemiology of human T-cell leukemia/lymphoma virus. *J. Infect. Dis.* 147:406–416.

16. Hino, S., Yamaguchi, K., Katamine, S., Sugiyama, H., Amagasaki, T., Kinoshita, K., Yoshida, Y., Doi, H., Tsuji, Y., and Miyamoto, T. (1985). Mother-to-child transmission of human T-cell leukemia virus type-I. *Jpn. J. Cancer Res.* 76:474–480.

17. Blattner, W., Nomura, A., Clark, J. W., Ho, G. Y. F., Nakao, Y., Gallo, R., and Robert-Guroff, M. (1986). Modes of transmission and evidence for viral latency from studies of human T-cell lymphotropic virus type I in Japanese migrant populations in Hawaii. *Proc. Natl. Acad. Sci. USA* 83:4895–4898.

18. Kajiyama, W. (1980). Intrafamilial transmission of adult T cell leukemia virus. *J. Infect. Dis.* 154:851–857.

19. Weiss, S. H., Saxinger, W. C., Ginzburg, H. M., Mundon, F. K., and Blattner, W. A. (1987). HTLV-1 and HIV Prevalence Among US Drug Abusers. *Proc. Am. Soc. Clin. Oncol.* 6:4.

20. Gessain, A., Barin, F., Vernant, J. C., Gout, O., Maurs, L., Calendar, A., and de The, G. (1985). Antibodies to human T-lymphotropic virus type-I in patients with tropical spastic paraparesis. *Lancet* 2:407–410.

21. Osame, M., Matsumoto, M., Usuku, K., Izumo, S., Ijichi, N., Amitani, H., Tara, M., and Igata, A. (1987). Chronic progressive myelopathy associated with elevated antibodies to human T-lymphotropic virus type I and adult T-cell leukemia-like cells. *Ann. Neurol.* 21:117–122.

22. Yoshida, M., Osame, M., Usuku, K., Matsumoto, M., and Igata, A. (1987). Viruses detected in HTLV-I-associated myelopathy and adult T-cell leukaemia are identical on DNA blotting (Letter). *Lancet* 1085–1086.

23. Seiki, M., Hattori, S., and Yoshida, M. (1982). Human adult T-cell leukemia virus: Molecular cloning of the provirus DNA and the unique terminal structure. *Proc. Natl. Acad. Sci. USA* 79:6899–6902.

24. Manazri, V., Wong-Staal, F., Franchini, G., Colombini, S., Gelmann, E. P., Oroszlan, S., and Gallo, R. C. (1983). Human T-cell leukemia-lymphoma virus (HTLV): Cloning of an integrated defective provirus and flanking cellular sequences. *Proc. Natl. Acad. Sci. USA* 80:1574–1578.

25. Chen, I. S. Y., McLaughlin, J., Gasson, J. C., Clark, S. C., and Golde, D. W. (1983). Molecular characterization of genome of a novel human T-cell leukaemia virus. *Nature* 305:502–508.

26. Gelmann, E. P., Franchini, G., Manzari, V., Wong-Staal, F., and Gallo, R. C. (1984). Molecular cloning of a unique human T-cell leukemia virus (HTLV–IIMo). *Proc. Natl. Acad. Sci. USA* 81:993–997.

27. Seiki, M., Hattori, S., Hirayami, Y., and Yoshida, M. (1983). Human adult T-cell leukemia virus: Complete nucleotide sequence of the provirus genome integrated in leukemia cell DNA. *Proc. Natl.. Acad. Sci. USA* 80: 3618–3622.

28. Haseltine, W. A., Sodroski, J. G., and Patarca, R. (1984). Structure and function of the genome of HTLV. *Current Topics Microbiol. Immunol.* 115:177–209.

29. Shimotohno, K., Takahashi, Y., Shimizyi, N., Golde, D. W., Chen, I. S. Y., Miwa, M., and Sugimara, T. (1985). Complete nucleotide sequence of an infectious clone of human T-cell leukemia virus type II: An open reading frame for the protease gene. *Proc. Natl. Acad. Sci. USA* 82:3101–3105.

30. Haseltine, W. A., Sodroski, J. G., Patarca, R., Briggs, ·D., Perkins, D., and Wong-Staal, F. (1984). Structure of 3′ terminal region of type II human T lymphotropic virus: Evidence for new coding region. *Science* 225:419–421.

31. Shimotohno, K., Wachsman, W., Takahashi, Y., Golde, D., Miwa, M., Sugimura, T., and Chen, I. S. Y. (1984). Nucleotide sequence of the 3′ region of an infectious human T-cell leukemia virus type II genome. *Proc. Natl. Acad. Sci. USA* 81:6657–6661.

32. Seiki, M., Hikikoshi, A., Taniguchi, T., and Yoshida, M. (1985). Expression of the pX gene of HTLV-I: General splicing mechanism in the HTLV family. *Science* 228:1532–1534.

33. Slamon, D. J., Shimotohno, K., Cline, M. J., Golde, D. W., and Chen, I. S. Y. (1984). Identification of the putative transforming protein of the human T-cell leukemia virus HTLV-I and HTLV-II. *Science* 226:61–65.

34. Lee, T. H., Coligan, J. E., Sodroski, J. G., Haseltine, W. A., Salahuddin, S. Z., Wong-Staal, F., Gallo, R. C., and Essex, M. (1984). Antigens encoded by the 3′-terminal region of human T-cell leukemia virus: Evidence for a functional gene. *Science* 226:57–61.

35. Haseltine, W. A., Sodroski, J., Patarca, R., Briggs, D., Perkins, D., and Wong-Staal, F. (1984). Structure of 3′ terminal region of type II human T lymphotropic virus: Evidence for new coding region. *Science* 225:419–421.

36. Goh, W. C., Sodroski, J. G., Rosen, C., Essex, M., and Haseltine, W. A. (1985). Subcellular localization of the product of the long open reading frame of human T-cell leukemia virus type I. *Science* 227:1227–1228.

37. Wachsman, W., Shimotohno, K., Clark, S. C., Golde, D. W., and Chen, I. S. Y. (1984). Expression of the 3′ terminal region of human T-cell leukemia viruses. *Science* 226:177–179.

38. Derse, D. (1987). Bovine leukemia virus transcription is controlled by a virus-encoded trans-acting factor and by cis-acting response elements. *J. Virol.* 61:2462–2471.

39. Slamon, D. J., Press, M. F., Souza, L. M., Murdock, D. C., Cline, M. J., Golde, D. W., Gasson, J. C., and Chen, I. S. Y. (1985). Studies of the putative transforming protein of the type I human T-cell leukemia virus. *Science* 228:1427–1430.

40. Sodroski, J., Rosen, C., Goh, W. C., and Haseltine, W. (1985). A transcriptional activator protein encoded by the x-lor region of the human T-cell leukemia virus. *Science* 228:1430–1434.

41. Seiki, M., Inoue, J., Takeda, T., and Yoshida, M. (1986). Direct evidence that p40x of human T-cell leukemia virus type I is a trans-acting transcriptional activator. *EMBO* 5(3):561–565.

42. Wachsman, W., Golde, D. W., Temple, P. A., Orr, E. C., Clark, S. C., and Chen, I. S. Y. (1985). HTLV x-gene product: requirement for the env methionine initiation codon. *Science* 228:1534–1537.

43. Adlovini, A., De Rossi, A., Feinberg, M. B., Wong-Staal, F., and Franchini, G. (1986). Molecular analysis of a deletion mutant provirus of type I human T-cell lymphotropic virus: Evidence for a doubly spliced x-lor mRNA. *Proc. Natl. Acad. Sci. USA* 83:38–42.

44. Kiyokawa, T., Seiki, Iwashita, S., Imagawa, K., Shimiza, F., and Yoshida, M. (1985). p27x-III and p21x-III, proteins encoded by the pX sequence of human T-cell leukemia virus type I. *Proc. Natl. Acad. Sci. USA* 82:8359–8363.

45. Nagashima, K., Yoshida, M., and Seiki, M. (1986). A single species of pX mRNA of juman T0cell leukemia virus type I encodes trans-activator p40x and two other phosphoproteins. *J. Virol.* 60:394–399.

46. Josephs, S. F., Wong-Staal, F., Manzari, V., Gallo, R. C., Sodroski, J. G., Trus, M. D., Perkins, D., Patarca, R., and Haseltine, W. A. (1984). Long terminal repeat structure of an American isolate of type I human T-cell leukemia virus. *Virology* 139:340–345.

47. Shimotohno, K., Miwa, M., Slamon, D. J., Chen, I. S. Y., Hoshing, H., Takano, M., Fujino, M., and Sugimura, T. (1985). Identification of new gene products coded from X regions of human T-cell leukemia viruses. *Proc. Natl. Acad. Sci. USA* 82:302–306.

48. Watanabe, T., Seiki, M., Tsujimoto, H., Miyoshi, I., Hayami, M., and Yoshida, M. (xxxx) Sequence homology of the simian retrovirus genome with human T-cell leukemia virus type I. *Virology* 144:59–65.

49. Sagata, N., Yasunaga, T., Tsuzuku-Kawamura, J., Ohishi, K., Ogawa, Y., and Ikawa, Y. (1985). Complete nucleotide sequence of the genome of bovine leukemia virus: Its evolutionary relationship to other retroviruses. *Proc. Natl. Acad. Sci. USA* 82:677–681.

50. Sagata, N., Yasunaga, T., and Igawa, Y. (1985). Two distinct polypeptides may be translated from a single spliced mRNA of the X genes of human T-cell leukemia and bovine leukemia viruses. *FEBS Lett.* 192:37–42.

51. Rice, N. R., Stephens, R. M., Couez, D., Deschamps, J., Kettmann, R., Burny, A., and Gilden, R. V. (1984). The nucleotide sequence of the *env* gene and post-*env* region of bovine leukemia virus. *Virology* 138:82–93.

52. Sodroski, J. G., Rosen, C. R., and Haseltine, W. A. (1984). Trans-acting transcriptional activation of the long terminal repeat of human T lymphotropic viruses in infected cells. *Science* 225:381–385.

53. Rosen, C. A., Sodroski, J. G., Willems, L., Kettmann, R., Campbell, K., Zaya, R., Burny, A., and Haseltine, W. A. (1986). The 3' region of bovine

leukemia virus genome encodes a *trans*-activator protein. *EMBO* 5(10): 2585–2589.

54. Derse, D., and Casey, J. W. (1986). Two elements in the bovine leukemia virus long terminal repeat that regulate gene expression. *Science* 231:1437–4411.

55. Pashkalis, H., Felber, B. K., and Pavlakis, G. N. (1986). *Cis*-acting sequences responsible for the transcriptional activation of human T-cell leukemia virus type I constitute a conditional enhancer. *Proc. Natl. Acad. Sci. USA* 83: 6558–6562.

56. Felber, B. K., Paskalis, H., Kleinman-Ewing, C., Wong-Staal, F., and Pavlakis, G. N. (1985). The pX protein of HTLV–I is a transcriptional activator of its long terminal repeats. *Science* 229:675–679.

57. Rosen, C. A., Sodroski, J. G., and Haseltine, W. A. (1985). Location of *cis*-acting regulatory sequences in the human T-cell leukemia virus type I long terminal repeat. *Proc. Natl. Acad. Sci. USA* 82:6502–6506.

58. Miwa, M., Takano, M., Teruuchi, T., and Shimotohno, K. (1986). Requirement of multiple copies of a 21-nucleotide sequence in the U3 regions of human T-cell leukemia virus type I and type II long terminal repeats for trans-acting activation of transcription. *Proc. Natl. Acad. Sci. USA* 83: 8112–8116.

59. Rosen, C. A., Park, R., Sodroski, J. G., and Haseltine, W. A. (1987). Multiple sequence elements are required for regulation of human T-cell leukemia virus gene expression. *Proc. Natl. Acad. Sci. USA* 84:4919–4923.

60. Brady, J., Jeang, K. T., Duvall, J., and Khoury, G. (1987). Identification of p40x-responsive regulatory sequences within the human T-cell leukemia virus type I long terminal repeat. *J. Virol.* 61:2175–2181.

61. Inoue, J. I., Yoshida, M., and Seiki, M. (1987). Transcriptional (p40x) and post-transcriptional (p27x-III) regulators are required for the expression and replication of human T-cell leukemia virus type I genes. *Proc. Natl. Acad. Sci. USA* 84:3653–3657.

62. Inoue, J-I., Seiki, M., and Yoshida, N. (1986). The second pX product p27 chi-III of HTLV-1 is required for *gag* expression. *FEBS Lett.* 209: 187–190.

63. Rosen, C. A., Sodroski, J. G., and Haseltine, W. A. (1985). The location of *cis*-acting regulatory sequences in the human t cell lymphotropic virus type III (HTLV–III/LAV) long terminal repeat. *Cell* 41:813–823.

64. Feinberg, M. B., Jarrett, R. F., Aldovini, A., Gallo, R. C., and Wong-Staal, F. (1986). HTLV–III expression and production involve complex regulation at the levels of splicing and translation of viral RNA. *Cell* 46:807–817.

65. Cullen, B. R. (1986). Trans-activation of human immunodeficiency virus occurs via a bimodel mechanism. *Cell* 46:973–982.

66. Seiki, M., Eddy, R., Shows, T. B., and Yoshida, M. (1983). Nonspecific integration of the HTLV provirus genome into adult T-cell leukaemia cells. *Nature* 309:640–642.

67. Yoshida, M., Seiki, M., Yamaguchi, K., and Takatsuki, K. (1984). Monoclonal integration of human T-cell leukemia provirus in all primary tumors

of adult T-cell leukemia suggests causative role of human T-cell leukemia virus in the disease. *Proc. Natl. Acad. Sci. USA* 81:2534–2537.

68. Hahn, B., Manzari, V., Colombini, S., Franchini, G., Gallo, R. C., and Wong-Staal, F. (1983). Common site of integration of HTLV in cells of three patients with mature T-cell leukaemia-lymphoma. Retraction of *Nature* 303:253–256; 305:340.

69. De Rossi, A. (1985). Clonal selection of T lymphocytes infected by cell-free human T-cell leukemia/lymphoma virus type I: Parameters of virus integration and expression. *Virology* 163:640–645.

70. Franchini, G., Mann, D. L., Popovic, M., Zicht, R. R., Gallo, R. C., and Wong-Staal, F. (1985). HTLV-I infection of T and B cells of a patient with adult T-cell leukemia-lymphoma (ATLL) and transmission of HTLV-I from B cells to normal T cells. *Leukemia Res.* 9:1305–1314.

71. Del Mistro, A., De Rossi, A., Aldovini, A., Salmi, R., and Chieco-Bianchi, L. (1986). Immortalization of human T lymphocytes by HTLV-I: Phenotypic characteristics of target cells and kinetics of virus integration and expression. *Leukemia Res.* 10:1109–1120.

72. Nerenberg, M., Hinrichs, S. H., Reynolds, R. K., Khoury, G., and Jay, G. (1987). The *tat* gene of human T-lymphotropic virus type 1 induces mesenchymal tumors in transgenic mice. *Science* 237:1324–1329.

73. Hinrichs, S. H., Nerenberg, M., Reynolds, R. K., Khoury, G., and Jay, G. (1987). A transgenic mouse model for human neurofibromatosis. *Science* 237:1340–1343.

74. Inoue, J., Seiki, M., Taniguchi, T., Tsuru, S., and Yoshida, M. (1986). Induction of interleukin 2 receptor gene expression by p40x encoded by human T-cell leukemia virus type I. *EMBO J.* 5:2883–2888.

75. Leonard, W. J., Depper, J. M., Crabtree, G. R., Rudikoff, S., Pumphgrey, J., Robb, R. J., Kronke, M., Svetlik, P. B., Peffer, N. J., Waldmann, T. A., and Greene, W. C. (1984). Molecular cloning and expression of cDNAs for the human interleukin-2 receptor. *Nature* 311:626–631.

76. Leonard, W. J., Depper, J. M., Kanehisa, M., Kronke, M., Peffer, N. J., Svetlik, P. B., Sullivan, M., and Greene, W. C. (1985). Structure of the human interleukin-2 receptor gene. *Science* 230:633–639.

77. Cross, S. L., Feinberg, M. B., Wolf, J. B., Holbrook, N. J., Wong-Staal, F., and Leonard, W. S. (1987). Regulation of the human interleukin-2 receptor alpha chain promoter: Activation of a nonfunctional promoter by the transactivator gene of HTLV-I. *Cell* 49:47–56.

78. Ho, D. D., Sarngadharan, M. G., Resnick, L., DiMarzo-Veronese, F., Rota, T. R., and Hirsch, M. S. (1985). Primary human T-lymphotropic virus type III infection. *Am. Intern. Med.* 103:880–883.

79. Shaw, G. M., Harper, M. E., Hahn, B. H., Epstein, L. G., Gajdusek, D. C., Price, R. W., Navia, B. A., Petito, C. K., O'Hara, C. J., Groopman, J. E., Cho, E., Oleske, J. M., Wong-Staal, F., and Gallo, R. C. (1985). HTLV-III infection in brains of children and adults with AIDS encephalopathy. *Science* 227:177.

80. Ho, D. D., Rota, T. R., Schooley, R. T., Kaplan, J. C., Allan, J. D., Groopman, J. E., Resnick, L., Felsenstein, D., Andrews, C. A., and Hirsch, M. S. (1985). Isolation of HTLV-III from cerebrospinal fluid and neural tissues of patients with neurologic syndromes related to the acquired immunodeficiency syndrome. *N. Engl. J. Med.* 313:1493–1497.

81. Resnick, L. (1985). Intra-blood-brain-barrier synthesis of HTLV-III-specific IgG in patients with neurologic symptoms associated with AIDS or AIDS-related complex. *N. Engl. J. Med.* 313:1498–1504.

82. Epstein, L. G. (1985). HTLV-III/LAV-like retrovirus particles in the brains of patients with AIDS encephalopathy. *AIDS Res.* 1:447–454.

83. Goldstick, L. (1985). Spinal cord degeneration in AIDS. *Neurology* 35: 103–106.

84. Anders, K. H., Guerra, W. F., Tomiyasa, U., Verity, M. A., and Vinters, H. V. (1986). The neuropathology of AIDS. UCLA experience and review. *Am. J. Pathol.* 124:537–538.

85. Curran, J. W. (1985). The epidemiology of AIDS: Current status and future prospects. *Science* 229:1352–1357.

86. Popovic, M. (1984). OKT-4 Antigen bearing molecule is a receptor for the human retrovirus HTLV-III. *Clin. Res.* 33:560A.

87. Klatzman, D., Barre-Sinoussi, F., Nugeyre, M. T., Dauguet, C., Vilmer, E., Griscell, C., Brun-Vesinet, F., Rouzioux, C., Gluckman, J. C., Chermann, J. C., and Montagnier, L. (1984). Selective tropism of lymphadenopathy associated virus (LAV) for helper-inducer T lymphocytes. *Science* 225: 59–63.

88. Fahey, J. Ll, Prince, H., Weaver, M., Groopman, J., Visscher, B., Schwartz, K., and Detels, R. (1984). Quantitative changes in T helper or T suppressor/cytotoxic lymphocyte subsets that distinguish acquired immune deficiency syndrome from other immune subset disorders. *Am. J. Med.* 76:695–100.

89. Redfield, R. R., Wright, D. C., and Tranont, E. C. (1986). The Walter Reed staging classification for HTLV-III/LAV infection. *N. Engl. J. Med.* 314(2): 131–132.

90. Barre-Sinoussi, F., Chermann, J. C., Rey, F., Nugeyre, M. T., Chamaret, S., Gruest, J., Dauguet, C., Axler-Blin, C., Brun-Vezinet, F., Rouzious, C., Rozenbaum, W., and Montaigner, L. (1983). Isolation of a T-lymphotropic retrovirus from a patient at risk for acquired immune deficiency syndrome (AIDS). *Science* 220:868–871.

91. Ho, D. D., Schooley, R. T., Rota, T. R., Kaplan, J. C., Flynn, T., Salahuddin, S. Z., Gonda, M. A., and Hirsch, M. S. (xxxx). Infection of monocyte/macrophages by human T lymphotropic virus type III. *Science* 226:451–453.

92. Gartner, S., Markovits, P., Markovits, D. M., Kaplan, M. H., Gallo, R. C., and Popovic, M. (1986). The role of mononuclear phagocytes in HTLV-III/LAV infection. *Science* 233:215–219.

93. Salahuddin, S. Z., Rose, R. M., Groopman, J. E., Markham, P. D., and Gallo, R. C. (1986). Human T lymphotropic virus type III infection of human alveolar macrophages. *Blood* 68:281–284.

94. Tschachler, E., Groh, V., Popovic, M., Mann, D. L., Konrad, K., Safai, B., Eron, L., di Marzo-Veronese, S., Wolff, K., and Stingl, G. (1987). Epidermal Langerhans cells—A target for HTLV-III/LAV infection. *J. Invest. Dermatol.* 88:233–237.

95. Montagnier, L., Gruest, J., Chamaret, C., Dauguet, C., Axler, D., Guetard, M. T., Nugeyre, Barre-Sinoussi, F., Chermann, J. C., Brunet, J. B., Klatzmann, D., and Gluckman, J. D. (1984). Adaptation of lymphadenopathy associated virus (LAV) to replication in EBV-transformed B lymphoblastoid cell lines. *Science* 225:63–66.

96. Chiodi, F., Fuerstenberg, S., Gudlund, M., Asjo, B., and Fenyo, E. M. (1987). Infection of brain-derived cells with the human immunodeficiency virus. *J. Virol.* 61:1244–1247.

97. Cheng-Meyer, C., Rutka, J. T., Rosenblum, M. L., Mchugh, T. Stittes, D. P., and Levy, J. A. (1987). Human immunodeficiency virus can productively infect cultured human glial cells. *Proc. Natl. Acad. Sci. USA* 84: 3526–3530.

98. Wiley, C. A., Schreier, R. D., Nelson, J. A., Lampert, P. W., and Oldstone, M. B. A. (1986). Cellular localization of human immunodeficiency virus infection within the brains of acquired immune deficiency syndrome patients. *Proc. Natl. Acad. Sci. USA* 83:7089–7093.

99. Donahue, R. E., Johnson, M. M., Zon, L. I., Clark, S. C., and Groopman, J. Suppression of in vitro haemoatopoiesis following human immunodeficiency virus infection. *Nature* 326:200–203.

100. Hammer, S. N., Gillis, J., Groopman, J. E., and Rose, R. M. (1986). In vitro modification of human immunodeficiency virus infection by granulocyte-macrophage colony-stimulating factor and gamma interferon. *Proc. Natl. Acad. Sci. USA* 83:8734–8738.

101. Harper, M. E., Marselle, L. M., Gallo, R. C., and Wong-Staal, F. (1986). Detection of lymphocyte expressing human T-lymphotropic virus type III in lymph nodes and peripheral blood from infected individuals by in situ hybridization. *Proc. Natl. Acad. Sci. USA* 83:772–776.

102. Koenig, S., Gendelman, H. E., Orenstein, J. M., Del Canto, M. C., Pezeshkpour, G. H., Yungbluth, M., Janotte, F., Arksoint, A., martin, M. A., and Fauci, A. S. (1986). Detection of AIDS virus in macrophages in brain tissue from AIDS patients with encephalopathy. *Science* 233:1089–1093.

103. Gallo, R. C., Salahuddin, S. Z., Popovic, M., et al. (1984). Frequent detection and isolation of cytopathic retroviruses (HTLV-III) from patients with AIDS and at risk for AIDS. *Science* 224:500–503.

104. Groopman, J. E., Salahuddin, S. Z., Sarngadharan, M. G., Markham, P. D., Gonda, M., Sliski, A., and Gallo, R. C. (1984). HTLV-III in saliva of people with AIDS-related complex and healthy homosexual men at risk for AIDS. *Science* 226:447–449.

105. Zagury, D., Bernard, J., Leibowitch, J., et al. (1984). HTLV-III in cells cultured from semen of two patients with AIDS. *Science* 226:449–457.

106. Salahuddin, S. Z., Markhana, P. D., Popovic, M., Sarngadharan, M. G., Orndorff, S., Fladagas, A., Patel, A., Gold, J., and Gallo, R. C. (1985).

Isolation of infectious human T-cell leukemia/lymphotropic virus type III (HTLV-III) from patients with acquired immunodeficiency syndrome (AIDS) or AIDS-related comples (ARC) and from healthy carriers: A study of risk groups and tissue sources. *Proc. Natl. Acad. Sci. USA* 82: 5530-5534.

107. Vogt, M. W., Witt, D. J., Craven, D. E., Byington, D. F., Schooley, R. T., and Hirsch, M. S. (1986). Isolation of HTLV-III/LAV from cervical secretions of women at risk for AIDS. *Lancet* 1:525-527.

108. Cooper, D. A., Gold, J., Maclean, P., Donovan, B., Finlayson, R., Barnes, T. G., and Michelmore, H. M. (1985). Acute AIDS retrovirus infection. Definition of a clinical illness associated with seroconversion. *Lancet* 1: 537:540.

109. Walker, B. D., Chakrabarti, S., Moss, B., Paradis, T. P., Flynn, T., Durno, A. G., Blumberg, R., Kaplan, J. C., Hirsch, M. S., and Schooley, R. T., (1987). HIV-specific cytotoxic T lymphocytes in seropositive individuals. *Nature* 328 (6128):345-348.

110. Kanki, P. J., Barin, P., M'Boup, S., Allan, J. S., Romet-Lemaine, J. L., Marlink, R., McLane, M. F., Lee, T. H., Arbeille, B., Denis, F., and Essex, M. (1986). New human T-lymphotropic retrovirus related to simian T-lymphotropic virus type III (STLV-IIIAGM). *Science* 232:238-243.

111. Kanki, P., and Essex, M. (1987). In: AIDS: *Modern Concepts inTherapeutic Challenges* (S. Broder, ed.). New York, Marcel Dekker, p. 63.

112. Clavell, F., Guetard, D., Brun-Vezinet, F., Chamaret, S., Rey, M., Santos-Ferreira, M. O., Laurent, A. G., Dauguet, C., Katlama, C., Rouzioux, C., Klatzmann, D., Champalimoud, J. L., and Montagnier, L. (1986). Isolation of a new human retrovirus from West African patients with AIDS. *Science* 233:343-346.

113. Kanki, P., McLane, M. F., King, N., Letvin, N. L., Hunt, R. D., Sehgal, P., Daniel, M. D., Desrosiers, R. C., and Essex, M. (1985). Serologic identification and characterization of a macaque T-lymphotropic retrovirus closely related to HTLV-III. *Science* 228:1199-1201.

114. Letvin, N., Daniel, M. D., Sehgal, P. K., Desrosiers, R. C., Hunt, R. D., Waldron, R. M., Mackey, J. J., Schmidt, D. K., Chalifoux, L. V., and King, N. W. (1985). Induction of AIDS-like disease in macaque monkeys with T-cell tropic retrovirus STLV-III. *Science* 230:71-73.

115. Ratner, L., Haseltine, W. A., Patarca, R., Livak, K. J., Starcich, B., Josephs, S. F., Doran, E. A.,Rafalski, J. A., Whitehorn, E. A., Baumeister, K., Ivanoff, L., Pette, Pearson, M. I., Lautenberger, J. A., Papas, T. S. Ghrayeb, J., Chang, N., Gallo, R. C., and Wong-Staal, F. (1985). Complete nucleotide sequence of the AIDS virus HTLV-III. *Nature* 313:277-284.

116. Wain-Hobson, S., Sonigo, P., Danos, O., Cole, S., and Alizon, M. (1985). Nucleotide sequence of the AIDS virus, LAV. *Cell* 40:9-18.

117. Sanchez-Pescador, R., Power, M. D., Barr, P. J., Steimer, K. S., Stempien, M. M., Brown-Shimer, S. L., Gee, W. W., Renard, A., Randolph, A., Levy, J. A., Dina, D., and Luciw, P. (1985). Nucleotide sequence and expression of an AIDS-associated retrovirus (ARV-2). *Science* 227:484-492.

118. Muesing, M. A. (1985). Nucleic acid structure and expression of the human AIDS/lymphadenopathy retrovirus. *Nature* 313:480–485.

119. Sodroski, J., Rosen, C., Wong-Staal, F., Salahuddin, S. Z., Popovic, M., Arya, S., Gallo, R. C., and Haseltine, W. A. (1985). Trans-acting transcriptional regulation of human T-cell leukemia virus type III long terminal repeat. *Science* 227:171–173.

120. Peterlin, B. M., Luciw, P. A., Barr, P. J., and Walker, M. D. (1986). Elevated levels of mRNA can account for the trans-activation of human immunodeficiency virus. *Proc. Natl. Acad. Sci. USA* 83:9734–9738.

121. Khoury, G., and Gross, X. X. (1986). Enhancer elements. *Cell* 33:313–316.

122. Jones, K. A., Kadonga, J. T., Luciw, P. A., and Tijan, R. (1986). Activation of the AIDS retrovirus promoter by the cellular transcription factor, Sp1. *Science* 23:755–759.

123. Okamoto, T., and Wong-Staal, F. (1986). Demonstration of virus-specific transcriptional activator(s) in cells infected with HTLV-III by an in vitro cell-free system. *Cell* 47:29–35.

124. Nabel, G., and Baltimore, D. (1987). An inducible transcription factor activates expression of human immunodeficiency virus in T cells. *Nature* 326:711–714.

125. Gendelman, H. E., Phelps, W., Feigenbaum, L., Ostrove, J. M., Adachi, A., Howley, P. M., Khoury, G., Ginsberg, H. S., and Martin, M. A. (1986). Trans-activation of the human immunodeficiency virus long terminal repeat sequence by DNA viruses. *Proc. Natl. Acad. Sci. USA* 83:9759–9763.

126. Mosca, J. D., Bednarik, D. P., Raj, N. B. K., Rosen, C. A., Sodroski, J. G., Haseltine, W. A., and Pitha, P. M. (1987). Herpes simplex virus type-1 can reactivate transcription of latent human immunodeficiency virus. *Nature* 325:67–70.

127. Jacks, T., Power, M. D., Masierz, F. R., Luciw, P. A., Barr, P. J., and Varmus, H. E. (1987). Characterization of ribosomal frameshifting in HIV-1 *gag-pol* expression. *Nature* 331:280–283.

128. Mous, J., Heimer, E. P., and LeGrice, S. F. (1988). Processing protease and reverse transcriptase from human immunodeficiency virus type I polyprotein in Escherichia coli. *J. Virol.* 62:1433–1436.

129. Veronese, F., Copland, P., DeVico, A., Rahman, R., Oroszlan, S., Gallo, R., and Sarangadharan, M. (1986). Characterization of highly immunogenic p66/p51 as the reverse transcriptase of HTLV-III/LAV. *Science* 231:1289–1293.

130. Allan, J. S., Coligan, J. E., Lee, T. H., Barin, F., Kanki, P. J., M'Boup, S., McLane, M. F., Groopman, J. E., and Essex, M. (1987). Immunogenic nature of a *pol* gene product of HTLV-III/LAV. *Blood* 69:331–333.

131. Allan, J. S., Coligan, J., Barin, F., McLane, M. F., Sodroski, J. G., Rosen, C. A., Haseltine, W. A., Lee, T. H., and Essex, M. (1985). Major glycoprotein antigens that induce antibodies in AIDS patients are encoded by HTLV-III. *Science* 228:1091–1094.

132. Robey, W. G., Safai, B., Oroszlan, S., Arthur, L. O., Gonda, M. A., Gallo, R. C., and Fischinger, P. J. (1985). Characterization of envelope and core structural gene products of HTLV-III with sera from AIDS patients. *Science* 228:593–595.

133. Barin, F., McLane, M. F., Allan, J. S., Lee, T. H., Groopman, J. E., and Essex, M. (1985). Virus envelope protein of HTLV-III represents major target antigen for antibodies in AIDS patients. *Science* 228:1094–1096.

134. Terwilliger, E., Sodroski, J. G., Rosen, C. A., and Haseltine, W. A. (1986). Effects of mutations within the 3' *orf* open reading frame region of HTLV-III/LAV on replication with cytopathogenicity. *J. Virol.* 60:754–760.

135. Kowalski, M., Potz, J., Basiripour, L., Dorfman, T., Goh, W. C., Terwilliger, E., Dayton, A., Rosen, C., Haseltine, W., and Sodroski, J. (1987). Functional regions of the envelope glycoprotein of human immunodeficiency virus type 1. *Science* 237:1351–1355.

136. Gelderblom, H. R. (1987). Fine structure of human immunodeficiency virus (HIV) and immunolocalization of structural proteins. *Virology* 156:171–176.

137. McDougal, J. S., Kennedy, M. S., Sligh, J. M., Cort, S. P., Mawle, A., and Nicholson, J. K. (1985). Binding of HTLV-III/LAV to T4+ T cells by a complex of the 110K viral protein and the T4 molecule. *Science* 231:382–385.

138. Dalgleish, A. G., Beverly, P. C., Clapham, P. R., Crawford, D.H., Greaves, M. F., and Weiss, R. (1985). The CD4 (T4) antigen is an essential component of the receptor for the AIDS retrovirus. *Nature* 312:763–767.

139. Lifson, J. D., Feinberg, M. B., Reyes, G. R.,Rabins, L., Banapour, B., Chakrabarti, S., Moss, B., Wong-Staal, F., Steimer, K. S., and Engleman, E. G. (1986). Induction of CD4-dependent cell fusion by the HTLV-III/LAV envelope glycoprotein. *Nature* 323:725–728.

140. Klatzmann, D. E., Champagne, E., Chamaret, S., Gurest, J., Guetard, D., Hercend, T., Gluckman, J., and Montagnier, L. (1985). T-lymphocyte T4 molecule behaves as the receptor for human retorvirus LAV. *Nature* 312:767–768.

141. McDougal, J. S., Nicholson, J. K., Cross, G. D., Cort, S. P., Kennedy, M. S., and Mawle, A. C. (1986). Binding of the human retrovirus HTLV-III/LAV/ARV/HIV to the CD4 (T4) molecule: Conformation dependence, epitope mapping, antibody inhibition, and potential for idiotypic mimicry. *J. Immunol.* 137(9):2937–2944.

142. Sattentau, Q. J., Dalgleish, A. G., Weiss, R. A., and Beverly, P. C. L. (1986). Epitopes of the CD4 antigen and HIV infeciton. *Science* 234:1120–1123.

143. Maddon, P. J., Dalgleish, A. G., McDougal, J. S., Clapham, P. R., Weiss, R. A., and Axel, R. (1986). The T4 gene encodes the AIDS virus receptor and is expressed in the immune system and the brain. *Cell* 47:333–348.

144. Daniel, M. D., Letvin, N. L., King, M. W., Kannagi, M., Sehgal, P. K., and Hunt, R. D. (1985). Isolation of T-cell tropic HTLV-III-like retrovirus from macaques. *Science* 228:1201–1204.

145. Nicholson, J. K., Cross, G. P., Callaway, C. S., and McDougal, J. S. (1986). In vitro infection of human monocytes with human T lymphotropic virus type III/lymphadenopathy-associated virus (HTLV–III/LAV). *J. Immunol.* 137:323–329.

146. Levy, J., Shimabukuro, J., McHugh, T., Casavant, C., Stites, D., and Oshiro, L. (1985). AIDS-associated retroviruses (ARV) can productively infect other cells besides human T helper cells. *Virology* 147:441.

147. Asjo, B., Ivhed, I., Gidlund, M., Fuerstenberg, S., Fenyo, E. M., Nilsson, K., and Wigzell, H. (1987). Susceptibility to infection by the human immunodeficiency virus (HIV) correlates with T4 expression in a parental monocytoid cell line and its subclones. *Virology* 157:359–365.

148. Adachi, A., Koenig, S., Gendelman, H., Daugherty, D., Gattoni-Celli, S., Fauci, A., and Martin, M. (1987). Productive, persistent infection of human colorectal cell lines with human immunodeficiency virus. *J. Virol.* 61:209–213.

149. Sodroski, J., Goh, W. C., Rosen, C., Campbell, K., and Haseltine, W. A. (1986). Role of the HTLV–III/LAV envelope in syncytium formation and cytopathicity. *Nature* 322(6078):470–474.

150. Lifson, J. D., Reyes, G. R., McGrath, M. S., Stein, B., and Engleman, E. G. (1986). AIDS retrovirus induced cytopathology: Giant cell formation and involvement of CD4 antigen. *Science* 232:1123–1127.

151. Rasheed, S., Gottlieb, A. A., and Garry, R. F. (1986). Cell killing by ultraviolet-inactivated human immunodeficiency virus. *Virology* 154(2):395–400.

152. Stein, B. S., Gowda, S. P., Lifson, J. D., Penhallow, R. C., Bensch, K. G., and Engleman, E. G. (1987). pH-independent HIV entry into CD4-positive T cells via virus envelope fusion to the plasma membrane. *Cell* 49:659–668.

153. Sharer, L. R. (1985). Multinucleated giant cells and HTLV–III in AIDS encephalopathy. *Hum. Pathol.* 16:760.

154. Brynes, R. K. (1983). Value of lymph node biopsy in unexplained lymphadenopathy in homosexual men. *JAMA* 250:1313–1317.

155. Lasky, L. A., Nakamura, G., Smith, D. H., Fennie, C., Shimasaki, C., Patzer, E., Berman, P., Gregory, T., and Capon, D. J. (1987). Delineation of a region of the human immunodeficiency virus type 1 gp120 glycoprotein critical for interaction with the CD4 receptor. *Cell* 50:975–985.

156. McCune, J. M., Rabin, L. B., Feinberg, M. B., Lieberman, M., Kosek, J. C., Reyes, G. R., and Weissman, I. L. (1988). Endoproteolytic cleavage of gp160 is required for the activation of human immunodeficiency virus. *Cell* 53:55–67.

157. Terwilliger, E., Sodroski, J. G., Rosen, C. A., and Haseltine, W. A. (1986). Effects of mutations within the 3' *orf* open reading frame region of human T-cell lymphotropic virus type III (HTLV–III/LAV) on replication and cytopathogenicity. *J. Virol.* 60:754–760.

158. Sarngadharan, M. G., Popovic, M., Bruch, L., Schupbach, J., and Gallo, R. C. (1984). Antibodies reactive with human T-lymphotropic retroviruses (HTLV–III) in the serum of patients with AIDS. *Science* 224:506–508.

159. Matthews, T.J. (1986). Restricted neutralization of divergent human T-lymphotropic virus type III isolates by antibodies to the major envelope glycoprotein. *Proc. Natl. Acad. Sci. USA* 83:9709-9713.

160. Palker, T. J., Matthews, T. J., Clark, M. E., Cianciolo, G. J., Randall, R. R. Langlois, A. J., White, G. C., Safai, G., Snyderman, R., Bolognesi, D. P., and Haynes, B. (1987). A conserved region at the COOH terminus of human immunodeficiency virus gp120 envelope protein contains an immunodominant epitope. *Proc. Natl. Acad. Sci. USA* 84:2479-2483.

161. Weiss, R. A., Clapham, P. R., Cheingsong-Popov, R., Dalgleish, A. G., Carne, C. A., Weller, I. V., and Tedder, R. S. (1985). Neutralization of human T-lymphotropic virus type III by sera of AIDS and AIDS-risk patients. *Nature* 316:69-74.

162. Sodroski, J., Goh, W. C., Rosen, C., Tartar, A., Portelle, D., Burny, A., and Haseltine, W. (1986). Replicative and cytopathic potential of HTLV-III/LAV with *sor* gene depletions. *Science* 231:1549-1553.

163. Fisher, A. G., Ratner, L., Mitsuya, H., Marselle, L. M., Harper, M. E., Broder, S., Gallo, R. C., and Wong-Staal, F. (1987). The *sor* gene of HIV-1 is required for efficient virus transmission in vitro. *Science* 237:888-893.

164. Strebel, K., Daugherty, D., Clouse, K., Cohen, D., Folks, T., and Martin, M. A. (1987). The HIV "A" (sor) gene product is essential for virus infectivity. *Nature* 328:728-730.

165. Wong-Staal, F., Chanda, P. K., and Ghrayeb, J. (1987). HIV The eighth gene. *AIDS Res. Hum. Retroviruses* 3:33-39.

166. Allan, J. S., Coligan, J. E., Lee, T. H., McLane, M. F., Kanki, P. S., Groopman, J. E., and Essex, M. (1985). A new HTLV-III/LAV encoded antigen detected by antibodies from AIDS patients. *Science* 230:810-813.

167. Haase, A. T. (1986). Pathogenesis of lentivirus infections. *Nature* 322:130-136.

168. Fisher, A. G., Ratner, L., Mitsuya, H., Marselle, L. M., Harper, M. E., Broder, S., Gallo, R. C., and Wong-Staal, F. (1986). Infectious mutants of HTLV-III with changes in the 3' region and markedly reduced cytopathic effects. *Science* 233:655-658.

169. Luciw, P. A., Cheng-Mayer, C., and Levy, J. A. (1987). Mutational analysis of the human immunodeficiency virus: The *orf*-B region down-regulates virus replication. *Proc. Natl. Acad. Sci. USA* 84:1434-1438.

170. Guy, B., Kieny, M. P., Riviere, Y., LePeuch, C., Dott, K., Girard, M., Montagnier, L., and Lewig, J. P. (1987). HIV F/3' *orf* encodes a phosphorylated GTP-binding protein resembling an oncogene product. *Nature* 330:266-269.

171. Ahman, N., and Venkatesan, S. (1988). Nef protein of HIV-1 is a transcriptional repressor of HIV-1 LTR. *Science* 241:1481-1485.

172. Arya, S. K., Guo, C., Joseph, S. F., and Wong-Staal, F. (1985). *Trans*-activator gene of human T-lymphotropic virus type III (HTLV-III). *Science* 229:69-73.

173. Ruben, S., Perkins, A., Purcell, R., Joung, K., Sia, R., Burghoff, R., Haseltine, W. A., and Rosen, C. A. (1989). Structural and functional characterization of human immunodeficiency virus *tat* protein. *J. Virol.* 63:1-8.

174. Goh, W. C., Rosen, C., Sodroski, J., Ho., D., and Haseltine, W. A. (1986). Identification of a protein encoded by the *trans* activator gene *tat*III of human T-cell lymphotropic retrovirus type III. *J. Virol.* 59:181-184.

175. Dayton, A. I., Sodroski, J. G., Rosen, C. A., Goh, W. C., and Haseltine, W. A. (1986). The *trans*-activator gene of the human T cell lymphotropic virus type III is required for replication. *Cell* 44:941-947.

176. Fisher, A. G., Feinberg, M. B., Josephs, S. F., Harper, M. E., Marselle, L. M., Reyes, G., Gonda, M. A., Aldovini,A., Debouk, C., Gallo, R. C., and Wong-Staal, F. (1986). The *trans*-activator gene of HTLV-III is essential for virus replication. *Nature* 320:367-371.

177. Wright, C. M., Felber, B. K., Paskalis, H., and Pavlakis, G. N. (1986). Expression and characterization of the *trans*-activator of HTLV-III/LAV virus. *Science* 234:988-992.

178. Hauber, J., and Cullen, B. (1988). Mutational analysis of the *trans*-activation-responsive region of the human immunodeficiency virus type I long terminal repeat. *J. Virol.* 62:673.

179. Sodroski, J., Patarca, R., Rosen, C., and Haseltine, W. A. (1985). Location of the *trans*-activating region on the genome of human T-cell lymphotropic virus type III. *Science* 229:74-77.

180. Frankel, A. D., Bredt, D. S., and Pabo, C. O. (1988). *Tat* protein from human immunodeficiency virus forms a metal-linked dimer. *Science* 240:70-73.

181. Sadaie, M. R., Rappaport, J., Benter, T., Josephs, S. F., Willis, R., and Wong-Staal, F. Missense Mutations in an Infectious Human Immunodeficiency Viral Genome. Proc. Natl. Acad. Sci. USA 85:9224-9228.

182. Goh, W. C., Sodroski, J. G., Rosen, C. A., and Haseltine, W. A. (1987). Expression of the *art* gene protein of human T-lymphotropic virus type III (HTLV-IIII/LAV) in bacteria. *J. Virol.* 61:633-637.

183. Sodroski, J. G., Goh, W. C., Rosen, C., Dayton, A., Terwilliger, E., and Haseltine, W. (1986). A second post-transcriptional *trans*-activator gene required for HTLV-III replication. *Nature* 321:412-417.

184. Dayton, A. I., Terwilliger, E. F., Potz, J., Kowalski, M. M. Sodroski, J. G., Haseltine, W. A. (1988). *Cis*-acting Sequences Responsive to the *rev* Gene Product of the Human Immunodeficiency Virus. Journal of Acquired Immune Deficiency Syndromes. 1:441-452.

185. Terwilliger, E. F., Sodroski, J. G., Haseltine, W. A., and Rosen, C. R. (1988). The *art* gene product of human immunodeficiency virus is required for replication. *J. Virol.* 62:655-658.

186. Knight, D. M., Flomerfelf, F. A., and Ghrayab, J. (1987). Expression of *art/trs* protein of HIV and study of its role in viral envelope synthesis. *Science* 236:837-840.

187. Rosen, C. R., Terwilliger, E. F., Dayton, A. I., Sodroski, J. G., and Haseltine, W. A. (1988). Intragenic *cis*-acting *art* gene-responsive sequences of the human immunodeficiency virus. *Proc. Natl. Acad. Sci. USA* 85:2071–2075.

188. Cohen, E., Terwilliger, E., Sodroski, J., and Haseltine, W. (1988). Identification of a Protein Encoded by the *vpu* Gene of HIV-1, *Nature* 334:532–534.

189. Strebel, K., Klimkait, T., and Martin, M. A. (1988). A Novel Gene of HIV-1, *vpu*, and its 16-kilodalton Product. *Science* 241:1221–1223.

190. Matsuda, Z., Chou, M., Matsuda, M., Huang, J., Chen, Y., Redfield, R., Mayer, K., Essex, M., and Lee, T. H. (1988). Human Immunodeficiency Virus Type 1 has an Additional Coding Sequence in the Central Region of the Genome. *Proc. Natl. Acad. Sci. USA* 85:6968–6972.

191. Rimsky, L., Hauber, J., Duovich, M., Malim, M., Langlois, A., Cullen, B., and Greene, W. (1988). Functional Replacement of the HIV-1 *rev* Protein by the HTLV–I *rex* Protein. *Nature* 335:738–740.

192. Greene, W., Leonard, W., Wano, Y., Svetlik, P., Pfeffer, P., Sodrosky, J., Rosen, C., Goh, W., and Haseltine, W. (1986). Trans-activator Gene of HTLV–II Induces IL-Z Receptor and IL-Z Cellular Gene Expression. *Science* 232:877-880.

193. Mervis, R., Ahmad, N., Lillehoj, E., Reum, M., Salazar, R., Chan, H., and Venkatesan, S. (1988). The *gag* Gene Products of HIV-1: Alignment within the *gag* Open Reading Frame, Identification of Post-translational Modifications, and Evidence for Alternative *gag* Precursors *J. Virol.* 62:3993–4002.

194. Kozarsky, K., Penman, M., Basiripoor, L., Haseltine, W., Sodroski, J., and Krieger, M. (1989). Glycosylation and Processing of the HIV-1 Envelope Protein. *J. Acquir. Immune Defic. Syndr.* 2:163–169.

195. Niederman, T., Thielan, B., and Ratner, L. (1989). HIV–I Negative Factor is a Transcriptional Silencer. *Proc. Natl. Acad. Sci. USA* 86:1128–1132.

196. Kao, S. Y., Calman, A. F., Luciw, P. A., and Peterlin, B. M. (1987). Antitermination of Transcription with the Long Terminal Repeat of HIV-1 by *tat* Gene Product. *Nature* 330:489–493.

197. Malim, M. H., Hauber, J., Fenrick, R., and Cullen, B. R. (1988). Immunodeficiency Virus *rev* Trans-activator Modulates the Expression of the Viral Regulatory Genes. *Nature* 335:181–183.

11

Immunology of AIDS and HIV Infection

Scott Koenig and Anthony S. Fauci *National Institute of Allergy and Infectious Diseases, National Institutes of Health, Bethesda, Maryland*

I. INTRODUCTION

The acquired immunodeficiency syndrome (AIDS) has created a challenge of unparalleled proportions for health care workers and scientists for the past 6 years since the initial patients affected with this disorder were described (1-3). As of April 1987, 33,000 cases of AIDS have been reported in the United States alone; this figure is projected to increase to 270,000 by the year 1991 (4).

The major advance in AIDS research has been identification of the etiological agent of this disorder, called lymphadenopathy associated virus (LAV) (5), human T-cell lymphotropic virus III (HTLV-III) (6), AIDS-related virus (ARV) (7), and, most recently, human immunodeficiency virus (HIV) (8). The isolation of HIV has fostered an intensive examination of the virological, immunological, and molecular properties of this virus in an effort to gain insight into the pathogenesis of AIDS. It has facilitated the identification of individuals at risk for the disease through serological screening (9,10) and viral isolation techniques. As a result, a major source of viral dissemination has been eliminated in countries where screening of prospective blood donors is carried out. About 3% of current cases of AIDS in the United States are a result of transmission of HIV through blood and blood products before this technology was available (11). In addition, individuals who are identified as having been exposed to HIV are educated in ways to curtail its sexual transmission.

Recently, another human T-cell tropic retrovirus, LAV-2, has been isolated and implicated as the cause of AIDS in certain individuals in western Africa (12). Structurally, this virus is closely related to a simian retrovirus called STLV-III (13-15) isolated from captive and wild primate species. Individuals infected with LAV-2 develop antibodies that strongly react with STLV-III but have minimal activity against HIV-1 proteins. Theoretically, HIV-1 and LAV-2 may be part of a larger group of human retroviruses that can cause AIDS; the majority of cases in the current epidemic, however, are clearly attributable to HIV-1. Another retrovirus, called HTLV-IV (16), had also been isolated from individuals in Senegal. It too is related to STLV-III (17) but, unlike LAV-2, has been isolated only from healthy individuals and has not been associated with an AIDS-like disorder. Comparison of the amino acid sequences and virological and immunological properties of the proteins from these different T-cell lymphotropic retroviruses should provide insight into those regions which govern the pathogenicity of this family of retroviruses in humans.

The ability to propagate these retroviruses in vitro provided a means to investigate the defects of the immune system observed in patients with AIDS. Following infection with HIV, individuals will develop progressive immunological abnormalities predisposing them to opportunistic infections (OI) (1-3,18,20) and unusual malignancies, most commonly, Kaposi's sarcoma (KS) (20,21). Moreover, the pathogens responsible for these OIs can also alter immune function independent of HIV (22). Thus, an observed depressed immune response in many AIDS patients, particularly those with OI, may be the result of an insidious infection with HIV plus infection with other organisms. The mechanisms responsible for the impaired immune response have not been fully defined, but some are clearly unique to HIV and others may be common to several pathogens. In vitro models allow for dissection of the complex mechanisms by which individual and combinations of pathogens affect the multiple components of the immune system.

HIV, like some other retroviruses, produce proteins that enhance their own replication (23). In addition, the in vivo propagation of HIV could potentially be fostered by interactions with a variety of other organisms which could promote HIV replication by similar or distinct mechanisms. Recent evidence suggests that HIV expression can be augmented by *trans*-acting products of herpes and papovavirus genes (24,25). Thus, in a patient with asymptomatic cytomegalovirus (CMV) viremia, HIV replication might be enhanced, resulting in more devastating immune compromise and clinical progression. When better antiviral agents are available, the treatment of HIV, CMV, and other viruses in asymptomatic and immunologically competent individuals might prevent the development of AIDS-associated disorders.

In this chapter, we will review observations related to immunological abnormalities detected in AIDS patients and individuals infected with HIV, summarize pertinent in vitro findings of the effects of HIV on immune function, examine immune responses to the HIV organism in patients, and present a model of immunopathogenesis.

II. T LYMPHOCYTE

A. T-Cell Subpopulations

One of the earliest laboratory abnormalities seen in patients with AIDS and in individuals in high-risk groups for this disorder was an alteration in the ratio of circulating T lymphocytes bearing the T4 and T8 phenotypes (26–36). The normal ratio of T4 to T8 cells of approximately 2 to 1 is markedly reduced, particularly in patients with more clinically advanced forms of the disease. These changes are due primarily to depletion of the T4 population. Generalized lymphopenia has been observed, but it usually occurs late in the clinical course. AIDS patients with OI, in general, have a more depressed T4-to-T8 ratio and lower T4 counts compared with patients with KS alone (35). Depressed ratios of T4 to T8 cells can be observed in the setting of a relatively normal T4 cell count due to increased numbers of T8 cells. This is seen most often in asymptomatic HIV-seropositive individuals and those with progressive generalized lymphadenopathy (PGL), AIDS-related complex (ARC), and milder forms of KS.

There is some indication that subpopulations of T4 cells may be preferentially eliminated in AIDS and AIDS-related conditions. The Leu 8^+ (TQ1) subset of T4 cells was found to be depressed in circulating lymphocytes of patients with lymphadenopathy and AIDS, whereas the Leu 8^- subset of T4 cells was reduced only in the AIDS patients (37,38). In lymph nodes from AIDS patients, paracortical populations of Leu 3^+8^+ cells were decreased, whereas Leu 3^+8^- cells within the germinal regions were maintained (39). The physiological and pathological significance of preferential loss of Leu 8^+ T4 cells in the different clinical groups is unknown. This population has been functionally associated with the induction of suppressor cells, and the complementary population (Leu 8^-) has been thought to mediate helper effects. Functional overlap exists, however, between these subpopulations, and other phenotypic markers (2H4 and 4B4) may more accurately delineate these functional subpopulations (40, 41).

Several mechanisms may account for these observations. The generation of increased numbers of T8 cells, which is seen in many healthy HIV-seropositive and ARC patients, could require induction by Leu 8^+ (or 4B4) T4 cells. Since the replication of HIV requires cellular activation (42–44, Figure 1), the preferential loss of the Leu 3^+ subset could occur in the setting of their activation during this induction process. Alternatively, these Leu 8^+T^+ cells may express a unique surface receptor in addition to CD4 which facilitates the preferential binding or uptake of HIV.

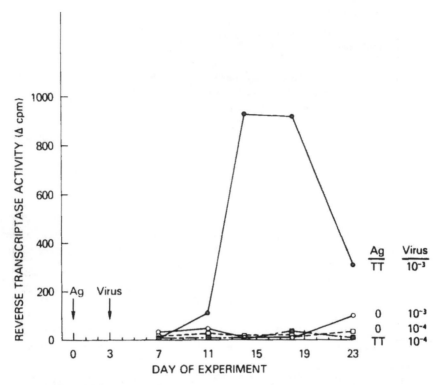

FIGURE 1   Effect of antigen-induced activation of PBMNC on production of reverse transcriptase (RT) after exposure to HTLV–III/LAV. PBMNC from a tetanus toxoid (TT) immune donor were incubated with TT (closed symbols) or medium alone (open symbols) from day 0 to day 3 and were exposed to HTLV–III/LAV [10^{-3} (circles) or 10^{-4} (squares) dilution of stock virus] or media alone on day 3. Each point represents the mean of duplicate observations with backgrounds subtracted. Backgrounds for RT counts produced by PBMNC not exposed to virus averaged 179.4 cpm (standard deviation = 63.6, range = 86–416). The blastogenic response (Δ cpm) of fresh (day 0) PBMNC to TT was 26,380 (mean of triplicate observations). [Adapted from Margolick et al. (44).]

B. In Vitro Infection of T4 Lymphocytes

The depletion of T4 cells from HIV-infected individuals occurs as a consequence of the binding of the viral envelope to the CD4 molecule which appears to be a receptor for HIV. Early studies indirectly showed the importance of this molecule as a receptor for HIV by demonstrating that HIV could be successfully propagated in T4 bearing normal lymphocytes and tumor lines (45–47) and that monoclonal antibodies to CD4 could prevent viral replication (48,49). The differential ability of these antibodies to block HIV infection indicated that not all regions of the CD4 molecule participate in HIV binding (50,51). More recent reports have demonstrated that the HIV envelope protein complexes with the CD4 molecule and can be immunoprecipitated with antibodies to HIV or the T4 molecule. HIV bound to T4 cells can block immunoprecipitation with mono-clonal antibodies to OKT4a, but not OKT4, suggesting that the epitope recognized by the former, but not the latter, anti-CD4 antibodies are involved in binding to the HIV envelope (51,52). The ability to infect cells with HIV from individuals not expressing the epitope recognized by OKT4 also suggests that this determinant is not required for viral binding (53,54). The central role of the CD4 molecule as a receptor for HIV is also supported by the observation that human-derived cell lines, which normally do not express CD4 and cannot be infected with HIV, may be rendered susceptible to infection by the transfection of the CD4 gene leading to expression of the CD4 molecule (55). Murine-derived cell lines transfected with the same CD4 protein are incapable of being infected with HIV, although binding to the viral envelope is achieved (55). Since murine cells that are transfected with cloned HIV DNA can support viral replication, other proteins may be required for viral uptake or propagation following binding to CD4.

C. Mechanisms of T-Cell Depletion

Binding of T4 to envelope proteins may have a role in the cytopathic effects of HIV. Cell lines that express the greatest density of T4 on their surfaces are the ones which appear to be most susceptible to cell death by HIV infection (52). T4 RNA expression is decreased in cell lines following infection with HIV. In addition, cells infected with HIV or with genetically engineered vaccinia virus producing recombinant HIV envelope proteins fuse with uninfected cells, which express CD4 molecules, form syncytia, and lyse (56,57). It is unclear whether this fusion process, or the formation of intracytoplasmic CD4-envelope complexes, or the production of virally induced cytotoxic factors is responsible for T4 cell depletion in vivo.

Depletion of the circulating pool of T4 lymphocytes in AIDS patients may reflect, in part, an inability of the bone marrow to adequately replenish precursor cells. This may be caused by cytotoxic or suppressive effects on marrow populations, depletion of precursor cells bearing the T4 phenotype, elimination of mature T4 cells, or absence of cytokines that regulate precursor cell differentiation. Little is known about differentiation of bone marrow cells in AIDS patients. A recent report showed that marrow aspirates from AIDS patients had decreased formation of colony-forming units of granulocytes-macrophages (CFU-GM) and burst-forming units-early erythroid (BFU-E) when cultured in the presence of sera from AIDS patients, but not in the presence of sera from seronegative individuals (58). Similar depression occurred when the cells were incubated with rabbit antisera raised to a recombinant *env* protein, suggesting that these nonlymphoid precursors are infected with HIV. HIV was recovered from one of four aspirates obtained from AIDS patients in that study.

Increased numbers of immature T8 lymphocytes with decreased ecto-5'-nucleotidase activity have been found in the peripheral blood of AIDS patients (59). Adenosine deaminase (ADA) production has been found to be elevated in circulating null cells of AIDS patients and may reflect an increase in an immature population, although ADA activity in T cells was unchanged in the same study (60). An increased population of non-CD4-bearing immature circulating cells (59,61) may reflect an unsuccessful attempt by the marrow to repopulate the peripheral blood with CD4 lymphocytes. Thymic dysplasia reported in AIDS patients could contribute to the failure to restore normal levels of T4 cells (62).

An alternative hypothesis for T-cell depletion has been proposed based on autoimmunity (63,64). It has been suggested that the HIV envelope mimics the physiological ligand for CD4, possibly a conserved stretch of Class II molecules. During the course of the normal immune response to the envelope protein, a cross-reactive autoimmune response to Class II–bearing cells may occur, as well as an antiidiotypic response to CD4-bearing cells. Consequently, CD4^{+}- and Class II–bearing cells may be eliminated by those responses. This theory provided some of the rationale for treating ADIS patients with cyclosporin (65).

D. T–Cell Function: Proliferation to Antigen and Mitogen

In association with T4 depletion, immunological functions of T4 and T8 cells are impaired. Multiple studies have reported increased spontaneous proliferative responses of T cells and depressed proliferative responses to T-cell mitogens and antigens in AIDS patients (1–3,30,31,33,35,66,67). The magnitude of these functional defects varies among individual patients, although, in general, proliferative responses to mitogens and antigens in more advanced AIDS patients, particularly those with OI, are more often depressed as compared with patients with KS, ARC, or PGL or asymptomatic individuals (35). The physiological mechanism for the depression in proliferative responses is not fully understood, although it appears that mitogen responses of unfractionated cell populations from AIDS patients can be reconstituted by increasing the relative proportion of T4 cells in culture (67). In contrast, proliferative responses to soluble antigens in purified T4-cell populations from AIDS patients remain depressed compared with cells derived from control individuals, implying that an intrinsic and selective defect in response to soluble antigen occurs in the T4 population. The site of the defect is unknown, but is not accountable for by an antigen presentation or processing defect. When monocytes from healthy control identical twins are used as a source for in vitro antigen presentation, T cells from these AIDS patients are still unresponsive to soluble antigen (S. Koenig, H. C. Lane, and A. S. Fauci, unpublished observations). Possible explanations for this observed defect include selective loss of a selective T4-cell population responding to soluble antigen, suppression of T4-cell function by soluble mediators or viral proteins without modulation of CD4 expression, or structural and/or functional changes in the antigen-specific receptor.

One study, utilizing crude preparations of viral proteins in vitro, showed inhibition of both antigen-stimulated antisheep red blood cell (SRBC) plaque-forming cell (PFC) production in peripheral blood lymphocyte (PBL) cultures and pokeweed mitogen (PWM)-induced PFC production when T cells were exposed to these preparations (68). These observations suggest a direct suppressive effect of the viral proteins on helper T cells, but production of soluble suppressor factors by T cells or macrophages is also possible.

The absence of a proliferative response to soluble antigen appears to be the earliest detectable change in immunological function in individuals infected with HIV, as many healthy HIV-seropositive individuals with circulating T4 cells in the normal range exhibit this diminution in antigen-specific responses (H. C. Lane, unpublished observations). The presence of this defect in an individual may be a harbinger of more severe immunological impairment and clinical deterioration.

Another in vitro immunological test, of unclear physiological importance, is the autologous mixed lymphocyte reaction, in which T4 cells proliferate in response to autologous non-T cells (69,70). This response has been shown to be depressed in both AIDS and ARC patients in the setting of either depressed or normal circulating T4 cell numbers and may be restored in some cases by the addition of IL-2. The mechanism responsible for the depressed proliferative response of these T4 cells may be similar to the defect in the response to soluble antigen.

Physiological immune functions associated with T4 lymphocytes are diverse since the population of cells expressing the CD4 molecule is comprised of multiple functional subpopulations of cells. Well-documented functions include "help" for the humoral immune response, "induction" of cells that mediate help and suppression, secretion of regulatory factors that act on lymphoid and nonlymphoid cells, and cell-mediated effector function by Class II-restricted cytotoxicity. Since T4 lymphocytes influence the funciton of other components of the immune system, their depletion or impairment would be expected to alter non-T4-cell-associated functions. For example, human lymphocyte antigen (HLA)-restricted cytotoxic T-cell responses against influenza and CMV are depressed early in the development of AIDS (71-74). Generation of Class I-restricted cytotoxicity by $CD8^+$ cells is dependent on intact T4-cell inducer function, and therefore these depressed T8 responses may be a consequence of T4-cell dysfunction. However, it is unknown whether depression of T8-lymphocyte cytotoxic function can be alleviated solely by T4-cell reconstitution or whether additional functional defects occur in the T8 population as well. In vitro T-cell-mediated cytotoxicity to CMV has been restored in cells derived from AIDS patients following incubation with interleukin-2 (IL-2) (75). This observation implies that T8-cytotoxic-cell function is intrinsically intact, and the lack of the proper inductive signals, possibly by T4 cells, may be responsible for depressed cytotoxic responses. This concept is also supported by the finding that not all T8 cytotoxic functions are impaired. Cytotoxic responses to alloantigens in AIDS patients appear to be relatively preserved (72). However, these alloreactive cells may be induced by an alternative pathway not requiring T4 cells, as has been demonstrated in the murine system (76), thus accounting for intact alloreactivity.

Common to the pathway of activation of T cells is the production of IL-2 and expression of IL-2 receptors (77). T-cell proliferation and function are dependent on the elaboration of these proteins. Conflicting results in different studies have documented both normal and decreased IL-2 receptor expression and IL-2 production of cells derived from AIDS patients following mitogenic stimulation (67,78–84). In one study, the decreased receptor expression, but not decreased IL-2 production, correlated in some of the clinical subgroups with the observed depression in cellular proliferation and T4-to-T8 ratios (80). There have been discrepencies in reports with regard to augmentation of mitogen-induced proliferative responses by the addition of exogenous IL-2, with some studies demonstrating reconstitution of responses (82,85) and others noting minor or no effects (70,79,86). In another study utilizing an antigen-stimulated system, impaired γ-interferon (IFN) generation by T cells could be enhanced 4.4- to 7.2-fold with supplemental recombinant (r)IL-2 (87) in cells from AIDS and ARC patients with decreased IL-2 production. Some of the difficulties in comparing these studies of IL-2 production and IL-2 receptor expression in HIV-seropositive individuals include differences in experimental design (e.g., both unfractionated and enriched cell populations have been used) and heterogeneity of clinical subjects. Insight into these questions may require use of in vitro models. Such an approach was used with the examination of HIV-infected cell lines for production of IL-2 messenger (m)RNA transcripts (88). HIV-infected tumor lines showed constituitive expression of mRNA which could be further induced with phytohemagglutinin (PHA) and phorbol-12 myristate acetate (PMA), thus suggesting normal IL-2 gene regulation in HIV-infected cells. Additional studies examining HIV-infected T4 cells derived from healthy individuals for IL-2 and IL-2 receptor mRNA and protein production are needed to determine whether this is a consistent observation in normal T4 cells infected with HIV. A recent study suggests a marked decrease in IL-2 mRNA expression, but preservation of IL-2 receptor mRNA production in normal T4 cells infected with HIV (H. C. Lane, personal communication). Mechanisms other than infection of these cells with HIV may be responsible for changes in IL-2 receptor expression, IL-2 production, and T4-cell function.

E. Suppressor Substances of T-Cell Function

Several groups have documented suppressive effects on T-cell responses by sera and plasma obtained from AIDS patients (89-92). One of the first studies showed that AIDS sera suppressed PHA and mixed-lymphocyte proliferative responses (MLR) of cells from healthy individuals (89). This suppressive activity of both types of proliferative responses was removed by absorbing the sera with SRBC; sera absorption with allogeneic cells removed suppressive activity in the MLR response. This suggested that the observed suppressor activity might be directed toward T-cell and major histocompatibility complex (MHC) determinants. In addition, some of these sera were able to suppress α-IFN-induced natural killer (NK) responses but not endogenous NK cytotoxicity.

Another report (90) described the presence of a substance in the plasma of AIDS and ARC patients that suppressed antigen- and mitogen-induced proliferative responses and IL-2 production and IL-2 receptor (Tac) expression in normal cells. In preliminary studies, this factor appeared to be greater than 70,000 molecular weight (MW) and not removed by passage over immunoglobulin (Ig) affinity columns.

A similar factor in the sera of AIDS patients which suppressed mitogen-induced IL-2 production by normal peripheral blood mononuclear cells (PBMC) was described and more thoroughly characterized (91). This activity was not removed by passage over anti-IgM or -IgG columns, was stable at pH 3 and 10 and at 60°C for 6 hr, and was not extractable by ether. It did not inhibit proliferation of tumor cell lines or prevent PHA-induced interferon production. Activity was not abrogated by the addition of IL-1 or indomethacin to cultures. Of interest, sera with comparable suppressive activity was found in one of five CMV-seropositive healthy males following experimental inoculation with CMV, suggesting that other pathogens may induce similar suppressive substances.

Suppressive factors derived from T cells of patients with AIDS have been produced by in vitro stimulation with concanvalin A (Con A) in the presence of monocytes (92). These factors could inhibit PWM-driven PFC responses and tetanus toxoid proliferative responses, but not Epstein-Barr virus (EBV)-induced proliferative responses. T-cell hybridomas, producing factors with similar suppressor activities, were also developed (93); the molecular weight of the major peak of suppressor activity appeared to be about 47,000 daltons. In contrast to certain sera which induced suppression of proliferative responses by inhibiting IL-2 production, these factors had no effect on the production of IL-2 by Jurkat cells or on the utilization of IL-2 by an IL-2-dependent T-cell clone.

There are conflicting reports regarding the presence of excessive suppressor activity in AIDS patients. One study, using unfractionated PBLs, reported the presence of suppressor cell activity in mitogen-induced proliferation assays in 12 of 21 AIDS patients when cells from these individuals were added in equal parts to normal control cells. However, this study did not attempt to control for the differences in the composition of the subpopulations of T cells added to these cultures (94). Another study examined the effects of PBLs and T cells from AIDS patients on PWM-induced Ig synthesis and found that these cells induced suppression that could be abrogated by irradiation. The addition of positively selected T8$^+$ cells from AIDS patients in a single concentration to normal PBLs suppressed Ig synthesis by 88% as compared with a mean suppression of 40% by T8$^+$ cells from normal controls. The authors conclude that OKT8$^+$ cells from AIDS patients have a greater suppressor activity on a per-cell basis compared to T8$^+$ cells from normals. A titration of T8$^+$ cells under similar experimental conditions, particularly in concentrations where normal T8$^+$ cells were not suppressive, might have strengthened these conclusions (94). When such an experiment was performed by other investigators, no significant differences in suppressor activity of T8$^+$ cells from two AIDS patients were noted in comparison to control T8$^+$ cells (66). Purified T8$^+$ cells from AIDS patients also did not appear to have greater suppressor activity in mitogen-induced proliferation assays (67). Thus, while HIV seropositive individuals may manifest increased suppressor activity in their unseparated PBLs because of increased numbers of T8$^+$ cells, purified populations of T8$^+$ cells do not appear to have greater suppressor effects.

III. MONOCYTES AND MACROPHAGES

While earlier studies of immune function in AIDS had focused on alterations in the T-cell compartment, subsequent findings suggested that monocytes and macrophages play a role in the pathogenesis of this disorder. It was first observed that promonocytic cell lines could be infected with HIV (45,96). The in vitro infection of these lines appeared to correlate with the expression of the CD4 molecule on the surface of the cells. The presence of CD4 protein on these tumor lines does not seem to be a consequence of their transformation, since normal circulating monocytes are found to express CD4 as well (97). The low level of CD4 expression on monocytes renders them susceptible to HIV infection in vivo. In vitro studies support this notion since circulating cells selected by adherence and expressing phenotypic characteristics of macrophages could be infected with HIV (98,99). Similarly, alveolar-derived macrophages appeared to be infected in vivo as detected by coculture techniques (100).

The in vitro (and possibly in vivo) propagation of particular HIV isolates may be dependent on their cell type of origin. Isolates obtained from adherent cells of brain and lung which were trypsin resistant were shown to be preferentially passaged through adherent macrophages derived from circulating monocytes (101). Conversely, isolates passaged through T cells maintained their tropism for this cell type. The mechanism responsible for this selectivity is unclear. Viral progeny acquire cellular-derived proteins, including those of the histocompatibility region on their envelope, during the process of budding, which may contribute to their subsequent cell association. In this regard, the LAV isolate could more readily infect B cells after initial passage through this cell type in coculture (102). There has been precedence for tissue tropism in particular strains of retroviruses. For instance, thymotropism has been observed for many strains of mink cell focus-forming murine leukemia viruses due to enhanced elements present in the long terminal repeat (LTR) segments of these viruses (103); similar events may be responsible for the tropism of some HIV strains for macrophages.

Direct evidence for in vivo infection of macrophages was demonstrated in examination of brain tissue from AIDS patients with encephalopathy. Earlier histological studies noted the appearance of multinucleated giant cells in brain specimens from these patients. By concomitant immunohistochemical and in situ hybridization techniques, and by transmission electron microscopy, cells of the monocyte/macrophages lineage were shown to be infected by HIV in brain tissue (104; Figure 2). Other studies performed at the same time supported this observation (105-107). To date, circulating monocytes have not been directly shown to be infected with HIV. However, reverse transcriptase activity could be demonstrated in cocultures of PHA-stimulated normal PBMC and T-cell-depleted (treatment with anti-T3) adherent cells obtained from PBMC of AIDS patients (98). Also, numbers of circulating T4+ monocytes were found to be depressed in patients with AIDS and lymphadenopathy as compared with healthy HIV sero-positive and seronegative subjects (108).

Abnormalities in some monocyte activities have been observed in cells obtained from HIV-infected individuals, but the majority of in vitro monocyte/ macrophage functions appear to be intact (109-117). Whether the altered functions that have been observed are a consequence of infection of these cells with HIV or reflect epiphenomena is unknown. Impaired in vitro chemotactic activity of circulating monocytes from AIDS patients and lymphadenopathy patients seems to be the most consistently observed defect (109,110). Impaired chemotaxis has been observed during other viral infections and in individuals with malignancy, and thus its clinical significance and association with HIV infection per se are unclear. In most studies, phagocytic activity (111), antibody-dependent cellular cytotoxicity (ADCC) (110), tumoricidal activity, and cytotoxicity against *Toxoplasma gondii* and *Chlamydia psittaci* in response to γ-IFN (112) and fungicidal activity against *Aspergillus fumigatus, Cryptococus neoformans*, and *Thermoascus crustaceus* (111) appeared to be intact in mono-cytes derived from individuals with AIDS or lymphadenopathy, although one group noted decreased candidacidal activity in their patients (115). IL-1 production, in general, appears to be preserved (116,117). Recent findings suggest that inhibitors of IL-1 are present in AIDS patients (117). A higher proportion

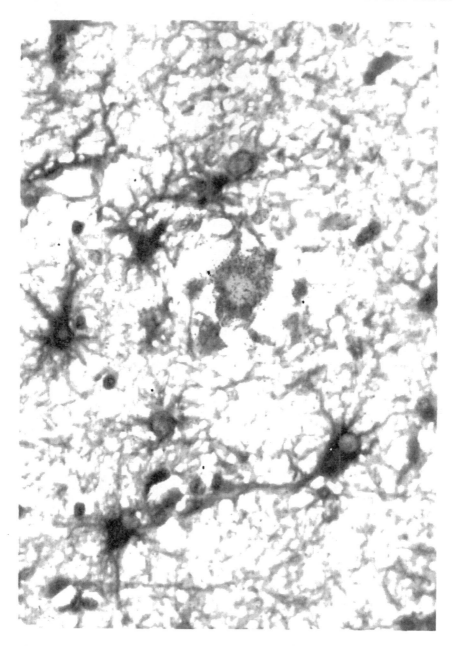

FIGURE 2 In situ hybridization with HIV RNA-specific probes on brain tissue. Astrocytes stained with the rabbit antibody to glial fibrillary acid protein surrounding an infected multinucleated giant cell. Silver grains indicate HIV RNA (hematoxylin and eosin, 100X). [Adapted from Koenig et al. (104).]

of lymphadenopathy patients compared to normal individuals had a low pro-liferative response to anti-T3 antibody, which could be augmented by the addition of normal allogeneic macrophages, suggesting a defect of accessory cell function in this T-cell-dependent response (118). Minimal abnormalities were seen in presentation of soluble antigen by circulating monocytes in a syngeneic system utilizing highly enriched populations derived from healthy identical twins (S. Koenig, H. C. Lane, and A. S. Fauci, unpublished observations). Similar observations were reported by other using Class II MHC-compatible presenting cells (114). Thus, although decreased numbers of circulating HLA-DR antigen-bearing monocytes (119) and Langerhans' cells in skin have been reported (120), antigen-presentation defects may not be a major problem in HIV infection.

IV. B CELLS

B-cell function and humoral immunity are altered in patients with AIDS. Patients characteristically develop polyclonal B-cell activation as manifested by hypergammaglobulinemia, spontaneous B-cell proliferation, and increased numbers of PFC (1–3,66,121–125). Markedly increased serum levels of IgG, IgA, and IgD, with inconsistent increases of IgM, have been noted in patients with AIDS (30,66,120,121). Elevation of IgM production is more frequently seen in children with AIDS (126). Increased spontaneous transformation of B cells by EBV, probably secondary to impaired T-cell and NK-cell surveillance, has been noted (127). This undoubtedly is associated with an increased incidence of B-cell lymphomas in these patients.

In contrast to their increased spontaneous activities, B cells from AIDS patients have impaired in vitro proliferative responses to antigens and mitogens (66). Likewise, PWM-induced Ig synthesis is decreased. In vivo specific antibody production is reduced in both primary and secondary immunizations of AIDS patients with protein and polysaccharide antigens (66,125).

In vitro studies have attempted to dissect the mechanisms of B-cell dys-function. Early studies demonstrated that EBV-transformed B cells can be infected with HIV (101), possibly through expression of a CD4 receptor, but nontransformed B cells have not been shown to be infectable with HIV, nor have HIV-infected B cells been identified in HIV-seropositive individuals. Despite the inability to infect these cells, HIV appears to affect in vitro B-cell functions in manners that recapitulate in vivo observations. Intact infectious HIV and disrup-ted viral particles will induce polyclonal activation with proliferation and Ig synthesis to levels comparable to those of other B-cell mitogens (128–130). Whether this polyclonal response is due to a direct effect on B cells (130; Figure 3) or is a T-cell-dependent response (68,129) is unresolved. The possibility that induction of B-cell proliferation is enhanced by the presence of T cells as a result

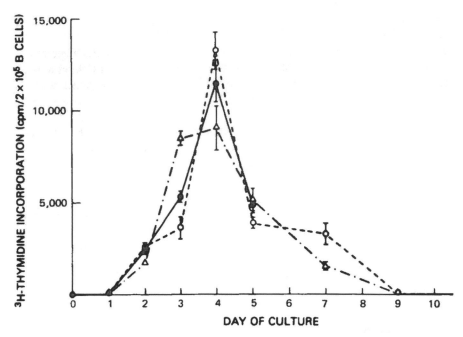

FIGURE 3 Kinetics of HIV-induced B-cell proliferation. Virus-containing supernatants were derived from the HIV-infected A3.01 cell line. The stock supernatant contained virus at a concentration of 10^4 infectious units per milliliter. The B cells (5×10^6) were centrifuged and the pellet was resuspended in 1 ml of virus-containing supernatant and incubated for 1 hr at $37°C$. The cells were then washed twice with RPMI 1640 and resuspended in RPMI 1640 containing 10% fetal calf serum (FCS) at 2×10^6 cells per milliliter. Portions (0.1 ml) were placed into round-bottom microtiter dishes already containing 0.1 ml of RPMI 1640 with 10% FCS in each well. The cultures were incubated at $37°C$ at 7% CO_2. On sequential days, the cultures were pulsed for 4 hr with 2 μCi of [^{3}H]thymidine. Proliferation is expressed as cpm per 2×10^5 B cells. The figure depicts three of six such experiments. Each point shows the average ($\pm$SEM) for counts from six wells. [Adapted from Schnittman et al. (130).]

of "presentation" to B cells of HIV antigen bound to T cells was not addressed in these studies.

In addition to the stimulatory effects of HIV on B cells in vitro, HIV was found to suppress EBV-induced PFC responses of purified B-cell populations (68). Since EBV is mitogenic for B cells in the absence of T cells or accessory cells, this suggests a direct effect of HIV on B cells. It would be of interest to see if CD4 molecules are induced in these short-term B-cell cultures stimulated with EBV, or if this activity is mediated through another receptor with a lower affinity for HIV.

V. NATURAL KILLER (NK) CELLS

NK cells comprise a circulating population consisting primarily of large granular lymphocytes that are capable of lysing virally infected cells and allogeneic and xenogeneic tumor cells. They are felt to have a central role in surveillance against spontaneously developing tumors and viral infections in vivo.

NK-cell function has been found to be frequently depressed in AIDS patients in comparison to healthy HIV-seronegative controls (75,131–135). Most laboratories have utilized the lysis of tumor cell lines such as K562 or U937 as a standard of NK function. The numbers of circulating NK cells do not appear to be significantly diminished in HIV-seropositive patients, and these cells are capable of binding target cells. Thus, depressed NK function appears to be secondary to a postbinding or "triggering" defect (136). NK activity of cells from AIDS patients may be restored to normal levels by incubation with IL-2. NK-cell function can also be enhanced by culture with Con A or with a phorbol ester and calcium ionophore (137). Therefore, NK cells from AIDS patients appear to be intrinsically intact to perform their cytolytic functions. A lack of IL-2 receptor expression or IL-2 production or release may account for the impaired function of NK cells in AIDS patients.

VI. MISCELLANEOUS IMMUNOLOGICAL DEFECTS

Other factors have been proposed to account for some of the immune defects seen in AIDS patients. Along with the hypergammaglobulinemia, circulating autoantibodies (138–142) and immune complexes (143,144) have been detected in patients with AIDS. Prior to the identification of HIV, some investigators had suggested that the specificities for some of these antibodies were for lymphocytes, particularly for those bearing OKT4 and OKT11 epitopes, and that these antibodies were responsible for CD4-cell depletion. However, these antibodies are present in low titers and their presence does not correlate with in vivo cell depletion.

The clinical consequences of elevated circulating immune complexes in AIDS patients are not entirely clear. Thrombocytopenia and peripheral neuropathies are often seen in HIV-infected individuals, and these conditions have been associated with the presence of immune complexes. However, an increased incidence of classic immune complex disease or vasculitis has not been reported in AIDS.

An elevated acid-labile interferon level was one of the early abnormalities detected in patients with AIDS but its clinical significance is unclear (145,146). One group has reported that impaired α-IFN production in vitro along with CD4-cell depletion in seropositive individuals is a marker for clinical progression of HIV-seropositive individuals to full-blown AIDS with opportunistic infections (137).

VII. IMMUNITY TO HIV

The life expectancy of an individual with the diagnosis of AIDS is approximately 2 years or less, depending on the initial clinical manifestations. While there appears to be a long latency period from an initial infection with HIV to the development of AIDS or ARC, not all individuals infected with HIV will necessarily become clinically symptomatic. By current estimates, approximately 30% of HIV-seropositive individuals will develop an AIDS-associated condition within 5 years of infection. It is unknown what proportion of patients will develop disease after longer periods of time.

The conditions that dictate the clinical sequelae following HIV infection are unclear. Some may be related to the virulence of a particular HIV isolate, the size of the initial inoculum of virus, or the types and nature of other pathogens coinfecting these individuals. However, a successful immunological response to HIV resulting in the elimination or containment of the virus undoubtedly will be a major determinant of the clinical outcome in an individual, although the conditions for development of effective immunity to HIV have not been established.

Immunological functions that may be naturally operative against HIV and HIV-infected cells include humoral responses (neutralizing and cytolytic antibodies) and cell-mediated responses (cytotoxic T cells, antibody-dependent cytotoxicity, and NK-cell activity).

Most HIV-infected individuals develop a humoral immune response to HIV and produce antibodies to multiple viral proteins (150,151). Usually, an individual will continue to produce HIV-specific antibody throughout his life, although loss of this antibody may occur late in the clinical course. Since HIV-specific antibodies are detectable in both healthy and symptomatic patients, the humoral immune response seems to be inadequate in preventing HIV propagation and clinical progression. Neutralizing antibodies have been detected in low titers in both AIDS patients and healthy seropositive individuals, suggesting that their presence in such low quantities may be clinically insignificant (152–155). In some cases, preferential neutralization of one of a pair of HIV isolates with serum from a single individual has been noted in in vitro assays (154,155). Neutralizing antibodies have been produced in animals immunized with recombinant envelope proteins expressed in both bacterial and mammalian vector systems and with purified native or nonglycosylated envelope proteins (156, 157). The neutralizing titers achieved were strain specific and in low titers, comparable to those detected in infected patients.

Other immunizing protocols may be necessary to induce neutralizing antibodies in higher titers. Fractionation of sera derived from patients with neutralizing activity on gp 120 affinity columns demonstrated neutralization activity in both adherent and effluent fractions (155). The flow-through fractions neutralized both homologous and divergent strains of HIV, while the absorbed fractions neutralized only homologous isolates. This suggests that either a low-affinity neutralizing antibody for gp 120 is contained in the effluent fraction or the antibody is directed against determinants of other viral proteins. In this regard, cross-reacting antibodies to $alpha_1$-thymosin and p17, a *gag*-associated protein, have been detected in seropositive individuals and have been suggested to manifest activity against HIV-infected cells (158). It appears that neutralizing antibodies are not raised against the amino-terminal half of p41 or gp120 since animals immunized with recombinant proteins representing these regions induced antibodies that could bind, but not neutralize, HIV (157).

Limited numbers of patients have been examined for neutralizing antibodies. The reports in the current literature may represent only a subgroup of HIV-seropositive individuals whose immune systems may be already compromised and who are destined to become clinically symptomatic. Identifying healthy HIV-seropositive individuals with unusually high titers of neutralizing antibodies with follow-up for HIV isolation and disease progression may provide insight into the clinical relevance of neutralizing antibodies. The development of clinical disease and persistent viremia in an individual with high titers of neutralizing antibody would clearly suggest that humoral immunity has only a limited role in inhibiting HIV propagation. One explanation for continued virus replication in the presence of neutralizing antibodies is that mutant HIV strains may emerge which display antigens that can no longer be recognized by the neutralizing antibodies. An in vitro model has demonstrated the selection of HIV isolates in cultures with neutralizing antibodies (159). Similarly, in some lentiviral diseases, especially equine infectious anemia, new viral strains are selected in a cyclical pattern following the production of neutralizing antibodies (160).

It is widely accepted that the cellular immune response to viruses, including retroviruses, is important in preventing the development of their associated diseases, particularly malignancies. For example, it has been shown recently in a Friend leukemia system that animals who developed cell-mediated immunity to viral proteins expressed in a recombinant form did not develop leukemia following challenge with Friend virus (161). In contrast, animals who developed only neutralizing antibodies were not protected. The nature of the cellular immunity to HIV-infected cells is only beginning to be characterized.

ADCC has been found in sera from HIV-seropositive individuals with the highest activity noted in sera derived from healthy seropositive individuals (162). Sera developed from AIDS patients with OI had the least activity, while sera from those with only KS had intermediate ADCC responses. Sera that contained antibodies directed toward 120 gp and 24 gp, as detected by Western blot analysis, mediated ADCC more consistently compared to sera that contained antibody only to the gp 120 (162). A recent study of a cohort of asymptomatic serpositive individuals followed for 3 years showed a decreased rate of progression to clinically overt disease in individuals with antibody to p24 (163). However, there was no correlation between antibody to this protein and the ability of the sera to neutralize HIV in vitro. It is possible that immune protection might be conferred in individuals with antibody to p24 through other immune mechanisms. The role of antibodies to proteins coded by the *gag* region requires further investigation.

NK-cell function against HIV-infected cells has been investigated using enriched populations of large granular lymphocytes (LGL). Minimal cytotoxic activity was detected in this cell population. However, cytotoxicity was markedly increased when cells were activated in the presence of IL-2 (164). Similarly, unfractionated PBMC from healthy HIV-seronegative individuals and from patients with AIDS could lyse HIV-infected target cells in an HLA-unrestricted manner if they were first stimulated with IL-2 (S. Koenig et al., submitted). This suggests that lymphokine-activated killer (LAK) cells could have a role in therapy in AIDS, as has been suggested for individuals with advanced malignancies (165).

Specific cytotoxic T-cell responses against HIV-infected target cells have been detected in the peripheral blood of some healthy HIV-seropositive individuals but not in HIV-seronegative individuals or in patients with AIDS (S. Koenig et al., submitted). However, cytotoxic T-cell responses to HIV-infected cells could be detected in two AIDS patients following syngeneic bone marrow transplantation and lymphocyte transfers from their healthy identical twins. In one of these individuals, this activity was consistently detectable over a 5-month period. Although bone marrow transplantation appears not to be feasible for the majority of AIDS patients because of the risks of immunosuppressive therapy for allogeneic transplants in this already immunologically compromised population, other forms of cellular adoptive therapy are possible.

Finally, the characterization of viral proteins necessary for generation of a cellular immune response to HIV will assist in the development of a vaccine for high-risk populations and possibly healthy seropositive individuals whose immunological functions are intact.

VIII. MODEL OF IMMUNOPATHOGENESIS

Our understanding of the in vivo pathogenesis of AIDS is still fragmented. However, based on our current knowledge of the clinical and research observations both in HIV and in other lentiviral diseases, the following may serve as a framework for examining the development of immune defects in AIDS.

An individual may be initially infected with HIV in several ways. First, infected lymphocytes and/or monocytes contained within semen, vaginal secretions, or blood may pass directly into capillaries of the newly infected individual. Alternatively, sexual transmission of HIV may occur through the direct infection of cells other than those of hematopoietic origin, including epithelium of rectum, colon, vagina, and cervix. These infected cells may then transmit the virus to the draining lymphoid tissue. Although there are no in vivo observations supporting this notion, tumor lines derived from rectum and colon have been infected with HIV in vitro (166).

These few infected and passaged cells would be primarily sequestered in lymphatic tissue, initiating a primary cellular and humoral immune response. In the course of this response, virions are passaged to other lymphocytes, dendritic cells, and monocytes by cellular contact and through phagocytosis of cellular debris. Viral antigenemia, with formation of antigen-antibody complexes, may occur in some individuals who develop a mononucleosis-like syndrome with a dermatitis and arthralgias.

After the primary HIV infection, a period of either true viral latency and/ or low-level chronic infection occurs in most individuals. A reservoir for HIV infection may be created when viral particles infect nonactivated lymphocytes and monocytes. Newly formed DNA transcripts of the virus integrated in the cellular genome could remain biologically silent until the cell receives an activation signal. This signal may be initiated by a specific antigen, lymphokine (e.g., IL-1, IL-2, γ-IFN), or by other pathogens tropic for the same cell type. In this regard, the isolation of HIV from patients' cells is best achieved by coculture with normal mononuclear cells stimulated with PHA in which these same lymphokines are generated. When latently infected cells receive a physiological activation signal, they will transcribe and translate viral RNA and proteins and new HIV progeny will be formed and released. Cell death may be the result of either direct cytopathic effects of the virus, decrease in production of IL-2 required to sustain cellular growth, fusion with uninfected CD4-bearing cells in the local environment, or through lysis by cytotoxic cells. A cycle for the perpetuation of HIV infection is established with the dissemination of new virions to other cells. New virions can be passed to other cells through direct contact or as free particles.

Infections with other organisms may precipitate HIV expression. Numerous pathogens, including CMV and mycobacteria, which coinfect HIV-infected individuals, are harbored by macrophages. These infected cells will release lymphokines that could activate lymphocytes latently infected with HIV and induce virus replication. In addition, the coinfection of latently infected monocytes and lymphocytes with other pathogens may contribute to viral propagation. The rate-limiting factor for lymphocyte depletion could be the activation of these latently infected cells. Thus, in the presence of a large opportunistic pathogen burden, a patient's time course for CD4-cell depletion, immunodeficiency, and clinical progression may be abbreviated. This may account for markedly lower rates of isolation of HIV from seropositive hemophiliacs and blood recipients as compared to sexually active homosexual men who may be infected with many other pathogens as well. Also, we have observed marked

HIV viral expression in brain specimens coinfected with other pathogens (e.g., progressive multifocal leukoencephalopathy [PML] due to JC virus and CMV) (104; S. Koenig, C. Fox, and A. S. Fauci, unpublished observation).

As a consequence of the loss of T4 cells, the function of many other cells that require induction by T4 cells is impaired. T8-cell-mediated cytotoxicity, Ig production by B cells, and possibly NK responses are some of the immune functions that are impaired by the T4-cell loss. While T4 depletion is a major factor contributing to the immunodeficiency of these patients, immunological dysfunction precedes the onset of detectable depletion of circulating lymphocytes. The failure of circulating T4 cells to respond to soluble protein antigens is an early detectable immunological defect whose cause is unclear. Stimulation of the CD4 molecule on cells by viral envelope proteins in the absence of concomitant antigen-specific stimulation of the T3-Ti receptor complex may render these cells refractory to further soluble antigen stimulation, but amenable to activation by an alternative pathway (e.g., by mitogens). This hypothesis is supported by the observation that some antibodies directed against CD4 inhibit their subsequent stimulation with anti-T3 (167). Alternatively, HIV envelope or virus, particularly during collaboration of HIV-infected macrophages and T cells in generation of an immune response, may block the physiological ligand (i.e., Class II molecule), which is necessary for the response to soluble antigens. Another possible mechanism is that increased quantities of soluble suppressor substances may be released as a consequence of polyclonal activation of the immune response or through stimulation of HIV-infected cells. Under these circumstances, proliferative responses to soluble antigens may be more affected than mitogenic responses.

An impaired specific humoral immune response may be a consequence of loss of T4 helper function plus coexisting polyclonal B-cell activation. The generation of polyclonal responses may occur by direct HIV stimulation, through T-cell products, and by other pathogens. In this regard a recently described recombinant protein called neuroleukin, which shares some amino acid homology with HIV, can also induce polyclonal activation of B cells (168).

Further delineation of the effects of HIV on the immune system will, it is hoped, be achieved through examination of the effects of purified HIV proteins and synthesized peptides with homology to established HIV sequences and through the use of nonlytic mutant forms of the virus. A better understanding of the early events causing immune compromise soon after HIV infection and before significant T-cell loss occurs and the role that other pathogens play in this process may help to guide therapeutic strategies for the healthy HIV-infected individual. If control of HIV replication can be achieved with agents that have efficacy against both HIV and DNA viruses, particularly against the herpes group, then HIV-seropositive individuals with relatively intact immune systems may have a better chance for long-term survival.

REFERENCES

1. Gottlieb, M. S., Schroff, R., Schanker, H. M., et al. (1981). *Pneumocystis carinii* pneumonia and mucosal candidiasis in previously healthy homosexual men: Evidence of a new acquired cellular immunodeficiency. *N. Engl. J. Med.* 305:1425-1431.

2. Masur, H., Michelis, M. A., Greene, J. B., et al. (1981). An outbreak of community-acquired *Pneumocystis carinii* pneumonia: Initial manifestations of cellular immune dysfunction. *N. Engl. J. Med.* 305:1431-1438.

3. Siegel, F. P., Lopez, C., Hammer, G. S., et al. (1981). Severe acquired immunodeficiency in male homosexuals, manifested by chronic perianal ulcerative herpes simplex lesions. *N. Engl. J. Med.* 305:1439-1444.

4. Institute of Medicine, National Academy of Sciences (1986). *Confronting AIDS*. Washington, DC, National Academy Press.

5. Barré-Sinoussi, F., Chermann, J. C., Rey, F., et al. (1983). Isolation of a T lymphotropic retrovirus from a patient at risk for acquired immune deficiency syndrome (AIDS). *Science* 220:868-871.

6. Popovic, M., Sarngadharan, M. G., Read, E., and Gallo, R. C. (1984). Detection, isolation, and continuous production of cytopathic retroviruses HTLV-III) from patients with AIDS and pre-AIDS. *Science* 224:497-500.

7. Levy, J. A., Hoffman, A. D., Kramer, S. M., Landis, J. A., Shimabukura, J. M., and Oshiro, L. S. (1984). Isolation of lymphocytopathic retroviruses from San Francisco patients with AIDS. *Science* 225:840-842.

8. Coffin, J., Haase, A., Levy, J. A., et al. (1986). Human immunodeficiency viruses. *Science* 232:697.

9. Sarngadharan, M. G., Popovic, M., Bruch, L. Schüpbach, J., and Gallo, R. C. (1984). Antibodies reactive with human T-lymphotropic retroviruses (HTLV-III) in the serum of patients with AIDS. *Science* 224:506-508.

10. Brun-Vezinet, F., Rouzioux, C., Barré-Sinoussi, F., et al. (1984). Detection of IgG antibodies to lymphadenopathy syndrome. *Lancet* 1:1253-1256.

11. Centers for Disease Control (1986). *Acquired Immunodeficiency Syndrome Weekly Surveillance Report.* Sept. 9, 1986. Atlanta, GA, Centers for Disease Control.

12. Brun-Vezinet, F., Rey, M. A., Katlama, C., et al. (1987). Lymphadenopathy-associated virus type 2 in AIDS and AIDS-related complex. *Lancet* 1:128-132.

13. Clavel, F., Guyader, M., Guetard, D., et al. (1986). Molecular cloning and polymorphism of the human immune deficiency virus type 2. *Nature* 324:691-695.

14. Kanki, P. J., McLane, M. F., King, N. W., Jr., Et al. (1985). Serologic identification and characterization of a macaque T-lymphotropic retrovirus closely related to HTLV-III. *Science* 228:119-121.

15. Kanki, P. J., Alroy, J., and Essex, M. (1985). Isolation of T-lymphotropic retrovirus related to HTLV-III/LAV from wild-caught African green monkeys. *Science* 230:951-954.

16. Kanki, P. J., Barin, F., M'Boup, S., et al. (1986). New human T-lympho-tropic retrovirus related to simian T-lymphotropic virus type III (STLV-IIIAGM). *Science* 232:238–243.

17. Kornfeld, H., Riedel, N., Viglianti, G. A., Hirsch, V., and Mullins, J. I. (1987). Cloning of HTLV-4 and its relation to simian and human immuno-deficiency viruses. 326:610–613.

18. Poon, M-C., Landay, A., Prasthofer, E. F., and Stagno, S. (1983). Acquired immunodeficiency syndrome with *Pneumocystis carinii* pneumonia and *Mycobacterium avium-intracellulare* infection in a previously healthy pa-tient with classic hemophilia. *Ann. Intern. Med.* 98:287–290.

19. Elliott, J. L., Hoppes, W. L., Platt, M. S., Thomas, J. G., Patel, I. P., and Gansar, A. (1983). The acquired immunodeficiency syndrome and *Myco-bacterium avium-intracellulare* bacteremia in a patient with hemophilia. *Ann. Intern. Med.* 98:290–293.

20. Hymesk, Cheung, T., Greene, J. B., et al. (1981). Kaposi's sarcoma in homosexual men: A report of eight cases. *Lancet* 2:598–600.

21. Fauci, A. S., Masur, H., Gelmann, E. P., et al. (1985). The acquired im-munodeficiency syndrome: An update. *Ann. Intern. Med.* 102:800–813.

22. Reinherz, E., O'Brien, C., Rosenthal, P., et al. (1980). The cellular basis for viral-induced immunodeficiency: Analysis by monoclonal antibodies. *J. Immunol.* 125:1264–1274.

23. Rosen, C. A., Sodroski, J. G., Goh, W. C., et al. (1986). Post-transcriptional regulation accounts for the *trans*-activation of the human T-lymphotropic virus type III. *Nature* 319:555–559.

24. Gendelman, H. E., Phelps, W., Feigenbaum, L., et al. (1986). *Trans*-activa-tion of the human immunodeficiency virus long terminal repeat sequence by DNA viruses. *Proc. Natl. Acad. Sci. USA* 83:9759–9763.

25. Mosca, J. D., Bednarik, D. P., Raj, N. B. K., et al. (1987). Herpes simplex virus type-q can reactivate transcription of latent human immunodeficiency virus. *Nature* 325:67–70.

26. Kornfeld, H., Van de Stouwe, R. A., Lange, M., et al. (1982). T-lympho-cyte subpopulations in homosexual men. *N. Engl. J. Med.* 307:729.

27. Goldsmith, J. C., Moseley, P. L., Monick, M., et al. (1983). T-lymphocyte subpopulation abnormalities in apparently healthy patients with hemo-philia. *Ann. Intern. Med.* 98:294–296.

28. Pinchign, A. J., Jeffries, D. J., Donaghy, M., et al. (1983). Studies of cellular immunity in male homosexuals in London. *Lancet* 2:126–130.

29. Detels, R., Schwartz, K., Visscher, B. R., et al. (1983). Relation between sexual practices and T-cell subsets in homosexually active men. *Lancet* 1: 609–611.

30. Ammann, A. J., Abrams, D., Conant, M., et al. (1983). Acquired immune dysfunction in homosexual men: Immunologic profiles. *Clin. Immunol. Immunopathol.* 27:315–325.

31. Rubinstein, A., Sicklick, M., Gupta, A., et al. (1983). Acquired immuno-deficiency with reverse T4/T8 ratios in infants born to promiscuous and drug-addicted mothers. *JAMA* 249:2350–2356.

32. Fahey, J. L., Prince, H., Weaver, M., et al. (1984). Quantitative changes in T helper or T suppressor/cytotoxic lymphocyte subsets that distinguish acquired immune deficiency syndrome from other immune subset disorders. *Am. J. Med.* 76:95–100.

33. Cavaille-Coll, M., Messiah, A., Klatzmann, D., et al. (1984). Critical analysis of T cell subset and function evaluation in patients with persistent generalized lymphadenopathy in groups at risk for AIDS. *Clin. Exp. Immunol.* 57: 511–519.

34. Biggar, R. J., Melbye, M., Ebbesen, P., et al. (1984). Low T-lymphocyte ratios in homosexual men. *JAMA* 251:1441–1446.

35. Lane, H. C., Masur, H., Gelmann, E. P., et al. (1985). Correlation between immunologic function and clinical subpopulations of patients with the acquired immune deficiency syndrome. *Am. J. Med.* 78:417–422.

36. Mittelman, A., Wong, G., Safai, B., Myskowski, P., Gold, J., and Koziner, B. (1985). Analysis of T cell subsets in different clinical subgroups of patients with the acquired immune deficiency syndrome. *Am. J. Med.* 78:951–956.

37. Nicholson, J. K. A., McDougal, J. S., Spira, T. J., Cross, G. D., Jones, B. M., and Reinherz, E. L. (1984). Immunoregulatory subsets of the T helper and T suppressor cell populations in homosexual men with chronic unexplained lymphadenopathy. *J. Clin. Invest.* 73:191–201.

38. Nicholson, J. K. A., McDougal, J. S., and Spira, T. J. (1985). Alterations of functional subsets of T helper and T suppressor cell populations in acquired immunodeficiency syndrome (AIDS) and chronic unexplained lymphadenopathy. *J. Clin. Immunol.* 5:269–274.

39. Wood, G. S., Burns, B. F., Dorfman, R. F., and Warnke, R. A. (1986). In situ quantitation of lymph node helper, suppressor, and cytotoxic T cell subsets in AIDS. *Blood* 67:596–603.

40. Morimoto, C., Letvin, N. L., Distaso, J. A. Aldrich, W. R., and Schlossman, S. F. (1985). The isolation and characterization of the human suppressor inducer T cell subset. *J. Immunol.* 134:1508.

41. Morimoto, C., Letvin, N. L., Boyd, A. W., Hagan, M., Brown, H. M., Kornacki, M. M., and Schlossman, S. F. (1985). The isolation and characterization of the human helper inducer T cell subset. *J. Immunol.* 134:3762.

42. Zagury, D., Bernard, J., Leonard, R., et al. (1985). Long-term cultures of HTLV–III-infected T cells: A model of cytopathology of T-cell depletion in AIDS. *Science* 2:850–853.

43. McDougal, J. S., Mawle, A., Cort, S. P., et al. (1985). Cellular tropism of the human retrovirus HTLV–III/LAV. *J. Immunol.* 135:3151–3162.

44. Margolick, J. B., Volkman, D. J., Folks, T. M., and Fauci, A. S. (xxxx). Amplification of HTLV–III/LAV infection by antigen-induced activation of T cells and direct suppression by virus of lymphocyte blastogenic responses.

45. Popovic, M., Read-Connole, E., and Gallo, R. C. (1984). T4 positive human neoplastic cell lines susceptible to and permissive for HTLV–III. *Lancet* 1: 79–80.

46. Klatzmann, D., Barré-Sinoussi, F., Nugeyre, M. T., et al. (1984). Selective tropism of lymphadenopathy associated virus (LAV) for helper-inducer T lymphocytes. *Science* 225:59–63.

47. Folks, T. M., Benn, S., Rabson, A., et al. (1985). Characterization of a continuous T-cell line susceptible to the cytopathic effects of the acquired immunodeficiency syndrome (AIDS)-associated retrovirus. *Proc. Natl. Acad. Sci. USA* 82:4539.

48. Klatzmann, D., Champagne, E., Chamaret, S., et al. (1984). T-lymphocyte T4 molecule behaves as the receptor for human retrovirus LAV. *Nature* 312:767–768.

49. Dalgleish, A. G., Beverly, C. L., Clapham, P. R., Crawford, D. H., Greaves, M. F., and Weiss, R. A. (1984). The CD4 (T4) antigen is an essential component of the receptor for the AIDS retrovirus. *Nature* 312:763–767.

50. Sattentau, Q. J., Dalgleish, A. G., Weiss, R. A., and Beverley, P. C. L. 91986). Epitopes of the CD4 antigen and HIV infection. *Science* 234:1120.

51. McDougal, J. S., Kennedy, M. S., Sligh, J. M., Cort, S. P., Mawle, A., and Nicholson, J. K. A. (1985). Binding of HTLV-III/LAV to T4$^+$ T cells by a complex of the 110K viral protein and the T4 molecule. *Science* 231:382–385.

52. Hoxie, J. A., Alpers, J. D., Rackowski, J. L., et al. (1986). Alterations in T4 (CD4) protein and mRNA synthesis in cells infected with HIV. *Science* 234:1123–1127.

53. Hoxie, J. A., Flaherty, L. E., Haggarty, B. S., and Rackowski, J. L. (1986). Infection of T4 lymphocytes by HTLV-III does not require expression of the OKT4 epitope. *J. Immunol.* 136:361–363.

54. Folks, T., Justement, S., Mitchell, S. R., Cupps, T. R., Katz, P., Maples, J., and Fauci, A. S. (1987). The T4 epitope is not required for a normal replicative cycle of human immunodeficiency virus. *J. Infect. Dis.* 155:592–593.

55. Maddon, P. J., Dalgleish, A. G., McDougal, J. S., Clapham, P. R., Weiss, R. A., and Axel, R. (1986). The T4 gene encodes the AIDS virus receptor and is expressed in the immune system and the brain. *Cell* 47:333–348.

56. Lifson, J. D., Feinberg, M. B., Reyes, G. R., et al. (1986). Induction of CD4-dependent cell fusion by the HTLV–III/LAV envelope glycoprotein. *Nature* 323:725–728.

57. Sodroski, J., Goh, W. C., Rosen, C., Campbell, K., and Haseltine, W. A. (1986). Role of the HTLV–III/LAV envelope in syncytium formation and cytopathicity. *Nature* 322:470–474.

58. Donahue, R. E., Johnson, M. M., Zon, L. I., Clark, S. C., and Groopman, J. E. (1987). Suppression of in vitro haematopoiesis following human immunodeficiency virus infection. *Nature* 326:200–205.

59. Winkelstein, A., Klein, R. S., Evans, T. L., Dixon, B. W., Holder, W. L., and Weaver, L. D. (1985). Defective in vitro T cell colony formation in the acquired immunodeficiency syndrome. *J. Immunol.* 134:151–156.

60. Murray, J. L., Loftin, K. C., Munn, C. G., Reuben, J. M., Mansell, P. W. A., and Hersh, E. M. (1985). Elevated adenosine deaminase and purine nucleoside phosphorylase activity in peripheral blood null lymphocytes from patients with acquired immune deficiency syndrome. *Blood* 65:1318-1323.

61. Tedder, T. F., Crain, M. J., Kubagawa, H., Clement, L. T., and Cooper, M. D. (1985). Evaluation of lymphocyte differentiation in primary and secondary immunodeficiency diseases. *J. Immunol.* 135:1786-1791.

62. Elie, R., Larouch, A. C., Arnoux, R., et al. (1983). Thymic dysplasia in acquired immunodeficiency syndrome. *N. Engl. J. Med.* 308:341-342.

63. Klatzman, D., and Montagnier, L. (1986). Approaches to AIDS therapy. *Nature* 319:10-11.

64. Ziegler, J. L., and Stites, D. P. (1986). Hypothesis: AIDS is an autoimmune disease directed at the immune system and triggered by a lymphotropic retrovirus. *Clin. Immunol. Immunopathol.* 41:305-313.

65. Walgate, R. (1985). Politics of premature French claim of cure. *Nature* 318:3.

66. Lane, H. C., Masur, H., Edgar, L. C., Whalen, G., Rook, A. H., and Fauci, A. S. (1983). Abnormalities of B-cell activation and immunoregulation in patients with the acquired immunodeficiency syndrome. *N. Engl. J. Med.* 309:453-458.

67. Lane, H. C., Depper, J. M., Greene, W. C., Whalen, G., Waldmann, T. A., and Fauci, A. S. (1985). Qualitative analysis of immune function in patients with the acquired immunodeficiency syndrome. *N. Engl. J. Med.* 313:79-84.

68. Pahwa, S., Pahwa, R., Good, R. A., Gallo, R. C., and Saxinger, C. (1986). Stimulatory and inhibitory influences of human immunodeficiency virus on normal B lymphocytes. *Proc. Natl. Acad. Sci. USA* 83:9124-9128.

69. Smolen, J. S., Bettelheim, P., Köller, U., et al. (1985). Deficiency of the autologous mixed lymphocyte reaction in patients with classic hemophilia treated with commercial factor VIII concentrate. *J. Clin. Invest.* 75:1828-1834.

70. Gupta, S., Gillis, S., Thornton, M., and Goldberg, M. (1984). Autologous mixed lymphocyte reaction in man. XIV. Deficiency of the autologous mixed lymphocyte reaction in acquired immune deficiency syndrome (AIDS) and AIDS related complex (ARC). In vitro effect of purified interleukin-1 and interleukin-2. *Clin. Exp. Immunol.* 58:395-401.

71. Shearer, G. M., Payne, S. M., Joseph, L. J., and Biddison, W. E. (1984). Functional T lymphocyte immune deficiency in a population of homosexual men who do not exhibit symptoms of acquired immune deficiency syndrome. *J. Clin. Invest.* 74:496-506.

72. Shearer, G. M., Salahuddin, S. Z., Markham, P. D., et al. (1985). Prospective study of cytotoxic T lymphocyte responses to influenza and antibodies to human T lymphotropic virus-III in homosexual men. *J. Clin. Invest.* 76:1699-1704.

73. Sheridan, J. F., Aurelian, L., Donnenberg, A. D., and Quinn, T. C. (1984). Cell-mediated immunity of cytomegalovirus (CMV) and herpes simplex

virus (HSV) antigens in the acquired immune deficiency syndrome: Interleukin-1 and interleukin-2 modify in vitro responses. *J. Clin. Immunol.* 4: 304–311.

74. Rook, A. H., Manischewitz, J. D., Frederick,W. R., et al. (1985). Deficient, HLA-restricted, cytomegalovirus-specific cytotoxic T cells and natural killer cells in patients with the acquired immunodeficiency syndrome. *J. Infect. Dis.* 152:627–628.

75. Rook, A. H., Masur, H., Lane, H. C., et al. (1983). Interleukin-2 enhances the depressed natural killer and cytomegalovirus-specific cytotoxic activities of lymphocytes from patients with the acquired immune deficiency syndrome. *J. Clin. Invest.* 72:398–403.

76. Mizuochi, T., Goldberg, H., Rosenberg, A. S., Climcher, L. H., Malek, T. R., and Singer, A. (1985). Both L3T4$^+$ and Lyt-2$^+$ helper cells initiate cytotoxic T lymphocyte responses against allogeneic major histocompatibility antigens but not against trinitrophenyl-modified self. *J. Exp. Med.* 162: 427–443.

77. Greene, W. C., Depper, J. M., Krönke, M., and Leonard, W. J. (1986). The human interleukin-2 receptor: Analysis of structure and function. *Immunol. Rev.* 92:29–48.

78. Tsuchiya, S., Imaizumi, M., Minegishi, M., Kono, T., and Tada, K. (1983). Lack of interleukin-2 production in a patient with OKT4$^+$ T-cell deficiency. *N. Engl. J. Med.* 308:1294.

79. Hauser, G. J., Bino, T., Rosenberg, H., Zakuth, V., Geller, E., and Spirer, Z. (1984). Interleukin-2 production and response to exogenous interleukin-2 in a patient with the acquired immune deficiency syndrome (AIDS). *Clin. Exp. Immunol.* 56:14–17.

80. Prince, H. E., Kermani-Arab, V., and Fahey, J. L. (1984). Depressed interleukin 2 receptor expression in acquired immune deficiency and lymphadenopathy syndromes. *J. Immunol.* 133:1313–1317.

81. Cavaille-Coll, M., Brisson, E., Klatzmann, D., and Gluckman, J-C., (1984). Exogenous interleukin-2 and mitogen responses in AIDS patients. *Lancet* 1:1245.

82. Alcocer-Varela, J., Alarcon-Segovia, D., and Abud-Mendoza, C. (1985). Immunoregulatory circuits in the acquired immune deficiency syndrome and related complex. Production of and response to interleukins 1 and 2, NK function and its enhancement by interleukin-2 and kinetics of the autologous mixed lymphocyte reaction. *Clin. Exp. Immunol.* 60:31–38.

83. Ebert, E. C., Stoll, D. B., Cassens, B. J., Lipshutz, W. H., and Hauptman, S. P. (1985). Diminished interleukin 2 production and receptor generation characterize the acquired immunodeficiency syndrome. *Clin. Immunol. Immunopathol.* 37:283–297.

84. Gupta, S. (1986). Study of activated T cells in man. *Clin. Immunol. Immunopathol.* 38:93–100.

85. Tsang, K. Y., Fudenberg, H. H., and Galbraith, G. M. P. (1984). In vitro augmentation of interleukin-2 production and lymphocytes with the Tac antigen marker in patients with AIDS. *N. Engl. J. Med.* 310:987.

86. Murray, J. L., Hersh, E. M., Reuben, J. M., Munn, C. G., and Mansell, P. W. A. (1985). Abnormal lymphocyte response to exogenous interleukin-2 in homosexuals with acquired immune deficiency syndrome (AIDS) and AIDS related complex (ARC). *Clin. Exp. Immunol.* 60:25–30.
87. Murray, H. W., Welte, K., Jacobs, J. L., Rubin, B. Y., Mertelsmann, R., and Roberts, R. B. (1985). Production of and in vitro response to interleukin 2 in the acquired immunodeficiency syndrome. *J. Clin. Invest.* 76:1959–1964.
88. Arya, S. K., and Gallo, R. C. (1985). Human T-cell growth factor (interleukin 2) and γ-interferon genes: Expression in human T-lymphotropic virus type III- and type I-infected cells. *Proc. Natl. Acad. Sci. USA* 82:8691–8695.
89. Cunningham-Rundles, S., Michelis, M. A., and Masur, H. (1983). Serum suppression of lymphocyte activation in vitro in acquired immunodeficiency disease. *J. Clin. Immunol.* 3:156–165.
90. Farmer, J. L., Gottlieb, A. A., and Nishihara, T. (1986). Inhibition of interleukin 2 production and expression of the interleukin 2 receptor by plasma from acquired immune deficiency syndrome patients. *Clin. Immunol. Immunopathol.* 38:235–243.
91. Siegel, J. P., Djeu, J. Y., Stocks, N. I., Masur, H., Gelmann, E.P., and Quinnan, G. V. (1985). Sera from patients with the acquired immunodeficiency syndrome inhibit production of interleukin-2 by normal lymphocytes. *J. Clin. Invest.* 75:1957–1964.
92. Laurence, J., Gottlieb, A. B., and Kunkel, H. G. (1983). Soluble suppressor factors in patients with acquired immune deficiency syndrome and its prodrome. *J. Clin. Invest.* 72:2072–2081.
93. Laurence, J., and Mayer, L. (1984). Immunoregulatory lymphokines of T hybridomas from AIDS patients: Constituitive and inducible suppressor factors. *Science* 225:66–69.
94. Hersh, E. M., Mansell, P. W. A., Reuben, J. M., et al. (1983). Leukocyte subset analysis and related immunological findings in acquired immunodeficiency disease syndrome (AIDS) and malignancies. *Diag. Immunol.* 1:168–173.
95. Benveniste, E., Schroff, R., Stevens, R. H., and Gottlieb, M. S. (1983). Immunoregulatory T cells in men with a new acquired immunodeficiency syndrome. *J. Clin. Immunol.* 3:359–367.
96. Levy, J. A., Shimabukuro, J., McHugh, T., Casavant, C., Stites, D., and Oshiro, L. (1985). AIDS-associated retrovirus (ARV) can productively infect other cells besides human T helper cells. *Virology* 147:441–448.
97. Moscicki, R. A., Amento, E. P., Krane, S. M., Kornick, S. M., and Colvin, R. B. (1983). Modulation of surface antigens of a human monocyte cell line, U937, during incubation with T lymphocyte-conditioned medium: detection of T4 antigen and its presence on normal blood monocytes. *J. Immunol.* 313:743–748.

98. Ho, D. D., Rota, T. R., and Hirsch, M. S. (1986). Infection of monocyte/ macrophage by human T lymphotropic virus type III. *J. Clin.* Invest. 77:1712–1715.

99. Nicholson, J. K. A., Cross, G. D., Callaway, C. S., and McDougal, J. S. (1986). In vitro infection of human monocytes with human T-lymphotropic virus type III/lymphadenopathy-associated virus (HTLV–III/LAV). *J. Immunol.* 137:323–329.

100. Salahuddin, S. Z., Rose, R. M., Groopman, J. E., Markham, P. D., and Gallo, R. C. (1986). Human T lymphotropic virus type II infection by human alveolar macrophages. *Blood* 86:281–284.

101. Gartner, S., Markovits, P., Markovits, D. M., et al. (1986). The role of mononuclear phagocytes in HTLV–III/LAV infection. *Science* 233:215–4.

102. Montagnier, L., Gruest, J., Chamaret, S., et al. (1984). Adaption of lymphadenopathy associated virus (LAV) to replication in EBV-transformed B lymphoblastoid cells lines. *Science* 225:63–66.

103. Celander, D., and Haseltine, W. A. (1984). Tissue-specific transcription preference as a determinant of cell tropism and leukaemogenic potential of murine retroviruses. *Nature* 312:159–162.

104. Koenig, S., Gendelman, H. E., Orenstein, J. M., et al. (1986). Detection of AIDS virus in macrophages in brain tissue from AIDS patients with encephalopathy. *Science* 233:1089–1093.

105. Wiley, C. A., Schrier, R. D., Nelson, J. A., Lampert, P. W., and Oldstone, M. B. A. (1986). Cellular localization of human immunodeficiency virus infection within the brains of acquired immunodeficiency syndrome patients. *Proc. Natl. Acad. Sci. USA* 83:7089–7093.

106. Gartner, S., Markovits, P., Markovits, D. M., Betts, R. F., and Popovic, M. (1986). Virus isolation from and identification of HTLV–III/LAV-producing cells in brain tissue from a patient with AIDS. *JAMA* 256:2365–2371.

107. Stoler, M. H., Eskin,T. A., Benn, S., Angerer, R. C., and Angerer, L. M. (1986). Human T-cell lymphotropic virus type III infection of the central nervous system. *JAMA* 256:2360–2364.

108. Rieber, P., and Riethmüller, G. (1986). Loss of circulating T4$^+$ monocytes in patients infected with HTLV–III. *Lancet* 1:270.

109. Smith, P. D., Ohura, K., Masur, K., Lane, H. C., and Fauci, A. S. (1984). Monocyte function in the acquired immune deficiency syndrome. *J. Clin. Invest.* 74:2121–2128.

110. Poli, G., Gottazzi, B., Acero, R., et al. (1985). Monocyte function in intravenous drug abusers with lymphadenopathy syndrome and in patients with acquired immunodeficiency syndrome: Selective impairment of chemotaxis. *Clin. Exp. Immunol.* 62:136–142.

111. Washburn, R. G., Tuazon, C. U., and Bennett, J. E. (1985). Phagocytic and fungicidal activity of monocytes from patients with acquired immunodeficiency syndrome. *J. Infect. Dis.* 151:565.

112. Murray, H. W., Gellene, R. A., Libby, D. M., Rothermel, C. D., and Rubin, B. Y. (1985). Activation of tissue macrophages from AIDS patients: In

vitro response of AIDS alveolar macrophages to lymphokines and interferon-γ. *J. Immunol.* 135:2374-2377.

113. Braun, D. P., and Harris, J. E. (1986). Abnormal monocyte function in patients with Kaposi's sarcoma. *Cancer* 57:1501-1506.

114. Hofmann, B., Odum, N., and Jakobsen, B. K. (1986). Immunological studies in the acquired immunodeficiency syndrome. *Scand. J. Immunol.* 23:669-678.

115. Estevez, M. E., Ballart, I. J., Diez, R. A., Planes, N., Scaglione, C., and Sen, L. (1986). Early defect of phagocyte cell function in subjects at risk for acquired immunodeficiency syndrome. *Scand. J. Immunol.* 24:215-221.

116. Kleinerman, E. S., Ceccorulli, L. M., Zwelling, L. A., et al. (1985). Activation of monocyte-mediated tumoricidal activity in patients with acquired immunodeficiency syndrome. *J. Clin. Oncol.* 3:1005-1012.

117. Enk, C., Gerstoft, J., Moller, S., and Remvig, L. (1986). Interleukin 1 activity in the acquired immunodeficiency syndrome. *Scand. J. Immunol.* 23:491-497.

118. Prince, H. E., Moody, D. J., Shubin, B. J., and Fahey, J. L. (1985). Defective monocyte function in acquired immune deficiency syndrome (AIDS): Evidence from a monocyte-dependent T-cell proliferative system. *J. Clin. Immunol.* 5:21-25.

119. Heagy, W., Kelley, V. E., Strom, T. B., et al. (1984). Decreased expression of human class II antigens on monocytes from patients with acquired immune deficiency syndrome. *J. Clin. Invest.* 74:2089-2096.

120. Belsito, D. V., Sanchez, M. R., Baer, R. L., Valentine, F., and Throbecke, G. J. (1984). Reduced Langerhans' cell Ia antigen and ATPase activity in patients with the acquired immunodeficiency syndrome. *N. Engl. J. Med.* 310:1279-1282.

121. Chess, Q., Daniels, J., North, E., and Macris, N. T. (1984). Serum immunoglobulin elevations in the acquired immunodeficiency syndrome (AIDS): IgG, IgA, IgM, and IgD. *Diag. Immunol.* 148-153.

122. Papadopoulos, N. M., and Frieri, M. (1984). The presence of immunoglobulin D in endocrine disorders and diseases of immunoregulation, including the acquired immunodeficiency syndrome. *Clin. Immunol. Immunopathol.* 32:248-252.

123. Anderson, K. C., Boyd, A. W., Fisher, D. C., et al. (1985). Isolation and functional analysis of human B cell populations. *J. Immunol.* 134:820-827.

124. Pahwa, S. G., Quilop, M. T. J., Lange, M., Pahwa, R. N., and Grieco, M. H. (1984). Defective B-lymphocyte function in homosexual men in relation to the acquired immunodeficiency syndrome. *Ann. Intern. Med.* 101:757-763.

125. Ammann, A. J., Schiffman, G., Abrams, D., Volberding, P., Ziegler, J., and Conant, M. (1984). B-cell immunodeficiency in acquired immune deficiency syndrome. *JAMA* 251:1447-1449.

127. Birx, D. L., Redfield, R. R., and Tosato, G. (1986). Defective regulation of Epstein-Barr virus infection in patients with acquired immunodeficiency

syndrome (AIDS) or AIDS-related disorders. *N. Engl. J. Med.* 314:874–879.

128. Pahwa, S., Pahwa, R., Saxinger, C., Gallo, R. C., and Good, R. A. (1985). Influence of the human T-lymphotropic virus/lymphadenopathy-associated virus on functions of human lymphocytes: Evidence for immunosuppressive effects and polyclonal B-cell activation by banded viral preparations. *Proc. Natl. Acad. Sci. USA* 82:8198–8202.

129. Yarchoan, R., Redfield, R. R., and Broder, S. (1985). Mechanisms of B cell activation in patients with acquired immunodeficiency syndrome and related disorders. *J. Clin. Invest.* 78:439–447.

130. Schnittman, S. M., Lane, H. C., Higgins, S. E., Folks, T., and Fauci, A. S. (1986). Direct polyclonal activation of human B lymphocytes by the acquired immune deficiency syndrome virus. *Science* 233:1084–1086.

131. Rook, A. H., Hooks, J. J., Quinnan, G. V., et al. (1985). Interleukin 2 enhances the natural killer cell activity of acquired immunodeficiency syndrome patients through a γ-interferon-independent mechanism. *J. Immunol.* 134:1503–1507.

132. Reddy, M. M., Chinoy, P., and Grieco, M. H. (1984). Differential effects of interferon-α_2a and interleukin-2 on natural killer cell activity in patients with acquired immune deficiency syndrome. *J. Biol. Res. Mod.* 3:379–386.

133. Lew, F., Tsang, P., Solomon, S., Selikoff, I. J., and Bekesi, J. G. (1984). Natural killer cell function and modulation of αIFN and IL2 in AIDS patients and prodromal subjects. *Clin. Lab. Immunol.* 14:115–121.

134. Creemers, P. C., Stark, D. F., and Boyko, W. J. (1985). Evaluation of natural killer cell activity in patients with persistent generalized lymphadenopathy and acquired immunodeficiency syndrome. *Clin. Immunol. Immunopathol.* 36:141–150.

135. Hersh, E. M., Gutterman, J. U., Spector, S., et al. (1985). Impaired in vitro interferon, blastogenic, and natural killer cell responses to viral stimulation in acquired immune deficiency syndrome. *Cancer Res.* 45:406–410.

136. Katzman, M., and Lederman, M. M. (1986). Defective postbinding lysis underlies the impaired natural killer activity in factor VIII-treated, human T lymphotropic virus type III seropositive hemophiliacs. *J. Clin. Invest.* 77:1057–1062.

137. Bonavida, B., Katz, J., and Gottlieb, M. (1986). Mechanism of defective NK cell activity in patients with acquired immunodeficiency syndrome (AIDS) and AIDS-related complex. *J. Immunol.* 137:1157–1163.

138. Pollack, M. S., Callaway, C., LeBlanc, D., et al. (1983). Lymphototoxic antibodies to non-HLA antigens in the sera of patients with acquired immunodeficiency syndrome (AIDS). In: *Non-HLA Antigens in Health, Aging, and Malignancy*, pp. 209–213. New York, Alan R Liss,

139. Kloster, B. E., Tomar, R. H., and Spira, T. J. (1984). Lymphotoxic antibodies in the acquired immune deficiency syndrome (AIDS). *Clin. Immunol. Immunopathol.* 30:330–335.

140. Williams, R. C., Jr., Masur, H., and Spira, T. J. (1984). Lymphocyte-reactive antibodies in acquired immune deficiency syndrome. *J. Clin. Immunol.* 4:118–123.

141. Dorsett, B., Cronin, W., Chuma, V., and Ioachim, H. L. (1985). Antilymphocyte antibodies in patients with the acquired immune deficiency syndrome. *Am. J. Med.* 78:621–626.

142. Tomar, R. H., John, P. A., Hennig, A. K., and Kloster, B. (1985). Cellular targets of antilymphocyte antibodies in AIDS and LAS. *Clin. Immunol. Immunopathol.* 37:37–47.

143. McDougal, J. S., Hubbard, M., and Nicholson, J. K. A. (1985). Immune complexes in the acquired immunodeficiency syndrome (AIDS). *J. Clin. Immunol.* 5:130–138.

144. Gupta, S., and Licorish, K. (1984). Circulating immune complexes in AIDS. *N. Engl. J. Med.* 310:1530–1531.

145. DeStefano, E., Friedman, R. M., Friedman-Kien, A. E., et al. (1982). Acid-labile human leukocyte interferon in homosexual men with Kaposi's sarcoma and lymphadenopathy. *J. Infect. Dis.* 146:451–455.

146. Buimovici-Klein, E., Lange, M., Klein, R. J., Grieco, M. H., and Cooper, L. Z. (xxxx). Long-term follow-up of serum-interferon and its acid-stability in a group of homosexual men. In: *AIDS Research*, pp. 99–108. Mary Ann Liebert, Inc.

147. Siegal, F. P., Lopez, C., Fitzgerald, P. A., et al. (1986). Opportunistic infections in acquired immune deficiency syndrome result from synergistic defects of both the natural and adaptive components of cellular immunity. *J. Clin. Invest.* 78:115–123.

150. Sarngadharan, M. F., Veronese, F. M., Lee, S., and Gallo, R. C. (1985). Immunological properties recognized by sera of patients with AIDS and AIDS-related complex and of asymptomatic carriers of HTLV–III infection. *Cancer Res.* 45:4574s–4677s.

151. Montagnier, L., Clavel, F., Krust, B., et al. (1985). Identification and antigenicity of the major envelope glycoprotein of lymphadenopathy-associated virus. *Virology* 141:283–289.

152. Robert-Guroff, M., Brown, M., and Gallo, R. C. (1985). HTLV–III-neutralizing antibodies in patients with AIDS and AIDS-related complex. *Nature* 316:72–74.

153. Weiss, R. A., Clapham, P. R., Cheingsong-Popov, R., et al. (1985). Neutralization of human T-lymphotropic virus type III by sera of AIDS and AIDS-risk patients. *Nature* 316:69–72.

154. Weiss, R. A., Clapham, P. R., Weber, J. N., Dalgleish, A. G., Lasky, L. A., and Berman, P. W. (1986). Variable and conerved neutralization antigens of human immunodeficiency virus. *Nature* 324:572.

155. Matthews, T. J., Langlois, A. J., Robey, W. G., et al. (1986). Restricted neutralization of divergent human T-lymphotropic virus type III isolates by antibodies to the major envelope glycoprotein. *Proc. Natl. Acad. Sci. USA* 83:9709–9713.

156. Lasky, L. A., Groopman, J. E., Fennie, C. W., et al. (1986). Neutralization

of the AIDS retrovirus by antibodies to a recombinant envelope glycoprotein. *Science* 233:209–212.

157. Putney, S. D., Matthews, T. J., Robey, W. G., et al. (1986). HTLV–III/ LAV-neutralizing antibodies to an *E. coli*-produced fragment of the virus envelope. *Science* 234:1392–1395.

158. Sarin, P. S., Sun, D. K., Thornton, A. H., Naylor, P. H., and Goldstein, A. L. (1986). Neutralization of HTLV–III/LAV replication by anti-serum to thymosine alpha I. *Science* 232:1135–1137.

159. Robert-Guroff, M., Reitz, M. S., Robey, W. G., and Gallo, R. C. (1986). In vitro generation of an HTLV–III variant by neutralizing antibody. *J. Immunol.* 137:3306–3309.

160. Montelaro, R. C., Parakh, B., Orrego, A., and Issel, C. J. (1984). Antigenic variation during persistent infection by equine infectious anemia virus, a retrovirus. *J. Biol. Chem.* 259:10539–10544.

161. Earl, P., Moss, B., Morrison, R. P., et al. (1986). T-lymphocyte primary and protection against Friend leukemia by vaccinia-retrovirus env gene recombinant. *Science* 234:728–731.

162. Rook, A. H., Lane, H. C., Folks, T., et al. (1987). Sera from HTLV–III/ LAV antibody-positive individuals mediate antibody-dependent cellular cytotoxicity against HTLV–III/LAV-infected T cells. *J. Immunol.* 138: 1064–1067.

163. Weber, J. N., Clapham, Weiss, R. A., et al. (1987). Human immunodeficiency virus infection in two cohorts of homosexual men: neutralizing sera and association of anti-gag antibody with prognosis. *Lancet* 1:119–121.

164. Ruscetti, F. W., Mikovits, J. A., Kalyanaraman, V. S., et al. (1986). Analysis of effector mechanism against HTLV-1 and HTLV–III/LAV infected lymphoid cells. *J. Immunol.* 136:3619–3624.

165. Rosenberg, S., Lotze, M., Muul, C., et al. (1985). Observations on the systemic administration of autologous lymphokine-activated killer cells and recombinant interleukin 2 to patients with metastatic cancer. *N. Engl. J. Med.* 313:1485–1492.

166. Adachi, A., Koenig, S., Gendelman, H. E., et al. (1987). Productive, persistent infection of human colorectal cell lines with human immunodeficiency virus. *J. Virol.* 61:209–213.

167. Bank, I., and Chess, L. (1985). Perturbation of the T4 molecule transmits a negative signal to T cells. *J. Exp. Med.* 162:1294–1303.

168. Gurney, M. E., Apatoff, B. R., Spear, G. T., et al. (1986). Neuroleukin: A lymphokine product of lectin-stimulated T cells. *Science* 234:574–581.

12
Biology of HIV-Related Viruses

Phyllis J. Kanki and Myron Essex   *Harvard School of Public Health, Boston, Massachusetts*

I. INTRODUCTION

It is now well recognized that AIDS is caused by an exogenous human retrovirus termed human immunodeficiency virus type 1 (HIV-1). The complexities of HIV-1 extend from the intricate genetic machinery that finely controls virus replication to the panorama of clinical and epidemiological features of this virus infection in humans. The identification of HIV-1 as the causative agent of AIDS propagated renewed interest in the search for related animal and human retroviruses. In the past 5 years two such new viruses have been identified, simian immunodeficiency virus (SIV) and human immunodeficiency virus type 2 (HIV-2). Early in 1985, a related virus, simian immunodeficiency virus (SIV, previously termed STLV-3) was found in immunodeficient macaque monkeys and healthy African green monkeys (1–3). Based on antigenic properties, SIV was demonstrated to be the closest known animal retrovirus to HIV-1. The existence of a primate relative of HIV-1 found in high numbers of naturally infected African primates suggested the possibility that humans might also be susceptible to infection with an SIV-related virus. In late 1985, such a virus was found in West African people; this virus has been designated human immunodeficiency virus type 2 (HIV-2) (4).

II. SIMIAN IMMUNODEFICIENCY VIRUS

At present, the simian immunodeficiency virus (SIV, previously termed simian T-lymphotropic virus type 3, STLV-3) represents one of the more important animal model systems for AIDS. SIV is the closest known animal retrovirus to HIV-1, based on antigenic, genetic and pathobiological features. The utility of this simian lentivirus model for the study of human AIDS viruses can be

317

summarized in the following areas: virological properties, host immune response, and disease pathogenesis.

A. Virology of SIV

SIV (STLV-3) was first described in captive macaques in 1985 (1) as a primate T-lymphotropic virus related to HIV-1. Evidence of SIV has also been described in African green monkeys, African sooty mangabeys, and captive macaques (2,3,5-9) The similarities of this virus to HIV-1 include: T-cell tropism, ultra-structural morphology, antigenic similarities and cross-reactivity, genomic organization, and 50-60% genetic homology (1-3,5-9). Importantly, SIV infection in macaques induces an immunodeficiency syndrome similar to that of human AIDS.

A number of virological issues important to our understanding of HIV viruses of humans may also be addressed in the SIV system. The role of genetic diversity in SIVs from various primate species and within one infected primate needs to be evaluated. At present, full nucleotide sequence is only available on two SIV isolates (10,11) with approximately 10% nucleotide difference. Restriction map analyses have yielded confusing results (12,13). The fact that SIV is not found in wild-caught Asian macaques (14) suggests the possibility that SIVmac results from infection from another primate species. Prior to evaluation of species-specific genetic diversity, the origin of the SIVmac must be determined.

SIV, similar to HIV-2, has a number of accessory genes with variable homology to HIV-1 genes; these include: *vif, tat, rev, nef,* and *vpr*. Functional evidence of *tat* activity has been shown for SIV, similar and cross-reactive with HIV-1 *tat* (15,16). A unique gene to SIV and HIV-2, termed *vpx*, appears to be expressed in the SIV virus and is immunogenic (17,18). Comparative analysis of these putative regulatory genes' functions and correlation to the biological effects of these viruses may provide important clues in identifying pathogenic determinants of HIV-1.

The existence of conserved epitopes in the major viral antigens of SIV and HIV was the mechanism by which these viruses were first discovered (1,2,4). Studies on the major envelope protein of HIV-1 have delineated a number of important functional domains that are highly conserved in SIV (10,19). Conserved epitopes on the envelope antigens of SIV and HIV-1 may be important as future vaccine candidates. These may be readily tested in the SIV model prior to testing in the HIV-1 system with chimpanzees or ultimately in humans.

B. Host Immunity to SIV

The identification of protective host immune factors in HIV infection is central to the development of an effective AIDS vaccine. SIV infection in Asian macaques is linked to immunodeficiency; however, SIV infection in African green monkeys appears to be innocuous (1-3,20). This apparent species-specific pathogenicity may provide important clues to viral-host mechanisms that result in protection from disease development. The West African HIV-2 virus that is more closely related to SIV than HIV-1 also appears to demonstrate a difference in relative pathogenicity (21-24).

The immune response of SIV-infected primates appears to parallel the immune response of people to HIV-1. The envelope antigens appear to be the most immunogenic in exposed primates; less frequently antibody response to *pol-* and/or *gag*-encoded antigens is seen (1-3,25). The humoral response in infected primates appears to follow similar kinetics to that of HIV-1 infected people, and antibodies are persistent throughout. Virus-neutralizing titers appear to be low and variable in infected primates (6,26,27), as in HIV-1-infected people. The role of cell-mediated responses in the SIV system has been less well studied to date.

C. Disease Pathogenesis in SIV

One of the most significant aspects of the SIV model is the ability of this virus to induce immunodeficiency in the macaque monkey (5,6,20,25). SIV-infected macaques have shown T4-cell diminution and dysfunction and death due to opportunistic infections (20,25). There are many similarities in the natural history of SIV-induced immunodeficiency. An animal model system that demonstrates a similar disease process in a primate host has obvious utility for drug and vaccine development. However, standardization of inoculation protocols and virus titration will be necessary prior to more general use of the SIV system for HIV therapeutics. Acutely pathogenic strains of SIV may be useful for certain types of studies; however, the normal long latency of SIV infection is one of the most striking parallels to the HIV system.

There is no doubt that the SIV system will play an important role in the continued research on HIV viruses. The close relationship of this simian virus to human AIDS viruses provides an excellent animal system for comparative studies. As standardization of SIV inoculation protocols in macaques is established, this system will provide an excellent mechanism for AIDS drug and vaccine testing.

III. HUMAN IMMUNODEFICIENCY VIRUS TYPE 2 (HIV-2)

In 1985, a new human T-lymphotropic retrovirus was discovered in Senegal, West Africa (4). We described antibody reactivity in healthy Senegalese that demonstrated strong antibodies to the *env, gag,* and *pol* antigens of SIV. These same samples when reacted with HIV-1 showed only weak cross-reactive antibodies to the *gag* and *pol* antigens. It was therefore recognized that these individuals had been exposed to a virus more closely related to SIV and more distantly related to the protoype AIDS virus, HIV-1 (4,21). Subsequently, Clavel and colleagues (28) demonstrated similar SIV antibody reactivity in two AIDS patients originating from West Africa. This new human retrovirus has now been termed human immunodeficiency virus type 2 (HIV-2) (29). Various strain names have been given to HIV-2, including: HTLV-4, LAV-2, SBL-6669, HTLV-4 (ST). It is now believed that these are all the same virus type. All HIV-2 strains thus far identified are serologically cross-reactive, and therefore serology-based studies are not thought to be strain-specific (28–32).

A. Virology of HIV-2

The antigenic relatedness of both SIV and HIV-2 to the prototype HIV virus prompted both the discovery and further classification of these related viruses (1,4). Subsequent genetic analysis has shown HIV-2 to be most closely related to SIV (less than 20% difference) and more distantly related to HIV-1 (approximately 50% difference) (10,16,33). Similar to HIV-1, these related viruses demonstrate tropism to the T4 lymphocyte (1,28,31,32). The overall genetic organization is similar in all three virus types with the exception of a unique open reading frame termed *vpx* found in both HIV-2 and SIV (10,16,33).

The major viral antigens of HIV-2 have been identified by immunoblot and radioimmunoprecipitation analysis; they are similar and cross-reactive with epitopes on the major viral antigens of HIV-1 (4,21,30). The *gag*-encoded products include p55, a myristylated precursor, a major core protein, p24–26, and an amino-terminal myristylated *gag* protein, p15 (4,21,30,34). The *pol*-encoded proteins, readily distinguished by immunoblot and radioimmunoprecipitation of virus preparations, include p64, p53, and p34 (endonuclease) (21,30,35). The most highly immunogenic antigens are the *env*-related glycoproteins, which include a precursor, gp160, the mature envelope protein, gp120, and the transmembrane gp32–40 (4,21,30). There appears to be polymorphism in the *env*-related glycoproteins, similar to what was reported for different strains of HTLV-1 (4,21,30,36).

The *gag* and *pol* genes are well conserved for both HIVs and SIV (10,11, 16). The *gag*- and *pol*-encoded proteins exhibit broad serological cross-reactivity. It is the presence of *gag* and *pol* antigens in various HIV-1 serological assays that enables the frequent detection of HIV-2 antibodies which are cross-reactive to those conserved epitopes.

The *env* genes of HIV-1 and HIV-2 show less conservation at a genetic and antigenic level, whereas HIV-2 and SIV show a high degree of conservation (4,10,16,30). Sequence analysis of several HIV-2 and SIV strains has demonstrated a stop codon in the middle of the open reading frame encoding the transmembrane protein (10,16,19); this corresponds to the smaller transmembrane protein size that is seen with certain HIV-2 isolates (4,21,36). HIV-2 isolates

SBL-6669 and HIV-2(ST) are reported to have two smaller-molecular-weight glycoproteins that are thought to correspond to two different-sized transmembrane proteins, a gp32 and a gp40 (31,36). It is not known whether these two glycoproteins are, in fact, transmembrane proteins. If so, the presence of two different-sized transmembrane proteins may indicate that these cell lines contain a mixture of virus strains or one virus strain capable of modulating expression of the transmembrane stop codon. It is still to be determined if the stop codon is seen in viruses in vivo or if it represents an in vitro artifact. If the former, it will be important to determine whether potential modulation of the stop codon can affect the functional properties of this virus. Nucleotide sequence comparison between HIV-1 and HIV-2 or SIV has revealed conserved regions scattered throughout the envelope gene (10,16,19). It is still not known whether these regions correspond to immunogenic domains capable of eliciting a cross-protective response to all virus types. In limited studies it has been shown that HIV-2-positive sera are capable of neutralizing various HIV-2 strains and additionally neutralizing some HIV-1 strains at lower titer, whereas the neutralizing response of HIV-1-infected individuals appeared to be type-specific (37).

The close antigenic relationship of HIV-1 and HIV-2 has created new problems for serological diagnosis. Currently employed HIV-1 immunoassays have shown variable ability in detecting HIV-2-antibody-positive samples depending on the testing format and antigen preparation (38–40). HIV-1 recombinant-based and/or competition-type assays are frequently more type-specific and therefore do not readily detect HIV-2-antibody-positive samples. Most of the commonly used, commercial, first-generation HIV-1 antibody ELISA assays will detect over 80% of HIV-2-positive samples (39,40). This probably results from the fact that HIV-2-antibody-positive sera frequently contain high-titer antibodies directed at epitopes which cross-react between the two virus types. At present, immunoblot and/or radioimmunoprecipitation analysis with both HIV-1 and HIV-2 virus types and the presence of antibodies to the *env* products are necessary to distinguish the virus infections.

B. Biology of HIV-2

HIV-2 was given its name to indicate its close relationship to HIV-1. This was based on similar cell tropism and antigenic and genetic properties (29). However, the comparative ability of HIV-2 to induce immunodeficiency is still under active study. Preliminary surveillance studies have already demonstrated significant rates of HIV-2 infection in West Africa. It is, therefore, important to determine the clinical significance of HIV-2 infection and evaluate its potential as a second AIDS-causing virus.

The health status of individuals from whom virus has been isolated is but one means of assessing the pathogenic potential of a virus. HIV-2 has been identified in healthy individuals and AIDS patients originating from West Africa

(4,21–23,28,31,41). Clavel and co-workers (41) have described 30 HIV-2-infected West African patients admitted for treatment to a hospital in Lisbon, Portugal; 6/30 West African patients did not show any symptomatology associated with immunosuppression. In clinical studies, described by Brun-Vezinet and colleagues (23), two of three HIV-2-positive AIDS cases were still alive and stable 3 years after the diagnosis of AIDS. This seemed to indicate a difference in the pathogenicity between HIV-1 and HIV-2. Unfortunately, it is difficult to adequately assess the pathogenic potential of any suspected agent with isolation and description of select disease patients alone.

In West Africa, seroepidemiological surveys of over 10,000 people, conducted in Senegal, Guinea, Guinea Bissau, Mauritania, Burkina Faso, Ivory Coast, Gambia, Cape Verde, and Benin, have demonstrated moderate to high rates of infection with HIV-2 in all countries surveyed except Mauritania (21,22, 32,38,42–44). For these studies, serological diagnosis was accomplished by immunoblot and/or radioimmunoprecipitation analysis with both HIV-1 and HIV-2 antigens. In general, the prevalence of HIV-2 was higher than that of HIV-1 in any West African country surveyed when control healthy or sexually active risk groups were examined. However, HIV-1 was more frequent in the few AIDS patients or suspect cases examined; these individuals were frequently of Central African origin or had a history of recent travel in Central Africa; (19/19 AIDS cases in Senegal from 1986-1987) (42,43).

In the countries surveyed the HIV-2 prevalence in healthy adult populations varied from 0.2 to 9.2%, whereas rates of HIV-1 were much lower. Limited studies conducted in select rural populations of Senegal have failed to demonstrate comparable rates (0%) of HIV-2 seropositives when compared to urban populations, indicating that HIV-2 infection may be more predominant in urban settings as compared to rural settings (42,43). Similar observations have been made with HIV-1 in Central African populations, where lower and more stable seroprevalences of HIV-1 are found in rural populations. (43).

HIV-2 prevalence was higher in sexually active risk groups such as female prostitutes and sexually transmitted disease patients when compared to healthy control populations (21,22,32,42,43). In some urban centers of West Africa the prevalence for HIV-2 in female prositititutes ranged from 15 to 64%. All individuals were healthy at the time of the serum sampling and were found to be without signs or symptoms of AIDS. The prevalence ratio of sexually active risk groups compared to control populations was 7.4 (X^2 = 297, $p < 0.001$). Therefore, HIV-2 appears to be sexually transmitted, like HIV-1, and in Africa this is thought to be primarily heterosexual transmission.

Large-scale seroepidemiological studies permit evaluation of large numbers of individuals of varying health status and the potential association of viral agent to disease. In our studies, there was no significant difference between the prevalence of HIV-2 in immunosuppressed individuals or AIDS patients when compared to healthy controls from the same geographical locales (22,38,42,43). As it is recognized that AIDS in West Africa is still relatively infrequent, we investigated the possibility that some other associated disease entity might better illustrate the potential immunosuppressive properties of HIV-2. It has been well recognized from numerous studies in Central Africa that tuberculosis is highly associated with HIV-1 infection and AIDS (45); 40–70% of tuberculosis cases examined in Zaire were shown to be HIV-1 seropositive (45). We therefore hypothesized that since tuberculosis was also endemic in West Africa, this population might also show an increased association with HIV-2 seropositivity. In Senegal, Ivory Coast, and Guinea Bissau (n = 345) there was no significant difference between the seroprevalance in control populations compared to that in tuberculosis patients (22). This is in contrast to what is seen in similar studies in Central Africa with HIV-1, indicating a significant difference in the patho-biology of HIV-2 from that of HIV-1.

In Burkina Faso and Ivory Coast significant rates of infection with both HIV-1 and HIV-2 are seen in risk populations (22,32,38). In these countries a number of individuals were found who possess antibodies with strong reactivity to the *env* antigens, particularly the gp120, of both HIV-1 and HIV-2 (21,22,23, 32,38,46). This type of serological profile may be indicative of either infection with an intermediate type of virus, double exposure to both HIV-1 and HIV-2, or unusually high-titer cross-reactive antibodies to both. Isolation of both HIV-1 and HIV-2 has not been demonstrated to date (22,32,38,46). It is therefore not known if some individuals may have been exposed to both viruses but maintain persistent infection with only one type. Alternatively, previous infection with one virus type may induce cross-protective immunity and/or interference for the second virus type. Prospective follow-up studies with virus isolation as well as clinical monitoring will determine if interference, double infection, or recom-bination with these viruses occurs and if so, the clinical consequences.

Preliminary studies on the hematological and immunological status of HIV-2-infected prostitutes in Senegal have been completed (24). Generalized lymph-adenopathy and clinical symptomatology of AIDS were not present. Comparison to seronegative prostitutes and seronegative surgery patients was made and sig-nificant elevations were seen in T8 lymphocytes. (p = 0.03), IgG (p = 0.001), and β_2-microglobulin (p = 0.03). The mean T4-lymphocyte count in seropositive prostitutes was normal but lower than in seronegative prostitutes (757 vs. 1179, p = 0.15), but this difference was not statistically significant and appeared to be correlated with age. No significant differences were noted between seronegative and seropositive prostitutes in blastogenic response to various mitogens. Anti-lymphocyte antibodies above background were not present in either population.

Clinical examination of over 62 HIV-2-infected West African prostitutes has failed to demonstrate an increase in AIDS-related signs or symptoms or generalized lymphadenopathy (24). This is in contrast to similar studies conducted on Nairobi prostitutes with HIV-1 infection, where 54% of the seropositive prostitutes were found to have generalized lymphadenopathy on physical examination (47). In Rwandese HIV-1 prostitutes, clinical examination revealed 83% generalized lymphadenopathy and 38% signs or symptoms suggestive of HIV-like disease (48). It therefore appears that the natural history and clinical course of HIV-2-infected individuals may vary significantly from those of HIV-1-infected persons. It is yet to be determined whether the biology of HIV-2 differs from that of HIV-1 only in latency and incubation period.

Despite the lack of association of HIV-2 with AIDS or related disease groups, it is still possible that this T-lymphotropic retrovirus is capable of inducing disease but perhaps with a long latency and decreased attack rate. The lack of abnormal clinical findings in HIV-2-infected individuals could be explained by a recent introduction of this retrovirus into the population. However, retrospective studies conducted in populations of Dakar, Senegal, in the mid-1970s demonstrate that HIV-2 was present at least 18 years ago (21). Furthermore, we have found that the age-specific seroprevalence to HIV-2 in seropositive female prostitutes in Dakar increases with age (32). The age-specific seroprevalence curve is indicative of an endemic virus present in the population for at least several generations. This is consistent with the relatively high rates of HIV-2 infection found in healthy adult control populations.

In summary, the results of epidemiological and prospective clinical studies indicate that HIV-2 is not identical to HIV-1 in pathogenicity, as was once reported. Further follow-up studies are necessary to define the natural history and clinical significance of HIV-2 infection. This will be critical to health policy decisions in many West African countries where HIV-2 infection is common as well as in other parts of the world where this virus is sure to become more widespread.

HIV-2 infection at the present time appears to be most concentrated in West Africa. Serological surveys conducted in other African nations, such as Zaire, Burundi, Tanzania, Zambia, Kenya, Cameroon, Congo, Equatorial Guinea, Chad, Ethiopia, Angola, and Malawi, failed to demonstrate evidence of HIV-2 infection (49–51), despite varying levels of HIV-1 infection. In many of these areas, the prevalence of HIV-1 infection and AIDS is quite high; however, there were no cases of HIV-2 seropositivity (49). Rare cases of HIV-2-infected individuals have been detected in Europe, usually in individuals with connections to West Africa (46,52–54). Limited studies conducted in the United States failed to identify HIV-2 individuals, although it has been recognized that the risk groups for infection with this virus may differ from that of HIV-1 (40,55).

Increased international travel will no doubt enhance the spread of HIV-2 outside of West Africa, where this virus is endemic. A better understanding of the clinical significance and natural history of HIV-2 infection will therefore be necessary for health policymakers worldwide. Prospective clinical studies on HIV-2 are necessary to assess the clinical significance of this second human retrovirus and determine its role in the development of AIDS. Over the past few years, our recognition and understanding of T-lymphotropic retroviruses of animals and people has increased dramatically. Our present data on the HIV-2/ SIV viruses indicate useful parallels and intriguing differences from the prototype AIDS virus. It is hoped that further comparative studies of these retroviruses will provide important direction in our studies to prevent and control the AIDS epidemic worldwide.

ACKNOWLEDGMENT

This research was supported by DAMD 17-87-C-7072 and NIH CA 37466, 18216, and FOD 630. PJK is a Fellow of the Leukemia Society of America.

REFERENCES

1. Kanki, P. J., McLane, M. F., King, N. W., Jr., Letvin, N. L., Hunt, R. D., Sehgal, P., Daniel, M. D., Desrosiers, R. C., and Essex, M. (1985). Serological identification and characterization of a macaque T-lymphotropic retrovirus closely related to HTLV–III. *Science* 1199-1201.
2. Kanki, P. J., Kurth, R. Becker, W., Dreesman, G., Mclane, M. F., and Essex, M. (1985). Antibodies to simian T-lymphotropic virus type III in African green monkeys and recognition of STLV–III Viral proteins by AIDS and related sera. *Lancet* 2:1330–1332.
3. Kanki, P. J., Alroy, J., and Essex, M. (1985). Isolation of T-lymphotropic retrovirus related to HTLV–III/LAV from wild-caught African green monkeys. *Science* 230:951–954.
4. Barin, F., M'Boup, S., Denis, F., Kanki, P., Allan, J. S., Lee, T. H., and Essex, M. (1985). Serological evidence for virus related to simian T-lymphotropic retrovirus III in residents of West Africa. *Lancet* 2:1387–1389.
5. Murphey-Corb, M., Martin, L. N., Rangan, S. R. S., Baskin, G. B., Gormus, B. J., Wolf, R. H., Andes, W. A., West, M., and Montelaro, R. C. (1986). Isolation of an HLTV–III related retrovirus from macaques with simian AIDS and its possible origin in asymptomatic mangabeys. *Nature* 321:435–437.
6. Fultz, P. N., McClure, H. M., Anderson, D. C., Swenson, R. B., Anand, R., and Srinivasan, A. (1986). Isolation of a T-lymphotropic retrovirus from naturally infected sooty mangabey monkeys (*Cercocebus atys*). *Proc. Natl. Acad. Sci. USA* 83:5286–5290.
7. Benveniste, R. E., Arthur, L. O., Tsai, C., Sowder, R., Copeland, T. D., Henderson, L. E., and Oroszlan, S. (1986). Isolation of a lentivirus from a macaque with lymphoma. Comparison with HTLV–III/LAV and other lentiviruses. *J. Virol.* 60:483–490.
8. Lowenstine, L.J., Pederson, N. C., Higgins, J., Pallis, K. C., Uyeda, A., Marx, P., Lerche, N. W., Munn, R. J., and Gardner, M. B. (1986). Seroepidemiologic

survey of captive Old World primates for antibodies to human and simian retroviruses, and isolation of a lentivirus from sooty mangabeys. *Int. J. Cancer* 38:563–574.

9. Ohta, Y., Masuda, T., Tsujimoto, H., Ishikawa, K., Kodama, T., Morikawa, S., Nakai, M., Honjo, S., and Hayami, M. (1988). Isolation of simian immunodeficiency virus from African green monkeys and seroepidemiologic survey of the virus in various non-human primates. *Int. J. Cancer* 41:115–122.

10. Franchini, G., Gurgo, C., Guo, H. G., Gallo, R. C., Collalti, E., Fargnoli, K. A., Hall, L. F., Wong-Staal, F., and Reitz, M. S. (1987). Sequence of simian immunodeficiency virus and its relationship to the human immunodeficiency viruses. *Nature* 328:539–543.

11. Chakrabati, L., Guyader, M., Alizon, M., Daniel, M. D., Desroisers, R. C., Tiollais, P., and Sonigo, P. (1987). Sequence of simian immunodeficiency virus from macaque and its relationship to other human and simian retroviruses. *Nature* 328:543–547.

12. Kornfeld, H., Riedel, N., Viglianti, G. A., Hirsch, V., and Mullins, J. (1987). Cloning of HTLV-4 and its relation to simian and human immunodeficiency viruses. *Nature* 326:610–613.

13. Kessler, H., Li, Y., Naidu, Y. M., Butler, C. V., Ochs, M. F., Jaenel, G., King, N. W., Daniel, M. D., and Desroisers, R. C. (1988). Comparison of simian immunodeficiency virus isolates. *Nature* 331:619–621.

14. Chou, M. J., Kanki, P. J., Lee, T. H., Yang, C. C., and Essex, M. (1989). Absence of natural SIV infection in healthy macaque monkeys, in press.

15. Arya, S. K., Beaver, B., Jagodzinski, L., Ensoli, B., Kanki, P. J., Albert, J., Fenyo, E-M., Biberfeld, G., Zagury, J. F., Laure, F., Essex, M., Norrby, E., Wong-Staal, F., and Gallo, R. C. (1987). New human and simian HIV-related retroviruses possess functional transactivator (*tat*) gene. *Nature* 328:548–550.

16. Guyader, M., Emerman, M., Sonigo, P., Clavel, F., Montagnier, L., and Alizon, M. (1987). Genome organization and transactivation of the human immunodeficiency virus type 2. *Nature* 326:662–669.

17. Yu, X. F., Ito, S., Essex, M., and Lee, T. H. (1988). A naturally immunogenic virion-associated protein specific for HIV-2 and SIV. *Nature* 335:262–265.

18. Franchini, V., personal communication.

19. Hirsch, V., Riedel, N., and Mullins, J. (1987). The genome organization of STLV-3 is similar to that of the AIDS virus except for a truncated transmembrane protein. *Cell* 49:307–319.

20. Letvin, N. L., Daniel, M. D., Sehgal, P. K., Desrosers, R. D., Hunt, R. D., Waldron, L. M., Mackey, J. J., Schmidt, D. K., Chalifoux, L. V., and King, N. W. (1985). Induction of AIDS-like disease in macaque monkeys with T-cell tropic retrovirus STLV–III. *Science* 230:71–73.

21. Kanki, P. J., Barin, F., M'Boup, S., Allan, J. S., Romet-Lemonne, J. L., Marlink, R., McLane, M. F., Lee, T-H., Arbeille, B., Denis, F., and Essex, M.

(1986). New human T-lymphotropic retrovirus related to simian T-lympho-tropic virus type III (STLV–IIIAGM). *Science* 232:238–243.

22. Kanki, P. J., M'Boup, S., Ricard, D., Barin, F., Denis, F., Boye, C., Sangare, L., Travers, K., Albaum, M., Marlink, R., Romet-Lemonne, J-L., and Essex, M. (1987). Human T-lymphotropic virus type 4 and the human immuno-deficiency virus in West Africa. *Science* 236:827–831.

23. Brun-Vezinet, F., Rey, M. A., Katlama, C., Girard, P. M., Roulot, D., Yeni, P., Clavel, F., Alizon, M., Gadelle, S., Madjar, J. J., Harzic, M., and Lenoble, L. (1987). Lymphadenopathy-associated virus type 2 in AIDS and AIDS-related complex. *Lancet* 1:128–132.

24. Marlink, R., Ricard, D., M'Boup, S., Kanki, P., Romet-Lemonne, J-L., N'Doye, I., Diop, K., Simpson, M-A., Greco, F., Chou, M-J., DeGruttola, V., Hseih, C-C., Boye, C., Barin, F., Denis, F., McLane, M-F., and Essex, M. (1988). Clinical, Hematologic, and Immunologic Cross-sectional evalua-tion of individuals exposed to human immunodeficiency virus type 2 (HIV-2). *AIDS Res. Hum. Retroviruses* 4(2):137–148.

25. Kanki, P. J., Eichberg, J., and Essex, M., Relevant Aspects of HIV-Related Viruses to Vaccine Development. *In*: Vaccines, (S. Putney, ed.), in press.

26. Kannagi, M., Kiyotaki, M., Desrosers, R. C., Reimann, K. A., King, N. W., Waldron, L. M., and Letvin, N. L. (1986). Humoral immune responses to T-cell tropic retrovirus simian T-lymphotropic virus type III in monkeys with experimentally induced acquired immune deficiency-like syndrome. *J. Clin. Invest.* 78:1229–1236.

27. Ho, D., and Kanki, P. J., unpublished data.

28. Clavel, F., Guetard, D., Brun-Vezinet, F., Chamaret, S., Rey, M-A., Santos-Ferreira, M. O., Laurent, A. G., Dauguet, C., Katlamà, C., Rouzioux, C., Klatzmann, D., Champalimaud, J. L., and Montagnier, L., (1986). Isolation of a new human retrovirus from West African patients with AIDS. *Science* 233:343–346.

29. Biberfeld, G., Brown, F., Esparza, J., Essex, M., Gallo, R. C., Montagnier, L., Najera, R., Risser, R., and Schild, G. (1987). Meeting report, WHO Working Group on Characterization of HIV-Related, Retroviruses: Criteria for characterization and proposal for a nomenclature system. *AIDS* 1:189–190.

30. Kanki, P. J., Essex, M., and Barin, F. (1987). Antigenic relationships between HTLV-3/LAV, STLV-3, and HTLV-4. In: *Vaccines 87* (R. Chanock, F. Brown, R. Lerner, and H. Ginsberg, eds.). Cold Spring Harbor, New York, Cold Spring Harbor Press.

31. Albert, J., Bredberg, U., Chiodi, F., Bottiger, B., Fenyo, E. M., Norrby, E., and Biberfield, G. (1987). A new human retrovirus isolate of West African origin (SBL-6669) and its relationship to HTLV–IV, LAV-II, and HTLV-IIIB. *AIDS Res. Hum. Retroviruses* 3:3–10.

32. Kanki, P. J. (1987). West African human retroviruses related to STLV-III. *AIDS* 1:141–145.

33. Franchini, G., Collalti, E., Arya, S. K., Fenyo, E. M., Biberfeld, G., Zagury, J. F., Kanki, P. J., Wong-Staal, G., and Gallo, R. C. (1987). Genetic analysis

of a new subgroup of human and simian T-lymphotropic retroviruses: HTLV–IV, LAV-2, SBL6669, and STLV–III AGM. *AIDS Res. Hum. Retroviruses* 3:11–17.

34. Allan, J. S., and Kanki, P. J., Identification of the 3'orf product in HIV-2 and SIV, in preparation.

35. Allan, J. S., Coligan, J. E., Lee, T-H., Barin, F., Kanki, P. J., M'Boup, S. Mclane, M. F., Groopman, J. E., and Essex, M. (1987). Immunogenic nature of a *pol* gene product of HTLV–III/LAV. *Blood* 69:331–333.

36. Kanki, P. J., Romet-Lemonne, J-L., M'Boup, S., and Essex, M. (1987). Antigenic properties of different HIV-2 isolates, presented at conference on Aids in Africa, Naples, Italy, 1987.

37. Weber, J. N., Clapham, P. R., Whitby, D., Tedder, R. S., and Weiss, R. A. (1987). Neutralization of African HIV–I and HIV-2, presented at conference on Aids in Africa, Naples, Italy, 1987.

38. Denis, F., Barin, F., Gershey-Damet, G., Rey, J-L., Lhuillier, M., Mounier, M., Leonard, G., Sangare, A., Goudeau, A., M'Boup, S., Essex, M., and Kanki, P. (1987). Prevalence of human T-lymphotropic retroviruses type III (HIV) and type IV in Ivory Coast. *Lancet* 1:408–411.

39. Denis, F., Leonard, G., Sangare, A., Gershey-Damet, G., Rey, J. L., Soro, B., Schmidt, D., Mounier, M., Verdier, M., Baillou, A., and Barin, F. (1989). Comparison of ten HIV enzyme immunoassays for detection of antibody to HIV-2 in West African sera. *J. Clin. Miciobiol.* in press.

40. Kanki, P. J., and Essex, M., in preparation.

41. Clavel, F., Mansinho, K., Chamaret, S., Guetard, D., Favier, V., Nina, J., Santos-Ferreira, M-O., Champalimaud, J-L., and Montagnier, L. (1987). Human immunodeficiency virus type 2 infection associated with AIDS in West Africa. *N. Engl. J. Med.* 316:1180–1185.

42. M'Boup, S. (1987). Summary of presentation for conference on Aids in Africa, Naples, Italy, 1987.

43. Kanki, P. J., B'Boup, S., Barin, F., Denis, F., Marlink, R., Romet-Lemonne, J-I., and Essex, M. The biology of HIV-1 and HIV-2 in Africa. In: *Aids in Africa*. Basel, S. Karger, in press.

44. Brun-Vezinet, F., Katlama, C., Ceuninck, D., Andrade, D., Dario, D., and Rey, M. A. (1987). Lymphadenopathy associated virus type 2 (LAV2)–Seroepidemiological study in Cape Verde Islands, presented at 3rd International Conference on AIDS, Washington, DC, 1987.

45. Quinn, T. C., Mann, J. M., Curran, J. W., and Piot, P. (1986). Aids in Africa: An epidemiologic paradigm. *Science* 234:955–963.

46. Rey, M. A., Girard, P. M., Harzic, M., Madjar, J. J., Brun-Vezinet, F., and Saimot, A. G. (1987). HIV-1 and HIV-2 double infection in French homosexual male with AIDS-related complex (Paris, 1985). *Lancet* 1:388–389.

47. Kreiss, J. K., Koech, D., Plummer, F. A., Holmes, K. K., Lightfoote, M., Piot, P., Ronald, A. R., Ndinya-Achola, J. O., D'Costa, L. J., Roberts, P., Ngugi, E. N., and Quinn, T. C. (1986). AIDS virus infection in Nairobi

prostitutes: Spread of the epidemic to East Africa. *N. Engl. J. Med.* 314: 414–418.

48. Van de Pierre, P., Clumeck, N., Carael, M., Nzabihimana, E., Robert-Guroff, M., De Mol, P., Freyens, P., Butzler, J.-P., Gallo, R. C., and Clumeck, N. (1985). Female prostitutes, a risk group for infection by the human T-cell lymphotropic virus type III. *Lancet* 2:524–527.

49. Kanki, P. J., Allan, J., Barin, F., Redfield, R., Clumeck, N., Quinn, T., Mowovindi, F., Thiry, L., Burny, A., Zagury, D., Petat, E., Kocheleff, P., Pascal, K., Lausen, I., Frederickson, B., Craighead, J., M'Boup, S., Denis, F., Curran, J. W., Mann, J., Francis, H., Albaum, M., Travers, K., McLane, M. F., Lee, T-H., and Essex, M. (1987). Absence of antibodies to HIV-2/ HTLV-4 in six Central African nations. *AIDS Res. Hum. Retroviruses* 3(3): 317–322.

50. Mhalu, F., Bredberg-Raden, U., Mbena, E., Pallangyo, K., Kiango, J., Mbise, R., Nyamuryekunge, K., and Biberfeld, G. (1987). Prevalence of HIV infection in healthy subjects and groups of patients in Tanzania. *AIDS* 1(4): 217–222.

51. Gurtler, L., Eberle, J., Deinhardt, F., Liomba, G. N., Ntaba, N. G., and Schmidt, A. J. (1987). Prevalence of HIV-1 in selected populations and areas of Malawi, presented at conference on Aids in Africa, Naples, Italy, 1987.

52. Couroce, A-M. (1987). HIV-2 in blood donors and in different risk groups in France. *Lancet* 1:1151.

53. Werner, A., Staszewski, S., Helm, E. B., Stille, W., Weber, K., and Kurth, R. (1987). HIV-2 (West Germany, 1984). *Lancet* 1:868–869.

54. Tedder, R. S., O'Connor, T. (1987). HIV-2 in UK. *Lancet* 1:869.

55. Schochetman, G., Schable, M. S., Goldstein, L. C., Epstein, J., and Zuck, T. F. (1987). Screening of U.S. populations for the presence of LAV-2, presented at 3rd International Conference on AIDS, Washington, DC, 1987.

13
Progress in the Therapy of Human Immunodeficiency Virus (HIV) Infections

Hiroaki Mitsuya and Samuel Broder   *National Cancer Institute, National Institutes of Health, Bethesda, Maryland*

I. INTRODUCTION

The acquired immune deficiency syndrome (AIDS) is a pandemic, lethal disease caused by the third known human T-lymphotropic virus (1), which has the capacity to replicate within critical cells of the immune system (in particular macrophages and T cells), leading to a profound destruction of certain T cells (1-6). AIDS was first recognized as a new clinical entity in 1981 (7,8). This was approximately 75 years after the identification of a transmissable agent that we would now call a retrovirus (9), 11 years after publication of the phenomoneon of reverse transcription as the biochemical basis for classifying such retroviruses (10,11), and 1 year after publication of the first formal proof that pathogenic human retroviruses of any kind exist (12). The number of cases of AIDS has increased substantially; as of July 1989 over 100,000 cases have been reported in the United States alone. In addition, it is estimated that in the United States alone over one million people may have been infected by the retrovirus that causes AIDS. It is not known how many of these people will develop a clinical illness related to their retroviral infection. Projections made in May 1988 estimate that 365,000 AIDS cases will have been diagnosed in the United States by the end of 1992 (13).

The development of effective therapies for AIDS is a crucial challenge from a scientific point of view. However, there is more to this challenge: Society is likely to judge the scientific community by what is accomplished or not accomplished in AIDS.

The available data indicate that the AIDS retrovirus is spread in two major ways: sexual activity (homosexual or heterosexual) and the administration of virus-bearing blood products (1,8). Maternal-fetal transmission can also occur (14). AIDS may have a long latency period, in some cases 5 years or more, and an individual who is otherwise well can be a source of infection for others.

AIDS was initially defined as the development of either an opportunistic infection or Kaposi's sarcoma (an unusual neoplasm that had previously been recognized to be associated with certain immunosuppressed states) in a person without a known cause for immunodeficiency (7). It soon became apparent that these patients had a cellular immunodeficiency characterized by an inexorable depletion of helper/inducer (CD4$^+$) T cells. Functionally, this was manifested as a loss of delayed-type cutaneous hypersensitivity reactions, a loss of certain in vitro T-cell proliferative responses, excessive immunoglobulin production by B cells perhaps due to abnormal regulation coupled with a T-cell-dependent mitogenic effect mediated by the virus (15), and a loss of in vitro cytotoxic T-cell responses (16). In addition, the opportunistic agents that most characteristically cause illness in this patient population (e.g., *Pneumocystis carinii, Mycobacterium avium, Candida albicans*, etc.) are normally controlled by cell-mediated (T-cell) immune responses. Once established, the immunodeficiency state is generally progressive and leads to death.

Shortly after the recognition that AIDS was caused by a pathogenic retrovirus, Gallo and his co-workers provided data that this virus could replicate within the brain (17), and there is a growing recognition that serious neurological diseases, ranging from peripheral neuropathy to fulminant dementia, can be caused by HIV infection with or without a clinically evident immunodeficiency (18,19). Perhaps infected macrophages play a role in the traffic of virus across the blood-brain barrier (20,21). It is still not known for sure how the virus enters the brain and how it brings about neurological damage, but it is a virtual certainty that successful therapeutic strategies must address the consequences of viral replication within the central nervous system.

II. HIV AS A RETROVIRUS

We can now turn to a discussion of potential strategies for treating AIDS and its related disorders. While many features in the life-cycle of HIV are not understood, enough is known to formulate specific strategies (see Table 1) (22). At the outset, it is worth stressing that effective therapy of HIV infections may well depend on a combination of strategies that attack multiple steps in viral replication and cytopathogenicity, in part because the emergence of drug-resistant strains might be less likely. The testing of new antiretroviral agents has been facilitated by the availability of rapid and sensitive in vitro screening systems which can determine whether a putative drug can inhibit the replication and T-cell killing activity of HIV (23) (Figure 1).

TABLE 1   Stages in the Replicative Cycle of a Pathogenic Human Retrovirus Which May Be Targets for Therapeutic Intervention

Stage	Potential intervention
Binding to target cell	Antibodies to the virus or cell receptor
Early entry into target cell	Drugs that block fusion or interfere with retroviral uncoating
Transcription of RNA to DNA by reverse transcriptase	Reverse transcriptase inhibitors
Degradation of viral RNA in an RNA–DNA hybrid	Inhibitors of RNase H activity
Integration of DNA into host genome	Drugs which inhibit *pol* gene-mediated "integrase" function
Expression of viral gene	"Antisense" constructs. inhibitors of the *tat*-III protein or *art/trs* protein; compounds that interfere with *gag-pol* frameshift
Viral component production and assembly	Myristylation, glycosylation, and protease inhibitors or modifiers
Budding of virus	Interferons

As discussed earlier, HIV belongs to a family of viruses known as retroviruses. In the era before the etiology of AIDS was recognized, one of the most notable features of such viruses was the capacity to induce neoplastic transformation in infected target cells; hence the expressions "RNA tumor virus" and "leukemia virus" appeared in the literature to denote this category of virus (24). However, HIV has not been shown to have a transforming capacity per se, although it is linked to the causation of cancer through its immunosuppressive activity. By definition such viruses must replicate through a DNA intermediate (i.e., at one step of their cycle of replication, genetic information flows from RNA to DNA, a reverse or "retro" direction). The viral DNA polymerase reverse transcriptase) that catalyzes this step is encoded by the *pol* gene of the virus, a gene that is conserved to a considerable extent in its amino acid and nucleotide sequences among all retroviruses (25). These viruses have as major structural components a core of genomic RNA, group-specific antigen (*gag*) proteins which play a role in the structure of the core and assembly of the virion, a lipid bilayer, and an outside envelope glycoprotein. HIV is the most complex retrovirus yet characterized (1,26–31) in that it contains at least six genes that were not known to exist in previously defined retroviruses. Essentially all of these genes have been expressed in bacteria (32), yeast (33),

NO DRUG

40 μM
DIDEOXYADENOSINE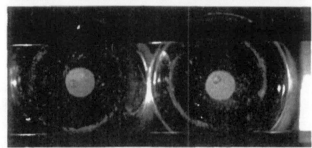

2 μM
DIDEOXYCYTIDINE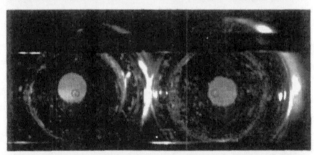

5 μM
AZIDO-THYMIDINE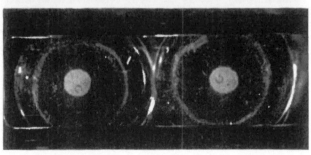

NO VIRUS   HTLV-III

and/or mammalian cells. The functions of several genes, including *vpr* and *vpu* are not yet fully understood. The functions of two genes, *tat* (28) and *rev*, previously designated as *art* (29) or *trs* (30), will be discussed later.

III. CELL BINDING AND ENTRY

The first step in infection of a cell by HIV is the binding to the target cell receptor. In the case of helper-inducer T cells, this receptor is thought to be on or near the CD-4 antigen (34,35). The CD-4-related epitope is thought to be responsible for the tropism of the virus for helper-inducer (CD-4$^+$) T cells, as a consequence of specific binding between the CD-4 molecule and portions of the envelope glycoprotein. This binding step may be vulnerable to attack by antibodies either to the virus or to the receptor, and in the future one can speculate that certain chemicals or small peptides could be designed to occupy the receptor and accomplish the same thing.

It mightbe worthwhile deviating from the chronology of the retroviral life-cycle briefly to address the cytopathogenicity of HIV. From one perspective, the CD-4 molecule and the envelope glycoprotein of the virus determine events both at the beginning and at the end of the life-cycle of HIV within susceptible T cells. In addition to a role in the receptor-mediated entry of the virus into

---

FIGURE 1 Rapid system for assessing antiretroviral activity against HIV. An immortalized CD4$^+$ T-cell clone (ATH8) was obtained by cloning a normal tetanus-toxoid-specific T-cell line in the presence of human T-lymphotropic virus type I (HTLV-I) (93). The ATH8 clone was selected for this study on the basis of its rapid growth and readily detectable sensitivity to the cytopathic effect of HIV (53). ATH8 cells (2 × 10^5) were exposed to HTLV-III$_B$ (3000 virus particles per cell) and cultured in test tubes in the presence or absence of 2$'$,3$'$-dideoxynucleosides. Since in this system the minimum cytopathic virus dose that destroys the culture is approximately five virus particles per cell (55), the conditions used in our system have been steered toward high estimated multiplicity of infection (MOI). By day 7 following exposure to the virus, ATH8 cells were almost completely killed by the virus. The cytopathic effect can be seen as a small disrupted pellet which contains debris of cells (top right). However, in the presence of 2$'$,3$'$-dideoxyadenosine, 2$'$,3$'$-dideoxycytidine, or 3$'$-azido-2$'$,3$'$-dideoxythimidine, ATH8 cells were completely protected and continued to grow, which can be seen as large cell pellets (second, third, and bottom right) comparable to a control virus-unexposed, drug-unexposed ATH8 population (top left). ATH8 cells exposed to only drug (second, third, and bottom left) formed large pellets comparable to control virus-unexposed and drug-unexposed cells (top left). This system is readily adaptable for rapid screening of new antiretroviral agents including neutralizing antibodies (23).

target T cells, the CD-4 receptor plays some role in the susceptibility of a target
T cell, once it begins to produce virus, to be killed by that virus. Precisely how
HIV destroys T cells in vivo is not known. The cytopathic effect of HIV is
thought to be mediated in part by an interaction between the CD-4 molecule
and the HIV envelope that brings about lethal cell-to-cell fusions (syncytia) or a
surface autofusion phenomenon that destroys the integrity of the cell membrane
(36). However, other factors in addition to the formation of syncytia are
thought to play a role in the cytopathic effects brought about by the virus.
Recent results raise the possibility that the carboxyl terminus of the envelope, a
region different from the portion that directly binds to the CD-4 molecule, is
important in the envelope-mediated destruction of T cells brought about by the
virus (37).

Conceivably, therapeutic agents could be designed to alter the properties
of the viral surface or target cell surface (e.g., the lipid composition) to reduce
viral infectivity or cytopathic effect. It is known that there is some variation in
the surface envelope from one viral isolate to another (38,39), and in fact, this
portion of the virus shows the greatest mutation rate. The range of possible
alterations in the envelope binding site is, however, most likely limited by the
need to bind to CD4 (which is relatively constant). An antibody directed against
this site on the envelope might bind to most strains of HIV and perhaps kill cells
as they begin to express envelope antigens so that spread of virus to uninfected
cells can be reduced. Thus, monoclonal antibodies to HIV could have a therapeutic role in patients with AIDS or related diseases (40). We have been able to
produce a complement-fixing human IgG$_k$ monoclonal antibody against the
major envelope glycoprotein of another pathogenic human T-lymphotropic
retrovirus (HTLV-I) (41), and similar approaches could be used to develop
human monoclonal antibodies against HIV. A potential difficulty of this approach, however, is that virally infected cells could make infectious cell-to-cell
contacts. Antibodies might not gain access to relevant epitopes under such
circumstances. Also, it has been shown that AIDS can occur in the face of what
in vitro appears to be neutralizing antibodies to HIV. Whether this occurs
because the titers of such antibodies are low or because such antibodies do not
block epitopes that mediate in vivo cytopathogenicity is a topic for ongoing
research (42,43). Finally, it might be relatively easy for the virus to undergo a
mutation that modifies the antigenicity profile of the virion, thereby evading
attack by antibodies.

After binding to a cell, HIV enters the target cell by an incompletely
defined mechanism, most likely by a fusion process. It is conceivable that drugs
that block this step could be developed. For example, calmodulin antagonists
block the entry of Epstein-Barr virus into B cells (44).

After a retrovirus enters a target cell, the virus loses its envelope-coat and
RNA is released into the cytoplasm (each virion conveys a dimer of two identical

genomic RNA subunits into the cell). Pharmacological agents that block this "uncoating" might be developed in future strategies for the experimental treatment of AIDS. It is likely that uncoated viral RNA is susceptible to degradation by cellular nucleases; so the process of reverse transcription must be accomplished promptly.

HIV uses a lysine transfer RNA as a primer and the *pol* gene product, reverse transcriptase (RT), to make a minus-strand DNA copy of the genome employing the viral RNA as a template (27,45). The same enzyme then catalyzes the production of a positive-strand DNA, and eventually, the genetic information encoded by the virus as a single strand of RNA is transcribed into a single strand of DNA (as an RNA-DNA hybrid) and then into a double-stranded DNA form. Retroviral DNA polymerase has an inherent RNase H activity that specifically degrades the RNA of an RNA-DNA hybrid (46). The use of reverse transcriptase inhibitors in patients with AIDS is predicated on the assumption that an ongoing state of viral replication occurs in this disease, and inhibition of viral replication would permit some regeneration of the immune system, or at least prevent further deterioration. Recent clinical results using antiviral chemotherapy support the notion that ongoing viral replciation is important in perpetuating the disease.

The HIV reverse transcriptase enzyme has been purified; it appears to exist as a p51 and a p66 molecule (47). Large quantities of the HIV reverse transcriptase should become readily available because the enzyme has been expressed in bacteria (32) and yeast (48) using recombinant DNA technology. Because of its unique role in retroviral replication and because a great deal is already known about it, RT is a high-priority target for antiviral therapy in AIDS (22,23,49–56). Several investigational drugs that are inhibitors of this enzyme are either already available, having been developed for the therapy of conventional viral diseases [e.g., phosphonoformate, (54)], or are being developed expressly for AIDS based on selection from in vitro screening systems for activity against HIV (23,53). We will return to this point again.

IV. INTEGRATION, LATENCY, AND REACTIVATION

Returning to the discussion of the retroviral life-cycle, the DNA copy of the virus may be circularized during or soon after its formation, the biological importance of circularized provirus is not clear, and it is possible that only the linear form is relevant. In the case of HIV, it is believed that the virus either can remain in an unintegrated form or can become integrated into the host cell genome. It is possible that chemicals could be developed to interfere with the viral "integrase" (thought to be a function of one of the *pol* gene products) which mediates this integration step. At some later time, perhaps after activation of the infected cell (57), the DNA is transcribed to mRNA and viral genomic

RNA using host RNA polymerases, and this RNA is then translated to form viral proteins, again using the biochemical apparatus of the host cell. However, within any given cell, it is thought that the retrovirus could remain latent, perhaps for the life of the cell. The capacity of pathogenic retroviruses to infect a target cell and remain replicatively silent until an activation event occurs, which then stimulates viral expression and virally mediated cell destruction, resembles certain aspects of lysogenic phage replication in bacteria.

It is now well accepted that HIV has a special regulatory gene (*tat*) coding for a diffusible protein that markedly enhances the production of other viral proteins (28). The *tat* gene and a related gene that uses an alternative, largely colinear reading frame (vide infra) are absolutely essential to viral replication. All of the known pathogenic human retroviruses have special transcriptional or posttranscriptional activating (*tat*) genes that act in *trans* as opposed to *cis* (i.e., through a mechanism that can affect genes not in direct proximity), but there are significant differences among them, and certain ideas about these *tat* genes are still evolving (58,59). One major effect of the *tat* protein is thought to be related to an increase of posttranscriptional events rather than effects on the transcription of DNA to RNA alone. Several features of this gene are virtually without precedent from studies of the genetic mechanisms of other retroviruses. For example, the *tat* gene is coded by two separated exons, and *tat* mRNA formation depends on a double-splicing mechanism not observed in retroviruses before the discovery of human T-lymphotropic viruses. The *tat* protein is thought to provide the virus with a positive feedback loop by which a viral product can amplify the production of new virions (28). The *tat* protein is small (86 amino acids), with a cluster of positively charged amino acids, and is thought to influence posttranscriptional events in the synthesis of other proteins by binding to critical regulatory sequences at the 5' end of mRNA. Possibly, drugs or other agents may be found that inhibit either the *tat* product itself, a crucial nucleic acid binding site for this protein, or both.

Employing the same mRNA that codes for *tat*, but using a different reading frame, a unique gene of HIV (denoted as the *rev* gene) produces a different small (116 amino acids) positively charged protein, which is thought to function as a second critical *trans*-acting factor in viral replication (29,30). Again, drugs that bind or inactivate this protein would be expected to inhibit viral replication. In the absence of this second regulatory factor, *gag*- and *env*-encoded protein synthesis is severely diminished. The probable mechanism was first described as an abrogation of negative regulatory effects on translation of viral mRNA encoding HIV structural proteins, and accordingly the gene was denoted as *art* (antirepression *trans*-activator) (29). Possibly, *tat* may control this gene, which in turn regulates the synthesis of *gag-pol* and *env*. A report by Feinberg et al. suggests that a different and rather novel mechanism might account for the regulatory activity of this *trans*-activating retroviral gene and that this gene negatively regulates the splicing of viral RNA to permit the preservation of *gag-pol* and *env* mRNA. Accordingly, they have denoted this gene as *trs* (*trans*-acting regulator of splicing) (30). According to this model, the absence of an adequate amount of the *trs*-encoded factor results in the presence of predominantly fully spliced 2-kb transcripts (e.g., *tat*, *rev*, and *nef* mRNA) and viral replication cannot proceed. If adequate amounts of the *rev* protein are available, the RNA splicing pattern shifts to permit the preservation of larger transcripts such as *gag-pol*, *env*, and genomic viral RNA, making it possible for replication to go ahead. Whether an *art* or a *trs* mechanism is involved, it is clear that this retrovirus has evolved an astonishingly complex system of genetic regulation. Perhaps this is because there is a race between viral replication within a cell and destruction of the cell that the virus has commandeered to reproduce itself, leaving no tolerance for inefficiency or improper timing in the synthesis of viral components. The complexity of the virus, however, could contribute to its defeat.

V. PROTEIN PRODUCTION AND ASSEMBLY

Continuing with our discussion of stages in the viral life-cycle that could be blocked, it is conceivable that a class of agents that interferes with the structure and function of retroviral mRNA in infected cells could have a role in AIDS. One drug, ribavirin, is believed to act as a guanosine analog that interferes with the 5' capping of viral mRNA in other viral systems, and perhaps this activity could be useful in retrovirally induced disorders (60). At present, no conclusive data support a role for ribavirin in the therapy of AIDS, but further research is necessary.

Another approach that may conceivably inhibit the translation of viral products would be the use of "antisense" oligodeoxynucleotides. Such approaches have been explored for a number of cellular and viral genes, including HIV in vitro (62). Basically, these approaches involve short sequences of DNA (or DNA that is chemically modified to enable better cell penetration and resistance to enzymatic degradation) whose base pairs are complementary to a vital segment of the viral genome. Such oligodeoxynucleotides could theoretically block expression of the viral genome through a kind of hybridization arrest of translation or possibly by interfering with the binding of a regulatory protein such as *tat*. Indeed, since the *tat* gene and the *rev* genes are expressed on the same mRNA transcript, it is possible that one antisense oligodeoxynucleotide could interrupt two functions that are vital for HIV replication. In principle, it might be possible to achieve the same goal by constructing an "antisense" virus (i.e., a retrovirus that has been genetically engineered to produce a stretch of mRNA that will bind to the messenger made by the wild-type virus). Such an approach could theoretically render certain target cells "genetically" resistant to HIV infection. While such an approach might pose serious complexities, there do not appear to be any fundamental technological barriers to implementing them.

The final stages in the replicative cycle of HIV involve crucial secondary processing of certain viral proteins by proteases (a function of one of the *pol* gene products) and myristylating and glycosylating enzymes (provided by the host) as a prelude to assembly of infectious virions. Thus, future strategies for the experimental treatment of AIDS might conceivably involve certain kinds of protease inhibitors or drugs which dampen or alter certain myristylation and glycosylation steps in the synthesis of viral components. Finally, retroviruses are released by a process of viral budding. Interferons may act to interfere with this stage of HIV replication (63).

While we have focused the discussion on how to suppress HIV, it might be worth noting that the virus could theoretically set off a chain of secondary events in vivo (autoimmune reaction, toxic lymphokine production, etc.) that is necessary for the expression of clinical disease. It is also intriguing to speculate that antiretroviral therapy coupled with bone marrow or lymphocyte replacement therapies might be successful in certain subsets of patients with HIV infections (64). However, we will not be able to discuss strategies for intervening against potential secondary events, or for adoptive cellular replacement, in this chapter.

VI. DIDEOXYNULCEOSIDES IN THE EXPERIMENTAL THERAPY OF HIV INFECTIONS

We now turn to a discussion of a broad family of 2′,3′-dideoxynulcoside analogs that can be metabolized to become potent inhibitors of HIV reverse transcriptase. These analogs can profoundly inhibit HIV replication and its capacity to destroy T-cell cultures at concentrations that are 10-20-fold lower than those that impair the proliferation and survival of target cells (53,55). From one perspective, these are not new chemicals and in several cases pioneering studies were accomplished over the past 20 years or so (65-74). Such compounds (as triphosphates) are familiar to molecular biologists as reagents for the Sanger DNA-sequencing procedure (75). However, their application as potenial antiretroviral chemotherapeutic agents in human beings will require an expansion of how these agents might have previously been categorized. They are of special interest because they prove that a simple chemical modification of the sugar moiety can predictably convert a normal substrate for nucleic acid synthesis into a potent compound with the capacity to inhibit the replication and cytopathic effect of HIV (55)

Certain relationships between the structure and activity of these nucleoside analogs are explored in Figure 2, which summarizes the capacity of five congeners of adenosine to act as inhibitors of HIV. In these experiments, the adenine portion of the nucleoside analog is kept constant; however, several simple changes in the sugar moiety are tested as variables. The reference agent is 2'-deoxyadenosine, a normal building block for DNA which, as expected, has no antiretroviral activity in this system. However, at the top Figure 2, one can see what happens when 2',3'-dideoxyadenosine is tested. It can be seen that a simple reduction (removal of the oxygen) at the 3'-carbon of the sugar converts the normal nucleoside at the 5'-carbon, creating 2',3',5'-trideoxyadenosine (bottom of the figure), nullifies the antiretroviral effect. Neither the 3'-deoxy configuration of adenosine (cordycepin), the *arabinosyl* derivative (Ara-A), nor 2',3'-b-epoxyadenosine exerted an antiretroviral effect. It is important to note that these experiments were performed under conditions of high estimated multiplicity of infection (23; see legend for Figure 1).

2',3'-Dideoxycytidine has been shown to have a virustatic effect on patients with AIDS and AIDS-related complex (ARC) (76), although this drug has a dose-dependent toxicity in the form of peripheral neuropathy, which limits its use in some patients. In this regard, it is worth stressing that durable and successful antiretroviral therapy may well depend upon the use of drug combinations. We have already administered an alternating regimen of AZT (vide infra) for 7 days and dideoxycytidine for 7 days to 16 patients in National Cancer Institute (76). Indeed, some patients are treated with this alternating regimen for more than 90 weeks without any intolerable side effects at this writing (R. Yarchoan et al. unpublished observation). It is noteworthy that ddC is the most potent antiretroviral dideoxynucleoside we have thus far tested on the basis of molarity. DdC can exert complete suppression of HIV replication at ½–1/5 the effective concentration of AZT in vitro (55) and the activity of ddC is more durable than that of AZT. Interestingly, ddC appears to be far more potent clinically both in terms of antiviral activity and toxicity as well than expected. This agent may have in vivo activity at 1/500–1/600 the effective daily dose of AZT. A large phase II/III study employing a low dose of ddC may define the role in the treatment of HIV infection. Another member of the dideoxynucleoside family, 2',3'-dideoxyniosine or ddI, has also been shown to suppress HIV replication in patients with AIDS and ARC (77), and thus far, ddI appears to be the least toxic of the nucleoside analogues we have tested as antiretroviral drugs. The most notable adverse effects which may possibly have been related to ddI administration included increases in serum uric acid, mild insomnia, headaches, and seizures. Further controlled studies to define the safety and efficacy of ddI are planned.

While several issues related to the antiretroviral effects of 2',3'-dideoxynucleosides are as yet not resolved, it appears that as 2',3'-dideoxynuucleosides are successively phosphorylated in the cytoplasm of a target cell to yield 2',3'-dideoxynucleoside-5'-triphosphates, they become analogs of the 2'-deoxynucleosides that are the natural substrates for cellular DNA polymerases *and* viral DNA

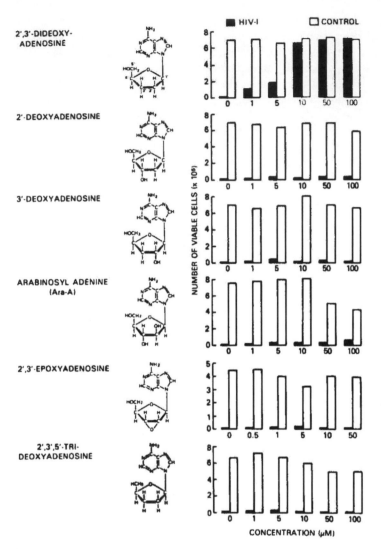

FIGURE 2 Protection of helper T cells against the cytopathic effect of HIV by adenosine congeners. ATH8 cells (2×10^5) were preexposed to polybrene, exposed to HTLV-III$_B$ (2000 virus particles per cell), resuspended in culture tubes (solid columns) inthe presence or absence of various amounts of adenosine congeners: $2',3'$-dideoxyadenosine, $2'$-deoxyadenosine, $3'$-deoxyadenosine (cordycepin), 9-β-D-arabinosyladenine, $2',3'$-β-epoxyadenosine, and $2',3',5'$-trideoxyadenosine. The primed numbers in the top refer to positions in the sugar moiety. Control cells (open columns) were not exposed to the virus. On day 5, the total viable cells were counted by dye exclusion method.

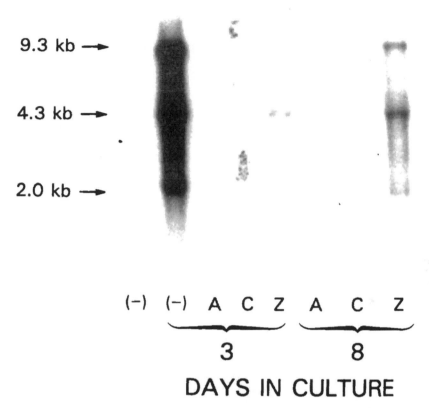

9.3 kb →

4.3 kb →

2.0 kb →

(−) (−) A C Z A C Z

3   8

DAYS IN CULTURE

FIGURE 3   Inhibition of HIV DNA synthesis in helper/inducer T cells exposed to the virus by 2′,3′-dideoxynucleosides. ATH8 cells (2×10^7) were exposed to HTLV–III$_B$ (1000 virus particles per cell) and cultured in the presence or absence of drugs. On days 3 and 8 after exposure to the virus, RNA was extracted and subjected to Northern blot hybridization using a ^{32}P-labeled antisense RNA transcript. On day 3, a substantial amount of viral RNA was detected (on day 4, ATH8 cells began to die from the HIV cytopathic effect, and therefore, on day 8, analyses of RNA from ATH8 cells could not be done). In contrast, when the cells were cultured in the presence of 50 μM dideoxyadenosine (A) and 2 μM dideoxycytidine (C), no viral RNA was detected throughout the study. The anti-HIV effect of 3′-azido-2′,3′-dideoxythymidine (AZT; Z) appears to be less durable in culture than that of dideoxyadenosine and dideoxycytidine. Note that at 10 μM AZT a small amount of viral RNA was detected on day 3, and the RNA expression further increased by day 8 in culture. Eventually these ATH8 cells were killed by HIV by day 15 of culture. In contrast, ATH8 cells protected by dideoxyadenosine or dideoxycytidine could grow continuously for more than 30 days (not shown, see Ref. 78).

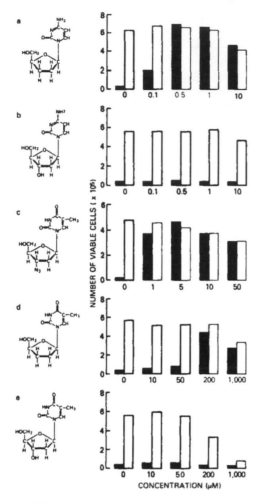

FIGURE 4 Inhibition of the cytopathic effect of HIV by 2′,3′-dideoxycytidine and 3′-azido-2′,3′-dideoxythymidine (AZT) against ATH8 cells. ATH8 cells (2×10^5) were preexposed to polybrene and exposed to HTLV-III$_B$ (2000 virus particles per cell) in culture tubes (solid columns) in the presence or absence of various concentrations of (a) 2′,3′-dideoxycytidine, (b) the normal nucleoside, 2′-deoxycytidine, (c) 3′-azido-2′,3′-dideoxythymidine, (d) 2′,3′-dideoxythimyidine, or (e) the normal nucleoside, 2′-deoxythymidine. Control cells (open columns) were similarly treated, but were not exposed to the virus. On day 5, total viable cells were counted. Note that ≥0.5 μM 2′,3′-dideoxycytidine and ≥1 μM 3′-azido-2′,3′-dideoxythymidine exhibited a strong antiviral effect, and 2′,3′-dideoxythymidine was protective at 200 μM [but this effect was lost by day 10 of culture (55)], while natural deoxynucleosides, 2′-deoxycytidine, and thymidine were not active against the virus. These results are corroborated by showing comparable effects on viral DNA, RNA, and p24 *gag* protein formation in target cells (data not shown).

polymerase (reverse transcriptase). (It is generally thought that nucleoside-5'-triphosphates do not cross cell membranes and are not active as drugs because of their ionic character and comparatively low lipophilicity.) Such analogs could compete with the binding of normal nucleotides to DNA polymerases, or they could be incorporated into DNA and bring about DNA chain termination because of normal $5' \rightarrow 3'$ phosphodiester linkages cannot be completed (see Figure 5). We have been able to show that, as expected, appropriately phosphorylated products of both pyrimidine and purine dideoxynulceoside analogs can serve as substrates for the HIV reverse transcriptase to elongate a DNA chain by one residue, after which the chain is terminated (Figure 6) (78).

Pyrimidine and purine dideoxynucleoside analogs can significantly inhibit the in vitro replication and pathogenic effects of a range of animal and human retroviruses, even when the pathogenic effect being monitored (transformation) requires only a single round of replication; moreover, with certain lentiviruses (a family of animal retroviruses related to HIV), these drugs can reduce the in vitro viral infectivity by more than five orders of magnitude (79). We have recently observed that 2',3'-dideoxyadenosine and 2',3'-dideoxycytidine suppress the infectivity and replication of human immunodeficiency virus II (HIV-II) and simian immunodeficiency virus (80). We have also learned that two dideoxynucleosides, 2',3'-dideoxycytidine and 3'-azido-2',3'-dideoxythymidine (AZT), can block the infectivity of human T-cell leukemia/lymphoma virus type I (HTLV-I) against helper/inducer T cells in vitro (81). The emergence of drug-resistant mutants, however, must always be considered in a discussion of why antiviral drug could fail. Indeed, it has recently been shown that AZT-insensitive HIV variants were isolated from patients with AIDS and ARC who had received AZT for 6 months or more (82).

It is important to stress that the crucial phosphorylation reactions are catalyzed by host cellular kinases, and therefore, caution must be used in extrapolating from experimental results from one species (or a cell type within a species) to another. If these kinases are lacking in the host cell, the retrovirus will appear resistant to these nucleoside analogs, but if the retrovirus is permitted to replicate in a different target cell which has the appropriate enzymes for anabolic phosphorylation, it will appear sensitive again. Similarly, we have observed that one dideoxynucleoside (2',3'-dideoxythymidine) behaves as a relatively poor substrate for human thymidine kinase (J. Balzarini, and S. Broder, unpublished data), and, at the same time, is the least potent among the dideoxynucleosides we have tested (55) (see Figure 4d). The substitution of an azido group at the 3'-carbon of 2',3'-dideoxythymidine (yielding AZT) produces a compound that is an excellent substrate for thymidine kinase and is a very potent inhibitor of HIV replication (vide infra). In this context, the lack of activity against HIV that was observed using 2',3',5'-trideoxyadenosine (bottom of Figure 2) probably relates to the unavailability of the 5'-site to undergo

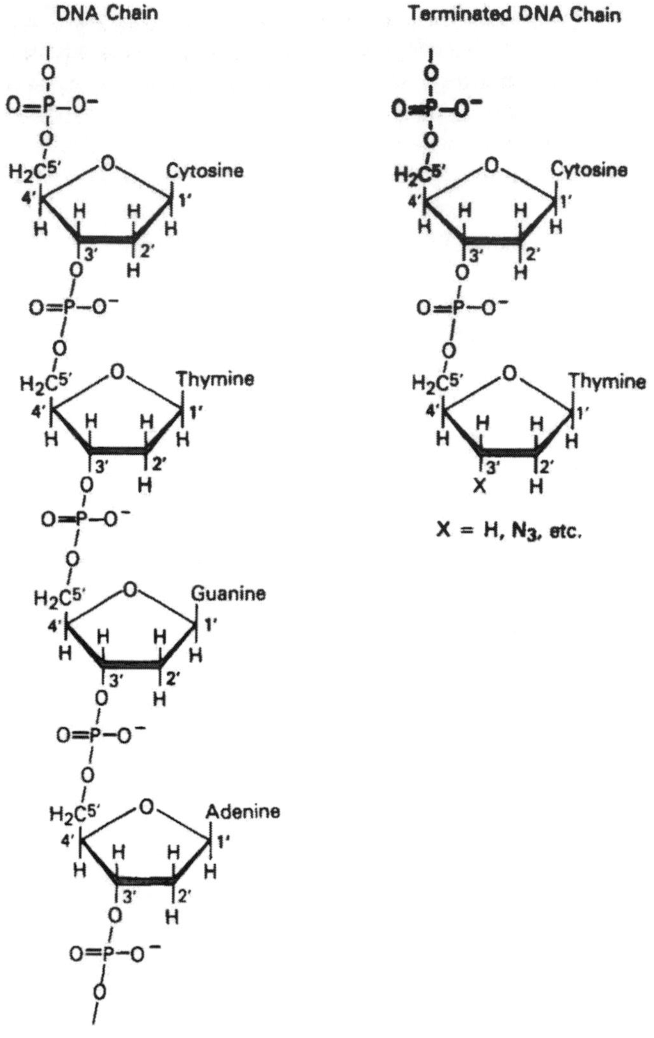

DNA Chain

Terminated DNA Chain

X = H, N₃, etc.

FIGURE 5 Possible mechanism of activity against HIV of $2',3'$-dideoxynucleosides and its derivatives as triphosphate products. When the $3'$-carbon of the deoxyribose is modified by certain substitutions (shown by X in right panel), it is not possible to form $5' \rightarrow 3'$ phosphodiester linkages that are necessary for DNA elongation in the replication of the virus from an RNA form to a DNA from. Reverse transcriptase of HTLV-III/LAV is much more sensitive to the inhibitory effect of $2',3'$-dideoxynucleotides than is mammalian DNA polymerase alpha (see text), which has key DNA synthetic and repair functions in the life of mammalian cells. It is thought that $2',3'$-dideoxynucleosides enter cells, where they are converted to a triphosphate form by cellular enzymes, utilized by reverse transcriptase, and act as DNA-chain terminators in retroviral DNA synthesis, although this does not have to be the only mechanism of the antiretroviral activity.

phosphorylation. To reemphasize, the retrovirus does not provide enzymes for phosphorylation, and therefore, the retrovirus (unlike herpes virus vs. certain antiviral drugs) cannot adopt a strategy of mutating a kinase gene to develop drug resistance.

Some data suggest that the viral DNA polymerase (reverse transcriptase) of HIV is much more susceptible to the inhibitory effects of these drugs as triphosphates than is mammalian DNA polymerase alpha (83,84), an enzyme that has key DNA synthetic and repair functions in the life of a cell. This parallels what had been learned in animal retroviral systems. Although, 2',3'-dideoxynucleotide triphosphates can inhibit mammalian DNA polymerase beta (a repair enzyme) and DNA polymerase gamma (a mitochondrial enzyme) (85), we have observed that dideoxynucleosides can suppress HIV replication and protect sensitive helper/inducer T-cell target cells in vitro for long periods without interfering with the function and growth of target T cells (55,78). As discussed earlier, our working explanation for the activity of these drugs against pathogenic retroviruses is that following anabolism to nucleoside-5'-triphosphates, they competitively inhibit reverse transcriptase or they bring about a selective chain termination as the RNA form of the virus attempts to make DNA copies of itself. The inherent RNase H activity of the viral DNA polymerase would be expected to degrade the viral RNA (46), giving the virus no second chance. Thus, one model for the antiretroviral activity of these compounds is that the viral DNA polymerase is more easily fooled into accepting the dideoxynulceotide than the mammalian enzyme counterpart, or the virus has less capacity than the host cell to repair the incorporation of the false nucleotide, or both. However, in some cases, such drugs may perturb normal nucleoside metabolism or exert other effects that influence antiretroviral activity and drug-related cellular toxicity. We will return to this point later.

VII. AZT

3'-Azido-2',3'-dideoxythymidine (AZT) is a special member of the dideoxynucleoside family that was synthesized over 20 years ago by Horwitz et al. (66) and shown to inhibit C-type murine retrovirus replication in vitro by Ostertag et al. more than 12 years ago (86). However, no application of the agent was found for the practice of medicine. We have recently explored this drug as an experimental agent against HIV infection (53,87). As shown in Figure 4c, AZT is a potent in vitro inhibitor of HIV replication and its cytopathic effect in susceptible target T cells, and it has an antiretroviral effect against widely divergent strains of HIV (53). The drug undergoes anabolic phosphorylation in human T cells to a nucleoside-5'-triphosphate (88), which can compete with thymidine-5'-triphosphate (TTP) and serve as a chain-terminating inhibitor of HIV reverse transcriptase.

In that sense, AZT parallels the other dideoxynucleosides. But it can also bring about several alterations of nucleoside metabolism within host cells,

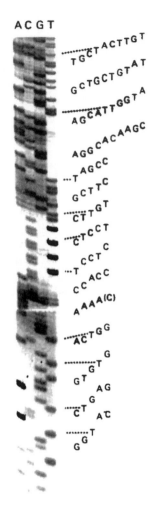

A C G T

...ᵍCᵀACTTGᵀ
TGᶜ

GᶜTGCTGᵀAᵀ

...AGᶜAᵀTᵀGGᵀᴬ

AGGᶜACAAGC

...Tᴬᴳᶜᶜ
GᶜTTᶜ

CᵀTᵀGᵀ

...ᶜTᶜCT

CCTᶜ

...TᶜᶜᴬCᶜ

Aᴬᴬᴬ(C)

...ᴬᶜTGᴳ

...Gᵀ

GᵀAᴳ

...ᶜTAᵀᶜ

...GᴳT

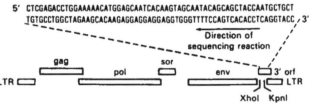

5′ CTCGAGACCTGGAAAAACATGGAGCAATCACAAGTAGCAATACAGCAGCTACCAATGCTGCT
TGTGCCTGGCTAGAAGCACAAGAGGAGGAGGAGGTGGGTTTTCCAGTCACACCTCAGGTACC 3′

Direction of
sequencing reaction

LTR   gag   pol   sor   env   3′ orf   LTR
                                   XhoI KpnI

FIGURE 6 Termination of viral DNA synthesized de novo by HIV reverse transcriptase. Each dideoxynuceoside-5′-triphosphate, dideoxyadenosine (A), dideoxycytidine (C), dideoxyguanosine (G), or dideoxythymidine (T), was incorporated into the growing DNA chain in place of the proper base and thus terminated the DNA chain. An M13 clone of Bam HI/Kpn I DNA fragment of a cloned HIV, HXB2, following annealing with an M13 universal primer, served as a template for nucleotide sequencing in a Sanger sequencing reaction using

which may be of clinical significance. For example, thymidine kinase catalyzes the formation of a monophosphate derivative of AZT, which then serves as a competitive inhibitor, with low V_{max}, of thymidylate kinase (an enzyme that catalyzes the second phosphorylation step to thymidine diphosphate), leading to the reduced formation of the third phosphorylation product, TTP (88). From this perspective, AZT has an interesting capacity to lower the level of its natural substrate competitor (TTP) for viral and cellular DNA polymerases. AZT can also reduce 2'-deoxycytidine triphosphate (dCTP) levels (88), but the mechanism of this effect is unknown. How these reductions of normal pyrimidine pools might affect the balance between antiretroviral activity and drug-induced cytotoxicity remains unknown. Nevertheless, pyrimidine starvation is likely to contribute to bone marrow suppression, one of the key side effects of the drug. This feature of the drug might lend itself to regimens that combine AZT with an agent that does not appear to deplete intracellular pyrimidine pools, such as dideoxycytidine (76). There are also other features that might have substantial clinical relevance, but remain unexplained at this time, including the capacity of acyclovir (a guanosine analog with potent activity against herpes viruses but rather weak antiretroviral activity per se) to potentiate the antiretroviral effect of AZT in vitro (22). Surbone and her coworkers have recently shown that the combination of AZT and acyclovir is well tolerated to patients with AIDS and ARC (89).

We began administering AZT to patients with AIDS and related conditions in a phase I (feasibility and toxicity-testing) trial in July 1985. In that study, it was found that the drug had good oral bioavailability and excellent penetration across the blood-brain barrier. Our conclusion from the phase I study (84) and extensions of that study (R. Yarchoan and S. Broder, unpublished data) is that during short courses of administration, AZT can bring about partial restoration of immune function and improve certain other clinical abnormalities associated with AIDS. The limiting toxicity observed was bone marrow suppression, predominantly in the form of anemia. In some adults and children with HIV-mediated dementia, the drug can cause neurologic improvements that are quite prompt and dramatic; however, the durability of the improvement is still not known (90). Moreover, the true frequency of neurological improvements cannot be determined until the appropriate controlled trials are concluded.

HIV reverse transcriptase purified by immunoaffinity chromatography with a monoclonal antibody against HIV reverse transcriptase (47). The DNA sequence matches the complementary base of the indicated nucleotide sequence of the 3'-orf gene of HXB2. The genome of HIV (tat-III, art/trs, and R genes are not shown) with the known nucleotide sequence of the Xho I/Kpn I fragment of HXB2 is illustrated at the bottom. (Reproduced from Ref. 78 by permission of the *Proceedings of the National Academy of Science of U.S.A.*)

AZT was tested in a randomized study of patients with fulminant AIDS and poor prognostic AIDS-related complex under double-blind, placebo-controlled conditions at several centers. The study was begun in February 1986 and approximately 140 patients were accrued to each arm shortly thereafter. By the end of September 1986 there were 19 deaths in the placebo arm and only one death in the AZT arm ($p < 0.001$). At the same time, patients in the drug arm showed other evidence of clinical and immunological improvements (91). However. it was also noted that AZT brings about a significant bone marrow suppression, which has been a serious side effect in patients with advanced AIDS (92). Nevertheless, the results at hand warrant the initiation of appropriate controlled studies to see whether therapy with AZT at an early point of retroviral infection can block the onset of AIDS in individuals who are at risk, who are otherwise free of symptoms, at the time of treatment. Furthermore, although AZT is not a cure for HIV infection, there is no doubt that this drug does prolong survival of certain patients.

Thus, AZT, a drug chosen on the basis of its selective in vitro antiviral effect against HIV, has been shown as a single agent to confer a clinical benefit in patients with advanced disease. In this respect, it represents a first step in developing practical chemotherapy against pathogenic human retroviruses. While the clinical results might be important in their own right, they are also important because they validate the idea that therapy for established retroviral infections is a feasible goal. This, in turn, should compel us to redouble our search for better and less toxic therapy.

VIII. CONCLUSION

The product(s) of each HIV gene represents a potential target of opportunity in developing new experimental therapies for diseases caused by this virus, and there are several steps related to the viral life-cycle which may be relevant for future therapeutic strategies. Curative therapy for diseases caused by pathogenic retroviruses will probably not be possible until the molecular biology of the virus and the structural chemistry of the key viral products are defined through further basic research, but it is not necessary to await new breakthroughs before implementing some strategies that we already know have clinical applications in patients with advanced disease. At this point, the question is no longer whether drugs that inhibit HIV can confer a clinical benefit in patients with AIDS and related diseases; the question now is how many such drugs will be found and what is the best way to apply candidate drugs for clinical trials.

ACKNOWLEDGMENTS

We thank Drs. Marvin S. Reitz, Ruth F. Jarrett, and Flossie Wong-Staal for their helpful discussions, Dr. David Barry for providing 3′-azido-2′,3′-dideoxythymidine, and Drs. Mika Popovic and Robert C. Gallo for providing HIV virions.

REFERENCES

1. Wong-Staal, F., and Gallo, R. C. (1985). Human T-lymphotropic retroviruses. *Nature* 317:395–403.
2. Barré-Sinoussi, F., Chermann, J. C., Rey, F., Nugeyre, M. T., Chamaret, S., Gruest, J., Dauguet, C., Axler-Blin, C., Vézinet-Brun, F., Rouxioux, C., Rozenbaum, W., and Montagnier, L. (1983). Isolation of a T cell lymphotropic virus from a patient at risk for acquired immunodeficiency syndrome (AIDS). *Science* 220:868–871.
3. Popovic, M., Sarngadharan, M. G., Read, E., and Gallo, R. C. (1984). Detection, isolation, and continuous production of cytopathic retrovirus (HTLV-III) from patients with AIDS and pre-AIDS. *Science* 224:497–500.
4. Gallo, R. C., Salahuddin, S. Z., Popovic, M., Shearer, G. M., Kaplan, M., Haynes, B. F., Palker, T. J., Redfield, R., Oleske, J., Safai, B., White, G., Foster, P., and Marcham, P. D. (1984). Frequent detection and isolation of cytopathic retroviruses (HTLV-III) from patients with AIDS and at risk for AIDS. *Science* 224:500–503.
5. Levi, J. A., Hoffman, A. D., Kramer, S. M., Lanois, J. A., Shimabukuro, J. M., and Oshiro, L. S. (1984). Isolation of lymphocytopathic retroviruses from San Francisco patients with AIDS. *Science* 225:840–842.
6. Coffin, J., Haase, A., Levy, J., Montagnier, L., Oroszlan, S., Teich, N., Temin, H., Toyoshima, K., Varmus, H., Vogt, P., and Weiss, R. (1986). What to call the AIDS virus? *Nature* 321:10.
7. Gottlieb, M. S., Schroff, R., Schanker, H. M., Weisman, J. D., Fan, P. T., Wolf, R. A., and Saxon, A. (1981). *Pneumocystis carinii* pneumonia and mucosal candidiasis in previously healthy homosexual men. Evidence of a new acquired cellular immunodeficiency. *N. Engl. J. Med.* 305:1425–1431.
8. Broder, S., and Gallo, R. C. (1984). A pathogenic retrovirus (HTLV-III) linked to AIDS. *N. Engl. J. Med.* 311:1292–1297.
9. Rous, P. (1911). *JAMA* 56:198.
10. Baltimore, D. (1970). RNA-dependent DNA polymerase in virions of RNA tumor viruses. *Nature* 226:1209–1211.
11. Temin, H., and Mizutani, S. (1970). RNA-dependent DNA polymerase in virions of Rous sarcoma virus. *Nature* 226:1211–1213.
12. Poiesz, B. J., Ruscetti, F. W., Gazdar, A. F., Bunn, P. A., Minna, J. D., and Gallo, R. C. (1980). Detection and isolation of type C retrovirus particles from fresh and cultured lymphocytes of a patient with cutaneous T-cell lymphoma. *Proc. Natl. Acad. Sci. USA* 77:7415–7419.

13. Morbidity and Mortality Weekly Report (1989) AIDS and human immuno-deficiency virus infection in the United States: 1988 Update. 38 (S-4): 1–39.

14. Parks, W. P., and Scott, G. B. (1987). An overview of pediatric AIDS: Approaches to diagnosis and outcome assessment. In: *AIDS: Modern Concepts and Therapeutic Challenges* (S. Broder, ed.), New York, Marcel Dekker, pp. 245–262.

15. Yarchoan, R., Redfield, R. R., and Broder, S. (1986). Mechanism of B cell activation in patients with acquired immunodeficiency syndrome and related disorders. *J. Clin. Invest.* 78:439–447.

16. Shearer, G. W., Salahuddin, S. Z., Markham, P. D., Joseph, L. J., Payne, S. M., Kriebel, P., Bernstein, D. C., Biddison, W. E., Sarngadharan, M. G., and Gallo, R. C. (1985). Prospective study of cytotoxic T lymphocyte responses to influenza and antibodies to human T lymphotropic virus-III in homosexual men. *J. Clin. Invest.* 76:1699–1704.

17. Shaw, G. M., Harper, M. E., Hahn, B. H., Epstein, L. G., Gajdusek, D. C., Price, R. W., Navia, B. A., Petito, C. K., O'Hara, C. J., Groopman, J. E., Cho, E.-S., Oleske, J. M., Wong-Staal, F., and Gallo, R. C. (1985). HTLV–III infection in brains of children and adults with AIDS encephalopathy. *Science* 227:177–182.

18. Ho, D. D., Rota, T. R., Schooley, R. T., Kaplan, J. C., Allan, J. D., Grrop-man, J. E., Resnick, L., Felsenstein, D., Andrews, C. A., and Hirsch, M. S. (1985). Isolation of HTLV-III from cerebrospinal fluid and neural tissue of patients with neurologic syndromes related to the acquired immunode-ficiency syndrome. *N. Engl. J. Med.* 313:1493–1497.

19. Navia, B. A., Jordan, B. D., and Price, R. W. (1986). The AIDS dementia complex. I. Clinical features. *Ann. Neurol.* 19:517–524.

20. Gartner, S., Markovits, P., Markovitz, D. M., Kaplan, M. H., Gallo, R. C., and Popovic, M. (1986). The role of mononuclear phagocytes in HTLV-III/LAV infection. *Science* 233:215–219.

21. Koenig, S., Gendelman, H. E., Orenstein, J. M., Dal Canto, M. C., Pezeshk-pour, G. H., Yungbluth, M., Janotta, F., Aksamit, A., Martin, M. A., and Fauci, A. S. (1986). Detection of AIDS virus in macrophages in brain tissue from AIDS patients with encephaloopathy. *Science* 233:1089–1093.

22. Mitsuya, H. and Broder, S. (1987) Strategies for antiviral therapy in AIDS. *Nature*, 325:773–778.

23. Mitsuya, H., Matsukura, M., and Broder, S. (1987). Rapid in vitro systems for assessing activity of agents against HTLV-III/LAV. In: *AIDS: Modern Concepts and Therapeutic Challenges* (S. Broder, ed.), pp. 303–333. New York, Marcel Dekker.

24. Gross, L. (1983). *Oncogenic Viruses*, 3rd ed. Oxford, Pergamon Press.

25. Toh, H., Hayashida, H., and Miyata, T. (1983). Sequence homology be-tween retroviral reverse transcriptase and putative polymerases of hepatitis and cauliflower mosaic virus. *Nature* 305:827–829.

26. Ratner, L., Haseltine, W., Patarca, R., Livak, K. J., Starcich, B., Josephs, S. F., Doran, E. R., Rafalski, J. A., Whitehorn, E. A., Baumeister, K.,

Ivanoff, L. K. J., Petteway, S. R., Pearson, M. L., Lautenberger, J. A., Papas, T. S., Ghrayeb, J., Chang, N. T., Gallo, R. A., and Wong-Staal, F. (1985). Complete nucleotide sequence of the AIDS virus, HTLV–III. *Nature* 313:277–284.

27. Wain-Hobson, S., Sonigo, P., Danos, O., Cole, S., and Alizon, M. (1985). Nucleotide sequence of the AIDS virus, LAV. *Cell* 40:9–17.

28. Sodroski, J., Rosen, C., Wong-Staal, F., Salahuddin, S. Z., Popovic, M., Arya, S., and Gallo, R. C. (1985). Trans-acting transcriptional regulation of human T-cell leukemia virus type III long terminal repeat. *Science* 227: 171–173.

29. Sodroski, J., Goh, W. C., Rosen, C., Dayton, A., Terwilliger, E., and Haseltine, W. A. (1986). A second post-transcriptional *trans*-activator gene required for HTLV–III replication. *Nature* 321:412–417.

30. Feinberg, M. B., Jarrett, R. F., Aldovini, A., Gallo, R. C., and Wong-Staal, F. (1986). HTLV–III expression and production involve complex regulation at the level of splicing and translation of viral RNA. *Cell* 46: 807–817.

31. Wong-Staal, F., Chandra, P. K., and Chrayeb, J. (1987). Human immunodeficiency virus type III: The eighth gene. *AIDS Res. Hum. Retroviruses* 3: 33–39.

32. Tanese, N., Sodroski, J., Haseltine, W. A., and Goff, S. P. (1986). Expression of reverse transcriptase activity of human T-lymphotropic type III (HTLV–III/LAV) in *Escherichia coli. J. Virol.* 59:743–745.

33. Kramer, R. A., Schaber, M. D., Skalka, A. M., Ganguly, K., Wong-Staal, F., and Reddy, E. P. (1986). HTLV–III *gag* protein is processed in yeast cells by the virus pol-protease. *Science* 231:1580–1584.

34. Dalgleish, A. G., Beverley, C. L., Calpman, P. R., Crawford, D. H., Greaves, M. F., and Weiss, R. A. (1985). The CD4 (T4) antigen is an essential component of the receptor for the AIDS retrovirus. *Nature* 312:763–767.

35. Klatzman, D., Champagne, E., Chamaret, S., Gruest, J., Guetard, D., Hercend, T., Gluckman, J. C., and Montagnier, L. (1984). T-lymphocyte T4 molecule behaves as the receptor for human retrovirus LAV. *Nature* 312:767–768.

36. Sodroski, J., Goh, W. C., Rosen, C., Campbell, K., and Haseltine, W. A. (1986). Role of HTLV–III/LAV envelope in syncytium formation and cytopathicity. *Nature* 322:470–474.

37. Fisher, A. G., Ratner, L., Mitsuya, H., Marselle, L. M., Harper, M. E., Broder, S., Gallo, R. C., and Wong-Staal, F. (1986). Infectious mutants of HTLV–III with changes in the 3' region and markedly reduced cytopathic effects. *Science* 655–659.

38. Hahn, B. H., Gonda, M. A., Shaw, G. M., Popovic, M., Hoxie, J. A., Gallo, R. C., and Wong-Staal, F. (1985). Genomic diversity of the acquired immune deficiency syndrome virus HTLV–III; Different viruses exhibit

39. Saag, M. S., Hahn, B. H., Gibbons, J., Li, Y., Parks, E. S., Parks, W. P., and Shaw, G. M. (1988) Extensive variation of human immunodeficiency virus type-1 *in vitro. Nature* 334: 449-444

greatest divergence in their envelope genes. *Proc. Natl. Acad. Sci. USA* 82: 4813–4817.

40. Lasky, L. A., Groopman, J. E., Fennie, C. W., Benz, P. M., Capon, D. J., Dowbenko, D. J., Nakamura, G. R., Nunes, W. M., Renz, M. E., and Berman, P. W. (1986). Neutralization of the AIDS retrovirus by antibodies to a recombinant envelope glycoprotein. *Science* 233:209–212.

41. Matsushita, S., Robert-Guroff, M., Trepel, J., Cossman, J., Mitsuya, H., and Broder, S. (1986). Human monoclonal antibody directed against an envelope glycoprotein of human T-cell leukemia virus type I. *Proc. Natl. Acad. Sci. USA* 83:2672–2676.

42. Weiss, R. A., Clapham, P. R., Cheingsong-Popov, R., Dagleish, A. G., Carne, C. A., Weller, V. D., and Tedder, R. S. (1985). Neutral antibodies of human T-lymphotropic virus type 3 in AIDS and AIDS-risk patients. *Nature* 316: 69–72.

43. Robert-Guroff, M., Brown, M., and Gallo, R. C. (1985). HTLV–III-neutralizing antibodies in patients with AIDS and AIDS-related complex. *Nature* 316:72–74.

44. Nemerow, G. R., and Cooper, N. R. (1984). Infection of B lymphocytes by a human herpes virus, Epstein-Barr virus, is blocked by calmodulin antagonists. *Proc. Natl. Acad. Sci. USA* 81:4955–4959.

45. Muesing, M. A., Smith, D. H., Cabradilla, C. D., Benton, C. V., Lasky, L. A., and Capon, D. J. (1985). Nucleic acid structure and expression of the human AIDS/lymphadenopathy retrovirus. *Nature* 313:450–458.

46. Collett, M. S., Dierks, P., Parsons, J. D., and Faras, J. T. (1978). RNase H hydrolysis of the 5′ terminus of the avian sarcoma virus genome during reverse transcription. *Nature* 272:181–183.

47. Di Marzo-Veronese, F., Copeland, T. D., DeVico, A. L., Rahman, R., Oroszlan, S., Gallo, R. C., and Sarngadharan, M. G. (1986). Characterization of highly immunogenic p66/p51 as the reverse transcriptase of HTLV–III/LAV. *Science* 231:1289–1291.

48. Barr, P. J., Power, M. D., Lee-Ng, C. T., Gibson, H. L., and Leciw, P. A. (1987). Expression of active human immunodeficiency virus reverse transcriptase in *Saccharomyces cerevisiae*. *Biotechnology* 5:486–489.

49. Mitsuya, H., Popovic, M., Yarchoan, R., Matsushita, S., Gallo, R. C., and Broder, S. (1984). Suramin protection of T cells in vitro against infectivity and cytopathic effect of HTLV–III. *Science* 226:172–174.

50. Mitsuya, H., Matsushita, S., Yarchoan, R., and Broder, S. (1985). Protection of T cells against infectivity and cytopathic effect of HTLV–III in vitro. In: *Retroviruses in Human Lymphoma/Leukemia* (M. Miwa et al., eds.), pp. 277–288, Utrecht, Japan Sci. Soc. Press, Tokyo/VNU Science Press.

51. Broder, S., Yarchoan, R., Collins, J. M., Lane, H. C., Markham, P. D., Mitsuya, H., Hoth, D. F., Gelmann, E., Groopman, J. E., Resnick, L., Gallo, R. C., Myers, C. E., and Fauci, A. S. (1985). Effects of suramin on HTLV–III/LAV infection presenting as Kaposi's sarcoma or AIDS-related complex:

clinical pharmacology and suppression of virus replication in vivo. *Lancet* 2:627–630.

52. Rosenbaum, W., Dormont, D., Spire, B., Vilmer, E., Gentilini, M., Griscelli, C., Montagnier, L., Barre-Sinoussi, F., and Chermann, J. C. (1985). Antimoniotungstate (HPA 23) treatment of three patients with AIDS and one with prodrome. *Lancet* 1:450–451.

53. Mitsuya, H., Weinhold, K. J., Furman, P. A., St. Clair, M. H., Lehrman, S. N., Gallo, R. C., Bolognesi, D., Barry, D. W., and Broder, S. (1985). 3'-Azido-3'-deoxythymidine (BW A509U): An antiviral agent that inhibits the infectivity and cytopathic effect of human T-lymphotropic virus type III/lymphadenopathy-associated virus in vitro. *Proc. Natl. Acad. Sci. USA* 82:7096–7100.

54. Sandstrom, E. G., Kaplan, J. C., Byington, R. E., and Hirsch, M. S. (1985). Inhibition of human T-cell lymphotorpic virus type III in vitro by phosphonoformate. *Lancet* 1:1480–1482.

55. Mitsuya, H., and Broder, S. (1986). Inhibition of the in vitro infectivity and and cytopathic effect of human T-lymphotropic virus type III/lymphadenopathy-associated virus (HTLV–III/LAV) by 2',3'-dideoxynucleosides. *Proc. Natl. Acad. Sci. USA* 83:1911–1915.

56. Anand, R., Moore, J., Feorinom, P., Curran, J., and Srinivasa, A. (1986). Rifabutine inhibits HTLV–III. *Lancet* 1:97–98.

57. Zagury, D., Bernard, J., Leonard, R., Cheynier, R., Feldman, M., Sarin, P. S., and Gallo, R. C. (1986). Long-term cultures of HTLV–III-infected T-cells: A model of cytopathology of T-cell depletion in AIDS. *Science* 231:850–853.

58. Cullen, B. R. (1986). *Trans*-activation of human immunodeficiency virus occurs via a bimodal mechanism. *Cell* 46:973–982.

59. Okamoto, T., and Wong-Staal, F. (1986). Demonstration of virus-specific transcriptional activator(s) in cells infected with HTLV–III by an in vitro cell-free system. *Cell* 47:29–35.

60. McCormick, J. B., Getchell, J. P., Mitchell, S. W., and Hicks, D. R. (1984). Ribavirin suppresses replication of lymphadenopathy-associated virus in culture of human adult lymphocytes. *Lancet* 2:1367–1369.

61. Zamenick, P. C., Goodchild, J., Taguchi, Y., and Sarin, P. (1986). Inhibition of replication and expression of human T-cell lymphotropic virus type III in cultured cells by exogenous synthetic oligonucleotide complimentar to viral RNA. *Proc. Natl. Acad. Sci. USA* 83:4143–4146.

62. Matsukura, M., Zon, G., Shinozuka, K., Robert-Guroff, M., Shimada, T., Stein, C., Mitsuya, H., Wong-Staal, F., Cohen, J. S., and Broder, S. (1989) Regulation of viral expression of human immunodeficiency virus *in vitro* by an antisense phosphorothioate ologodeoxynucleotide against *rev* (*art/trs*) in chronically infected cells. *Proc. Natl. Acad. Sci. USA* 86:4244–4248.

63. Ho, D. D., Hartshorn, K. L., Rota, T. R., Andrews, C. K., Kaplan, J. C., Shooley, R. T., and Hirsch, M. S. (1985). Recombinant human interferon alpha suppressed HTLV–III replication in vitro. *Lancet* 1:602–604.

64. Bolognesi, D. P., and Fischinger, P. J. (1985). Prospects for treatment of human retrovirus-associated diseases. *Cancer Res.* 45:4700s–4705s.

65. Robins, M. J., and Robins, R. K. (1964). The synthesis of 2′,3′-dideoxyadenosine from 2′-deoxyadenosine. *J. Am. Chem. Soc.* 86:3585–3586.

66. Horwitz, J. P., Chua, J., and Noel, M. (1964). Nucleosides V. The monomesylates of 1-(2′-deoxy-β-D-lyxofuranosyl)thymine. *J. Org. Chem.* 29: 2076–2078.

67. Robins, M. J., McCarthy, J. R., and Robins, R. K. (1966). *Biochemistry* 5: 224–231.

68. Doering, A. M., Jansen, M., and Cohen, S. S. (1966). Polymer synthesis in killed bacteria: Lethality of 2′,3′-dideoxyadenosine. *J. Bact.* 92:565–574.

69. Horwitz, J. P., Chua, J., Noel, M., and Donatti, J. T. (1967). Nucleosides. XI. 2′,3′-dideoxycytidine. *J. Org. Chem.* 32:817–818.

70. Atkinson, M. R., Deutscher, M. P., Kornberg, A., Ruseel, A. F., and Moffat, J. G. (1969). Enzymatic synthesis of Deoxyribonucleic acid. XXXIV. *Biochemistry*, 8:4897–4904.

71. Toji, L., and Cohen, S. S. (1970). Termination of deoxyribonucleic acid in *Escherichia coli* by 2′,3′-dideoxyadenosine. *J. Bact.* 103:323–328.

72. Smoller, D., Molineux, I., and Baltimore, D. (1971). Direction of polymerization by the avian myeloblastosis virus deoxyribonucleic acid polymerase. *J. Biol. Chem.* 246:7697–7700.

73. Faras, A. J., Taylor, J. M., Levinson, W. E., Goodman, H. M., and Bishop, J. M. (1973). RNA-directed DNA polymerase of Rous sarcoma virus: Initiation of synthesis with 70S viral RNA as template. *J. Mol. Biol.* 79: 163–183.

74. Lin, T. S., and Prusoff, W. H. (1978). Synthesis and biological activity of several amino analogs of thymidine. *J. Med. Chem.* 21:109–112.

75. Sanger, F., Nicklen, S., and Coulson, A. R. (1977). DNA sequencing with chain-terminating inhibitors. *Proc. Natl. Acad. Sci. USA* 74:5463–5467.

76. Yarchoan, R., Perno, C. F., Thomas, R. V., Klecker, R. W., Allain, J.-P., Willis, R. J., McAtee, N., Fischl, M. A., Mitsuya, H., Pluda, J. M., Lawlee, T. J., Leuther, M., Safai, B., Collins, J. M., Myers, C.E., and Broder, S. (1988). Phase I studies of 2′,3′-dideoxycytidine in severe human immunodeficiency virus infection as a single agent and alternating with Zidovudine (AZT). *Lancet* 1:76–80.

77. Yarchoan, R., Mitsuya, H., Thomas, R. V., Pluda, J.M., Hartman, N. R., Perno, C.-F., Marczyk, K. S., Allain, J.-P., Johns, D. G., and Broder, S. 1989. *In vivo* activity against HIV and favorable toxicity profile of 2′,3′-dideoxyinosine. *Science* (in press).

78. Mitsuya, H., Jarrett, R. F., Matsukura, M., Veronese, F., Devico, A. L., Sarngadharan, M. G., Johns, D. G., Reitz, M. S., and Broder, S. (1987). Long-term inhibition of HTLV–III/LAV DNA synthesis and RNA expression in T-cells protected by 2′,3′-dideoxynucleosides. *Proc. Natl. Acad. Sci. USA* 84:2033–2037.

79. Dahlberg, J. E. Mitsuya, H., Broder, S., Blam, S. B., and Aaronson, S. A. (1987). Broad spectrum antiretroviral activity of 2′,3′-dideoxynucleosides. *Proc. Natl. Acad. Sci. USA* 84:2469–2473.

80. Mitsuya, H. and Broder, S. 1988. Inhibition of infectivity and replication of HIV-2 and SIV in helper T-cells by 2',3'-dideoxynucleosides *in vitro. AIDS Res. Hum. Retrov.* 4:107–113.

81. Matsushita, S., Mitsuya, H., Reitz, M. S., and Broder, S. (1987). Pharmacological inhibition of human T-lymphotropic virus type 1 (HTLV-I). *J. Clin. Invest.,* in press.

82. Larder, B. A., Darby, G., and Richman, D. D. 1989. HIV with reduced sensitivity to zidovudine (AZT) isolated during prolonged therapy. *Science,* 243:1731–1734.

83. Mitsuya, H., Dahlberg, J. E., Spigelman, Z., Matsushita, S., Jarrett, R. F., Matsukura, M., Currens, M. J., Aaronson, S. A., Reitz, M. S., McCaffrey, R. S., and Broder, S. (1987). *Proc. UCLA Symposium on Human Retroviruses, Cancer and AIDS,* in press.

84. Starnes, M. C., and Cheng, Y-C. (1987). Cellular metabolism of 2',3'-dideoxycytidine, a compound active against human immunodeficiency virus in vitro. *J. Biol. Chem.* 226:988–991.

85. Waqar, M. A., Evans, M. J., Manly, K. F., Hughs, R. G., and Huberman, J. A. (1984). Effects of 2',3'-dideoxynucleosides on mammalian cells and viruses. *J. Cell. Physiol.* 121:402–408.

86. Ostertag, W., Roesler, G., Krieg, C. J., Kind, J., Cole, T., Crozier, T., Gaedicke, G., Steinherider, G., Kluge, N., and Dube, S. (1974). Induction of Endognous virus and of thymidine kinase by bromodeoxyuridine in cell cultures transformed by Friend virus. *Proc. Natl. Acad. Sci. USA* 71:4980–4985.

87. Yarchoan, R., Klecker, R. W., Weinhold, K. J., Markham, P. D., Lyerly, H. K., Durack, D. T., Gelmann, E., Lehrman, S. N., Blum, R. M., Baryy, D. W., Shearer, G. M., Fischl, M. A., Mitsuya, H., Gallo, R. C., Collins, J. M., Bolognesi, D. P., Myers, C. E., and Broder, S. (1986). Administration of 3'-azido-3'-deoxythymidine, An inhibitor of HTLV–III/LAV replication, to patients with AIDS or AIDS-related complex. *Lancet* 1:575–580.

88. Furman, P. A., Fyfe, J. A., St. Clair, M. H., Weinhold, K., Rideout, J. L., Freeman, G. A., Nusinoff-Lehrman, S., Bolognesi, D., Broder, S., Mitsuya, H., and Barry, D. W. (1986). Phosphorylation of 3'-azido-3'-deoxythymidine and selective interaction of the 5'-triphosphate with human immunodeficiency virus reverse transcriptase. *Proc. Natl. Acad. Sci. USA* 83:8333–8337.

89. Surbone, A., Yarchoan, R., McAtee, N., Blum, R., Maha, M., Allain, J.-P., Thomas, R. V., Mitsuya, H., Lehrman, S., Leuther, M., Pluda, J. M., Jacobsen, F. K., Kessler, H. A., Myers, C. E., and Broder, S. (1988) Treatment of the acquired immunodeficiency syndrome (AIDS) and AIDS-related complex with a regimen of 3'-azido-2',3'-dideoxythymidine (azidothymidine or zidovudine) and acyclovir. *Ann. Int. Med.* 108:534–540.

90. Yarchoan, R., Berg, G., Brouwers, P., Spitzer, A. R., Fischl, M. A., Thomas, R. V., Schmidt, P., Safai, B., Perno, C. F., Myers, C. E., and Broder, S. (1987). Preliminary observation of the response of HTLV–III/LAV-associated neurological disease to the administration of 3'-azido-3'-deoxythymidine. *Lancet* 1:132–135.

91. Fischl, M. A., Richman, D. D., Grieco, M. H., Gottlieb, M. S., Volberding, P. A., Laskin, S. L., Leedom, J. M., Allan, J. D., Mildvan, D., Schooley, R. T., Jackson, G. G., Durack, D. T., King, D., and AZT Collaborative Group. (1987). The efficacy of 3'-azido-3'-deoxythimidine (azidotdhymidine) in the treatment of patients with AIDS and AIDS-related complex: A double-blind placebo-controlled trial. *N. Engl. J. Med.*, (in press).

92. Richman, D. D., Fischl, M. A., Greico, M. H., Gottlieb, M. S., Volberding, P. A., Laskin, O. L., Leedom, J. M., Groopman, J. M., Mildvan, D., Hirsch, M. S., Jackson, G. G., Durack, D. T., Nusinoff-Lehrman, S., and the AZT collaborative working group. (1987) The toxicity of azidothymidine (AZT) in the treatment of patients with AIDS and AIDS-related complex. A double blind, placebo-controlled trial. *New Engl. J. Med.*, 317: 192-197.

93. Mitsuya, H., Guo, H-G., Cossman, J., Megson, M., Reitz, M. S., and Broder, S. (1984). Functional properties of antigen-specific T cells infected by human T-cell leukemia-lymphoma virus (HTLV-I). *Science* 225:1484-1486.

14

The Clinical Spectrum of HIV Infection

R. Jan Gurley and **Jerome E. Groopman** *New England Deaconess Hospital, Harvard Medical School, Boston, Massachusetts*

I. INTRODUCTION

The first descriptions of the acquired immunodeficiency syndrome were chillingly accurate reports of what was becoming increasingly recognized as a new disease entity. In the December 10, 1981 issue of the *New England Journal of Medicine*, three separate articles focused on all of the major features of the syndrome (1–3). In the abstract of "*Pneumocystis carinii* Pneumonia and Mucosal Candidiasis in Previously Healthy Homosexual Men: Evidence of a New Acquired Cellular Immunodeficiency," the authors write:

> Four previously healthy homosexual men contracted *Pneumocystis carinii* pneumonia, extensive mucosal candidiasis, and multiple viral infections. In three of the patients these infections followed prolonged fevers of unknown origin. In all four cytomegalovirus was recovered from secretions. Kaposi's sarcoma developed in one patient eight months after he presented with esophageal candidiasis. All patients were anergic and lymphopenic; they had no lymphocyte proliferative responses to soluble antigens, and their responses to phytohemagglutinin were markedly reduced. Monoclonal-antibody analysis of peripheral-blood T-cell subpopulations revealed virtual elmination of the Leu-3+ helper/inducer subset, and increased percentage of the Leu-2+ suppressor/cytotoxic subset, and an increased percentage of cells bearing the thymocyte-associated antigen T10.

Following the recognition of this disease entity, a mere 2 years passed before the isolation of the etiological agent, a retrovirus now called human immunodeficiency virus (HIV) (4–7). Much has been learned, not only about the biology of the virus, but also about the clinical spectrum of disease manifestations and complications. The original descriptions of the acquired immunodeficiency syndrome are now recognized as the end result of HIV infection in its terminal stages. Prior to that event, the disease may pass through a variety of stages, varying from the clinically silent asymptomatic carrier state to differing types of immune aberrations to the AIDS dementia complex. Several recent epidemiological studies have also attempted to decipher the probability that an infected individual will progress to develop symptoms. This chapter focuses on what is now known about the clinical consequences of HIV infection.

II. ACUTE HIV INFECTION

Original descriptions of the acquired immunodeficiency syndrome were limited to persons of specific, well-defined high-risk groups. These included homosexual men, IV drug abusers, hemophiliacs, transfusion recipients, and Haitian immigrants to the United States. Because the disease was transmissable by blood and blood products, the etiological agent was known to be a nonfilterable substance, probably a virus (8–12). Additional support for this hypothesis was based on the observation that other blood-borne viral diseases, such as hepatitis B, are common in these high-risk groups. Evaluations of contact spread of the disease, or point source exposure to a group of persons, or infection following a known exposure were difficult to document initially. Problems included multiple exposures for most individuals in high-risk groups (e.g., many homosexual contacts, usage of pooled sera from thousands of donors for hemophiliacs), initial lack of methods for documenting the presence of virus or antibody to virus, and the believed long latency between exposure and development of clinical AIDS. Recently, however, several studies have attempted to document known seroconversion from isolated exposures to the virus. In three of these papers (13–15), a new clinical syndrome, known as acute HIV infection or primary HTLV-III infection, has been described associated with initial exposure to the virus.

Primary HIV infection is described as an acute febrile illness resembling influenza or mononucleosis. Incubation time until the onset of symptoms varied from 6 days to 7 weeks. In the retrospective study by Cooper et al., symptoms of the illness in 11 patients lasted up to 14 days (14). In the prospective study of three patients by Ho et al., symptoms lasted up to three and a half weeks (15). Seroconversion for antibodies to the virus followed from 3 to 12 weeks from the time of the presumed exposure, although, in the study by Ho, viral isolation was possible in two of three patients prior to seroconversion. Seroconversion was detected by immunofluorescence antibody testing, Western blot analysis, or radioimmunoprecipitation methods. The following clinical features were cited most commonly among the 14 patients in the two studies reporting symptoms:

Signs and symptoms	Percentage (number)
Fever/sweats	100 (14)
Myalgias/arthralgias	100 (14)
Malaise/lethargy	71 (10)
Lymphadenopathy	64 (9)
Sore throat	64 (9)
Anorexia/nausea/vomiting	57 (8)
Maculopapular rash	57 (8)
Diarrhea	36 (5)

Other symptoms included a stiff neck (2), severe shooting pains (2), urticaria (1), sacral hyperesthesia (1), major weight loss (1), abdominal cramps (1), and desquamation of the palms and soles (1). Laboratory abnormalities included mild leukopenia and lymphopenia, with a few cases of documented T4:T8*inversion following the acute illness. Ho et al. also documented cerebrospinal fluid analyses consistent with a lymphocytic meningitis in the two patients presenting with headaches and stiff neck. In their study, acute and convalescent titers for Epstein-Barr virus and cytomegalovirus were negative in all three patients, and additional cultures of blood, cerebrospinal fluid, stool, and stool examinations for ova and parasites were also negative.

The documented seroconversions in the same study were also analyzed in terms of the time sequence of antibody formation to the various viral components. Antibodies to gp120, gp160, and p24 developed prior to other viral antigens. Antibody to p41 (transmembrane protein) and other antigens developed approximately 4 weeks later.

In general, primary HIV infection resembles acute infections with other viral agents, such as Epstein-Barr virus or cytomegalovirus. The presence of a maculopapular rash or lymphocytic meningitis, however, is also consistent with the diagnosis. Besides its importance in the differential diagnosis of a prolonged acute febrile illness, acute HIV infection should be recognized as an early presentation of infection by HIV in patients who may be seronegative. Virus isolation from blood was possible in two of three patients prior to seroconversion. Patients in high-risk groups who experience symptoms consistent with acute HIV infection should be aggressively counseled regarding sexual behavior, blood donations, needle sharing, and pregnancy, even if antibody tests for HIV are negative. Such modifications in behavior should remain in effect for a minimum of 12 weeks following presumed exposure, in spite of negative antibody tests. Serological titers for other possible agents may also be helpful, although these results would not generally be available for many weeks. Recognition of acute HIV infection may also be useful in the future for initiating suppressive antiretroviral therapy early in the course of the disease prior to the

---

* "T" as used throughout this chapter is equivalent to the terminology "CD."

establishment of immunological abnormalities. However, 1 of the 12 patients reviewed retrospectively by Cooper and colleauges did not report any symptoms or illness between exposure and seroconversion. Ho and colleagues also state that they have documented HIV seroconversion in three other persons who were asymptomatic in the 3 months preceding their seroconversion. These findings suggest that primary infection with HIV may also be subclinical.

III. IDIOPATHIC THROMBOCYTOPENIC PURPURA

In addition to acute HIV infection, a number of other disorders are believed to be sequelae of HIV infection prior to, or in addition to, the development of the acquired immunodeficiency syndrome. One of these is the development of idiopathic thrombocytopenic purpura, whose increased incidence has recently been described in sexually active homosexual men, homosexual men with AIDs, and hemophiliacs (16-18). Idiopathic thrombocytopenic purpura (ITP), in its classic form unassociated with HIV infection, is an autoimmune disorder more common in women than men, in which the patient produces antibodies directed against platelets. In addition to the production of specific, antiplatelet antibodies, thrombocytopenic purpura has also been described as a consequence of a non· specific increase in IgG production. In these circumstances, the destruction of the platelets is considered an "epiphenomenon." Circulating immune complexes may bind nonspecifically to the platelet Fe receptor, with subsequent clearing of the antibody-coated platelets by macrophage/monocytes. Drug-induced haptenation of platelets, with antibody production directed against the platelets, has been described as another mechanism for the development of ITP. With each of these processes antibody binding results in an increased rate of platelet destruction by the reticuloendothelial system, leading to thrombocytopenia. Treatment for ITP is directed primarily toward suppressing the immune responses of the patient. This may include steroid therapy to suppress antibody production and/ or splenectomy to prevent the increased destruction of antibody-coated platelets in the spleen. Diagnosis of the disorder includes the presence of thrombocytopenia, elevated levels of platelet-bound IgG, increased bone-marrow megakaryocytes, and negative antinuclear antibody titers. The initial report of the increased incidence of ITP among homosexual men was by Morris, et al., in 1982 (16). They reported severe ITP in 11 sexually active homosexual men, all of whom (even with prior evidence of immunodeficiency as characterized by T4:T8 ratios of <I.O) responded to steroid therapy and/or splenectomy with significant increases in their platelet counts. In a follow-up paper in 1984, the authors described an additional 33 cases of ITP in homosexual men (17). Ten of the thirty-three patients had AIDS, as defined by the presence of Kaposi's sarcoma or an opportunistic infection. In analyzing possible mechanisms for ITP in homosexual men, the authors found that circulating immune complexes were increased in 88% of homosexual patients, but

were normal in five controls. They also found that platelet eluates from 12 of ·15 patients with autoimmune thrombocytopenia purpura were capable of binding to platelets at a mean titer of 1:8, while only 1 of 10 platelet eluates from homosexuals with ITP reacted with platelets at a titer of 1:2. They also found no difference between their patients with AIDS and those without. They postulated that ITP among homosexual men was related to immune-complex-mediated destruction. Stricker et al., however, in their study of 30 homosexual men with ITP, identified an antiplatelet antibody in the serum of 29 of the 30 men studied (18). In addition to their 30 patients, none of whom had AIDS, they also studied a separate group of 16 who had lymphadenopathy or AIDS with normal platelet counts. Of these 16 patients, platelet-associated immunoglobulin was measured in two with lymphadenopathy and three with AIDS, despite normal platelet counts. Similar antiplatelet antibody activity was not found in the serum from 30 nonhomosexual patients with either ITP or nonimmune thrombocytopenia. The serum antibody reacts with a similar molecular-weight antigen associated with cultured herpes simplex virus. These authors concluded that ITP in homosexual men is mediated by an autoimmune antibody directed against a component of the platelet membrane.

While the association between HIV infection, ITP, and a newly identified antibody to the platelet membrane is far from established, the possibility that HIV may play a role in the pathogenesis of an autoimmune disorder is intriguing. Elevated levels of circulating antibodies were part of the initial descriptions of the acquired immunodeficiency syndrome. Patients with AIDS-related syndromes and increased immunoglobulin levels have also been described. It is, of course, also tempting to imagine B-cell "deregulation" (with elevated and/or autoimmune antibody production) as a consequence of HIV acting as a mitogen for B-cell proliferation.

IV. PERSISTENT GENERALIZED LYMPHADENOPATHY

Among the first clinical disorders initially recognized as a "prodrome" to AIDS is the condition known as persistent generalized lymphadenopathy (PGL) (19, 20). Although not recognized by some as a distinctly different disorder from the AIDS-related complex (see below), PGL, as described, can be considered a milder, perhaps longer-lasting stage in the spectrum of HIV infection. Persistent generalized lymphadenopathy is defined as the presence of lymphadenopathy of at least 3 months' duration involving two or more extrainguinal sites in the absence of any illnesses or drugs known to cause lymphadenopathy. Patients may also exhibit low-grade fevers (intermittent or continuous) or night sweats. However, patients with unexplained wasting, diarrhea, or thrush, or with evidence of other opportunistic infections are regarded as having a more advanced form of HIV infection. Patients with PGL may also have other constitutional symptoms, such as fatigue, muscle pains, decreased libido, or rashes. Patients also are frequently found to have evidence of impaired cutaneous delayed hypersensitivity (cell-mediated immunity) on the basis of a lack of delayed cutaneous reactivity to one or more recall antigens. Laboratory abnormalities often include anemia, neutropenia, thrombocytopenia, and lymphopenia. Patients frequently also have elevated levels of serum globulins. Immunoelectrophoresis shows a polyclonal elevation of IgG, IgA, and/or IgM. Phenotypic analyses of subsets generally show a marked reduction in circulating T4+ cells and a relative increase in the number of T8+ cells, leading to a reversal of the T4/T8 ratio. Normal T4/T8 ratios are 2-3:1, while up to 95% of patients with PGL have T4/T8 ratios of less than 1. Patients may also have marked decreases in T-cell function as exhibited by stimulation response studies, lymphocyte proliferation studies, and/or natural killer activity.

Lymph nodes in patients with PGL reveal a number of changes (21). There is often exuberant follicular proliferation with infection with HIV. With progression to the acquired immunodeficiency syndrome, lymph node biopsies show regression and marked involution of nodes. Between the early florid follicular hyperplasia of PGL and the marked follicular regression of AIDS lies a transition state with variable mixtures of hyperplasia and involution within the same node or in different nodes.

Histologically, lymph node changes associated with PGL, while not specific for that disorder, have several interesting features. In normal non-HIV-infected reactive nodes, nodal architecture consists predominantly of follicular (generally B-cell) and interfollicular (generally T-cell) zones. The follicles can be further divided into mantle and germinal center areas, with marked differentiation by immunoglobulin staining of IgD in the mantle areas. In PGL, a similar distribution is maintained, but there is a marked decrease in mantle size, with attenuation of this compartment. The explosive follicular hyperplasia of PGL can result in the disruption of nodal architecture, with subsequent invagination of mantle zone areas into the germinal centers. This disruption, or follicle "lysis," not only results in a mixing of the normally compartmentalized mantle and germinal center areas of the nodes, but is also thought to contribute to the partial destruction of the delicate dendritic cells of the germinal centers. These cells, called the dendritic reticulum cells (DRC), form a woven network thought to be important in the capture of antigen and antigen presentation. With progression of the disease, isolated clusters of the DRC can be seen, leading to the "burnt-out" small clusters of dendritic cells of AIDS.

The distribution of helper-inducer (Leu 3/T4 positive) cells in normal reactive nodes is predominantly paracortical, while those few T cells observed within the germinal centers are almost exclusively Leu 3+ helper/inducer cells. In PGL, there is a decrease in the overall number of these cells in all the nodal compartments, especially marked in the paracortical areas. With progression to

AIDS, the decrease in number is even more extreme, with very few, isolated cells staining positive. Multinucleated cells have also been variably reported in the nodes of patients with PGL, their occurrence being commonly associated with viral infections. The phenotypic origin of these cells, while not yet fully defined, is believed by some to be of T-cell origin, corresponding to the observed formation in vitro of T-cell syncytia with HIV infection.

In normal, reactive nodes, the number of suppressor/cytotoxic (Leu 2, T8-staining cells) T cells is usually a small percentage of the T cells in the para-cortical areas and only rarely are these cells found in germinal centers. In PGL, T suppressor/cytotoxic cells are not only increased in number, but are also found within germinal centers. Similar to what is found in the patient's peripheral blood, there is a marked inversion of the T4/T8 ratio in lymph nodes. With progression to AIDS, even the number of suppressor/cytotoxic cells is decreased, with a resulting overall paucity of lymphocytes. While many of these lymph node features are not specific to PGL and have been described in other conditions, the overall changes are quite striking and seem to occur more frequently than reported for other types of infections (41). In contrast, the marked inversion of the T4/T8 ratio and the unusual presence of suppressor/cytotoxic cells within germinal centers do seem unique to PGL. Clinically, patients in whom lymphadenopathy regresses probably represent further progression of disease with an overall depletion of lymphocytes. Conversely, patients in whom continued gradual enlargement of nodes occurs, particularly single, isolated nodes, should be suspected of harboring a malignancy and aggressively evaluated for the possibility of lymphoma or Kaposi's sarcoma (see below). Patients in high-risk groups who present with lymphadenopathy should also be evaluated to rule out other possible causes of nodal swelling, including the presence of an opportunistic infection, such as mycobacterium avium which may be an initial presentation of AIDS. Notwithstanding the possible occurrence of primary HIV infection (associated with an initial exposure to the virus), PGL is commonly thought to be an important first indication of the progression of disease from the asymptomatic, seropositive state to a condition representing the first derangement of the immune system of sufficient magnitude to cause symptoms.

V. AIDS-RELATED COMPLEX

AIDS-related complex (ARC) may be viewed as a more severe form of HIV infection than persistent generalized lymphadenopathy. In both the Centers for Disease Control and the Walter Reed systems for staging of HIV infection, a distinction is made between those patients exhibiting only chronic lymphadeno-pathy and those who have gone on to develop further symptoms, such as oral candidiasis, complete anergy to delayed hypersensitivity testing to recall antigens, watery diarrhea for 2 or more weeks, or sustained weight loss of more than

10% of body weight (22,23). Such symptoms distinguish patients considered to have AIDS-related complex from those with PGL. A separate classification is probably valid, considering that the development of these symptoms is particularly ominous with regard to further progression to AIDS (see below). Such symptoms are believed to indicate severe immunological deterioration secondary to HIV infection. Laboratory abnormalities may antedate the development of symptoms, but are not sufficient alone for the diagnosis of ARC. While there may be considerable variability, most patients with ARC demonstrate severely depleted total numbers of circulating T4 cells, a marked reversal of the T4/T8 ratio, and, occasionally, anergy on skin testing. Leukopenia, thrombocytopenia, and/or anemia with polyclonal hypergammaglobulinemia is also frequently seen.

VI. THE ACQUIRED IMMUNODEFICIENCY SYNDROME

The diagnosis of the acquired immunodeficiency syndrome (AIDS) is ultimately a clinical one. The hallmark of the disease is the presence of an infection or neoplasm, of a type usually found only in patients who are severely immunocompromised, without other reasons for immunosuppression. The list of infections and malignancies accepted as diagnostic of AIDS by the Centers for Disese Control is shown in Table 1. While prior laboratory examinations may show derangements of the immune system or exposure to HIV, it is the onset of an "opportunistic" infection (or neoplasm, which is discussed below) which heralds the point at which the patient's immune system is no longer competent. With HIV infection, once a patient's immune system is sufficiently compromised to exhibit the symptoms of AIDS, the disease is generally fatal.

Recognition of the disorder was possible, in part, because of the prior experience of physicians with other types of immunocompromised states. Patients who are born with hereditary immunodeficiency conditions, who are intentionally immunosuppressed pharmacologically (e.g., for organ transplantation), or who become immunocompromised as a side effect of other severe diseases (such as leukemia) were all known prior to the AIDS epidemic to be susceptible to infections with unusual organisms. Many of these organisms do not cause disease in healthy hosts. These infections came to be known as "opportunistic," to signify their variable pathogenicity, depending on the immune state of the host. Based on studies particularly of individuals with selective hereditary disorders of either B-cell or T-cell function, different types of opportunistic infections are indicative of either "cell-mediated" (T-cell and monocyte) disorders, or "humoral" (B-cell) disorders. Types of infections that appear to correlate with impairment of cell-mediated immune function include protozoal infections (especially *Pneumocystis carinii*), fungal infections (such as *candida* and *cryptococcus*), and reactivated, systemic infections of previously latent viral and bacterial organisms (such as herpes simplex, cytomegalovirus,

TABLE 1 Centers for Disease Control Surveillance Definition of AIDS

1. Diagnosed disease process that is at least moderately predictive of a defect in cell-mediated immunity occurring in persons with no known cause for diminished resistance
2. Diseases diagnostic for AIDS include:
   Kaposi's sarcoma
   Pneumocystic carinii pneumonia
   Serious opportunistic infections

	Aspergillosis
	Candidiasis
Pneumonia	*Cryptoccus*
Meningitis	Cytomegalovirus
Encephalitis	Nocardiosis
	Strongyloidosis
	Toxoplasmosis
	Zygomycosis
	Atypical mycobacteriosis
Esophagitis	*Candida*, cytomegalovirus, herpes simplex virus, progressive multifocal leukoencephalopathy
	Chronic (>4 weeks) cryptosporidiosis
	Severe, chronic (>5 weeks) mucocutaneous herpes simplex

3. Diagnostic only if HIV-positive antibody or culture
   a. Disseminated histoplasmosis
   b. Isosporiasis with chronic (>1 month) diarrhea
   c. Bronchial or pulmonary candidiasis
   d. Non-Hodgkin's lymphoma or high-grade, B-cell type
   e. Kaposi's sarcoma in patients older than 60 years
4. Miscellaneous changes
   a. Chronic lymphoid interstitial pneumonitis in child (<13 years of age) unless proven HIV negative
   b. Lymphoreticular malignancy occurring more than 3 months after other AIDS diagnostic disease will not alter a patient's AIDS diagnosis
   c. Patients with any disease processes that are HIV antibody negative and have normal immune studies will not be diagnosed as AIDS cases irrespective of disease process

and *Mycobacterium tuberculosis*). Impairment of humoral function has classically been recognized by the host's vulnerability to infection by pyogenic organisms.

Opportunistic infections in AIDS patients apparently may reflect the different types of immune deficits. However, over 60% of the patients who develop opportunistic infections as the initial presentation of AIDS will manifest *P. carinii* pneumonia (combined statistics from Ref. 24–26). Protozoal infections, of which *P. carinii* is but one, are the most common AIDS-related opportunistic infections. Symptoms of *Pneumocystis* pneumonia at presentation generally include fever, nonproductive cough, dyspnea at rest or on exertion, and chest tightness. The time course of symptoms may range from several days to months and the onset may be insidious or rapidly progressive. Five to ten percent of patients will have an apparently normal chest x-ray, although arterial blood gas analyses usually show a marked decrease in oxygenation. As with other opportunistic infections seen in AIDS patients, *P. carinii* infections are characterized by an aggressive clinical course, frequent resistance to standard therapy, and high rate of recurrence.

Toxoplasma gondii, another protozoal infection, is the most common cause of central nervous system infection in AIDS patients. The clinical presentation often includes focal or diffuse seizures, changes in personality or cognition, or focal neurological deficits. Serological testing to antibodies to toxoplasmosis is not useful for diagnosis. CT scans are helpful for diagnosis if characteristic ring-enhancing lesions are seen, but such lesions may also be seen with central nervous system lymphoma.

Another, less common protozoal infection seen in AIDS patients is *Cryptosporidium*, a coccidian parasite. This parasite in normal hosts may produce a self-limited diarrheal illness. In AIDS patients, however, *Cryptosporidium* infection generally results in sustained and profuse diarrhea with malabsorption and wasting. Identification of the organism requires careful analysis of stool by a sugar flotation test. No effective therapy has been found to date, so that supportive measures are employed with fluid replacement and control of bowel motility. Another coccidian parasite, *Isospora belli*, may also be seen in AIDS. This organism produces a severe diarrhea clinically indistinguishable from that of *Cryptosporidium*. Therapy is also supportive, as with *Cryptosporidium*.

Fungal infections are frequent manifestations of AIDS. In patients with ARC, oral candidiasis (thrush) is strongly predictive of subsequent development of AIDS and is one of the criteria for distinguishing ARC from persistent generalized lymphadenopathy. However, only invasive or systemic infections with *Candida* (not simply thrush) fulfill the surveillance definition of AIDS. Esophageal candidiasis is often observed, but disseminated *Candida*, particularly fungemia, is unusual in AIDS. Esophageal candidiasis is often suspected clinically if pain with swallowing develops, as nearly all patients with esophageal candidiasis also have oral thrush. Various therapies have proven efficacious against this disorder.

Another fungal infection in AIDS patients is *Cryptococcus neoformans* (27). In nearly three-quarters of cases, the meninges lining the brain are the sole site of infection by this organism. Cryptococcal meningitis generally presents with fever, severe headache, and mental status changes. Cerebral spinal fluid examination usually shows elevated protein with or without decreased glucose. Other fungal infections, such as *histoplasmosis, coccidioidomycosis,* and *aspergillosis*, have been reported in AIDS patients, although the number of cases is small. There also appears to be a high rate of relapse with these infections following therapy.

Viral infections are common in AIDS patients. Infections with herpes simplex virus are frequent in both ARC and AIDS patients. Severe cutaneous herpes infection lasting for longer than 4 weeks is one diagnostic criterion for AIDS. This infection may be the initial presentation and/or an ongoing problem throughout the course of the illness. Herpes simplex virus may also involve the oral mucosa and face and occasionally results in encephalitis, myelitis, or pneumonia. Herpes zoster infection is also frequent in patients with ARC or AIDS, although there are few data on the frequency of disseminated disease.

Cytomegalovirus (CMV) is another frequent viral pathogen in AIDS (28). It is often difficult to distinguish between mere colonization and invasive disease. At bronchoscopy, for example, more than one-third of patients with interstitial pneumonia have CMV cultured from the lavage fluid; a much smaller proportion have histiological evidence of tissue invasion. In preliminary studies, the presence of CMV at bronchoscopy had no clear effect on survival in patients with *P. carinii* pneumonia. CMV is, however, significantly associated with symptomatic retinitis, adrenalitis, colitis, and encephalitis in AIDS. Therapy may be difficult owing to drug toxicities and there is a high rate of relapse.

Other viral infections, such as hepatitis B, hepatitis A, non-A/non-B hepatitis, and Epstein-Barr virus, are all frequent in populations at high risk for HIV infection. The clinical outcome of such infections is similar to that seen in individuals not infected with HIV. Much speculation exists, however, as to their possible role in immunopathogenesis and subsequent clinical progression of immune deficiency.

Bacterial organisms, while not originally recognized as important pathogens in AIDS, have recently been described in a number of settings. These bacterial agents may not only represent possible reactivated infections (as with *Mycobacterium tuberculosis*), but may also signify impairment of humoral immunity (29,30). Polsky et al. described 18 episodes of bacterial pneumonia in 13 patients with AIDS (30). Fourteen of the eighteen episodes were caused by *Hemophilus influenza* or *Streptococcus pneumoniae*, organisms usually found in

patients with B-cell defects, chronic pulmonary lung disease, or splenic dysfunction. Such pyogenic infections, while relatively uncommon in AIDS patients,
often relapse. Immunization with pneumococcal vaccine does not appear to
elicit a normal protective response in AIDS patients.

Mycobacterium avium intracellulare (MAI) is a frequent organism and can
be isolated from sputum, blood, urine, and feces (31). Its clinical significance is
controversial. Because MAI may occur with other opportunistic infections and/
or neoplasia, it may be difficult to attribute fever, weight loss, or gastrointestinal
dysfunction directly to the organism. Nonetheless, in some patients MAI does
appear to account for these symptoms, as well as hepatosplenomegaly and bone
marrow failure. MAI is generally resistant to antituberculous therapy.

M. tuberculosis, another bacterial organism, is being seen with increased
frequency in HIV-infected individuals (32). This is particularly true among
persons from areas endemic for tuberculosis, such as Haiti or Africa. In patients
with AIDS, *M. tuberculosis* is often disseminated and may even be recovered
from blood. Some AIDS patients may present with extrapulmonary tuberculosis,
particularly cervical lymphadenopathy reminiscent of scrofula. Response to standard antituberculous therapy, however, is generally good.

Salmonella infections are also often disseminated, with bacteremia and/or
persistent enteritis. The rate of recurrence after standard antibiotic therapy is
quite high, sometimes occurring depsite maintenance therapy.

While most AIDS patients initially present with an opportunistic infection,
about 40% present with malignancy. Kaposi's sarcoma is the most frequent
AIDS-related neoplasm, with a lower prevalence of non-Hodgkin's lymphoma.
Prior to the AIDS epidemic, Kaposi's sarcoma, a rare disorder, was observed
among fairly well-defined patient groups. These included elderly men of Ash-
kenazic Jewish or Mediterranean origin (whose predisposition for the tumor has
remained undefined), Central Africans, and immunosuppressed patients.
Transplant patients, for example, develop Kaposi's sarcoma at an incidence of
400-500 times that of the general population (21). Anecdotal reports of resolu-
tion of the tumor in renal allograft recipients following discontinuation of
immunosuppressive therapy further supported a relationship between this
malignancy and cell-mediated immunity. Kaposi's sarcoma is a vascular tumor,
with a well-characterized histopathology. Controversy exists as to the exact cell
of origin, but it is likely to be endothelial. The tumor manifests itself as firm,
purplish plaques, papules, or nodules which may be round or elliptical. Occasionally lesions are linear and follow cutaneous lymphatics. Lesions on the face and
in the oral cavity are particularly common. Patients present with these mucocutaneous lesions or lymph node involvement, while visceral lesions, although
clinically silent, occur in nearly one-half the patients. Pulmonary Kaposi's sarcoma, however, is usually very clinically aggressive and occasionally may be con-
fused with *P. carinii* pneumonia, as both present with fever and interstitial infil-

trates on chest x-ray. The prognosis is extremely poor, with survival measured in several weeks to a few months. Most striking is the variable natural history of Kaposi's sarcoma in AIDS. Some patients have a remarkably indolent course, without disease progression for months to years. Other patients have rapidly advancing lesions that may extend over the skin and involve the viscera. At the time of initial diagnosis, prediction of the clinical course may be difficult, but factors that indicate disease progression and shortened survival time include a history of prior opportunistic infection, constitutional symptoms, absolute T-helper count below $100/mm^3$, and pulmonary involvement.

Therapy in AIDS-related Kaposi's sarcoma is controversial owing to the variable natural history of the illness and concern that treatment (such as chemo therapy) could exacerbate the underlying immune deficiency. Interferon alpha therapy has resulted in a response rate of 20-40%, with a subset of about 5% of patients surviving for 3 years or more without development of opportunistic infections (33). It is unclear, however, whether this outcome is a direct result of interferon therapy or represents the natural history of the disease in these individuals.

Kaposi's sarcoma also shows a variable distribution pattern among AIDS patients. The tumor occurs in nearly one-half of homosexual males, but is found in only about 10% of heterosexual AIDS cases. Although there has been considerable speculation that infections, such as CMV, or environmental toxins, such as recreational use of volatile nitrates, could contribute to the development of the neoplasm, there are few compelling data to identify a pathogenic role for these factors.

Another neoplasm frequently found in AIDS patients is malignant lymphoma (34). Lymphoma in AIDS occurs in two major forms—non-Hodgkin's disease (the most common) and Hodgkin's disease. Malignant lymphoma has been reported prior to the advent of AIDS as occurring with a much greater frequency in other immunosuppressed populations. Transplant recipients have a 30-40 times greater incidence of malignant lymphoma than the general population (21). Patients with certain types of congenital immunodeficiencies have a much greater risk of malignant lymphoma, with the non-Hodgkin's type predominating. Non-Hodgkin's lymphoma comprises 40-70% of all tumors in patients with Wiskott-Aldrich syndrome, ataxia-telangiectasia, combined variable immune deficiency, or severe combined immune deficiency. There are striking similarities between the lymphomas described in these congenitally immunocompromised patients and AIDS patients and those occurring in transplant recipients. These similarities include an increased proportion of very aggressive large-cell or "histiocytic"-type tumor of B-cell origin, common extranodal involvement, and frequent involvement of the brain and central nervous system. In addition, there have been described a variety of atypical proliferations, which, although classified as lymphomas, display polyclonal markers when studied for immunoglobulin expression and gene rearrangement. These tumors, interestingly, may regress or diminish in size following reversal of the immunosuppression in transplant patients (35). There is also a well-recognized association between Epstein-Barr virus and the lymphomas occurring in these subgroups of immunosuppressed patients (36).

Malignant lymphomas in AIDS patients manifest many similarities to those described in other immunocompromised patients. Non-Hodgkin's lymphoma of B-cell origin predominates and is now one of the diagnostic criteria for AIDS. Histopathologically, the B-cell lymphomas are often of the large-cell, undifferentiated, or immunoblastic sarcoma type. The rate of growth of these lymphomas may be quite rapid, with clinical doubling of size in weeks, reminiscent of African Burkitt's lymphoma. Pathologically, high-grade, aggressive, Burkitt's-like lymphoma has been reported as occurring in about half of malignant lymphomas in HIV-infected individuals. Rare "plasmacytoid" cell types have also been reported both in AIDS patients and in other immunodeficiency states (37). Patients with AIDS and malignant lymphoma also exhibit an association with Epstein-Barr virus infection. This association has been seen, not only serologically, but also by detection of the EBV genome by hybridization techniques within the tumor (38). Characteristic reciprocal gene translocations are also frequently found in both Burkitt's lymphoma and some of the non-Hodgkin's lymphomas in AIDS patients. A number of theories exist as to the possible role of EBV in the development of malignant lymphomas, but proof for it as an etiological agent is still lacking.

AIDS patients with lymphoma tend to present with more advanced stages of the disease. Extranodal involvement occurs in 85–95% of cases, central nervous system involvement in approximately 40% of cases, and bone marrow involvement in approximately 30% of cases (21). This contrasts sharply with the usual central nervous system and bone marrow involvement of 2% and 7.5%, respectively, in the non-AIDS lymphoma patients. The advanced stage, aggressive tumor type, and prior compromised state of the host may all be contributing factors to the extremely poor prognosis and poor response to therapy. Relapses occur frequently and the mean survival is generally 6–8 months.

Hodgkin's disease, although not yet accepted as a diagnostic criterion for AIDS, may also present with increased frequency in HIV-infected patients. Similar to the non-Hodgkin's type of lymphoma, the tumor tends to be advanced at presentation, with frequent extralymphatic involvement of the skin, liver, and bone marrow. In contrast to classic Hodgkin's disease, involvement of the mesenteric nodes is common. Although the tumor in HIV-infected homosexual men responds well to conventional combination chemotherapy, prolonged pancytopenia and complications with secondary opportunistic infections are frequent. Often these complications limit the achievement of complete remission and result in death.

VII. NEUROLOGICAL MANIFESTATIONS OF AIDS

Patients with the acquired immunodeficiency syndrome frequently develop neurological complications (39). Approximately 30% of patients with AIDS have clinically apparent central nervous system (CNS) dysfunction and approximately 80% have neuropathological abnormalities at postmortem examination (40). In addition to infectious and neoplastic involvement of the central nervous system secondary to immunosuppression, a number of distinct neurological syndromes have been described which are believed to be a consequence of primary HIV infection of the CNS. These include a progressive dementia (41-43), vacuolar myelopathy (40), and peripheral neuropathy (39,44,45). Chronic and acute meningitides have also been described occurring at the time of seroconversion in the absence of other detectable infectious processes (15,44). Patients with AIDS may also experience neurological vascular accidents in the form of hemorrhage or stroke (39). Neurological complications of the acquired immunodeficiency syndrome may therefore be classified into four major groups: (a) infectious, (b) neoplastic (including both primary CNS tumors and central nervous system involvement secondary to a systemic primary tumor), (c) vascular, and (d) disorders believed secondary to primary HIV infection of the nervous system.

Of the four major groupings of neurological complications of AIDS, the two major diagnostic subgroups in four large patient series were a progressive dementing illness (also known as a subacute encephalitis) and infections involving the CNS (see Table 2) (39,42,44,46). Together, these two diagnostic subgroups accounted for almost 70% of patients with neurological disorders. In these series' 214 patients with neurological complications of AIDS, 22% of patients were found to have an opportunistic infection of the central nervous system, 10% had a CNS malignancy (almost all of which were lymphomas), and 46% had progressive dementia or subacute encephalitis. Other categories included: metabolic encephalopathies (8%), myelopathies (2%), peripheral neuropathies (5%), and "unknown" (8%). Vascular complications, including sagittal sinus thrombosis and nonbacterial thrombotic emboli, accounted for 3%.

Infections of the central nervous system frequently occur in AIDS patients. Toxoplasmosis and cryptococcal meningitis are common, as discussed previously. Other infectious processes include *Candida* infections, progressive multifocal leukoencephalopathy, and varicella-zoster virus encephalopathy. Evidence of *M. avium intracellulare* and/or cytomegalovirus infection is common, but their role as pathogens remains controversial. Infections of the central nervous system generally have an acute or subacute presentation, often accompanied by focal neurological findings on physical examination. Similar clinical findings may occur in patients with central nervous system malignancies (with either primary or metastatic involvement). Differentiation between infection or

Table 2   Complications of AIDS

Study	No. of patients	No. without CNS involvement	Toxo	MAI/MTB	Cryptococcus	Candida	PMLE	VZV encephalitis	Total infections	SE	Lymphoma	Peripheral neuropathy	Myelopathy	Metabolic encephalopathy	Hemorrhage	SAG sinus thrombosis	Nonbacterial thrombotic endocard.	Total vascular complications	Unexplained meningitis	Unexplained mass lesion	Other[a]
Snider et al.	50		5	3	2	1	2			18	7	8			3		2	5	4	3	
Ho et al.	45	9	1	1						16	1	3	4					1	9	1	4
Nielsen et al.	40	9	5		2		1	3		19	3			3							
Navia et al.	121	24	16			3	3			46	10			15		1		1			
Total	256																				
Total with CNS complications	214		27	4	4	4	6	3	48	99	22	11	4	18	3	1	2	7	13	4	4
% of total with CNS complications			13	2	2	2	3	1	22	46	10	5	2	8	1	0.5	1	3	6	2	2

Toxo = toxoplasmosis; MAI = *Mycobacterium avium intracellulare*; MTB = *Mycobacterium tuberculosis*; PMLE = progressive multifocal leukoencephalopathy; VZV = varicella zoster virus; SE = subacute encephalitis, also known as AIDS dementia complex; SAG = sagittal.

aOther = psychosis (1), meningoencephalitis (1), neurosyphilis (1), and astrocytoma (1)

neoplasm may require cerebrospinal fluid culture and cytology, CT scans, and/or brain biopsy.

Many patients with AIDS or ARC develop neuropsychiatric abnormalities, particularly cognitive dysfunction, memory loss, and mild dementia (42). These symptoms and signs may antedate the onset of AIDS or may occur in the absence of other criteria for the diagnosis of AIDS. Patients frequently manifest marked cortical atrophy on CT scan. Early manifestations of HIV infection may include behavioral or motor changes, such as leg weakness, ataxia, loss of fine motor corrdination, apathy, withdrawal, or frank psychoses. This disorder, termed the AIDS dementia complex, is progressive in most patients. In the final stages, this neurological degeneration frequently results in profound dementia, incontinence, paraplegia, and mutism. This clinical picture, which often occurs without other infectious or neoplastic processes, has led many to believe that the disorder results from primary HIV infection of the central nervous system. Isolation of the virus from the CSF or from CNS tissue has been achieved in patients with AIDS dementia complex, acute aseptic meningitis associated with seroconversion, with AIDS or ARC and unexplained chronic meningitis, and in patients with myelopathy and peripheral neuropathy (44,47,48). Intra-blood-brain-barrier synthesis of anti-HIV IgG has also been documented (49). Molecular hybridization techniques have revealed HIV sequences in the brains of patients with AIDS and encephalopathy (47,50). Controversy still exists, however, as to which cells in brain and spinal cord are infected with HIV (50-52). Macrophage-monocytes may be the source of HIV in the CNS, although many people believe that the virus may possess a distinct neurotropism, much like lentiviruses (Y. Koyannagi et al., *Science*, May 15, 1987).

VIII. EFFECTS OF HIV HEMATOPOIESIS

Patients with AIDS or ARC may also exhibit a number of unexplained hematological abnormalities. In its most severe form, there may be pancytopenia, with marked decreases in erythroid, myeloid, megakaryocytic, and lymphoid cell lines. Patients who do not manifest such marked abnormalities may nonetheless have an increased sensitivity to the myelotoxic effects of certain therapuetic agents. Bone marrow suppression is the most common limiting hematological toxicity of both chemotherapeutic agents and antiretroviral drugs such as azido-thymidine. Widely used antibiotics, such as Fansidar and trimethoprim-sulfa-methoxazole, also appear to have limiting hematological toxicities at lower doses than that found in the general population. Seventy percent of patients with AIDS are anemic, 60% (in the later stages of the disease) are leukopenic, and approximately 40% are thrombocytopenic (21). Lymphopenia, one of the early signs of HIV infection, is seen in approximately 85% of AIDS patients (53-55).

While some patients' thrombocytopenia may be explained on the basis of an autoimmune phenomenon (as discussed previously), few data are available as to the basis of the hematological disorders commonly seen. Recently, however, virus has been isolated from cultures of bone marrow progenitors from AIDS patients. Immunoglobulin purified from HIV-infected sera has also been shown to inhibit hematopoiesis in marrow from HIV-infected donors but not normal donors (56). HIV, therefore, may have an inhibitory effect on bone marrow, perhaps secondary to infection of stem cells. Initial studies of the hematological effects of recombinant granulocyte-macrophage colony-stimulating factor treatment of leukopenic AIDS patients demonstrate a clear dose-response relationship with normalization of the circulating neutrophil and monocyte counts (57). Further work should help elucidate the role of HIV in bone marrow suppression and the clinical utility of therapy with hematopoietic growth factors.

IX. EPIDEMIOLOGICAL PROSPECTS FOR THE FUTURE

Current estimates of the number of HIV-infected persons in the United States range from 1 to 2 million. Several studies have addressed the natural history of HIV infection (24–26,59,60). However, attempts to determine the number of people who will develop AIDS from a population of infected individuals have proven difficult for a variety of reasons. The time since HIV infection for most people is unknown. Factors affecting the development of disease versus persistence of an ayrmptomatic carrier state are still undefined. Until fairly recently, only the terminal stages of the disease were recognized. Several groups have now developed useful staging classifications for HIV infection (22,23,48). With such staging systems, evaluation for progression of HIV disease (advancing stage) has been reported to be as high as 80%. In three different prospective studies of high-risk-group patients, a total of 5722 people were followed; 3569 entered as seronegative and 2153 as seropositive (24–26). None of the seropositive patients had AIDS. Of the seropositive individuals, one study of 1835 included 637 who were asymptomatic, 1094 with persistent generalized lymphadenopathy, and 104 with ARC. Estimates for the approximate yearly incidence of AIDS ranged from 2.6% (in asymptomatic seropositive individuals) to 10.1% (with no clinical differentiation between groups). Variations in estimates may reflect the duration of the time since seroconversion. For instance, high-risk individuals in New York are believed to have been exposed to HIV earlier than populations in other cities. Differences in estimates may also reflect variation in clinical status among patients. The prospective study by Polk et al. found that 8.6% of patients with ARC progressed to AIDS during their mean follow-up time of 15 months, while only 2.6% of asymptomatic individuals progressed to AIDS during the same time span (26). Incidence data for seroconversion reflect exposure to virus and may be a good indicator of behavioral changes among high-risk groups. While such studies are valuable in providing guidelines for long-range estimates of the magnitude of the epidemic, much information is still needed to delineate the natural history of HIV infection over a longer period of time in larger populations.

Since the initial description of the illness 8 years ago, much has been learned about the etiology, transmission, and clinical manifestations of AIDS. The relatively recent recognition of the epidemic, however, also limits our ability to predict the ultimate outcome for the majority of infected individuals. Future clinical observation may also reveal an even wider spectrum of disease than that presently defined.

REFERENCES

1. Gottlieb, M. S., Schroff, R., Schanker, H. M., et al. (1981). *Pneumocystic carinii* pneumonia and mucosal candidiasis in previously healthy homosexual men: Evidence of a new acquired cellular immunodeficiency. *N. Engl. J. Med.* 305:1425–1431.
2. Masur, H., Michelis, M. A., Greene, J. B., et al. (1981). An outbreak of community-acquired *Pneumocystis carinii* pneumonia: Initial manifestation of cellular immune dysfunction. *N. Engl. J. Med.* 305:1431–1438.
3. Siegal, F. P., Lopez, C., Hammer, G. S., et al. (1981). Severe acquired immunodeficiency in male homosexuals, manifested by chronic perianal ulcerative herpes simplex lesions. *N. Engl. J. Med.* 305:1439–1444.
4. Barre-Sinoussi, F., Chermann, J. C., Rey, F., et al. (1983). Isolation of a T-lymphotropic retrovirus from a patient at risk for acquired immune deficiency syndrome (AIDS). *Science* 220:868–871.
5. Gallo, R., Salahuddin, S. Z., Popovic, M., et al. (1984). Frequent detection and isolation of cytopathic retroviruses (HTLV–III) from patients with AIDS and at risk for AIDS. *Science* 224:500–503.
6. Popovic, M., Sarngadharan, M., Read, E., and Gallo, R. (1984). Detection, isolation and continuous production of cytopathic retroviruses (HTLV–III) from patients with AIDS and pre-AIDS. *Science* 224:497–500.
7. Levy J. A., Hoffman, A. D., Kramer, S. M., et al. (1984). Isolation of lymphocytopathic retroviruses from San Francisco patients with AIDS. *Science* 225:840–842.
8. Jett, J. R., Kuritsky, J. N., Katzmann, J. A., and Homburger, H. A. (1983). Acquired immunodeficiency syndrome associated with blood-product transfusions. *Ann. Intern. Med.* 99:621–624.
9. Curran, J. W., Lawrence, D. N., Jaffe, H., et al. (1984). Acquired immunodeficiency syndrome (AIDS) associated with transfusions. *N. Engl. J. Med.* 310:69–75.
10. Groopman, J. E., Salahuddin, S. Z., Sarngadharan, M. G., et al. (1984). Virologic studies in a case of transfusion-associated AIDS. *N. Engl. J. Med.* 311:1419–1422.
11. Church, J. A., and Isaac, H. (1984). Transfusion-associated acquired immune deficiency syndrome in infants. *J. Pediatr.* 105:731–737.

12. Feorino, P. M., Jaffe, H. W., Palmer, E., et al. Transfusion-associated acquired immunodeficiency syndrome: Evidence for persistent infection in blood donor. *N. Engl. J. Med.* 312:1293–1296.

13. Editorial (1984). Needlestick transmission of HTLV–III from a patient infected in Africa. *Lancet* 2:451–453.

14. Cooper, D. A., Gold, J., MacLean, P., et al. (1985). Acute AIDS retrovirus infection. Definition of a clinical illness associated with seroconversion. *Lancet* 1:437–540.

15. Ho, D. D., Sarngadharan, M. G., Resnick, L., et al. (1985). Primary human T-lymphotropic virus type III infection. *Ann. Intern. Med.* 103:880–883.

16. Morris, L., Distenfeld, A., Amorosi, E., and Kirkpatkin, S. (1982). Autoimmune thrombocytopenic purpura in homosexual men. *Ann. Intern. Med.* 96:714–717.

17. Walsh, C., Nardi, M., and Kirkpatkin, S. (1984). On the mechanism of thrombocytopenic purpura in sexually active homosexual men. *N. Engl. J. Med.* 311:635–639.

18. Stricker, R., Abrams, D., Corash, L., and Shuman, M. (1985). Target platelet antigen in homosexual men with immune thrombocytopenia. *N. Engl. J. Med.* 313:1375–1380.

19. Persistent genralized lymphoadenopathy among homosexual males. (1982). *Morbid. Mortal. Week. Rev.* 31:249–291.

20. Metroka, C. E., Cunningham-Rundles, S., Pollack, M. S., et al. (1983). Generalized lymphadenopathy in homosexual men. *Ann. Intern. Med.* 99: 585–591.

21. Harawi, S., and O'Hara, C. (in press). *The Pathology and Pathophysiology of AIDS.* Chapman and Hall, London.

22. Fahey, J. L., Prince, H., Weaver, M., et al. (1984). Quantitative changes in T helper or T suppressor/cytotoxic lymphocyte subsets that distinguish acquired immune deficiency syndrome from other immune disorders. *Am. J. Med.* 76:95–100.

23. CDC classification system for human T-lymphotropic virus type III/lymphadenopathy associated virus infections. (1986). *Morbid. Mortal. Week. Rev.* 35:334–339.

24. Redfield, R., Wright, D., and Tramont, E. (1986). The Walter Reed staging classification for HTLV–III/LAV infection. *N. Engl. J. Med.* 314:131–132.

25. Mathur-Wagh, U., Mildvan, D., and Senie, R. (1985). Follow-up at 4½ years on homosexual men with generalized lymphadenopathy. *N. Engl. J. Med.* 313:1524–1543.

26. Goedert, J. J., Biggar, R. J., Weiss, S. H., et al. (1986). Three-year incidence of AIDS in five cohorts of HTLV–III-infected risk group members. *Science* 231:992–995.

27. Eng, R. H. K., Bishburg, E., Smith, S. M., and Kapila, R. (1986). Crytpococcal infections in patients with the acquired immunodeficiency syndrome. *Am. J. Med.* 81:19–23.

28. Ho, M. (1982). *Cytomegalovirus: Biology and Infection.* New York, Plenum, p. 309.

29. Hawkins, C. C., Gold, J. W. M., Whimbey, E., et al. (1986). Mycobacterium avium complex infections in patients with the acquired immunodeficiency syndrome. *Ann. Intern. Med.* 105:184-188.

30. Sunderam, G., McDonald, R. J., Maniatis, T., et al. (1986). Tuberculosis as a manifestation of the acquired immunodeficiency syndrome (AIDS). *JAMA* 256:362-366.

31. White, S., Tsou, E., Waldhorn, R., and Katz, P. (1985). Life-threatening bacterial pneumonia in homosexuals with laboratory features of AIDS. *Chest* 87:486-488.

32. Polsky, B., Gold, J. W. M., Whibergy, E., et al. (1986). Bacterial pneumonia in patients with the acquired immunodeficiency syndrome. *Ann. Intern. Med.* 104:38-41.

33. Groopman, J. E. (1986). Therapy of epidemic Kaposi's sarcoma. *Semin. Hematol* 23:14-19.

34. Ziegler, J. L., Beckstead, J. A. Volberding, P. A., et al. (1984). Non-Hodgkin's lymphoma in 90 homosexual men. Relation to generalized lymphadenopathy and acquired immunodeficiency syndrome. *N. Engl. J. Med.* 311:565-570.

35. Starzl, T. E., Porter, K. A., Iwatsuki, S., et al. (1984). Reversibility of lymphomas and lymphoproliferative lesions developing under cyclosporin-steroid therapy. *Lancet* 1:583-587.

36. Hanto, D. W., Gajl-Peczalska, K. J., Frizzera, G., et al. (1983). Epstein-Barr virus (EBV) induced polyclonal and monoclonal B-cell lymphoproliferative diseases occurring after renal transplantation. *Ann. Surg.* 3:356-369.

37. Israel, A. M., Koziner, B., and Straus, J. (1983). Plasmacytoma and the acquired immunodeficiency syndrome. *Ann. Intern. Med.* 99:635-636.

38. Petersen, J. M., Tubbs, R. R., Savage, R. A., et al. (1985). Small noncleaved B cell Burkitt-like lymphoma chromosome t (8:14) translocation and Epstein-Barr virus nuclear-associated antigen in a homosexual man with acquired immune deficiency syndrome. *Am. J. Med.* 78:141-148.

39. Snider, W. G., Simpson, D. M., Nielsen, S., et al. (1983). Neurological complications of acquired immune defieincy syndrome: Analysis of 50 patients. *Ann. Neurol.* 14:403-418.

40. Ho, D. D., Rota, T. R., Schooley, R. T., et al. (1985). Isolation of HTLV-III from cerebrospinal fluid and neural tissues of patients with neurologic syndromes related to the acquired immunodeficiency syndrome. *N. Engl. J. Med.* 313:1493-1497.

41. Nielson, S., Petito, C., Urmacher, C., and Posner, J. (1984). Subacute encephalitis in acquired immune deficiency syndrome: A postmortem study. *Am. J. Clin. Pathol.* 82:678-682.

42. Navia, B., Jordon, B., and Price, R. (1986). The AIDS dementia complex: I. Clinical Features. *Ann. Neurol* 19:517-524.

43. Navia, B., Cho, E. S., Petito, C., and Price, R. (1986). The AIDS dementia complex: II. Neuropathology. *Ann. Neurol.* 19:525-535.

44. Petito, C. K., Navia, B. A., Cho, E. S., et al. (1985). Vacuolar myelopathy pathologically resembling subacute combined degeneration in patients with acquired immunodeficiency syndrome. *N. Engl. J. Med.* 312:874–879.

45. Lipkin, W., Parry, G., Kiprov, D., and Abrams, D. (1985). Inflammatory neuropathy in homosexual men with lymphadenopathy. *Neurology* 35: 1479–1483.

46. Levy. J., Shimabukuro, J., Hollander, H., Mills, J., and Kaminsky, L. (1985). Isolation of AIDS-associated retroviruses from cerebrospinal fluid and brains of patients with neurological symptoms. *Lancet* 2:586–588.

47. Shaw, G. M., Harper, M. E., Hahn, B. H., et al. (1985). HTLV–III infection in brains of children and adults with AIDS encephalopathy. *Science* 227: 177–182.

48. Resnick, L., diMarzo-Veronese, F., Schupbach, J., et al. (1985). Intra-blood-brain-barrier synthesis of HTLV–III-specific IgG in patients with neurologic symptoms associated with AIDS or AIDS-related complex. *N. Engl. J. Med.* 313:1498–1504.

49. Gabuzda, D. H., Ho, D. D., di la Monte, S. M., et al. (1986). Immunohisto-chemical identification of HTLV–III antigen in brains of patients with AIDS. *Ann. Neurol.* 20:289–295.

50. Wiley, C., Schrier, R., Nelson, J., Lampert, P., and Oldstone, M. (1986). Cellular localization of human immunodeficiency virus infection within the brains of acquired immune deficiency syndrome patients. *Proc. Natl. Acad. Sci. (USA)* 83:7089–7093.

51. Koenig, S., Gendelman, H. E., Orenstein, J. M., et al. (1986). Detection of AIDS virus in macrophages in brain tissue from AIDS patients with encephalopathy. *Science* 233:1089–1093.

52. Koyanagi, Y., MIles, S., Mitsuyasu, R. J., et al. (1987). Dual infection of the central nervous system by AIDS viruses with distinct cellular tropisms. *Science* 236:819–822.

53. Spivak, J., Selonick, S., and Quinn, T. (1983). Acquired immune deficiency syndrome and pancytopenia. *JAMA* 250:3084–3087.

54. Spivak, J., Bender, B., and Quinn, T. (1984). Hematologic abnormalities in the acquired immune deficiency syndrome. *Am. J. Med.* 77:224–228.

55. Zon, L., Arkin, C., and Groopman, J. E. (1987). Haemotologic manifesta-tions of the human immunodeficiency virus (HIV). *Br. J. Haematol.* 66: 251–256.

56. Donahue, R. E., Johnson, M. M., Zon, L. I., Clark, S. C., and Groopman, J. E. (1987). In vitro suppression of haematopoiesis after human immuno-deficiency virus infection. *Nature* 326:200–203.

57. Groopman, J. E., Mitsuyasu, R. T., DeLeo, M. J., Oette, D. H., and Golde, D. W. (1987). Effect of recombinant human granulocyte-macrophage colony-stimulating factor on myelopoiesis in the acquired immunodefi-ciency syndrome. *N. Engl. J. Med.* 317:593–595.

58. Jaffe, H. W., Darrow, W. W., Echenberg, D. F., et al. (1985). The acquired immunodeficiency syndrome in a cohort of homosexual men: A six-year follow-up study. *Ann. Intern. Med.* 103:210–214.

59. Schwartz, K., Visscher, B. R., Detels, R., et al. (1985). Immunological changes in lymphadenopathy virus positive and negative symtomless male homosexuals: Two years of observation. *Lancet* 2:831–832.
60. Salahudin, S., Groopman, J. E., Markham, P., Sarngadharan, M., Redfield, R., and McLane, M. (1984). HTLV-III in symptom free seronegative persons. *Lancet* 2:1418–1420.
61. Haverkos, H., Gottlieb, M., Killen, J., and Edelman, R. (1985). Classification of HTLV-III/LAV-related diseases. *J. Infect. Dis.* 152:1095.

15

Vaccine Strategies for AIDS

Dani P. Bolognesi   *Duke University Medical Center, Durham, North Carolina*

I. INTRODUCTION

The immune response that develops subsequent to infection with human immunodeficiency virus (HIV) consists of both humoral and cellular elements that, when tested in vitro can inhibit virus infection and syncytium formation and lyse virus-infected target cells. If such activities were operative in vivo, one would expect them to be effective in suppressing virus replication and virus-induced cytopathic effects. Antiviral immunity may thus represent one of the primary host defense mechanisms responsible for the protracted asymptomatic phase of the disease which, in most patients, can last several years. An important question to be answered for development of vaccine strategies against HIV is whether this response might be an effective barrier to de novo infection. Stated otherwise, if the immune response can indeed control the virus even for a limited period of time, the issue would not be whether or not it is possible to mount a protective immune response against HIV, but that it is necessary to establish it prior to or, possibly, during the very early stages of infection in order for it to be effective.

Information relevant to this issue is becoming available, and although the emerging picture is far from clear, one can begin to visualize certain avenues to pursue that could provide more definitive answers to this question.

II. NATURAL HISTORY OF HIV INFECTION IN HUMANS

Both preventive and interventive strategies against HIV must take into account what might be termed the natural history of infection in humans, a working model for which is depicted in Figure 1a. The essential concepts can be summarized as follows:

383

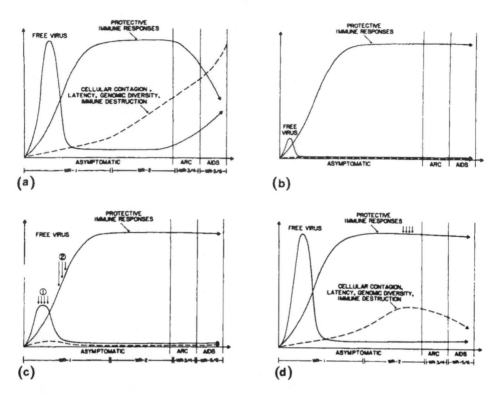

FIGURE 1 Hypothetical course of virus replication, immune responses, and pathogenic processes upon infection with HIV (a). (b) Results in an individual who managed to overcome an HIV infection, possibly as a result of a subclinical infectious dose. (c) Possible points for early intervention with chemical (1) or immunological (2) means in cases where exposure is known or suggested. (d) An approach to intervention prior to the decline of immune defenses, as discussed in the case of NK effector cells.

1. Exposure to the virus in the absence of any protective barriers would result in its replication in susceptible cells.
2. Within a variable period of time, which depends on several factors including the infectious dose, the route of infection, and the competency of the individual to respond to the virus, an immune defense is established which begins to curtail active viral replication.*

---

*The immune response may not be the only factor to impact on virus replication. The accessory genes of HIV which control virus replication may themselves interfere with off virus expression, and latency could be maintained until appropriate activating signals come into play.

3. As the immune response intensifies, viral activity is generally suppressed throughout the asymptomatic phase of the infection. However, several unique properties of HIV—notably its ability to spread by cell-to-cell transmission, its capacity to exist in latent form, the propensity of the virus to diversify in the face of an immune response, and its potential for suppressing immune function by direct infection or even by indirect means altogether—gradually tip the balance in its favor.
4. As virus begins to reemerge with failing immune responses, clinical symptoms become manifest and gradually progress.
5. Profound immune defects follow which render the individual defenseless against other pathogens, resulting in the full syndrome of AIDS.

There are a few bits of evidence in support of these general concepts which deserve attention. First, the initial phase of virus replication has been noted in a few patients through detection of viral antigens in the circulation prior to the onset of seropositivity (1). Also, some individuals experience an influenza-like illness accompanied by swollen lymph nodes during this period which resolves within a matter of days or weeks (2). Precisely when seroconversion occurs after the initial infection is unknown, but in some cases this can be a considerably later (3). Nonetheless, that a vigorous immune response to the virus is present in many asymptomatic individuals is well documented. Meanwhile, virus becomes difficult to isolate during this phase and free viral antigen is not detectable in the circulation (1). One might, then, consider this period to represent an "eclipse phase" for active virus replication. By contrast, as clinical symptoms become more severe and immune dysfunction is evident, HIV antigenemia reappears (4) and the efficiency of virus isolation increases (5,6). A prognostic marker for the development of these events is a decline in antibodies to one of the core proteins (p24) of HIV (7).

If this general premise is correct, it would highlight the importance of determining the course of the infection in an individual following exposure to the virus and using this information to devise both pre- and postexposure prophylaxis. For instance, if the immune response that develops is indeed effective in suppressing the virus, the precise characterization of its working parts should be defined. Such information would be useful for defining essential components of a vaccine regimen.

Assuming that an effective response to the virus can be generated but that, in all likelihood, it occurs too late under natural circumstances, it may also be possible through various means to stimulate the process to develop earlier and perhaps more vigorously. This would have important implications for early intervention postexposure to the virus. Stated otherwise, the amplitude of virus replication might thereby be significantly reduced and the overall degree of infection lowered and possibly limited to a subclinical level. Such manipulations might influence the subsequent rate of progression of the infection, perhaps eliminating it altogether. By reducing the infectious virus pool, one would also curtail the spread of the virus to other individuals. Ample precedents exist for early immunological intervention in retrovirus infections of animals with antiviral antibodies (8,9) and chemical antiviral agents as well (10-13). All of these studies stress the importance of interfering with viral replication during the early phase of infection.

III. CHARACTERISTICS OF THE NATURAL IMMUNE RESPONSE AGAINST HIV

A protective immune response against HIV infection would have to deal not only with the virus, but also with the infected cells that define its reservoirs. This is because infection can be transmitted as either free or cell-associated virus. Even small amounts of virus that escape the initial host defenses could establish residence in susceptible target cells. This may present a significant problem when dealing with a virus that integrates its genome in the host cell. It therefore follows that both the humoral and cell-mediated immunity must be active in an effective antiviral response.

Antibodies are likely to be the primary defense against free virus. Two facets of humoral immunity to be noted are: (1) the secretory immune response associated with epithelial surfaces, which may represent the initial barrier to natural infection; and (2) the systemic response, which is responsible for antibodies in the circulation. Both have been documented for HIV. Very little is known about secretory immunity to the virus except that antiviral IgA is present in secretions, notably saliva (14). However, evidence that secretory IgA exhibits biological activity against the virus or the infected cells is not available to date.

A strong systemic response in the form of IgG is manifest in most patients with HIV infection. Antibodies are present that are able to neutralize the infectivity of all HIV substrains in vitro and are thus broadly cross-reactive (15,16). This is quite important given the genomic diversity among HIV in the viral envelope gene and the propensity of the virus to mutate even during infection of a single patient (17). An important question which remains unanswered is whether virus neutralization measured in vitro actually occurs in vivo.

A search has also been made for antibodies that are cytotoxic for virus-infected cells in the presence of complement. While antibodies that readily bind to the surface of virus-infected cells are plentiful, no cytotoxic reactivity could be demonstrated in two independent studies (18,19). In marked contrast, one does find unusually high titers of antibodies (10^2-10^5) that cooperate with Fc receptor-bearing cells in vitro to destroy virus-infected targets by a process termed antibody-dependent cell cytotoxicity (ADCC) (19). These antibodies are directed largely at the outer envelope glycoprotein (gp120) of the virus (20). Moreover, a measurable portion of the ADCC activity is directed against an epitope representing a highly conserved region located at the very C' terminus of gp120, partially explaining why such antibodies can mediate lysis of cells infected with widely divergent strains of HIV (20). More recent studies have identified a similar activity in the circulation of HIV infected individuals (21). These cells bear the phenotypic markers of NK cells including Fc receptors (21). Studies are in progress to verify that they are armed with anti-HIV antibodies.

The more classical form of cell-mediated immunity, namely HIV-specific cytotoxic lymphocytes (CTL), has also been described (22,23). These have been documented in seropositive individuals and shown to lyse HIV-infected cells in an MHC-restricted fashion (22). Moreover, it could be shown that the activity in patients is directed against the envelope components of HIV as well as internal antigens of the virus (22,24).

Precisely how HIV-specific ADCC relates to CTL in quantitative or qualitative terms remains to be established, but their combined presence along with neutralizing antibodies speaks in favor of the presence of a functional immune response against the virus, at least within the asymptomatic phase of the disease.

IV. IMMUNE RESPONSES TO SELECTED EPITOPES OF THE HIV ENVELOPE

The humoral and cellular immune responses against HIV described above are from individuals at various stages of infection and disease progression. Only rarely, however, could this be determined in a single case where temporal knowledge of exposure to the virus in relation to seroconversion could be documented. Such information is now available from a laboratory worker who became infected with the HTLV-III$_B$ strain of HIV (19). One notes, early after seroconversion, the emergence of virus-neutralizing antibodies, antibodies that inhibit syncytium formation as well as those mediating ADCC against virus-infected target cells. The ADCC activity is broadly cross-reactive among diverse HIV isolates from the outset, whereas neutralizing antibodies are highly isolate-specific and remain so for about 18 months. Thereafter, the latter responses begin to broaden resembling the pattern noted in individuals infected long term under natural conditions. (Table 1).

Chimpanzees that have been infected with HIV also develop immune responses to the virus similar to those seen in humans. There is perhaps one exception in that chimps are able to mount a cytotoxic antibody response to virus-infected cells whereas humans rarely, if ever, do so (18,19). The sequence of the antiviral responses in relation to infection approximates the following: Humoral antibodies detectable by sensitive binding assays appear between 4 and 6 weeks after infection. These are directed against both the envelope and internal antigens of the virus. Neutralizing antibodies develop 8-12 weeks postinoculation at which time they exhibit a type- or strain-specific response. Eventually, this response begins to broaden, but the peak activity is considerably

TABLE 1 Anti-HIV Activities of Serum Samples From a Laboratory Worker Infected With the III$_B$ Strain of HIV

Month: after initial sample	ADCC[a]	Inhibition of fusion (III$_B$)[b]	Blockade of fusion inhibited by RP135[c]	Inhibition of fusion (RF)[d]	Blockade of gp120/CD4 binding[e]
0	13.9	0	NT[f]	0	0
5	21.9	89	85	0	2
6	24.8	83	73	0	0
11	28.0	81	88	NT[f]	4
12	25.5	100	63	0	12
13	24.6	100	6	0	28
17	37.9	100	0	5	41
23	32.1	100	0	38	67

[a]Percent lysis of cells infected with the III$_B$ strain of HIV at a 1/200 serum dilution. Lysis also ob served on RF infected cell.
[b]Inhibition of cell fusion between MOLT-4 cells infected with the III$_B$ strain of HIV and uninfected MOLT-4 cells (Matthews et al, 1987).
[c]Blockade of b by the RP135 peptide sequence derived from the BH10 clone of the III$_B$ HIV isolate
[d]Inhibition of fusion between cells infected with the RP isolate of HIV and uninfected MOLT-4 cells
[e]Percent inhibition of ^{125}I gp120 (III$_B$) binding to CD4 at a serum dilution of 1:250 (Skinner et a 1988a).
[f]Not tested.

lower and is achieved several months to a year after the type-specific response (26).

In an effort to better understand the nature and significance of these immune responses, a number of studies have been done to map the respective target epitopes of the virus. One of these, on the exterior envelope glycoprotein of the virus, represents a dominant neutralizing epitope (27,28). It is a hyper-variable sequence of 24 amino acids situated near the middle of gp120 which exists within a 35-amino-acid loop linked at its base by a disulfide bond between two cysteine residues (Figure 2). Immune responses to this region are highly specific for the immunizing antigen of the replicating virus (27,28). A single amino acid change in critical areas of this loop can abrogate susceptibility to neutralizing sera (29).

As noted previously, humans or chimps infected with HIV initially mount a type-specific neutralizing response which can be shown to be directed to this region of the envelope of the infecting virus. However, this response broadens and is soon capable of cross-neutralizing most HIV isolates. The kinetics of this broadening preclude a response to the emergence of a large number of highly divergent variants. More likely is that it signals the presence of antibodies to regions of the envelope which are conserved among HIV. Foremost among these are the domains of gp120 that bind to the HIV receptor, the CD4 molecule. A segment comprising 28 amino acids, located toward the C' terminus of gp120 (Figure 2) is thought to be involved in this interaction (30). This region, which also exists as a disulfide-linked loop, is highly conserved among HIV (30). Antibodies to this domain are able to block the binding of gp120 to CD4 and similar activities are found in humans (31). Of considerable interest is that the emergence of the broad cross-neutralizing activity in the infected laboratory worker coincides with a rising titer of antibodies that block gp120/CD4 binding (Table 1). Although this strongly implicates this region of the molecule as a target for broadly neutralizing antibodies, one must keep in mind other conserved regions of gp120 (32,33) and gp41 (33) as being additional potential targets for such activities.

A critical function for virus infectivity and pathogenicity is the phenomenon of fusion. In HIV, this occurs subsequent to virus binding to CD4 through a complex process thought to involve both gp120 and the transmembrane envelope glycoprotein, gp41 (34). It can also occur between virus-infected cells exhibiting gp120 and gp41 on their surface and uninfected cells bearing CD4.

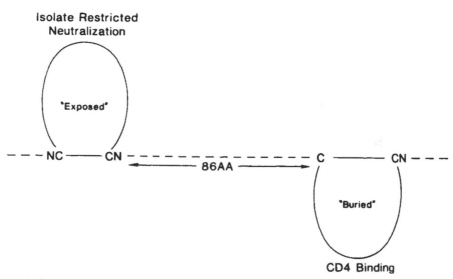

FIGURE 2 Schematic of two loop structures in HIV gp120 which relate to important biological functions of the virus. The hypervariable loop is between cysteine 296 and 331. The conserved loop is between cysteine 418 and 445.

The latter process eventually results in multinucleated giant cell formation and represents a likely mechanism for cell-to-cell transmission of the virus.

Based on current knowledge, antibodies that inhibit fusogenic activities related to HIV appear to be represented by three distinct types. Certain monoclonal antibodies directed against gp41 as well as monoclonal and polyclonal antibodies to certain peptides from gp41 (33) can inhibit cell fusion (R. Weiss, personal communication). Although not conclusively proven, the close correlation between the emergence of antibodies that block the binding of gp120 to CD4 and the appearance of antibodies in infected chimps and humans (which inhibit syncytium formation in a broad fashion) points to this region of gp120 as the likely target. Finally, and perhaps most surprising, antibodies directed against the hypervariable loop of gp120, which is the immunodominant target for neutralizing antibodies, are also very effective at inhibiting fusion formation, albeit in a highly isolate-restricted fashion (27,28). Moreover, they do so without having an effect on the binding of the virus to its receptor (31), raising the possibility that this region of gp120 is fundamentally involved in the process of virus entry into the cell. As suggested by the studies of Kowalski et al. (34), the hypervariable loop may be involved in the interaction of gp120 with gp41. It is remarkable that HIV has chosen a strategy where an apparently critical domain is associated with hypervariable elements and is still able to remain functional. The hypervariability would appear to be a defense mechanism against immune attack, but other explanations are possible.

Insofar as antibodies that mediate ADCC are concerned, only limited information is presently available. The kinetics of appearance of such antibodies in infected animals and humans indicates that the respective epitopes are distinct from those responsible for neutralization, syncytium formation, and gp120 binding to CD4 (Table 1). In the case of HIV-specific CTL, these have been detected after immunization of animals with components of the HIV envelope as well as in infected chimps and humans. Their targets can include both conserved and variable (35,36) regions of the HIV envelope and, as noted earlier, various internal virus antigens (22,24).

V. SELECTED ISSUES CONCERNING PREVENTIVE STRATEGIES FOR HIV INFECTION

Traditionally, successful vaccines against other pathogens have rested on their ability to induce immunological memory such that greatly elevated or anamnestic responses are achieved upon contact with the invading organism. The success of this response requires that effector mechanisms representing both the humoral and cellular arms of the immune system be activated to mediate a direct attack on the pathogen. Antibodies to specific antigens are secreted from plasma cells which either require T-cell help (T-cell-dependent epitopes) or are independent of it (B-cell-dependent epitopes). In a general sense, antibodies must recognize regions on native structures of the pathogen or the cells it infects in order to be effective. On the other hand, antigen recognition by the cellular arm of the immune system often requires that the antigenic structures be appropriately processed and presented in association with the major histocompatibility complex (MHC) for recognition by helper T cells and to serve as targets for cytotoxic T cells. It follows that the proper presentation of desired antigenic targets in HIV will be needed in order to optimize both the recognition and effector phases of the response.

Insofar as a more complete identification of epitopes is required to induce protective immunological memory against HIV, one must first draw from experience in other areas. Based on studies in influenza virus models (37), it appears that different T-cell epitopes are utilized to generate antibodies to a particular target antigen of the virus. There also appears to be a considerable variation among individuals to which epitopes of a given molecule are recognized, possibly regulated by allotype restriction. Finally, a concept has emerged from recent work with the hepatitis virus core antigen which points to the possible presence of universal T-cell epitopes that might be needed to enhance recognition of other antigenic sites (38). Much less is known about the nature of T-cell-independent antigens and their role in protection against infection. Some of these may constitute "discontinuous" epitopes which are formed after folding of the primary protein sequences. In this regard, the extensive glycosylation of the HIV envelope glycoprotein may contribute to the definition of such sites.

Any approaches toward this end will have to account for the extensive heterogeneity of HIV and its apparent ability to diversify in the face of an immune response. In this regard, one has to ask which, if any, of the target epitopes discussed thus far are likely to represent effective targets for immune attack. Based on the strong humoral and cellular antiviral responses to one of the hypervariable regions of gp120, one must consider this a potential target. The dilemma, of course, is that this would only be effective when and if the entire spectrum of infective strains that harbor variations in this region is known. Because only limited amino acid substitutions in subregions of this sequence are able to influence the neutralizing epitopes, the possibility of a cocktail of immunogens that is not prohibitive in number and which can induce a sufficiently broad response to account for HIV variants existing in the population remains a conceptual goal.

Approaches to enhance the immunogenicity of conserved regions of the envelope of HIV are also badly needed. On the basis of studies in animals immunized with gp120 or gp160, it appears that the epitopes responsible for the broader response may be hidden and require novel modes of presentation. Other studies also point to the dependency of conformation of the binding region of gp120 to CD4 (30) and the apparent requirement for the presence of the carbohydrate side chains (39). These are all obstacles that may hinder the formation of the types of antibodies that would be required to successfully compete with the high-affinity binding of gp120 to CD4. Parenthetically, the hypervariable neutralization target is not conformation dependent, can be synthesized as a linear peptide, is independent of glycosylation, and has no apparent receptor that antibodies need to compete with.

A note of caution about the potential efficacy of naturally occurring human antibodies which exhibit reactivities against conserved regions of HIV has recently been raised (Prince et al., in preparation). In this study, large quantities of human antibodies selected for high titers of broadly cross-reactive neutralizing and fusion-blocking antibodies failed to prevent infection of chimpanzees following challenge with a single isolate (HTLV-IIIB) of HIV. Curiously, there were no neutralizing or fusion-blocking antibodies in the immunoglobulin pool which were specifically directed against the hypervariable region of the challenge virus. Thus, one reason for failure of this trial could be attributed to the lack of efficacy of the broadly cross-reactive antibodies. It remains unexplained why such antibodies can exhibit biological activity in vitro, yet not reflect this in vivo.

As noted previously, not only the content of the immunogen but its mode of presentation is of critical importance. It is generally believed, for instance, that CTL are best induced when a replicating vector is utilized which permits antigen processing and presentation in association with MHC during recognition and effector phases of the response. Thus, vaccinia or other apathogenic viral recombinants bearing the envelope and/or internal antigens of HIV would represent an excellent vehicle for this, and this approach has already demonstrated promise (40,41). However, chimpanzees immunized with vaccinia recombinants bearing HIV *env* were not protected against virus challenges in spite of the presence of HIV-specific CTL (35). Notably lacking were neutralizing antibodies of sufficient titer. However, recent studies in humans using a protocol involving similar vaccinia HIV *env* recombinants as a primer followed by boosting with killed autologous vaccinia-infected cells generated an anamnestic response against HIV which exhibited strong cellular and humoral anti-HIV responses (42).

It remains to be seen whether viral subunits, peptides, or live recombinant vectors can be presented to the immune system in such a manner as to effectively induce the full spectrum of immunity needed to protect against HIV. If subunits will emerge as respectable immunogens in this regard, their proven ability to induce antibodies that neutralize virus infectivity, mediate ADCC, and inhibit syncytial formation will permit cocktails of individual fragments which define the critical epitopes for humoral and cell-mediated immunity to be assembled in a combined regimen.

All these studies bear heavily on the strategies needed to design a vaccine for HIV. They highlight the value of selecting appropriate regions of the virus structural components for inclusion in the immunogen, as both primary and stimulating epitopes. They also illustrate the importance of antigen presentation, perhaps the inclusion of multiple epitopes in a single subunit or particle, in order to obtain the most effective response.

VI. POSTEXPOSURE INTERVENTION

As noted previously, one can make a reasonably good argument for the presence of an effective immune response against the virus following exposure. However, this would appear to occur too late since considerable replication of the virus could take place during the interval prior to the onset. This period of time is likely to be quite variable, but in some cases may represent months and even as much as a year (3). Hence, there would be ample opportunity for HIV to establish residence in many of its potential target cells in various sanctuaries throughout the body, most notably the brain (43). Therefore, it stands to reason that any form of antiviral intervention during this period would have great potential benefit, as suggested in Figure 1. The target populations that comes immediately to mind in this regard are individuals who are known to have been or are likely to be exposed to the virus, including, among others, infants born to infected mothers or health care and laboratory workers who have experienced a hazardous contact with virus-containing material.

Treatment of such individuals with anti-HIV immunoglobulin (HIV-Ig) prepared from pools of blood selected on the basis of high neutralizing titers against HIV is already being contemplated. However, the recent failure to protect chimpanzees against challenge with HIV with such material makes one question the value of this approach. Clearly, much more needs to be understood with regard to both the protective and harmful elements in the Ig preparations. Some of the antibodies could actually potentiate infection of cells bearing Fc receptors as occurs with other viruses (44). Alternative approaches might include more specific antisera to recognize viral targets ideally in the form of monoclonal antibodies. Finally, one must consider antiidiotype antibodies to CD4 which bind to gp120 and might prevent the association of the virus with its receptors (45). Likely to be more effective are fragments of CD4 that contain the binding site to HIV gp120 because these might bind much more tightly to the virus.

There is also the likelihood that a window exists after virus exposure where the individual is already seropositive but immune responses have not optimized either in strength or specificity. Active immunization with viral components has been suggested as a means to stimulate such responses (46). Immunotherapy through administration of immune globulin may also be of benefit. The obvious difficulty with these situations is that each individual will have a different response to the virus, making it important to be able to chart the respective parameters for each patient (see Figure 1).

There are also ample opportunities for intervention with antiviral chemotherapeutic agents either alone or in combination with immunological
treatments in each of the situations outlined above. In animal studies, administration of AZT either concomitantly with type-C retroviruses or even 1-2
weeks postinfection (for a period of 1-7 days) prevented establishment of
leukemias in both murine (11,12) and feline (13) retrovirus models. Moreover, in
the cat, animals that repelled the virus subsequently developed an immune
response against the agent which was sufficient to resist a subsequent lethal virus
challenge. AZT is thus functioning as a *chemical vaccine*. In this context, it
demonstrates an effect similar to that observed in mice and cats treated with
hyperimmune IgG to the viral envelope glycoprotein, which, after clearing the
virus, stimulated the animals to mount an autologous protective response (47). A
combination of AZT and HIV-Ig during the early period might prove even more
effective.

In addition, the use of antiviral agents together with immunostimulants
is clearly worth considering. Such approaches are currently being formulated,
and one example was alluded to above when IL-2 and/ or gamma interferon
could be used to stimulate NK cell activity prior to their natural decline in
HIV-infected individuals (Figure 1d). This must be done in the presence of AZT,
or similar agents, since immune activation alone would enhance virus expression
(48).

In general, as in most other biomedical problems, the earlier one can inter-
vene, the more favorable the outcome for the infected individual. Clearly,
studies with AZT and other antiviral agents are moving from the direction of
treating only the late stage of AIDS to treating ARC and now even to treating
the asymptomatic individual. If that experience alone proves favorable, the
prospect for effective results with combined therapy with antiviral antibodies
and/or immunomodulators may be good and the outcomes might approach what
is depicted in Figure 1c and 1d.

VII. CONCLUDING REMARKS

This discussion has dealt with a few of the issues relating to development of
vaccine and early interventive strategies against HIV infection. The overall
message is that several opportunities may avail themselves to retard and possibly
halt transmission of the virus from individual to individual as well as within a
given individual exposure. It is now being recognized that not all humans exposed to HIV develop an active infection. Moreover, although the progression to
disease following an established infection is alarmingly high, it has not reached
100%. This may be indicative that, in some cases, the host is able to control the
infection, which could also depend on the infecting dose and/or its rouge (Figure
1b). As noted earlier, it also appears as if a favorable immune response develops

post exposure which is capable of holding the virus in check for long periods of time (years). These considerations strongly suggest that any measures that are effective in reducing the virus burden, particularly early in infection (Figure 1c) before it "seeds" the organism, will be beneficial. Reduction of the infecting agent to a "subclinical" level as has apparently occurred in animals infected with other retroviruses following treatment with hyperimmune sera or AZT, supports this notion.

Clearly, the problems associated with development of an effective vaccine against HIV are prodigious. The issues discussed herein highlight the potential value of identifying appropriate regions of the virus structural components for inclusion in vaccine candidates, both as primary and stimulating epitopes. They also illustrate the importance of antigen presentation, perhaps the inclusion of multiple epitopes in a single subunit or particle in order to obtain the most effective response. In addition to questions relating to definition of the optimal targets for immune attack are those which come under the heading of vaccinology, namely, efficacy, safety, formulation, adjuvenation, large-scale production, and so forth. In order to resolve such problems, animal models for testing are critical and currently represent the major bottleneck in this area of research. A most important advance in this field will be development of in vivo model systems where the animals are plentiful, easy to manage, able to support infection by HIV and, ideally, manifest disease syndromes. If this is not possible with HIV, one must draw from the related SIV models in small nonhuman primates which are currently being intensely investigated.

ACKNOWLEDGMENTS

A number of colleagues are responsible for the work described in this report: (a) Duke University Medical Center: Drs. Thomas J. Matthews, Kent J. Weinhold, Alphonse Langlois, Barton F. Haynes, Thomas Palker, Michael Skinner, Kim Lyerly, and Chet Nastala; (b) Repligen Corporation: Drs. James Rusche, Kashi Javaherian, and Scott Putney; (c) National Cancer Institute: Dr. William Blattner, Dr. Flossie Wong-Staal, Dr. Robert Gallo; (d) New York Blood Center: Dr. Alfred Prince; (e) Southwest Research Foundation: Dr. Jorg Eichberg.

REFERENCES

1. Allain, J-P., Laurian, Y., Paul, D. A., and Senn, D. (1986). Serological markers in early stages of human immunodeficiency virus infection in haemophiliacs. *Lancet* 2:1233–1236.
2. Redfield, R. R., Wright, D. C., and Tramont, E. C. (1986). The Walter Reed staging classification for HTLV–III/LAV infection. *N. Engl. J. Med.* 314: 131–132.

3. Ranki, A., Krohn, M., Allain, J-P., Franchini, G., Valle, S-L., Antonen, J., Leuther, M., and Krohn, K. (1987). Long latency precedes overt seroconversion in sexually transmitted human immunodeficiency virus infection. *Lancet* 2:589–593.

4. Pedersen, C., Nielsen, C. M., Vestergaard, B. F., Gerstoft, J., Krogsgaard, K., and Nielsen, J. O. (1987). Temporal relation of antigenaemia and loss antibodies to core antigens to development of clinical disease in HIV infection. *Br. Med. J.* 295:567–569.

5. deWolf, F., Goudsmit, J., Paul, D. A., Lange, J. M. A., Hooijkaas, C., Schellekens, P., Coutinho, R. A., and van der Noordaa, J. (1985). Risk of AIDS related complex and AIDS in homosexual men with persistent HIV antignaemia. *Br. Med. J.* 295:569–572.

6. Redfield, R. R. (1987), personal communication.

7. Weber, J. N., Clapham, P. R., Weiss, R. A., et al. (1987). Human immunodeficiency virus infection in two corhorts of homosexual men: Neutralizing sera and association of anti-*gag* antibody with prognosis. *Lancet* 1:119–122.

8. Schäfer, W., and Bolognesi, D. (1977). Mammalian C-type oncornaviruses: Relationships between viral structural and cell-surface antigens and their possible significance in immunological defense mechanisms. *Contemp. Topics Immunobiol.* 6:127–167.

9. Thiel, H. J. L., Schwarz, H., Fischinger, P., Bolognesi, D., and Schäfer, W. (1987). Role of antibodies to murine leukemia virus p15E transmembrane protein in immunotherapy against AKR leukemia: A model for studies in human acquired immunodeficiency syndrome. *Proc. Natl. Acad. Sci. USA* 84:5893–5897.

10. Wu, A. M., Ting, R. C., and Gallo, R. C. (1973). RNA-directed DNA polymerase and virus-induced leukemia in mice. *Proc. Natl. Acad. Sci. USA* 70:1298–1302.

11. Ruprecht, L. G., O'Brien, G., Rossoni, L. D., and Nusinoff-Lehrman, S. (1986). Suppression of mouse viremia and retroviral disease by 3′-azido-3′deoxythymidine. *Nature* 323:467–469.

12. Sharpe, A. H., Jaenisch, R. O., and Ruprecht, R. (1987). A rapid model for neurovirulence and transplacental antiviral therapy. *Science* 236:1671–1674.

13. Tavares, L., Roneker, C., Johnston, K., Nusinoff-Lehrman, S., and de Noronha, F. (1987). 3′-Azido-3′-deoxythymidine in feline leukemia virus-infected cats: A model for therapy and prophylaxis of AIDS. *Cancer Res.* 47:3190–3194.

14. Archibald, D. W., Barr, C. E., Torosian, J. P., McLane, M. F., and Essex, M. (1987). Secretroy IgA antibodies to human immunodeficiency virus in the parotid saliva of patients with AIDS and AIDS related complex. *J. Infect. Dis.* 155:793–796.

15. Robert-Guroff, M., Brown, M., and Gallo, R. C. (1985). HTLV–III neutralizing antibodies in patients with AIDS and AIDS related complex. *Nature* 316:72–74.

16. Matthews, T. J., Langlois, A. J., Robey, W. J., Chang, N. T., Gallo, R. C., Fischinger, P. J., and Bolognesi, D. P. (1986). Restricted neutralization of divergent human T-lymphotropic virus type III isolates by antibodies to the major envelope glycoprotein. *Proc. Natl. Acad. Sci. USA* 83:9709–9713.

17. Hahn, B. H., Shaw, G. M., Taylor, M. E., Redfield, R. R., Markham, P. D., Salahuddin, S. F., Wong-Staal, F., Gallo, R. C., Parks, E. S., and Parks, W. P. (1986). Genetic variation in HTLV-III/LAV over time in patients with AIDS or at risk for AIDS. *Science* 232:1548–1553.

18. Nara, P. L., Robey, W. G., Gonda, M. A., Cater, S. G., and Fischinger, P. J. (1987). Absence of cytotoxic antibody to human immunodeficiency virus-infected cells in humans and its induction in animals after infection or immunization with purified envelope glycoprotein gp120. *Proc. Natl. Acad. Sci. USA* 84:3797–3801.

19. Lyerly, H. K., Reed., D. L., Matthews, T. J., Langlois, A. J., Ahearne, P. M., Petteway, S. R, Bolognesi, D. P., and Weinhold, K. J. (1988). Anti-gp120 antibodies from HIV seropositive individuals mediate broadly reactive anti-HIV ADCC. *AIDS Res. Hum. Retroviruses* 3(4):409–422.

20. Lyerly, H. K., Matthews, T. J., Langlois, A. J., Bolognesi, D. P., and Weinhold, K. J. (1987). Human T-cell lymphotropic virus IIIB glycoprotein (gp120) bound to CD4 determinants on normal lymphocytes and expressed by infected cells serves as targets for immune attack. *Proc. Natl. Acad. Sci. USA* 84:4601–4605.

21. Weinhold, K. J., Lyerly, H. K., Matthews, T. J., Tyler, D. S. Ahearne, P. M., Stine, K. C., Langlois, A. J., Durack, D. T., and Bolognesi, D. P. (1988). Cellular anti-gp120 cytolytic reactivities in HIV-1 seropositive individuals. *Lancet* 902–904.

22. Walker, B. D., Chakrabarti, S., Moss, B., Paradis, T. J., Flynn, T., Durno, A. G., Blumberg, R. S., Kaplan, J. C., Hirsch, M. S., and Schooley, R. T. (1987). HIV-specific cytotoxic T lymphocytes in seropositive individuals. *Nature* 328:345–348.

23. Plata, F., Autran, B., Matrins, L. P., Wain-Hobson, S., Raphael, M., Mayaud, C., Denis, M., Guillon, J., and Debre, P. (1987). AIDS virus-specific cytotoxic T lymphocytes in lung disorders. *Nature* 328:348–351.

24. Walker, B. D., Flexner, C., Paradis, T. J., Fuller, T. C., Hirsch, M. S., Schooley, R. T., and Moss, B. (1988). HIV-1 reverse transcriptase is a target for cytotoxic T lymphocytes in infected individuals. *Science* 240;64–66.

25. Weiss, S. H., Goedert, J. J., Gartner, S., Popovic, M.,Waters, D., Markham, P., Di Marzo Veronese, F., Gail, M. T., Barkley, W. E., Gibbons, J., Gill, F. A., Leuther, M., Shaw, G. M., Gallo, R. C., and Blattner, W. A. (1988). Risk of human immunodeficiency virus (HIV-1) infection among laboratory workers. *Science* 239:68–71.

26. Nara, P. L., Robey, W. G., Arthur, L. O., Asher, D. M., Wolff, A. V., Gibbs, Jr., C. J., Gajdusek, C., and Fischinger, P. J. (1987). Persistent infection of chimpanzees with human immunodeficiency virus: Serological responses and properties of reisolated viruses. *J. Virol.* 61:3173–3180.

27. Palker, T. J., Clark, M. E., Langlois, A. J., Matthews, T. J., Weinhold, K. J., Randall, R. R., Bolognesi, D. P., and Haynes, B. S. (1988). Type-specific neutralization of the human immunodeficiency virus with antibodies to env-encoded synthetic peptides. *Proc. Natl. Acad. Sci. USA* 85:1932–1936.

28. Rusche, J. R., Javaherian, K., McDanal, C., Petro, J., Lynn, D. L., Grimaila, R., Langlois, A., Gallo, R. C., Arthur, L. O., Fischinger, P. J., Bolognesi, D. P., Putney, S. D., and Matthews, T. J. (1988). Antibodies that inhibit fusion of human immunodeficiency virus-infected cells bind a 24-amino acid sequence of the viral envelope, gp120. *Proc. Natl. Acad. Sci. USA* 85: 3198–3202.

29. Looney, D. J., Fischer, A. G., Putney, S. D., Rusche, J. R., Redfield, R. R., Gallo, R. C., and Wong-Staal, F. (1988). Type-restricted neutralization of molecular clones of human immunodeficiency virus. *Science* 241:357–359.

30. Lasky, L. A., Nakamura, G., Smith, D. H., Fennie, C., Shimasaki, C., Patzer, E., Berman, P., Gregory, T., and Capon, D. J. (1987). Delineation of a region of the human immunodeficiency virus type 1 gp120 glycoprotein critical for interaction with the CD4 receptor. *Cell* 50:975–985.

31. Skinner, M. A., Langlois, A. J., McDanal, C. B., Bolognesi, D. P., and Matthews, T. J. (1988). Neutralizing antibodies to an immunodominant envelope sequence do not prevent gp120 binding to CD4. *J. Virology* 62: 4195–4200.

32. Ho, D. D., Kaplan, J. C., Rackauskas, I. E., and Gurney, M. E. (1988). Second conserved domain of gp120 is important for HIV infectivity and antibody neutralization. *Science* 239:1021–1023.

33. Chanh, T. L., Dreesman, G. R., and Kanda, P. (1986). Induction of anti-HIV neutralising antibodies by synthetic peptides. *Eur. Mol. Biol. Org. J.* 5:3065–3071.

34. Kowalski, M., Potz, J., Basiripour, L., Dorfman, T., Goh, W. C., Terwilliger, E., Dayton, A., Rosen, C., Haseltine, W., and Sodroski, J. (1987). Functional regions of the envelope glycoprotein of human immunodeficiency virus type 1. *Science* 237:1351–1355.

35. Hu, S-L., Fultz, P. N., McClure, H. M., Wichberg, J. W., Thomas, E. K., Zarling, J., Signhal, M. C., Kosowski, S. G., Swenson, R. B., Anderson, D. C., and Todar, G. (1987). Effect of immunization with a vaccinia-HIV *env* recombinant on HIV infection of chimpanzees. *Nature* 328:721–723.

36. Takahashi, H., Cohen, J., Hosmalin, A., Cease, K. B., Houghten, R., Cornette, J. L., DeLisi, C., Moss, B., Germain, R. N., and Berzofsky, J. A. (1988). An immunodominant epitope of the human immunodeficiency virus envelope glycoprotein gp160 recognized by class I major histocompatibility complex molecule-restricted murine cytotoxic T lymphocytes. *Proc. Natl. Acad. Sci. USA* 85:3105–3109.

37. Townsend, A. R. M., Gotch, F. M., and Davey, J. (1985). Cytotoxic T-cells recognize fragments of influenza nucleoprotein. *Cell* 42:457–467.

38. Milich, D. R., Hughes, J. L., McLachlan, A., Thornton, G. B., and Moriarty, A. (1987). Hepatitis B synthetic immunogen comprised of nucleocapsid T-cell sites and an envelope B-cell epitope. *Proc. Natl. Acad. Sci. USA* 85:

cell sites and an envelope B-cell epitope. *Proc. Natl. Acad. Sci. USA* 85: 1610–1614.

39. Matthews, T. J., Weinhold, K. J., Lyerly, H. K., Langlois, A. J., Wigzell, H., and Bolognesi, D. P. (1987). Interaction between the human T-cell lymphotropic virus type III$_B$ envelope glycoprotein gp120 and the surface antigen CD4: Role of carbohydrate in binding and cell fusion. *Proc. Natl. Acad. Sci. USA* 84:5424–5428.

40. Zarling, J. M., Eichberg, W., Moran, P. A., McClure, J., Sridhar, P., and Hu, S. (1987). Proliferative and cytotoxic T cells to AIDS virus glycoproteins in chimpanzees immunized with a recombinant vaccinia virus expressing AIDS virus envelope glycoproteins. *J. Immunol.* 139:988–990.

41. Chakrabarti, S., Robert-Guroff, M., Wong-Staal, F., Gallo, R. C., and Moss, B. (1987). Expression of the HTLV-III envelope gene by a recombinant vaccinia virus. *Nature* 320:535–537.

42. Zagury, D., Bernard, J., Cheynier, R., Desportes, I., Leonard, R., Fouchard, M., Reveil, B., Ittele, D., Zirimwabagangabo, L., Mbayo, K., Wane, J., Salaun, J-J., Goussard, B., Dechazal, L., Burny, A., Nara, P., and Gallo, R. C. (1987). A group specific anamnestic immune reaction against HIV-1 induced by a candidate vaccine against AIDS. *Nature* 332:728–731.

43. Shaw, G. M., Harper, M. E. Hahn, B. H., Epstein, L. G., Gajdusek, D. C., Price, R. W., Navia, B. A., Petito, C. K., O'Hara, C. J., Groopman, J. E., Cho, E., Oleske, J. M., Wong-Staal, F., and Gallo, R. C. (1985). HTLV-III infection in brains of children and adults with AIDS encephalopathy. *Nature* 227:177–182.

44. Gollins, S. W., and Porterfield, J. S. (1986). A new mechanism for the neutralization of enveloped viruses by antiviral antibody. *Nature* 321:244–246.

45. Chanh, T. C., Dreesman, G. R., and Kennedy, R. C. (1987). Monoclonal anti-idiotypic antibody mimics the CD4 receptor and binds human immunodeficiency virus. *Proc. Natl. Acad. Sci. USA* 84:3891–3895.

46. Salk, J. (1987). Prospects for the control of AIDS by immunizing seropositive individuals. *Nature* 327:473–476.

47. Iglehart, J. D., Weinhold, K. J., Ward, E. C., Matthews, T. J., Langlois, A. J., Schafer, W., and Bolognesi, D. P. (1983). Prospects for the immunological management of lethal tumors. *Cancer Invest.* 1(5):409.

48. Zagury, D., Bernard, J., Leonard, R., Cheynur, R., Feldman, M., Sann, P. S., and Gallo, R. C. (1985). Long-term cultures of HTLV-III-infected T-cells: A model of cytopathology of T-cell depletion in AIDS. *Science* 231:850–853.

Index

Printed and bound by CPI Group (UK) Ltd, Croydon, CR0 4YY

30/10/2024

01781142-0001